Climate Change 2014
Mitigation of Climate Change

Summary for Policymakers

Technical Summary

Part of the Working Group III Contribution to the Fifth Assessment Report of the Intergovernmental Panel on Climate Change

Edited by

Ottmar Edenhofer
Working Group III Co-Chair
Potsdam Institute for
Climate Impact Research

Ramón Pichs-Madruga
Working Group III Co-Chair
Centro de Investigaciones de la
Economía Mundial

Youba Sokona
Working Group III Co-Chair
South Centre

Jan C. Minx
Head of TSU

Ellie Farahani
Head of Operations

Susanne Kadner
Head of Science

Kristin Seyboth
Deputy Head of Science

Anna Adler
Team Assistant

Ina Baum
Project Officer

Steffen Brunner
Senior Economist

Patrick Eickemeier
Scientific Editor

Benjamin Kriemann
IT Officer

Jussi Savolainen
Web Manager

Steffen Schlömer
Scientist

Christoph von Stechow
Scientist

Timm Zwickel
Senior Scientist

Working Group III Technical Support Unit

ISBN 978-92-9169-142-5

Figure SPM.4 as originally included in the digital version of this publication contained an error. This error is now corrected in this publication after having completed, in January 2015, the relevant procedures under the IPCC Protocol for Addressing Errors in IPCC Assessment Reports, Synthesis Reports, Special Reports or Methodological Reports.

The designations employed and the presentation of material on maps do not imply the expression of any opinion whatsoever on the part of the Intergovernmental Panel on Climate Change concerning the legal status of any country, territory, city or area or of its authorities, or concerning the delimitation of its frontiers or boundaries.

Cover photo:
Shanghai, China, aerial view © Ocean/Corbis

Dedication photo:
Elinor Ostrom © dpa

Foreword, Preface, Dedication and In Memoriam

Foreword

Climate Change 2014: Mitigation of Climate Change is the third part of the Fifth Assessment Report (AR5) of the Intergovernmental Panel on Climate Change (IPCC)—Climate Change 2013/2014—and was prepared by its Working Group III. The volume provides a comprehensive and transparent assessment of relevant options for mitigating climate change through limiting or preventing greenhouse gas (GHG) emissions, as well as activities that reduce their concentrations in the atmosphere.

This report highlights that despite a growing number of mitigation policies, GHG emission growth has accelerated over the last decade. The evidence from hundreds of new mitigation scenarios suggests that stabilizing temperature increase within the 21st century requires a fundamental departure from business-as-usual. At the same time, it shows that a variety of emission pathways exists where the temperature increase can be limited to below 2 °C relative to pre-industrial level. But this goal is associated with considerable technological, economic and institutional challenges. A delay in mitigation efforts or the limited availability of low carbon technologies further increases these challenges. Less ambitious mitigation goals such as 2.5 °C or 3 °C involve similar challenges, but on a slower timescale. Complementing these insights, the report provides a comprehensive assessment of the technical and behavioural mitigation options available in the energy, transport, buildings, industry and land-use sectors and evaluates policy options across governance levels from the local to the international scale.

The findings in this report have considerably enhanced our understanding of the range of mitigation pathways available and their underlying technological, economic and institutional requirements. The timing of this report is thus critical, as it can provide crucial information for the negotiators responsible for concluding a new agreement under the United Nations Framework Convention on Climate Change in 2015. The report therefore demands the urgent attention of both policymakers and the general public.

As an intergovernmental body jointly established in 1988 by the World Meteorological Organization (WMO) and the United Nations Environment Programme (UNEP), the IPCC has successfully provided policymakers with the most authoritative and objective scientific and technical assessments, which are clearly policy relevant without being policy prescriptive. Beginning in 1990, this series of IPCC Assessment Reports, Special Reports, Technical Papers, Methodology Reports and other products have become standard works of reference.

This Working Group III assessment was made possible thanks to the commitment and dedication of many hundreds of experts, representing a wide range of regions and scientific disciplines. WMO and UNEP are proud that so many of the experts belong to their communities and networks.

We express our deep gratitude to all authors, review editors and expert reviewers for devoting their knowledge, expertise and time. We would like to thank the staff of the Working Group III Technical Support Unit and the IPCC Secretariat for their dedication.

We are also thankful to the governments that supported their scientists' participation in developing this report and that contributed to the IPCC Trust Fund to provide for the essential participation of experts from developing countries and countries with economies in transition.

We would like to express our appreciation to the government of Italy for hosting the scoping meeting for the IPCC's Fifth Assessment Report, to the governments of Republic of Korea, New Zealand and Ethiopia as well as the University of Vigo and the Economics for Energy Research Centre in Spain for hosting drafting sessions of the Working Group III contribution and to the government of Germany for hosting the Twelfth Session of Working Group III in Berlin for approval of the Working Group III Report. In addition, we would like to thank the governments of India, Peru, Ghana, the United States and Germany for hosting the AR5 Expert meetings in Calcutta, Lima, Accra, Washington D.C., and Potsdam, respectively. The generous financial support by the government of Germany, and the logistical support by the Potsdam Institute for Climate Impact Research (Germany), enabled the effective operation of the Working Group III Technical Support Unit. This is gratefully acknowledged.

We would particularly like to thank Dr. Rajendra Pachauri, Chairman of the IPCC, for his direction and guidance of the IPCC and we express our deep gratitude to Professor Ottmar Edenhofer, Dr. Ramon Pichs-Madruga, and Dr. Youba Sokona, the Co-Chairs of Working Group III for their tireless leadership throughout the development and production of this report.

M. Jarraud
Secretary-General
World Meteorological Organization

A. Steiner
Executive Director
United Nations Environment Programme

Preface

The Working Group III contribution to the Fifth Assessment Report (AR5) of the Intergovernmental Panel on Climate Change (IPCC) provides a comprehensive and transparent assessment of the scientific literature on climate change mitigation. It builds upon the Working Group III contribution to the IPCC's Fourth Assessment Report (AR4) in 2007, the Special Report on Renewable Energy Sources and Climate Change Mitigation (SRREN) in 2011 and previous reports and incorporates subsequent new findings and research. The report assesses mitigation options at different levels of governance and in different economic sectors. It evaluates the societal implications of different mitigation policies, but does not recommend any particular option for mitigation.

Approach to the assessment

The Working Group III contribution to the AR5 explores the solution space of climate change mitigation drawing on experience and expectations for the future. This exploration is based on a comprehensive and transparent assessment of the scientific, technical, and socio-economic literature on the mitigation of climate change.

The intent of the report is to facilitate an integrated and inclusive deliberation of alternative climate policy goals and the different possible means to achieve them (e.g., technologies, policies, institutional settings). It does so through informing the policymakers and general public about the practical implications of alternative policy options, i.e., their associated costs and benefits, risks and trade-offs.

During the AR5 cycle, the role of the Working Group III scientists was akin to that of a cartographer: they mapped out different pathways within the solution space and assessed potential practical consequences and trade-offs; at the same time, they clearly marked implicit value assumptions and uncertainties. Consequently, this report may now be used by policymakers like a map for navigating the widely unknown territory of climate policy. Instead of providing recommendations for how to solve the complex policy problems, the report offers relevant information that enables policymakers to assess alternative mitigation options.

There are four major pillars to this cartography exercise:

Exploration of alternative climate policy goals: The report lays out the technological, economic and institutional requirements for stabilizing global mean temperature increases at different levels. It informs decision makers about the costs and benefits, risks and opportunities of these, acknowledging the fact that often more than one path can lead to a given policy goal.

Transparency over value judgments: The decision which mitigation path to take is influenced by a series of sometimes disputed normative choices which relate to the long-term stabilization goal itself, the weighing of other social priorities and the policies for achieving the goal. Facts are often inextricably interlinked with values and there is no purely scientific resolution of value dissent. What an assessment can do to support a rational public debate about value conflicts is to make implicit value judgments and ethical viewpoints as transparent as possible. Moreover, controversial policy goals and related ethical standpoints should be discussed in the context of the required means to reach these goals, in particular their possible consequences and side-effects. The potential for adverse side-effects of mitigation actions therefore requires an iterative assessment approach.

Multiple objectives in the context of sustainable development and equity: A comprehensive exploration of the solution space in the field of climate change mitigation recognizes that mitigation itself will only be one objective among others for decision makers. Decision makers may be interested in pursuing a broader concept of well-being. This broader concept also involves the sharing of limited resources within and across countries as well as across generations. Climate change mitigation is discussed here as a multi-objective problem embedded in a broader sustainable development and equity context.

Risk management: Climate change mitigation can be framed as a risk management exercise. It may provide large opportunities to humankind, but will also be associated with risks and uncertainties. Some of those may be of a fundamental nature and cannot be easily reduced or managed. It is therefore a basic requirement for a scientific assessment to communicate these uncertainties, wherever possible, both in their quantitative and qualitative dimension.

Scope of the report

During the process of scoping and approving the outline of the Working Group III contribution to the AR5, the IPCC focused on those aspects of the current understanding of the science of climate change mitigation that were judged to be most relevant to policymakers.

Working Group III included an extended framing section to provide full transparency over the concepts and methods used throughout the report, highlighting their underlying value judgments. This includes an improved treatment of risks and risk perception, uncertainties, ethical questions as well as sustainable development.

The exploration of the solution space for climate change mitigation starts from a new set of baseline and mitigation scenarios. The entire scenario set for the first time provides fully consistent information on radiative forcing and temperature in broad agreement with the information provided in the Working Group I contribution to the AR5. The United Nations Framework Convention on Climate Change requested the IPCC to provide relevant scientific evidence for reviewing the 2 °C goal as well as a potential 1.5 °C goal. Compared to the AR4 the report therefore assesses a large number of low stabilization scenarios broadly consistent with the 2 °C goal. It includes policy scenarios that investigate the impacts of delayed and fragmented international mitigation efforts and of restricted mitigation technologies portfolios on achieving specific mitigation goals and associated costs.

The WGIII contribution to the AR5 features several new elements. A full chapter is devoted to human settlements and infrastructures. Governance structures for the design of mitigation policies are discussed on the global, regional, national and sub-national level. The report closes with a novel chapter about investment needs and finance.

Structure of the report

The Working Group III contribution to the Fifth Assessment report is comprised of four parts:

    Part I: Introduction (Chapter 1)
    Part II: Framing Issues (Chapters 2–4)
    Part III: Pathways for Mitigating Climate Change (Chapters 5–12)
    Part IV: Assessment of Policies, Institutions and Finance (Chapters 13–16)

Part I provides an introduction to the Working Group III contribution and sets the stage for the subsequent chapters. It describes the 'Lessons learned since AR4' and the 'New challenges for AR5'. It gives a brief overview of 'Historical, current and future trends' regarding GHG emissions and discusses the issues involved in climate change response policies including the ultimate objective of the UNFCCC (Article 2) and the human dimensions of climate change (including sustainable development).

Part II deals with framing issues that provide transparency over methodological foundations and underlying concepts including the relevant value judgments for the detailed assessment of climate change mitigation policies and measures in the subsequent parts. Each chapter addresses key overarching issues (Chapter 2: Integrated Risk and Uncertainty Assessment of Climate Change Response Policies; Chapter 3: Social, Economic and Ethical Concepts and Methods; Chapter 4: Sustainable Development and Equity) and acts as a reference point for subsequent chapters.

Part III provides an integrated assessment of possible mitigation pathways and the respective sectoral contributions and implications. It combines cross-sectoral and sectoral information on long-term mitigation pathways and short- to mid-term mitigation options in major economic sectors. Chapter 5 (Drivers, Trends and Mitigation) provides the context for the subsequent chapters by outlining global trends in stocks and flows of greenhouse gases (GHGs) and short-lived climate pollutants by means of different accounting methods that provide complementary perspectives on the past. It also discusses emissions drivers, which informs the assessment of how GHG emissions have historically developed. Chapter 6 (Assessing Transformation Pathways)

analyses 1200 new scenarios generated by 31 modelling teams around the world to explore the economic, technological and institutional prerequisites and implications of mitigation pathways with different levels of ambition. The sectoral chapters (Chapter 7–11) and Chapter 12 (Human Settlements, Infrastructure and Spatial Planning) provide information on the different mitigation options across energy systems, transport, buildings, industry, agriculture, forestry and other land use as well as options specific to human settlements and infrastructure, including the possible co-benefits, adverse side-effects and costs that may be associated with each of these options. Pathways described in Chapter 6 are discussed in a sector-specific context.

Part IV assesses policies across governance scales. Beginning with international cooperation (Chapter 13), it proceeds to the regional (Chapter 14), national and sub-national levels Chapter 15) before concluding with a chapter that assesses cross-cutting investment and financing issues (Chapter 16). It reviews experience with climate change mitigation policies — both the policies themselves and the interactions among policies across sectors and scales — to provide insights to policymakers on the structure of policies which best fulfill evaluation criteria such as environmental and economic effectiveness, and others.

The assessment process

This Working Group III contribution to the AR5 represents the combined efforts of hundreds of leading experts in the field of climate change mitigation and has been prepared in accordance with the rules and procedures established by the IPCC. A scoping meeting for the AR5 was held in July 2009 and the outlines for the contributions of the three Working Groups were approved at the 31st Session of the Panel in November 2009. Governments and IPCC observer organizations nominated experts for the author teams. The team of 235 Coordinating Lead Authors and Lead Authors plus 38 Review Editors selected by the Working Group III Bureau, was accepted at the 41st Session of the IPCC Bureau in May 2010. More than 170 Contributing Authors provided draft text and information to the author teams at their request. Drafts prepared by the authors were subject to two rounds of formal review and revision followed by a final round of government comments on the Summary for Policymakers. More than 38,000 written comments were submitted by more than 800 expert reviewers and 37 governments. The Review Editors for each chapter monitored the review process to ensure that all substantive review comments received appropriate consideration. The Summary for Policymakers was approved line-by-line and the underlying chapters were then accepted at the 12th Session of IPCC Working Group III from 7–11 April 2014 in Berlin.

Acknowledgements

Production of this report was a major effort, in which many people from around the world were involved, with a wide variety of contributions. We wish to thank the generous contributions by the governments and institutions involved, which enabled the authors, Review Editors and Government and Expert Reviewers to participate in this process.

Writing this report was only possible thanks to the expertise, hard work and commitment to excellence shown throughout by our Coordinating Lead Authors and Lead Authors, with important assistance by many Contributing Authors and Chapter Science Assistants. We would also like to express our appreciation to the Government and Expert Reviewers, acknowledging their time and energy invested to provide constructive and useful comments to the various drafts. Our Review Editors were also critical in the AR5 process, supporting the author teams with processing the comments and assuring an objective discussion of relevant issues.

We would very much like to thank the governments of the Republic of Korea, New Zealand and Ethiopia as well as the University of Vigo and the Economics for Energy Research Centre in Spain, that, in collaboration with local institutions, hosted the crucial IPCC Lead Author Meetings in Changwon (July 2011), Wellington (March 2012), Vigo (November 2012) and Addis Ababa (July 2013). In addition, we would like to thank the governments of India, Peru, Ghana, the United States and Germany for hosting the Expert Meetings in Calcutta (March 2011), Lima (June 2011), Accra (August 2011), Washington D.C. (August 2012), and Potsdam (October 2013), respectively. Finally, we express our appreciation to the Potsdam Institute for Climate Impact Research (PIK) for welcoming our Coordinating Lead Authors on their campus for a concluding meeting (October 2013).

We are especially grateful for the contribution and support of the German Government, in particular the Bundesministerium für Bildung und Forschung (BMBF), in funding the Working Group III Technical Support Unit (TSU). Coordinating this funding, Gregor Laumann and Sylke Lenz of the Deutsches Zentrum für Luft- und Raumfahrt (DLR) were always ready to dedicate time and energy to the needs of the team. We would also like to express our gratitude to the Bundesministerium für Umwelt, Naturschutz, Bau und Reaktorsicherheit (BMUB) for the good collaboration throughout the AR5 cycle and the excellent organization of the 39th Session of the IPCC — and 12th Session of IPCC WGIII — particularly to Nicole Wilke and Lutz Morgenstern. Our thanks also go to Christiane Textor at Deutsche IPCC Koordinierungsstelle for the good collaboration and her dedicated work. We acknowledge the contribution of the Ministry for Science, Technology and Environment (CITMA) of the Republic of Cuba, the Cuban Institute of Meteorology (INSMET) and the Centre for World Economy Studies (CIEM) for their

support as well as the United Nations Economic Commission for Africa (UNECA) and its African Climate Policy Centre (ACPC).

We extend our gratitude to our colleagues in the IPCC leadership. The Executive Committee strengthened and facilitated the scientific and procedural work of all three working groups to complete their contributions: Rajendra K. Pachauri, Vicente Barros, Ismail El Gizouli, Taka Hiraishi, Chris Field, Thelma Krug, Hoesung Lee, Qin Dahe, Thomas Stocker, and Jean-Pascal van Ypersele. For his dedication, leadership and insight, we specially thank IPCC chair Rajendra K. Pachauri.

The Working Group III Bureau — consisting of Antonina Ivanova Boncheva (Mexico), Carlo Carraro (Italy), Suzana Kahn Ribeiro (Brazil), Jim Skea (UK), Francis Yamba (Zambia), and Taha Zatari (Saudi Arabia) — provided continuous and thoughtful advice throughout the AR5 process. We would like to thank Renate Christ, Secretary of the IPCC, and the Secretariat staff Gaetano Leone, Jonathan Lynn, Mary Jean Burer, Sophie Schlingemann, Judith Ewa, Jesbin Baidya, Werani Zabula, Joelle Fernandez, Annie Courtin, Laura Biagioni, Amy Smith and Carlos Martin-Novella, Brenda Abrar-Milani and Nina Peeva, who provided logistical support for government liaison and travel of experts from developing and transitional economy countries. Thanks are due to Francis Hayes who served as the conference officer for the Working Group III Approval Session.

Graphics support by Kay Schröder and his team at Daily-Interactive Digitale Kommunikation is greatly appreciated, as is the copy editing by Stacy Hunt and her team at Confluence Communications, the layout work by Gerd Blumenstein and his team at Da-TeX, the index by Stephen Ingle and his team at WordCo and printing by Matt Lloyd and his team at Cambridge University Press. PIK kindly hosted and housed the TSU offices.

Last but not least, it is a pleasure to acknowledge the tireless work of the staff of the Working Group III Technical Support Unit. Our thanks go to Jan Minx, Ellie Farahani, Susanne Kadner, Kristin Seyboth, Anna Adler, Ina Baum, Steffen Brunner, Patrick Eickemeier, Benjamin Kriemann, Jussi Savolainen, Steffen Schlömer, Christoph von Stechow, and Timm Zwickel, for their professionalism, creativity and dedication to coordinate the report writing and to ensure a final product of high quality. They were assisted by Hamed Beheshti, Siri Chrobog, Thomas Day, Sascha Heller, Ceren Hic, Lisa Israel, Daniel Mahringer, Inga Römer, Geraldine Satre-Buisson, Fee Stehle, and Felix Zoll, whose support and dedication are deeply appreciated.

Sincerely,

Ottmar Edenhofer
IPCC WG III CO-Chair

Ramon Pichs-Madruga
IPCC WG III CO-Chair

Youba Sokona
IPCC WG III CO-Chair

Dedication

Elinor Ostrom
(7 August 1933 – 12 June 2012)

We dedicate this report to the memory of Elinor Ostrom, Professor of Political Science at Indiana University and Nobel Laureate in Economics. Her work provided a fundamental contribution to the understanding of collective action, trust, and cooperation in the management of common pool resources, including the atmosphere. She launched a research agenda that has encouraged scientists to explore how a variety of overlapping policies at city, national, regional, and international levels can enable humankind to manage the climate problem. The assessment of climate change mitigation across different levels of governance, sectors and regions has been a new focus of the Working Group III contribution to AR5. We have benefited greatly from the vision and intellectual leadership of Elinor Ostrom.

In Memoriam

Luxin Huang (1965–2013)
Lead Author in Chapter 12 on Human Settlements, Infrastructure and Spatial Planning

Leon Jay (Lee) Schipper (1947–2011)
Review Editor in Chapter 8 on Transport

Luxin Huang contributed to Chapter 12 on Human Settlements, Infrastructure and Spatial Planning. During this time, he was the director of the Department of International Cooperation and Development at the China Academy of Urban Planning and Design (CAUPD) in Beijing, China, where he worked for 27 years. The untimely death of Luxin Huang at the young age of 48 has left the Intergovernmental Panel on Climate Change (IPCC) with great sorrow.

Lee Schipper was a leading scientist in the field of transport, energy and the environment. He was looking forward to his role as review editor for the Transport chapter when he passed away at the age of 64. Schipper had been intimately involved with the IPCC for many years, having contributed as a Lead Author to the IPCC's Second Assessment Report's chapter on Mitigation Options in the Transportation Sector. The IPCC misses his great expertise and guidance, as well as his humorous and musical contributions.

Both researchers were dedicated contributors to the IPCC assessment process. Their passing represents a deep loss for the international scientific community. Luxin Huang and Lee Schipper are dearly remembered by the authors and members of the IPCC Working Group III.

Contents

Summary for Policymakers

SPM

Summary
for Policymakers

Drafting Authors:

Ottmar Edenhofer (Germany), Ramón Pichs-Madruga (Cuba), Youba Sokona (Mali), Shardul Agrawala (France), Igor Alexeyevich Bashmakov (Russia), Gabriel Blanco (Argentina), John Broome (UK), Thomas Bruckner (Germany), Steffen Brunner (Germany), Mercedes Bustamante (Brazil), Leon Clarke (USA), Felix Creutzig (Germany), Shobhakar Dhakal (Nepal/Thailand), Navroz K. Dubash (India), Patrick Eickemeier (Germany), Ellie Farahani (Canada), Manfred Fischedick (Germany), Marc Fleurbaey (France), Reyer Gerlagh (Netherlands), Luis Gómez-Echeverri (Colombia/Austria), Sujata Gupta (India/Philippines), Jochen Harnisch (Germany), Kejun Jiang (China), Susanne Kadner (Germany), Sivan Kartha (USA), Stephan Klasen (Germany), Charles Kolstad (USA), Volker Krey (Austria/Germany), Howard Kunreuther (USA), Oswaldo Lucon (Brazil), Omar Masera (México), Jan Minx (Germany), Yacob Mulugetta (Ethiopia/UK), Anthony Patt (Austria/Switzerland), Nijavalli H. Ravindranath (India), Keywan Riahi (Austria), Joyashree Roy (India), Roberto Schaeffer (Brazil), Steffen Schlömer (Germany), Karen Seto (USA), Kristin Seyboth (USA), Ralph Sims (New Zealand), Jim Skea (UK), Pete Smith (UK), Eswaran Somanathan (India), Robert Stavins (USA), Christoph von Stechow (Germany), Thomas Sterner (Sweden), Taishi Sugiyama (Japan), Sangwon Suh (Republic of Korea/USA), Kevin Chika Urama (Nigeria/UK/Kenya), Diana Ürge-Vorsatz (Hungary), David G. Victor (USA), Dadi Zhou (China), Ji Zou (China), Timm Zwickel (Germany)

Draft Contributing Authors

Giovanni Baiocchi (UK/Italy), Helena Chum (Brazil/USA), Jan Fuglestvedt (Norway), Helmut Haberl (Austria), Edgar Hertwich (Austria/Norway), Elmar Kriegler (Germany), Joeri Rogelj (Switzerland/Belgium), H.-Holger Rogner (Germany), Michiel Schaeffer (Netherlands), Steven J. Smith (USA), Detlef van Vuuren (Netherlands), Ryan Wiser (USA)

This Summary for Policymakers should be cited as:

IPCC, 2014: Summary for Policymakers. In: *Climate Change 2014: Mitigation of Climate Change. Contribution of Working Group III to the Fifth Assessment Report of the Intergovernmental Panel on Climate Change* [Edenhofer, O., R. Pichs-Madruga, Y. Sokona, E. Farahani, S. Kadner, K. Seyboth, A. Adler, I. Baum, S. Brunner, P. Eickemeier, B. Kriemann, J. Savolainen, S. Schlömer, C. von Stechow, T. Zwickel and J.C. Minx (eds.)]. Cambridge University Press, Cambridge, United Kingdom and New York, NY, USA.

Table of Contents

SPM

SPM

SPM.1   Introduction

The Working Group III contribution to the IPCC's Fifth Assessment Report (AR5) assesses literature on the scientific, technological, environmental, economic and social aspects of mitigation of climate change. It builds upon the Working Group III contribution to the IPCC's Fourth Assessment Report (AR4), the Special Report on Renewable Energy Sources and Climate Change Mitigation (SRREN) and previous reports and incorporates subsequent new findings and research. The report also assesses mitigation options at different levels of governance and in different economic sectors, and the societal implications of different mitigation policies, but does not recommend any particular option for mitigation.

This Summary for Policymakers (SPM) follows the structure of the Working Group III report. The narrative is supported by a series of highlighted conclusions which, taken together, provide a concise summary. The basis for the SPM can be found in the chapter sections of the underlying report and in the Technical Summary (TS). References to these are given in square brackets.

The degree of certainty in findings in this assessment, as in the reports of all three Working Groups, is based on the author teams' evaluations of underlying scientific understanding and is expressed as a qualitative level of confidence (from very low to very high) and, when possible, probabilistically with a quantified likelihood (from exceptionally unlikely to virtually certain). Confidence in the validity of a finding is based on the type, amount, quality, and consistency of evidence (e.g., data, mechanistic understanding, theory, models, expert judgment) and the degree of agreement.[1] Probabilistic estimates of quantified measures of uncertainty in a finding are based on statistical analysis of observations or model results, or both, and expert judgment.[2] Where appropriate, findings are also formulated as statements of fact without using uncertainty qualifiers. Within paragraphs of this summary, the confidence, evidence, and agreement terms given for a bolded finding apply to subsequent statements in the paragraph, unless additional terms are provided.

SPM.2   Approaches to climate change mitigation

Mitigation is a human intervention to reduce the sources or enhance the sinks of greenhouse gases. Mitigation, together with adaptation to climate change, contributes to the objective expressed in Article 2 of the United Nations Framework Convention on Climate Change (UNFCCC):

> *The ultimate objective of this Convention and any related legal instruments that the Conference of the Parties may adopt is to achieve, in accordance with the relevant provisions of the Convention, stabilization of greenhouse gas concentrations in the atmosphere at a level that would prevent dangerous anthropogenic interference with the climate system. Such a level should be achieved within a time frame sufficient to allow ecosystems to adapt naturally to climate change, to ensure that food production is not threatened and to enable economic development to proceed in a sustainable manner.*

Climate policies can be informed by the findings of science, and systematic methods from other disciplines. [1.2, 2.4, 2.5, Box 3.1]

---

[1]   The following summary terms are used to describe the available evidence: limited, medium, or robust; and for the degree of agreement: low, medium, or high. A level of confidence is expressed using five qualifiers: very low, low, medium, high, and very high, and typeset in italics, e.g., *medium confidence*. For a given evidence and agreement statement, different confidence levels can be assigned, but increasing levels of evidence and degrees of agreement are correlated with increasing confidence. For more details, please refer to the guidance note for Lead Authors of the IPCC Fifth Assessment Report on consistent treatment of uncertainties.

[2]   The following terms have been used to indicate the assessed likelihood of an outcome or a result: virtually certain 99–100 % probability, very likely 90–100 %, likely 66–100 %, about as likely as not 33–66 %, unlikely 0–33 %, very unlikely 0–10 %, exceptionally unlikely 0–1 %. Additional terms (more likely than not >50–100 %, and more unlikely than likely 0–<50 %) may also be used when appropriate. Assessed likelihood is typeset in italics, e.g., *very likely*.

SPM

Sustainable development and equity provide a basis for assessing climate policies and highlight the need for addressing the risks of climate change.[3] Limiting the effects of climate change is necessary to achieve sustainable development and equity, including poverty eradication. At the same time, some mitigation efforts could undermine action on the right to promote sustainable development, and on the achievement of poverty eradication and equity. Consequently, a comprehensive assessment of climate policies involves going beyond a focus on mitigation and adaptation policies alone to examine development pathways more broadly, along with their determinants. [4.2, 4.3, 4.4, 4.5, 4.6, 4.8]

Effective mitigation will not be achieved if individual agents advance their own interests independently. Climate change has the characteristics of a collective action problem at the global scale, because most greenhouse gases (GHGs) accumulate over time and mix globally, and emissions by any agent (e.g., individual, community, company, country) affect other agents.[4] International cooperation is therefore required to effectively mitigate GHG emissions and address other climate change issues [1.2.4, 2.6.4, 3.2, 4.2, 13.2, 13.3]. Furthermore, research and development in support of mitigation creates knowledge spillovers. International cooperation can play a constructive role in the development, diffusion and transfer of knowledge and environmentally sound technologies [1.4.4, 3.11.6, 11.8, 13.9, 14.4.3].

Issues of equity, justice, and fairness arise with respect to mitigation and adaptation.[5] Countries' past and future contributions to the accumulation of GHGs in the atmosphere are different, and countries also face varying challenges and circumstances, and have different capacities to address mitigation and adaptation. The evidence suggests that outcomes seen as equitable can lead to more effective cooperation. [3.10, 4.2.2, 4.6.2]

Many areas of climate policy-making involve value judgements and ethical considerations. These areas range from the question of how much mitigation is needed to prevent dangerous interference with the climate system to choices among specific policies for mitigation or adaptation [3.1, 3.2]. Social, economic and ethical analyses may be used to inform value judgements and may take into account values of various sorts, including human wellbeing, cultural values and non-human values [3.4, 3.10].

Among other methods, economic evaluation is commonly used to inform climate policy design. Practical tools for economic assessment include cost-benefit analysis, cost-effectiveness analysis, multi-criteria analysis and expected utility theory [2.5]. The limitations of these tools are well-documented [3.5]. Ethical theories based on social welfare functions imply that distributional weights, which take account of the different value of money to different people, should be applied to monetary measures of benefits and harms [3.6.1, Box TS.2]. Whereas distributional weighting has not frequently been applied for comparing the effects of climate policies on different people at a single time, it is standard practice, in the form of discounting, for comparing the effects at different times [3.6.2].

Climate policy intersects with other societal goals creating the possibility of co-benefits or adverse side-effects. These intersections, if well-managed, can strengthen the basis for undertaking climate action. Mitigation and adaptation can positively or negatively influence the achievement of other societal goals, such as those related to human health, food security, biodiversity, local environmental quality, energy access, livelihoods, and equitable sustainable development; and vice versa, policies toward other societal goals can influence the achievement of mitigation and adaptation objectives [4.2, 4.3, 4.4, 4.5, 4.6, 4.8]. These influences can be substantial, although sometimes difficult to quantify, especially in welfare terms [3.6.3]. This multi-objective perspective is important in part because it helps to identify areas where support for policies that advance multiple goals will be robust [1.2.1, 4.2, 4.8, 6.6.1].

---

[3]   See WGII AR5 SPM.

[4]   In the social sciences this is referred to as a 'global commons problem'. As this expression is used in the social sciences, it has no specific implications for legal arrangements or for particular criteria regarding effort-sharing.

[5]   See FAQ 3.2 for clarification of these concepts. The philosophical literature on justice and other literature can illuminate these issues [3.2, 3.3, 4.6.2].

Climate policy may be informed by a consideration of a diverse array of risks and uncertainties, some of which are difficult to measure, notably events that are of low probability but which would have a significant impact if they occur. Since AR4, the scientific literature has examined risks related to climate change, adaptation, and mitigation strategies. Accurately estimating the benefits of mitigation takes into account the full range of possible impacts of climate change, including those with high consequences but a low probability of occurrence. The benefits of mitigation may otherwise be underestimated (*high confidence*) [2.5, 2.6, Box 3.9]. The choice of mitigation actions is also influenced by uncertainties in many socio-economic variables, including the rate of economic growth and the evolution of technology (*high confidence*) [2.6, 6.3].

The design of climate policy is influenced by how individuals and organizations perceive risks and uncertainties and take them into account. People often utilize simplified decision rules such as a preference for the status quo. Individuals and organizations differ in their degree of risk aversion and the relative importance placed on near-term versus long-term ramifications of specific actions [2.4]. With the help of formal methods, policy design can be improved by taking into account risks and uncertainties in natural, socio-economic, and technological systems as well as decision processes, perceptions, values and wealth [2.5].

SPM.3   Trends in stocks and flows of greenhouse gases and their drivers

Total anthropogenic GHG emissions have continued to increase over 1970 to 2010 with larger absolute decadal increases toward the end of this period (*high confidence*). Despite a growing number of climate change mitigation policies, annual GHG emissions grew on average by 1.0 gigatonne carbon dioxide equivalent ($GtCO_2eq$) (2.2 %) per year from 2000 to 2010 compared to 0.4 $GtCO_2eq$ (1.3 %) per year from 1970 to 2000 (Figure SPM.1).[6,7] Total anthropogenic GHG emissions were the highest in human history from 2000 to 2010 and reached 49 (±4.5) $GtCO_2eq$/yr in 2010. The global economic crisis 2007/2008 only temporarily reduced emissions. [1.3, 5.2, 13.3, 15.2.2, Box TS.5, Figure 15.1]

CO_2 emissions from fossil fuel combustion and industrial processes contributed about 78 % of the total GHG emission increase from 1970 to 2010, with a similar percentage contribution for the period 2000–2010 (*high confidence*). Fossil fuel-related CO_2 emissions reached 32 (±2.7) $GtCO_2$/yr, in 2010, and grew further by about 3 % between 2010 and 2011 and by about 1–2 % between 2011 and 2012. Of the 49 (±4.5) $GtCO_2eq$/yr in total anthropogenic GHG emissions in 2010, CO_2 remains the major anthropogenic GHG accounting for 76 % (38±3.8 $GtCO_2eq$/yr) of total anthropogenic GHG emissions in 2010. 16 % (7.8±1.6 $GtCO_2eq$/yr) come from methane (CH_4), 6.2 % (3.1±1.9 $GtCO_2eq$/yr) from nitrous oxide (N_2O), and 2.0 % (1.0±0.2 $GtCO_2eq$/yr) from fluorinated gases (Figure SPM.1). Annually, since 1970, about 25 % of anthropogenic GHG emissions have been in the form of non-CO_2 gases.[8] [1.2, 5.2]

---

[6]   Throughout the SPM, emissions of GHGs are weighed by Global Warming Potentials with a 100-year time horizon (GWP_{100}) from the IPCC Second Assessment Report. All metrics have limitations and uncertainties in assessing consequences of different emissions. [3.9.6, Box TS.5, Annex II.9, WGI SPM]

[7]   In this SPM, uncertainty in historic GHG emission data is reported using 90 % uncertainty intervals unless otherwise stated. GHG emission levels are rounded to two significant digits throughout this document; as a consequence, small differences in sums due to rounding may occur.

[8]   In this report, data on non-CO_2 GHGs, including fluorinated gases, are taken from the EDGAR database (Annex II.9), which covers substances included in the Kyoto Protocol in its first commitment period.

Total Annual Anthropogenic GHG Emissions by Groups of Gases 1970–2010

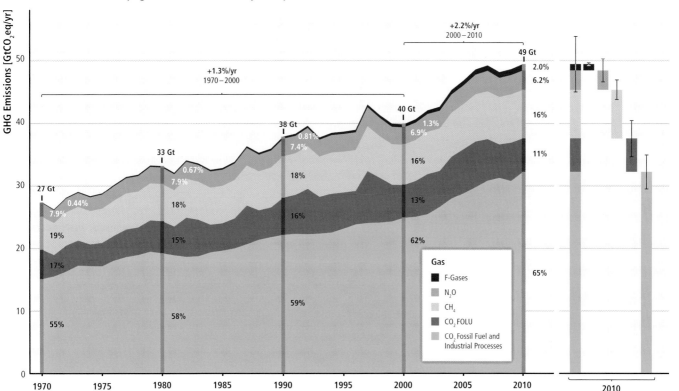

SPM

Figure SPM.1| Total annual anthropogenic GHG emissions (GtCO$_2$eq/yr) by groups of gases 1970–2010: CO$_2$ from fossil fuel combustion and industrial processes; CO$_2$ from Forestry and Other Land Use (FOLU); methane (CH$_4$); nitrous oxide (N$_2$O); fluorinated gases[8] covered under the Kyoto Protocol (F-gases). At the right side of the figure GHG emissions in 2010 are shown again broken down into these components with the associated uncertainties (90 % confidence interval) indicated by the error bars. Total anthropogenic GHG emissions uncertainties are derived from the individual gas estimates as described in Chapter 5 [5.2.3.6]. Global CO$_2$ emissions from fossil fuel combustion are known within 8 % uncertainty (90 % confidence interval). CO$_2$ emissions from FOLU have very large uncertainties attached in the order of ±50 %. Uncertainty for global emissions of CH$_4$, N$_2$O and the F-gases has been estimated as 20 %, 60 % and 20 %, respectively. 2010 was the most recent year for which emission statistics on all gases as well as assessment of uncertainties were essentially complete at the time of data cut-off for this report. Emissions are converted into CO$_2$-equivalents based on GWP$_{100}$[6] from the IPCC Second Assessment Report. The emission data from FOLU represents land-based CO$_2$ emissions from forest fires, peat fires and peat decay that approximate to net CO$_2$ flux from FOLU as described in Chapter 11 of this report. Average annual growth rate over different periods is highlighted with the brackets. [Figure 1.3, Figure TS.1]

About half of cumulative anthropogenic CO$_2$ emissions between 1750 and 2010 have occurred in the last 40 years (*high confidence*). In 1970, cumulative CO$_2$ emissions from fossil fuel combustion, cement production and flaring since 1750 were 420±35 GtCO$_2$; in 2010, that cumulative total had tripled to 1300±110 GtCO$_2$. Cumulative CO$_2$ emissions from Forestry and Other Land Use (FOLU)[9] since 1750 increased from 490±180 GtCO$_2$ in 1970 to 680±300 GtCO$_2$ in 2010. [5.2]

Annual anthropogenic GHG emissions have increased by 10 GtCO$_2$eq between 2000 and 2010, with this increase directly coming from energy supply (47 %), industry (30 %), transport (11 %) and buildings (3 %) sectors (*medium confidence*). **Accounting for indirect emissions raises the contributions of the buildings and industry sectors** (*high confidence*). Since 2000, GHG emissions have been growing in all sectors, except AFOLU. Of the 49 (±4.5) GtCO$_2$eq emissions in 2010, 35 % (17 GtCO$_2$eq) of GHG emissions were released in the energy supply sector,

9   Forestry and Other Land Use (FOLU)—also referred to as LULUCF (Land Use, Land-Use Change, and Forestry)—is the subset of Agriculture, Forestry and Other Land Use (AFOLU) emissions and removals of GHGs related to direct human-induced land use, land-use change and forestry activities excluding agricultural emissions and removals (see WGIII AR5 Glossary).

24 % (12 GtCO$_2$eq, net emissions) in AFOLU, 21 % (10 GtCO$_2$eq) in industry, 14 % (7.0 GtCO$_2$eq) in transport and 6.4 % (3.2 GtCO$_2$eq) in buildings. When emissions from electricity and heat production are attributed to the sectors that use the final energy (i.e. indirect emissions), the shares of the industry and buildings sectors in global GHG emissions are increased to 31 % and 19 %[7], respectively (Figure SPM.2). [7.3, 8.2, 9.2, 10.3, 11.2]

Globally, economic and population growth continue to be the most important drivers of increases in CO$_2$ emissions from fossil fuel combustion. The contribution of population growth between 2000 and 2010 remained roughly identical to the previous three decades, while the contribution of economic growth has risen sharply (*high confidence*). Between 2000 and 2010, both drivers outpaced emission reductions from improvements in energy intensity (Figure SPM.3). Increased use of coal relative to other energy sources has reversed the long-standing trend of gradual decarbonization of the world's energy supply. [1.3, 5.3, 7.2, 14.3, TS.2.2]

Without additional efforts to reduce GHG emissions beyond those in place today, emissions growth is expected to persist driven by growth in global population and economic activities. Baseline scenarios, those without additional mitigation, result in global mean surface temperature increases in 2100 from 3.7 °C to 4.8 °C compared to pre-industrial levels[10] (range based on median climate response; the range is 2.5 °C to 7.8 °C when including climate uncertainty, see Table SPM.1)[11] (*high confidence*). The emission scenarios collected for this assessment represent full radiative forcing including GHGs, tropospheric ozone, aerosols and albedo change. Baseline scenarios (scenarios without explicit additional efforts to constrain emissions) exceed 450 parts per million (ppm) CO$_2$eq by 2030 and reach CO$_2$eq concentration levels between 750 and more than 1300 ppm CO$_2$eq by 2100. This is similar to the range in atmospheric concentration levels between the RCP 6.0 and RCP 8.5 pathways in 2100.[12] For comparison, the CO$_2$eq concentration in 2011 is estimated to be 430 ppm (uncertainty range 340–520 ppm).[13] [6.3, Box TS.6; WGI Figure SPM.5, WGI 8.5, WGI 12.3]

---

[10] Based on the longest global surface temperature dataset available, the observed change between the average of the period 1850–1900 and of the AR5 reference period (1986–2005) is 0.61 °C (5–95 % confidence interval: 0.55–0.67 °C) [WGI SPM.E], which is used here as an approximation of the change in global mean surface temperature since pre-industrial times, referred to as the period before 1750.

[11] The climate uncertainty reflects the 5th to 95th percentile of climate model calculations described in Table SPM.1.

[12] For the purpose of this assessment, roughly 300 baseline scenarios and 900 mitigation scenarios were collected through an open call from integrated modelling teams around the world. These scenarios are complementary to the Representative Concentration Pathways (RCPs, see WGIII AR5 Glossary). The RCPs are identified by their approximate total radiative forcing in year 2100 relative to 1750: 2.6 Watts per square meter (W/m²) for RCP2.6, 4.5 W/m² for RCP4.5, 6.0 W/m² for RCP6.0, and 8.5 W/m² for RCP8.5. The scenarios collected for this assessment span a slightly broader range of concentrations in the year 2100 than the four RCPs.

[13] This is based on the assessment of total anthropogenic radiative forcing for 2011 relative to 1750 in WGI, i.e. 2.3 W/m², uncertainty range 1.1 to 3.3 W/m². [WGI Figure SPM.5, WGI 8.5, WGI 12.3]

Greenhouse Gas Emissions by Economic Sectors

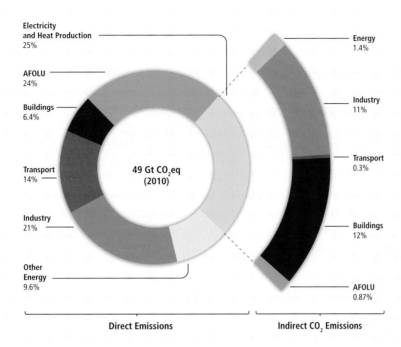

Figure SPM.2 | Total anthropogenic GHG emissions (GtCO₂eq/yr) by economic sectors. Inner circle shows direct GHG emission shares (in % of total anthropogenic GHG emissions) of five economic sectors in 2010. Pull-out shows how indirect CO_2 emission shares (in % of total anthropogenic GHG emissions) from electricity and heat production are attributed to sectors of final energy use. 'Other Energy' refers to all GHG emission sources in the energy sector as defined in Annex II other than electricity and heat production [A.II.9.1]. The emissions data from Agriculture, Forestry and Other Land Use (AFOLU) includes land-based CO_2 emissions from forest fires, peat fires and peat decay that approximate to net CO_2 flux from the Forestry and Other Land Use (FOLU) sub-sector as described in Chapter 11 of this report. Emissions are converted into CO_2-equivalents based on GWP₁₀₀[6] from the IPCC Second Assessment Report. Sector definitions are provided in Annex II.9. [Figure 1.3a, Figure TS.3 upper panel]

Decomposition of the Change in Total Annual CO₂ Emissions from Fossil Fuel Combustion by Decade

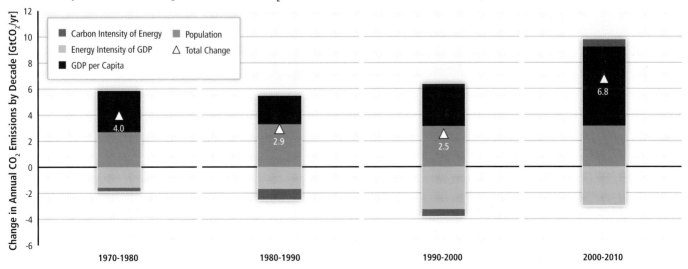

Figure SPM.3 | Decomposition of the change in total annual CO_2 emissions from fossil fuel combustion by decade and four driving factors: population, income (GDP) per capita, energy intensity of GDP and carbon intensity of energy. The bar segments show the changes associated with each factor alone, holding the respective other factors constant. Total emissions changes are indicated by a triangle. The change in emissions over each decade is measured in gigatonnes of CO_2 per year [GtCO₂/yr]; income is converted into common units using purchasing power parities. [Figure 1.7]

SPM.4 Mitigation pathways and measures in the context of sustainable development

SPM.4.1 Long-term mitigation pathways

There are multiple scenarios with a range of technological and behavioral options, with different characteristics and implications for sustainable development, that are consistent with different levels of mitigation. For this assessment, about 900 mitigation scenarios have been collected in a database based on published integrated models.[14] This range spans atmospheric concentration levels in 2100 from 430 ppm CO_2eq to above 720 ppm CO_2eq, which is comparable to the 2100 forcing levels between RCP 2.6 and RCP 6.0. Scenarios outside this range were also assessed including some scenarios with concentrations in 2100 below 430 ppm CO_2eq (for a discussion of these scenarios see below). The mitigation scenarios involve a wide range of technological, socioeconomic, and institutional trajectories, but uncertainties and model limitations exist and developments outside this range are possible (Figure SPM.4, upper panel).
[6.1, 6.2, 6.3, TS.3.1, Box TS.6]

Mitigation scenarios in which it is *likely* **that the temperature change caused by anthropogenic GHG emissions can be kept to less than 2 °C relative to pre-industrial levels are characterized by atmospheric concentrations in 2100 of about 450 ppm CO_2eq** (*high confidence*). Mitigation scenarios reaching concentration levels of about 500 ppm CO_2eq by 2100 are *more likely than not* to limit temperature change to less than 2 °C relative to pre-industrial levels, unless they temporarily 'overshoot' concentration levels of roughly 530 ppm CO_2eq before 2100, in which case they are *about as likely as not* to achieve that goal.[15] Scenarios that reach 530 to 650 ppm CO_2eq concentrations by 2100 are *more unlikely than likely* to keep temperature change below 2 °C relative to pre-industrial levels. Scenarios that exceed about 650 ppm CO_2eq by 2100 are *unlikely* to limit temperature change to below 2 °C relative to pre-industrial levels. Mitigation scenarios in which temperature increase is *more likely than not* to be less than 1.5 °C relative to pre-industrial levels by 2100 are characterized by concentrations in 2100 of below 430 ppm CO_2eq. Temperature peaks during the century and then declines in these scenarios. Probability statements regarding other levels of temperature change can be made with reference to Table SPM.1. [6.3, Box TS.6]

Scenarios reaching atmospheric concentration levels of about 450 ppm CO_2eq by 2100 (consistent with a *likely* **chance to keep temperature change below 2 °C relative to pre-industrial levels) include substantial cuts in anthropogenic GHG emissions by mid-century through large-scale changes in energy systems and potentially land use** (*high confidence*). Scenarios reaching these concentrations by 2100 are characterized by lower global GHG emissions in 2050 than in 2010, 40 % to 70 % lower globally,[16] and emissions levels near zero GtCO₂eq or below in

---

GHG Emission Pathways 2000-2100: All AR5 Scenarios

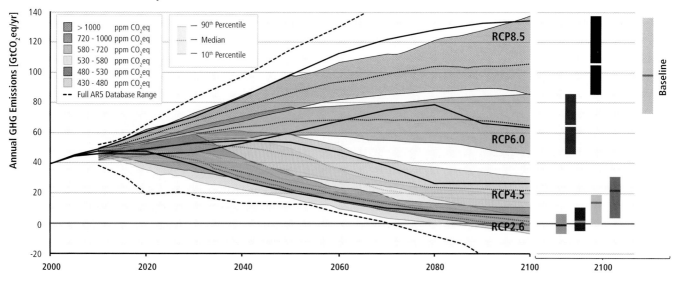

Associated Upscaling of Low-Carbon Energy Supply

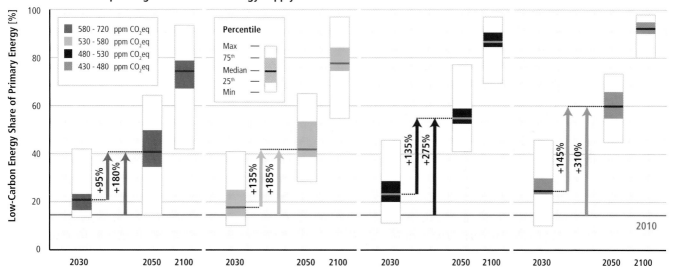

Figure SPM.4 | Pathways of global GHG emissions (GtCO₂eq/yr) in baseline and mitigation scenarios for different long-term concentration levels (upper panel) [Figure 6.7] and associated upscaling requirements of low-carbon energy (% of primary energy) for 2030, 2050 and 2100 compared to 2010 levels in mitigation scenarios (lower panel) [Figure 7.16]. The lower panel excludes scenarios with limited technology availability and exogenous carbon price trajectories. For definitions of CO₂-equivalent emissions and CO₂-equivalent concentrations see the WGIII AR5 Glossary.

2100. In scenarios reaching about 500 ppm CO_2eq by 2100, 2050 emissions levels are 25 % to 55 % lower than in 2010 globally. In scenarios reaching about 550 ppm CO_2eq, emissions in 2050 are from 5 % above 2010 levels to 45 % below 2010 levels globally (Table SPM.1). At the global level, scenarios reaching about 450 ppm CO_2eq are also characterized by more rapid improvements in energy efficiency and a tripling to nearly a quadrupling of the share of zero- and low-carbon energy supply from renewables, nuclear energy and fossil energy with carbon dioxide capture and storage (CCS), or bioenergy with CCS (BECCS) by the year 2050 (Figure SPM.4, lower panel). These scenarios describe a wide range of changes in land use, reflecting different assumptions about the scale of bioenergy production, afforestation, and reduced deforestation. All of these emissions, energy, and land-use changes vary across regions.[17] Scenarios reaching higher concentrations include similar changes, but on a slower timescale. On the other hand, scenarios reaching lower concentrations require these changes on a faster timescale. [6.3, 7.11]

Mitigation scenarios reaching about 450 ppm CO_2eq in 2100 typically involve temporary overshoot of atmospheric concentrations, as do many scenarios reaching about 500 ppm to about 550 ppm CO_2eq in 2100. Depending on the level of the overshoot, overshoot scenarios typically rely on the availability and widespread deployment of BECCS and afforestation in the second half of the century. The availability and scale of these and other Carbon Dioxide Removal (CDR) technologies and methods are uncertain and CDR technologies and methods are, to varying degrees, associated with challenges and risks (*high confidence*) (see Section SPM.4.2).[18] CDR is also prevalent in many scenarios without overshoot to compensate for residual emissions from sectors where mitigation is more expensive. There is uncertainty about the potential for large-scale deployment of BECCS, large-scale afforestation, and other CDR technologies and methods. [2.6, 6.3, 6.9.1, Figure 6.7, 7.11, 11.13]

Estimated global GHG emissions levels in 2020 based on the Cancún Pledges are not consistent with cost-effective long-term mitigation trajectories that are at least *about as likely as not* to limit temperature change to 2 °C relative to pre-industrial levels (2100 concentrations of about 450 to about 500 ppm CO_2eq), but they do not preclude the option to meet that goal (*high confidence*). Meeting this goal would require further substantial reductions beyond 2020. The Cancún Pledges are broadly consistent with cost-effective scenarios that are *likely* to keep temperature change below 3 °C relative to preindustrial levels. [6.4, 13.13, Figure TS.11]

Delaying mitigation efforts beyond those in place today through 2030 is estimated to substantially increase the difficulty of the transition to low longer-term emissions levels and narrow the range of options consistent with maintaining temperature change below 2 °C relative to pre-industrial levels (*high confidence*). Cost-effective mitigation scenarios that make it at least *about as likely as not* that temperature change will remain below 2 °C relative to pre-industrial levels (2100 concentrations of about 450 to about 500 ppm CO_2eq) are typically characterized by annual GHG emissions in 2030 of roughly between 30 $GtCO_2$eq and 50 $GtCO_2$eq (Figure SPM.5, left panel). Scenarios with annual GHG emissions above 55 $GtCO_2$eq in 2030 are characterized by substantially higher rates of emissions reductions from 2030 to 2050 (Figure SPM.5, middle panel); much more rapid scale-up of low-carbon energy over this period (Figure SPM.5, right panel); a larger reliance on CDR technologies in the long-term; and higher transitional and long-term economic impacts (Table SPM.2, orange segment). Due to these increased mitigation challenges, many models with annual 2030 GHG emissions higher than 55 $GtCO_2$eq could not produce scenarios reaching atmospheric concentration levels that make it *about as likely as not* that temperature change will remain below 2 °C relative to pre-industrial levels. [6.4, 7.11, Figures TS.11, TS.13]

---

[17] At the national level, change is considered most effective when it reflects country and local visions and approaches to achieving sustainable development according to national circumstances and priorities. [6.4, 11.8.4, WGII SPM]

[18] According to WGI, CDR methods have biogeochemical and technological limitations to their potential on the global scale. There is insufficient knowledge to quantify how much CO_2 emissions could be partially offset by CDR on a century timescale. CDR methods carry side-effects and long-term consequences on a global scale. [WGI SPM.E.8]

Table SPM.1 | Key characteristics of the scenarios collected and assessed for WGIII AR5. For all parameters, the 10th to 90th percentile of the scenarios is shown.[1,2] [Table 6.3]

CO₂eq Concentrations in 2100 [ppm CO₂eq] — Category label (concentration range)[9]	Subcategories	Relative position of the RCPs[5]	Cumulative CO₂ emissions[3] [GtCO₂] 2011–2050	2011–2100	Change in CO₂eq emissions compared to 2010 in [%][4] 2050	2100	2100 Temperature change [°C][7]	Likelihood of staying below temperature level over the 21st century[8] 1.5°C	2.0°C	3.0°C	4.0°C
< 430	Only a limited number of individual model studies have explored levels below 430 ppm CO₂eq										
450 (430–480)	Total range[1,10]	RCP2.6	550–1300	630–1180	−72 to −41	−118 to −78	1.5–1.7 (1.0–2.8)	More unlikely than likely	Likely		
500 (480–530)	No overshoot of 530 ppm CO₂eq		860–1180	960–1430	−57 to −42	−107 to −73	1.7–1.9 (1.2–2.9)		More likely than not		
	Overshoot of 530 ppm CO₂eq		1130–1530	990–1550	−55 to −25	−114 to −90	1.8–2.0 (1.2–3.3)	Unlikely	About as likely as not		
550 (530–580)	No overshoot of 580 ppm CO₂eq		1070–1460	1240–2240	−47 to −19	−81 to −59	2.0–2.2 (1.4–3.6)			Likely	
	Overshoot of 580 ppm CO₂eq		1420–1750	1170–2100	−16 to 7	−183 to −86	2.1–2.3 (1.4–3.6)		More unlikely than likely[12]		Likely
(580–650)	Total range	RCP4.5	1260–1640	1870–2440	−38 to 24	−134 to −50	2.3–2.6 (1.5–4.2)				
(650–720)	Total range	RCP4.5	1310–1750	2570–3340	−11 to 17	−54 to −21	2.6–2.9 (1.8–4.5)		Unlikely	More likely than not	
(720–1000)	Total range	RCP6.0	1570–1940	3620–4990	18 to 54	−7 to 72	3.1–3.7 (2.1–5.8)	Unlikely[11]		More unlikely than likely	
>1000	Total range	RCP8.5	1840–2310	5350–7010	52 to 95	74 to 178	4.1–4.8 (2.8–7.8)	Unlikely[11]	Unlikely	Unlikely	More unlikely than likely

[1] The 'total range' for the 430–480 ppm CO₂eq scenarios corresponds to the range of the 10th–90th percentile of the subcategory of these scenarios shown in Table 6.3.

[2] Baseline scenarios (see SPM.3) fall into the >1000 and 720–1000 ppm CO₂eq categories. The latter category also includes mitigation scenarios. The baseline scenarios in the latter category reach a temperature change of 2.5–5.8°C above preindustrial in 2100. Together with the baseline scenarios in the >1000 ppm CO₂eq category, this leads to an overall 2100 temperature range of 2.5–7.8°C (range based on median climate response: 3.7–4.8°C) for baseline scenarios across both concentration categories.

[3] For comparison of the cumulative CO₂ emissions estimates assessed here with those presented in WGI, an amount of 515 [445–585] GtC (1890 [1630–2150] GtCO₂), was already emitted by 2011 since 1870 [Section WGI 12.5]. Note that cumulative emissions are presented here for different periods of time (2011–2050 and 2011–2100) while cumulative emissions in WGI are presented as total compatible emissions for the RCPs (2012–2100) or for total compatible emissions for remaining below a given temperature target with a given likelihood [WGI Table SPM.3, WGI SPM.E.8].

[4] The global 2010 emissions are 31% above the 1990 emissions (consistent with the historic GHG emission estimates presented in this report). CO₂eq emissions include the basket of Kyoto gases (CO₂, CH₄, N₂O as well as F-gases).

[5] The assessment in WGIII involves a large number of scenarios published in the scientific literature and is thus not limited to the RCPs. To evaluate the CO₂eq concentration and climate implications of these scenarios, the MAGICC model was used in a probabilistic mode (see Annex II). For a comparison between MAGICC model results and the outcomes of the models used in WGI, see Sections WGI 12.4.1.2 and WGI 12.4.8 and 6.3.2.6. Reasons for differences with WGI SPM Table.2 include the difference in reference year (1986–2005 vs. 1850–1900 here), difference in reporting year (2081–2100 vs 2100 here), set-up of simulation (CMIP5 concentration driven versus MAGICC emission-driven here), and the wider set of scenarios (RCPs versus the full set of scenarios in the WGIII AR5 scenario database here).

[6] Temperature change is reported for the year 2100, which is not directly comparable to the equilibrium warming reported in WGIII AR4 [Table 3.5, Chapter 3]. For the 2100 temperature estimates, the transient climate response (TCR) is the most relevant system property. The assumed 90% range of the TCR for MAGICC is 1.2–2.6°C (median 1.8°C). This compares to the 90% range of TCR between 1.2–2.4°C for CMIP5 [WGI 9.7] and an assessed *likely* range of 1–2.5°C from multiple lines of evidence reported in the WGI AR5 [Box 12.2 in Section 12.5].

[7] Temperature change in 2100 is provided for a median estimate of the MAGICC calculations, which illustrates differences between the emissions pathways of the scenarios in each category. The range of temperature change in the parentheses includes in addition the carbon cycle and climate system uncertainties as represented by the MAGICC model [see 6.3.2.6 for further details]. The temperature data compared to the 1850–1900 reference year was calculated by taking all projected warming relative to 1986–2005, and adding 0.61°C for 1986–2005 compared to 1850–1900, based on HadCRUT4 [see WGI Table SPM.2].

[8] The assessment in this table is based on the probabilities calculated for the full ensemble of scenarios in WGIII using MAGICC and the assessment in WGI of the uncertainty of the temperature projections not covered by climate models. The statements are therefore consistent with the statements in WGI, which are based on the CMIP5 runs of the RCPs and the assessed uncertainties. Hence, the likelihood statements reflect different lines of evidence from both WGs. This WGI method was also applied for scenarios with intermediate concentration levels where no CMIP5 runs are available. The likelihood statements are indicative only [6.3], and follow broadly the terms used by the WGI SPM for temperature projections: *likely* 66–100%, *more likely than not* >50–100%, *about as likely as not* 33–66%, and *unlikely* 0–33%. In addition the term *more unlikely than likely* 0–<50% is used.

[9] The CO₂-equivalent concentration includes the forcing of all GHGs including halogenated gases and tropospheric ozone, as well as aerosols and albedo change (calculated on the basis of the total forcing from a simple carbon cycle/climate model, MAGICC).

[10] The vast majority of scenarios in this category overshoot the category boundary of 480 ppm CO₂eq concentrations.

[11] For scenarios in this category no CMIP5 run [WGI Chapter 12, Table 12.3] as well as no MAGICC realization [6.3] stays below the respective temperature level. Still, an *unlikely* assignment is given to reflect uncertainties that might not be reflected by the current climate models.

[12] Scenarios in the 580–650 ppm CO₂eq category include both overshoot scenarios and scenarios that do not exceed the concentration level at the high end of the category (like RCP4.5). The latter type of scenarios, in general, have an assessed probability of *more unlikely than likely* to stay below the 2°C temperature level, while the former are mostly assessed to have an *unlikely* probability of staying below this level.

SPM

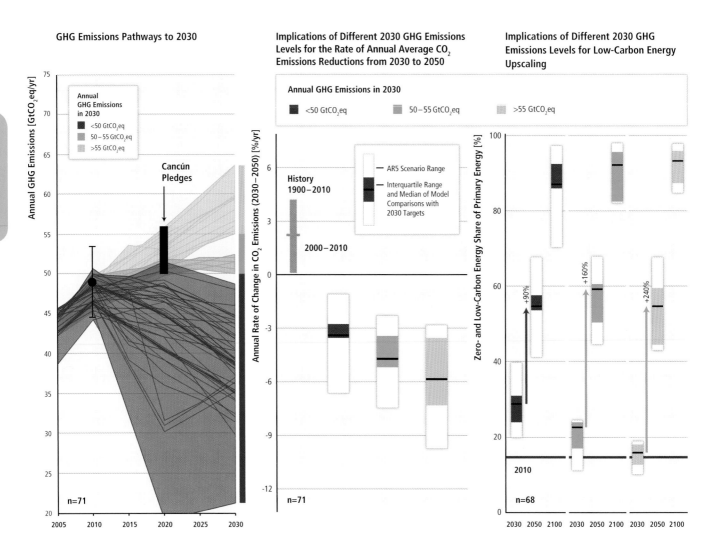

Figure SPM.5 | The implications of different 2030 GHG emissions levels (left panel) for the rate of CO₂ emissions reductions from 2030 to 2050 (middle panel) and low-carbon energy upscaling from 2030 to 2050 and 2100 (right panel) in mitigation scenarios reaching about 450 to about 500 (430 – 530) ppm CO₂eq concentrations by 2100. The scenarios are grouped according to different emissions levels by 2030 (coloured in different shades of green). The left panel shows the pathways of GHG emissions (GtCO₂eq/yr) leading to these 2030 levels. The black bar shows the estimated uncertainty range of GHG emissions implied by the Cancún Pledges. The middle panel denotes the average annual CO₂ emissions reduction rates for the period 2030–2050. It compares the median and interquartile range across scenarios from recent intermodel comparisons with explicit 2030 interim goals to the range of scenarios in the Scenario Database for WGIII AR5. Annual rates of historical emissions change between 1900 – 2010 (sustained over a period of 20 years) and average annual emissions change between 2000 – 2010 are shown in grey. The arrows in the right panel show the magnitude of zero and low-carbon energy supply up-scaling from 2030 to 2050 subject to different 2030 GHG emissions levels. Zero- and low-carbon energy supply includes renewables, nuclear energy, fossil energy with carbon dioxide capture and storage (CCS), and bioenergy with CCS (BECCS). Note: Only scenarios that apply the full, unconstrained mitigation technology portfolio of the underlying models (default technology assumption) are shown. Scenarios with large net negative global emissions (>20 GtCO₂/yr), scenarios with exogenous carbon price assumptions, and scenarios with 2010 emissions significantly outside the historical range are excluded. The right-hand panel includes only 68 scenarios, because three of the 71 scenarios shown in the figure do not report some subcategories for primary energy that are required to calculate the share of zero- and low-carbon energy. [Figures 6.32 and 7.16; 13.13.1.3]

Table SPM.2 | Global mitigation costs in cost-effective scenarios[1] and estimated cost increases due to assumed limited availability of specific technologies and delayed additional mitigation. Cost estimates shown in this table do not consider the benefits of reduced climate change as well as co-benefits and adverse side-effects of mitigation. The yellow columns show consumption losses in the years 2030, 2050, and 2100 and annualized consumption growth reductions over the century in cost-effective scenarios relative to a baseline development without climate policy. The grey columns show the percentage increase in discounted costs[2] over the century, relative to cost-effective scenarios, in scenarios in which technology is constrained relative to default technology assumptions.[3] The orange columns show the increase in mitigation costs over the periods 2030–2050 and 2050–2100, relative to scenarios with immediate mitigation, due to delayed additional mitigation through 2030.[4] These scenarios with delayed additional mitigation are grouped by emission levels of less or more than 55 $GtCO_2$eq in 2030, and two concentration ranges in 2100 (430–530 ppm CO_2eq and 530–650 ppm CO_2eq). In all figures, the median of the scenario set is shown without parentheses, the range between the 16th and 84th percentile of the scenario set is shown in the parentheses, and the number of scenarios in the set is shown in square brackets.[5] [Figures TS.12, TS.13, 6.21, 6.24, 6.25, Annex II.10]

	Consumption losses in cost-effective scenarios[1]				Increase in total discounted mitigation costs in scenarios with limited availability of technologies				Increase in medium- and long-term mitigation costs due to delayed additional mitigation until 2030			
	[% reduction in consumption relative to baseline]			[percentage point reduction in annualized consumption growth rate]	[% increase in total discounted mitigation costs (2015–2100) relative to default technology assumptions]				[% increase in mitigation costs relative to immediate mitigation]			
2100 Concentration [ppm CO_2eq]	2030	2050	2100	2010–2100	No CCS	Nuclear phase out	Limited Solar/Wind	Limited Bioenergy	≤ 55 $GtCO_2$eq		>55 $GtCO_2$eq	
									2030–2050	2050–2100	2030–2050	2050–2100
450 (430–480)	1.7 (1.0–3.7) [N: 14]	3.4 (2.1–6.2)	4.8 (2.9–11.4)	0.06 (0.04–0.14)	138 (29–297) [N: 4]	7 (4–18) [N: 8]	6 (2–29) [N: 8]	64 (44–78) [N: 8]	28 (14–50) [N: 34]	15 (5–59)	44 (2–78) [N: 29]	37 (16–82)
500 (480–530)	1.7 (0.6–2.1) [N: 32]	2.7 (1.5–4.2)	4.7 (2.4–10.6)	0.06 (0.03–0.13)	N/A	N/A	N/A	N/A				
550 (530–580)	0.6 (0.2–1.3) [N: 46]	1.7 (1.2–3.3)	3.8 (1.2–7.3)	0.04 (0.01–0.09)	39 (18–78) [N: 11]	13 (2–23) [N: 10]	8 (5–15) [N: 10]	18 (4–66) [N: 12]	3 (–5–16) [N: 14]	4 (–4–11)	15 (3–32) [N: 10]	16 (5–24)
580–650	0.3 (0–0.9) [N: 16]	1.3 (0.5–2.0)	2.3 (1.2–4.4)	0.03 (0.01–0.05)	N/A	N/A	N/A	N/A				

[1] Cost-effective scenarios assume immediate mitigation in all countries and a single global carbon price, and impose no additional limitations on technology relative to the models' default technology assumptions.

[2] Percentage increase of net present value of consumption losses in percent of baseline consumption (for scenarios from general equilibrium models) and abatement costs in percent of baseline GDP (for scenarios from partial equilibrium models) for the period 2015–2100, discounted at 5 % per year.

[3] No CCS: CCS is not included in these scenarios. Nuclear phase out: No addition of nuclear power plants beyond those under construction, and operation of existing plants until the end of their lifetime. Limited Solar/Wind: a maximum of 20 % global electricity generation from solar and wind power in any year of these scenarios. Limited Bioenergy: a maximum of 100 EJ/yr modern bioenergy supply globally (modern bioenergy used for heat, power, combinations, and industry was around 18 EJ/yr in 2008 [11.13.5]).

[4] Percentage increase of total undiscounted mitigation costs for the periods 2030–2050 and 2050–2100.

[5] The range is determined by the central scenarios encompassing the 16th and 84th percentile of the scenario set. Only scenarios with a time horizon until 2100 are included. Some models that are included in the cost ranges for concentration levels above 530 ppm CO_2eq in 2100 could not produce associated scenarios for concentration levels below 530 ppm CO_2eq in 2100 with assumptions about limited availability of technologies and/or delayed additional mitigation.

Estimates of the aggregate economic costs of mitigation vary widely and are highly sensitive to model design and assumptions as well as the specification of scenarios, including the characterization of technologies and the timing of mitigation (*high confidence*). Scenarios in which all countries of the world begin mitigation immediately, there is a single global carbon price, and all key technologies are available, have been used as a cost-effective benchmark for estimating macroeconomic mitigation costs (Table SPM.2, yellow segments). Under these assumptions, mitigation scenarios that reach atmospheric concentrations of about 450 ppm CO_2eq by 2100 entail losses in global consumption— not including benefits of reduced climate change as well as co-benefits and adverse side-effects of mitigation[19]—of 1 % to 4 % (median: 1.7 %) in 2030, 2 % to 6 % (median: 3.4 %) in 2050, and 3 % to 11 % (median: 4.8 %) in 2100 relative to consumption in baseline scenarios that grows anywhere from 300 % to more than 900 % over the century. These numbers

[19] The total economic effect at different temperature levels would include mitigation costs, co-benefits of mitigation, adverse side-effects of mitigation, adaptation costs and climate damages. Mitigation cost and climate damage estimates at any given temperature level cannot be compared to evaluate the costs and benefits of mitigation. Rather, the consideration of economic costs and benefits of mitigation should include the reduction of climate damages relative to the case of unabated climate change.

Co-Benefits of Climate Change Mitigation for Air Quality
Impact of Stringent Climate Policy on Air Pollutant Emissions (Global, 2005-2050)

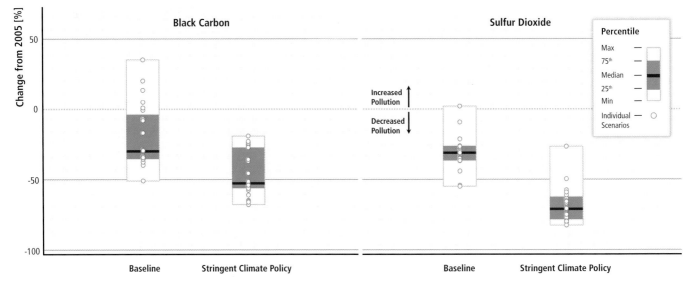

Figure SPM.6 | Air pollutant emission levels for black carbon (BC) and sulfur dioxide (SO$_2$) in 2050 relative to 2005 (0=2005 levels). Baseline scenarios without additional efforts to reduce GHG emissions beyond those in place today are compared to scenarios with stringent mitigation policies, which are consistent with reaching about 450 to about 500 (430–530) ppm CO$_2$eq concentrations by 2100. [Figure 6.33]

correspond to an annualized reduction of consumption growth by 0.04 to 0.14 (median: 0.06) percentage points over the century relative to annualized consumption growth in the baseline that is between 1.6 % and 3 % per year. Estimates at the high end of these cost ranges are from models that are relatively inflexible to achieve the deep emissions reductions required in the long run to meet these goals and/or include assumptions about market imperfections that would raise costs. Under the absence or limited availability of technologies, mitigation costs can increase substantially depending on the technology considered (Table SPM.2, grey segment). Delaying additional mitigation further increases mitigation costs in the medium- to long-term (Table SPM.2, orange segment). Many models could not achieve atmospheric concentration levels of about 450 ppm CO$_2$eq by 2100 if additional mitigation is considerably delayed or under limited availability of key technologies, such as bioenergy, CCS, and their combination (BECCS). [6.3]

Only a limited number of studies have explored scenarios that are *more likely than not* to bring temperature change back to below 1.5 °C by 2100 relative to pre-industrial levels; these scenarios bring atmospheric concentrations to below 430 ppm CO$_2$eq by 2100 (*high confidence*). Assessing this goal is currently difficult because no multi-model studies have explored these scenarios. Scenarios associated with the limited number of published studies exploring this goal are characterized by (1) immediate mitigation action; (2) the rapid upscaling of the full portfolio of mitigation technologies; and (3) development along a low-energy demand trajectory.[20] [6.3, 7.11]

Mitigation scenarios reaching about 450 to about 500 ppm CO$_2$eq by 2100 show reduced costs for achieving air quality and energy security objectives, with significant co-benefits for human health, ecosystem impacts, and sufficiency of resources and resilience of the energy system; these scenarios did not quantify other co-benefits or adverse side-effects (*medium confidence*). These mitigation scenarios show improvements in terms of the sufficiency of resources to meet national energy demand as well as the resilience of energy supply, resulting in energy systems that are less vulnerable to price volatility and supply disruptions. The benefits from reduced impacts to

---

[20] In these scenarios, the cumulative CO$_2$ emissions range between 680 and 800 GtCO$_2$ for the period 2011–2050 and between 90 and 310 GtCO$_2$ for the period 2011–2100. Global CO$_2$eq emissions in 2050 are between 70 and 95 % below 2010 emissions, and they are between 110 and 120 % below 2010 emissions in 2100.

health and ecosystems associated with major cuts in air pollutant emissions (Figure SPM.6) are particularly high where currently legislated and planned air pollution controls are weak. There is a wide range of co-benefits and adverse side-effects for additional objectives other than air quality and energy security. Overall, the potential for co-benefits of energy end-use measures outweighs the potential for adverse side-effects, whereas the evidence suggests this may not be the case for all energy supply and AFOLU measures. [WGIII 4.8, 5.7, 6.3.6, 6.6, 7.9, 8.7, 9.7, 10.8, 11.7, 11.13.6, 12.8, Figure TS.14, Table 6.7, Tables TS.3–TS.7; WGII 11.9]

There is a wide range of possible adverse side-effects as well as co-benefits and spillovers from climate policy that have not been well-quantified (*high confidence*). Whether or not side-effects materialize, and to what extent side-effects materialize, will be case- and site-specific, as they will depend on local circumstances and the scale, scope, and pace of implementation. Important examples include biodiversity conservation, water availability, food security, income distribution, efficiency of the taxation system, labour supply and employment, urban sprawl, and the sustainability of the growth of developing countries. [Box TS.11]

Mitigation efforts and associated costs vary between countries in mitigation scenarios. The distribution of costs across countries can differ from the distribution of the actions themselves (*high confidence*). In globally cost-effective scenarios, the majority of mitigation efforts takes place in countries with the highest future emissions in baseline scenarios. Some studies exploring particular effort-sharing frameworks, under the assumption of a global carbon market, have estimated substantial global financial flows associated with mitigation for scenarios leading to 2100 atmospheric concentrations of about 450 to about 550 ppm CO_2eq. [4.6, 6.3.6, 13.4.2.4; Box 3.5; Table 6.4; Figures 6.9, 6.27, 6.28, 6.29]

Mitigation policy could devalue fossil fuel assets and reduce revenues for fossil fuel exporters, but differences between regions and fuels exist (*high confidence*). Most mitigation scenarios are associated with reduced revenues from coal and oil trade for major exporters (*high confidence*). The effect of mitigation on natural gas export revenues is more uncertain, with some studies showing possible benefits for export revenues in the medium term until about 2050 (*medium confidence*). The availability of CCS would reduce the adverse effect of mitigation on the value of fossil fuel assets (*medium confidence*). [6.3.6, 6.6, 14.4.2]

SPM.4.2 Sectoral and cross-sectoral mitigation pathways and measures

SPM.4.2.1 Cross-sectoral mitigation pathways and measures

In baseline scenarios, GHG emissions are projected to grow in all sectors, except for net CO_2 emissions in the AFOLU sector[21] (*robust evidence, medium agreement*). Energy supply sector emissions are expected to continue to be the major source of GHG emissions, ultimately accounting for the significant increases in indirect emissions from electricity use in the buildings and industry sectors. In baseline scenarios, while non-CO_2 GHG agricultural emissions are projected to increase, net CO_2 emissions from the AFOLU sector decline over time, with some models projecting a net sink towards the end of the century (Figure SPM.7).[22] [6.3.1.4, 6.8, Figure TS.15]

---

[21]   Net AFOLU CO_2 emissions include emissions and removals of CO_2 from the AFOLU sector, including land under forestry and, in some assessments, CO_2 sinks in agricultural soils.

[22]   A majority of the Earth System Models assessed in WGI project a continued land carbon uptake under all RCPs through to 2100, but some models simulate a land carbon loss due to the combined effect of climate change and land-use change. [WGI SPM.E.7, WGI 6.4]

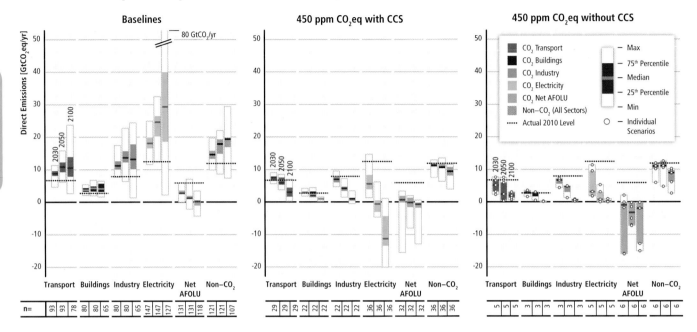

Figure SPM.7 | Direct emissions of CO_2 by sector and total non-CO_2 GHGs (Kyoto gases) across sectors in baseline (left panel) and mitigation scenarios that reach around 450 (430–480) ppm CO_2eq with CCS (middle panel) and without CCS (right panel). The numbers at the bottom of the graphs refer to the number of scenarios included in the range which differs across sectors and time due to different sectoral resolution and time horizon of models. Note that many models cannot reach about 450 ppm CO_2eq concentration by 2100 in the absence of CCS, resulting in a low number of scenarios for the right panel. [Figures 6.34 and 6.35]

Infrastructure developments and long-lived products that lock societies into GHG-intensive emissions pathways may be difficult or very costly to change, reinforcing the importance of early action for ambitious mitigation (*robust evidence, high agreement*). This lock-in risk is compounded by the lifetime of the infrastructure, by the difference in emissions associated with alternatives, and the magnitude of the investment cost. As a result, lock-in related to infrastructure and spatial planning is the most difficult to reduce. However, materials, products and infrastructure with long lifetimes and low lifecycle emissions can facilitate a transition to low-emission pathways while also reducing emissions through lower levels of material use. [5.6.3, 6.3.6.4, 9.4, 10.4, 12.3, 12.4]

There are strong interdependencies in mitigation scenarios between the pace of introducing mitigation measures in energy supply and energy end-use and developments in the AFOLU sector (*high confidence*). The distribution of the mitigation effort across sectors is strongly influenced by the availability and performance of BECCS and large scale afforestation (Figure SPM.7). This is particularly the case in scenarios reaching CO_2eq concentrations of about 450 ppm by 2100. Well-designed systemic and cross-sectoral mitigation strategies are more cost-effective in cutting emissions than a focus on individual technologies and sectors. At the energy system level these include reductions in the GHG emission intensity of the energy supply sector, a switch to low-carbon energy carriers (including low-carbon electricity) and reductions in energy demand in the end-use sectors without compromising development (Figure SPM.8). [6.3.5, 6.4, 6.8, 7.11, Table TS.2]

Mitigation scenarios reaching around 450 ppm CO_2eq concentrations by 2100 show large-scale global changes in the energy supply sector (*robust evidence, high agreement*). In these selected scenarios, global CO_2 emissions from the energy supply sector are projected to decline over the next decades and are characterized by reductions of 90 % or more below 2010 levels between 2040 and 2070. Emissions in many of these scenarios are projected to decline to below zero thereafter. [6.3.4, 6.8, 7.1, 7.11]

Final Energy Demand Reduction and Low-Carbon Energy Carrier Shares in Energy End-Use Sectors

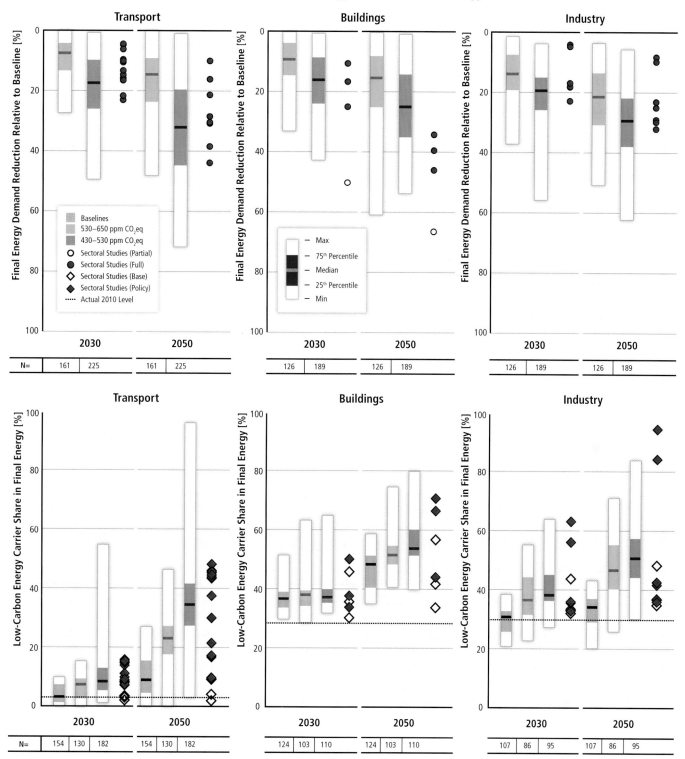

Figure SPM.8 | Final energy demand reduction relative to baseline (upper row) and low-carbon energy carrier shares in final energy (lower row) in the transport, buildings, and industry sectors by 2030 and 2050 in scenarios from two different CO_2eq concentration categories compared to sectoral studies assessed in Chapters 8–10. The demand reductions shown by these scenarios do not compromise development. Low-carbon energy carriers include electricity, hydrogen and liquid biofuels in transport, electricity in buildings and electricity, heat, hydrogen and bioenergy in industry. The numbers at the bottom of the graphs refer to the number of scenarios included in the ranges which differ across sectors and time due to different sectoral resolution and time horizon of models. [Figures 6.37 and 6.38]

Efficiency enhancements and behavioural changes, in order to reduce energy demand compared to baseline scenarios without compromising development, are a key mitigation strategy in scenarios reaching atmospheric CO$_2$eq concentrations of about 450 to about 500 ppm by 2100 (*robust evidence, high agreement*). Near-term reductions in energy demand are an important element of cost-effective mitigation strategies, provide more flexibility for reducing carbon intensity in the energy supply sector, hedge against related supply-side risks, avoid lock-in to carbon-intensive infrastructures, and are associated with important co-benefits. Both integrated and sectoral studies provide similar estimates for energy demand reductions in the transport, buildings and industry sectors for 2030 and 2050 (Figure SPM.8). [6.3.4, 6.6, 6.8, 7.11, 8.9, 9.8, 10.10]

Behaviour, lifestyle and culture have a considerable influence on energy use and associated emissions, with high mitigation potential in some sectors, in particular when complementing technological and structural change[23] (*medium evidence, medium agreement*). Emissions can be substantially lowered through changes in consumption patterns (e.g., mobility demand and mode, energy use in households, choice of longer-lasting products) and dietary change and reduction in food wastes. A number of options including monetary and non-monetary incentives as well as information measures may facilitate behavioural changes. [6.8, 7.9, 8.3.5, 8.9, 9.2, 9.3, 9.10, Box 10.2, 10.4, 11.4, 12.4, 12.6, 12.7, 15.3, 15.5, Table TS.2]

SPM.4.2.2   Energy supply

In the baseline scenarios assessed in AR5, direct CO$_2$ emissions from the energy supply sector are projected to almost double or even triple by 2050 compared to the level of 14.4 GtCO$_2$/year in 2010, unless energy intensity improvements can be significantly accelerated beyond the historical development (*medium evidence, medium agreement*). In the last decade, the main contributors to emission growth were a growing energy demand and an increase of the share of coal in the global fuel mix. The availability of fossil fuels alone will not be sufficient to limit CO$_2$eq concentration to levels such as 450 ppm, 550 ppm, or 650 ppm. (Figure SPM.7) [6.3.4, 7.2, 7.3, Figures 6.15, TS.15]

Decarbonizing (i.e. reducing the carbon intensity of) electricity generation is a key component of cost-effective mitigation strategies in achieving low-stabilization levels (430–530 ppm CO$_2$eq); in most integrated modelling scenarios, decarbonization happens more rapidly in electricity generation than in the industry, buildings, and transport sectors (*medium evidence, high agreement*) (Figure SPM.7). In the majority of low-stabilization scenarios, the share of low-carbon electricity supply (comprising renewable energy (RE), nuclear and CCS) increases from the current share of approximately 30 % to more than 80 % by 2050, and fossil fuel power generation without CCS is phased out almost entirely by 2100 (Figure SPM. 7). [6.8, 7.11, Figures 7.14, TS.18]

Since AR4, many RE technologies have demonstrated substantial performance improvements and cost reductions, and a growing number of RE technologies have achieved a level of maturity to enable deployment at significant scale (*robust evidence, high agreement*). Regarding electricity generation alone, RE accounted for just over half of the new electricity-generating capacity added globally in 2012, led by growth in wind, hydro and solar power. However, many RE technologies still need direct and/or indirect support, if their market shares are to be significantly increased; RE technology policies have been successful in driving recent growth of RE. Challenges for integrating RE into energy systems and the associated costs vary by RE technology, regional circumstances, and the characteristics of the existing background energy system (*medium evidence, medium agreement*). [7.5.3, 7.6.1, 7.8.2, 7.12, Table 7.1]

Nuclear energy is a mature low-GHG emission source of baseload power, but its share of global electricity generation has been declining (since 1993). Nuclear energy could make an increasing contribution to low-carbon energy supply, but a variety of barriers and risks exist (*robust evidence, high agreement*). Those include:

---

[23]   Structural changes refer to systems transformations whereby some components are either replaced or potentially substituted by other components (see WGIII AR5 Glossary).

operational risks, and the associated concerns, uranium mining risks, financial and regulatory risks, unresolved waste management issues, nuclear weapon proliferation concerns, and adverse public opinion (*robust evidence, high agreement*). New fuel cycles and reactor technologies addressing some of these issues are being investigated and progress in research and development has been made concerning safety and waste disposal. [7.5.4, 7.8, 7.9, 7.12, Figure TS.19]

GHG emissions from energy supply can be reduced significantly by replacing current world average coal-fired power plants with modern, highly efficient natural gas combined-cycle power plants or combined heat and power plants, provided that natural gas is available and the fugitive emissions associated with extraction and supply are low or mitigated (*robust evidence, high agreement*). In mitigation scenarios reaching about 450 ppm CO_2eq concentrations by 2100, natural gas power generation without CCS acts as a bridge technology, with deployment increasing before peaking and falling to below current levels by 2050 and declining further in the second half of the century (*robust evidence, high agreement*). [7.5.1, 7.8, 7.9, 7.11, 7.12]

Carbon dioxide capture and storage (CCS) technologies could reduce the lifecycle GHG emissions of fossil fuel power plants (*medium evidence, medium agreement*). While all components of integrated CCS systems exist and are in use today by the fossil fuel extraction and refining industry, CCS has not yet been applied at scale to a large, operational commercial fossil fuel power plant. CCS power plants could be seen in the market if this is incentivized by regulation and/or if they become competitive with their unabated counterparts, for instance, if the additional investment and operational costs, caused in part by efficiency reductions, are compensated by sufficiently high carbon prices (or direct financial support). For the large-scale future deployment of CCS, well-defined regulations concerning short- and long-term responsibilities for storage are needed as well as economic incentives. Barriers to large-scale deployment of CCS technologies include concerns about the operational safety and long-term integrity of CO_2 storage as well as transport risks. There is, however, a growing body of literature on how to ensure the integrity of CO_2 wells, on the potential consequences of a pressure build-up within a geologic formation caused by CO_2 storage (such as induced seismicity), and on the potential human health and environmental impacts from CO_2 that migrates out of the primary injection zone (*limited evidence, medium agreement*). [7.5.5., 7.8, 7.9, 7.11, 7.12, 11.13]

Combining bioenergy with CCS (BECCS) offers the prospect of energy supply with large-scale net negative emissions which plays an important role in many low-stabilization scenarios, while it entails challenges and risks (*limited evidence, medium agreement*). These challenges and risks include those associated with the upstream large-scale provision of the biomass that is used in the CCS facility as well as those associated with the CCS technology itself. [7.5.5, 7.9, 11.13]

SPM.4.2.3   **Energy end-use sectors**

Transport
The transport sector accounted for 27% of final energy use and 6.7 GtCO$_2$ direct emissions in 2010, with baseline CO$_2$ emissions projected to approximately double by 2050 (*medium evidence, medium agreement*). This growth in CO_2 emissions from increasing global passenger and freight activity could partly offset future mitigation measures that include fuel carbon and energy intensity improvements, infrastructure development, behavioural change and comprehensive policy implementation (*high confidence*). Overall, reductions in total transport CO_2 emissions of 15–40% compared to baseline growth could be achieved in 2050 (*medium evidence, medium agreement*). (Figure SPM.7) [6.8, 8.1, 8.2, 8.9, 8.10]

Technical and behavioural mitigation measures for all transport modes, plus new infrastructure and urban redevelopment investments, could reduce final energy demand in 2050 by around 40% below the baseline, with the mitigation potential assessed to be higher than reported in the AR4 (*robust evidence, medium agreement*). Projected energy efficiency and vehicle performance improvements range from 30–50% in 2030 relative to 2010 depending on transport mode and vehicle type (*medium evidence, medium agreement*). Integrated urban planning, transit-oriented development, more compact urban form that supports cycling and walking, can all lead to modal shifts as can, in the longer term, urban redevelopment and investments in new infrastructure such as high-speed rail systems that reduce short-haul air travel demand (*medium evidence, medium agreement*). Such mitigation measures are challenging, have uncertain outcomes, and could reduce transport GHG emissions by 20–50 % in 2050 compared to baseline (*limited evidence, low agreement*). (Figure SPM.8 upper panel) [8.2, 8.3, 8.4, 8.5, 8.6, 8.7, 8.8, 8.9, 12.4, 12.5]

Strategies to reduce the carbon intensities of fuel and the rate of reducing carbon intensity are constrained by challenges associated with energy storage and the relatively low energy density of low-carbon transport fuels (*medium confidence*). Integrated and sectoral studies broadly agree that opportunities for switching to low-carbon fuels exist in the near term and will grow over time. Methane-based fuels are already increasing their share for road vehicles and waterborne craft. Electricity produced from low-carbon sources has near-term potential for electric rail and short- to medium-term potential as electric buses, light-duty and 2-wheel road vehicles are deployed. Hydrogen fuels from low-carbon sources constitute longer-term options. Commercially available liquid and gaseous biofuels already provide co-benefits together with mitigation options that can be increased by technology advances. Reducing transport emissions of particulate matter (including black carbon), tropospheric ozone and aerosol precursors (including NO_x) can have human health and mitigation co-benefits in the short term (*medium evidence, medium agreement*). [8.2, 8.3, 11.13, Figure TS.20, right panel]

The cost-effectiveness of different carbon reduction measures in the transport sector varies significantly with vehicle type and transport mode (*high confidence*). The levelized costs of conserved carbon can be very low or negative for many short-term behavioural measures and efficiency improvements for light- and heavy-duty road vehicles and waterborne craft. In 2030, for some electric vehicles, aircraft and possibly high-speed rail, levelized costs could be more than USD100/tCO_2 avoided (*limited evidence, medium agreement*). [8.6, 8.8, 8.9, Figures TS.21, TS.22]

Regional differences influence the choice of transport mitigation options (*high confidence*). Institutional, legal, financial and cultural barriers constrain low-carbon technology uptake and behavioural change. Established infrastructure may limit the options for modal shift and lead to a greater reliance on advanced vehicle technologies; a slowing of growth in light-duty vehicle demand is already evident in some OECD countries. For all economies, especially those with high rates of urban growth, investment in public transport systems and low-carbon infrastructure can avoid lock-in to carbon-intensive modes. Prioritizing infrastructure for pedestrians and integrating non-motorized and transit services can create economic and social co-benefits in all regions (*medium evidence, medium agreement*). [8.4, 8.8, 8.9, 14.3, Table 8.3]

Mitigation strategies, when associated with non-climate policies at all government levels, can help decouple transport GHG emissions from economic growth in all regions (*medium confidence*). These strategies can help reduce travel demand, incentivise freight businesses to reduce the carbon intensity of their logistical systems and induce modal shifts, as well as provide co-benefits including improved access and mobility, better health and safety, greater energy security, and cost and time savings (*medium evidence, high agreement*). [8.7, 8.10]

Buildings

In 2010, the buildings sector[24] accounted for around 32 % final energy use and 8.8 $GtCO_2$ emissions, including direct and indirect emissions, with energy demand projected to approximately double and CO_2 emissions to increase by 50–150 % by mid-century in baseline scenarios (*medium evidence, medium agreement*). This energy demand growth results from improvements in wealth, lifestyle change, access to modern energy services and adequate housing, and urbanisation. There are significant lock-in risks associated with the long lifespans of buildings and related infrastructure, and these are especially important in regions with high construction rates (*robust evidence, high agreement*). (Figure SPM.7) [9.4]

---

[24] The buildings sector covers the residential, commercial, public and services sectors; emissions from construction are accounted for in the industry sector.

Recent advances in technologies, know-how and policies provide opportunities to stabilize or reduce global buildings sector energy use by mid-century (*robust evidence, high agreement*). For new buildings, the adoption of very low energy building codes is important and has progressed substantially since AR4. Retrofits form a key part of the mitigation strategy in countries with established building stocks, and reductions of heating/cooling energy use by 50–90 % in individual buildings have been achieved. Recent large improvements in performance and costs make very low energy construction and retrofits economically attractive, sometimes even at net negative costs. [9.3]

Lifestyle, culture and behaviour significantly influence energy consumption in buildings (*limited evidence, high agreement*). A three- to five-fold difference in energy use has been shown for provision of similar building-related energy service levels in buildings. For developed countries, scenarios indicate that lifestyle and behavioural changes could reduce energy demand by up to 20 % in the short term and by up to 50 % of present levels by mid-century. In developing countries, integrating elements of traditional lifestyles into building practices and architecture could facilitate the provision of high levels of energy services with much lower energy inputs than baseline. [9.3]

Most mitigation options for buildings have considerable and diverse co-benefits in addition to energy cost savings (*robust evidence, high agreement*). These include improvements in energy security, health (such as from cleaner wood-burning cookstoves), environmental outcomes, workplace productivity, fuel poverty reductions and net employment gains. Studies which have monetized co-benefits often find that these exceed energy cost savings and possibly climate benefits (*medium evidence, medium agreement*). [9.6, 9.7, 3.6.3]

Strong barriers, such as split incentives (e. g., tenants and builders), fragmented markets and inadequate access to information and financing, hinder the market-based uptake of cost-effective opportunities. Barriers can be overcome by policy interventions addressing all stages of the building and appliance lifecycles (*robust evidence, high agreement*). [9.8, 9.10, 16, Box 3.10]

The development of portfolios of energy efficiency policies and their implementation has advanced considerably since AR4. Building codes and appliance standards, if well designed and implemented, have been among the most environmentally and cost-effective instruments for emission reductions (*robust evidence, high agreement*). In some developed countries they have contributed to a stabilization of, or reduction in, total energy demand for buildings. Substantially strengthening these codes, adopting them in further jurisdictions, and extending them to more building and appliance types, will be a key factor in reaching ambitious climate goals. [9.10, 2.6.5.3]

Industry

In 2010, the industry sector accounted for around 28 % of final energy use, and 13 $GtCO_2$ emissions, including direct and indirect emissions as well as process emissions, with emissions projected to increase by 50–150 % by 2050 in the baseline scenarios assessed in AR5, unless energy efficiency improvements are accelerated significantly (*medium evidence, medium agreement*). Emissions from industry accounted for just over 30 % of global GHG emissions in 2010 and are currently greater than emissions from either the buildings or transport end-use sectors. (Figures SPM.2, SPM.7) [10.3]

The energy intensity of the industry sector could be directly reduced by about 25 % compared to the current level through the wide-scale upgrading, replacement and deployment of best available technologies, particularly in countries where these are not in use and in non-energy intensive industries (*high agreement, robust evidence*). Additional energy intensity reductions of about 20 % may potentially be realized through innovation (*limited evidence, medium agreement*). Barriers to implementing energy efficiency relate largely to initial investment costs and lack of information. Information programmes are a prevalent approach for promoting energy efficiency, followed by economic instruments, regulatory approaches and voluntary actions. [10.7, 10.9, 10.11]

Improvements in GHG emission efficiency and in the efficiency of material use, recycling and re-use of materials and products, and overall reductions in product demand (e.g., through a more intensive use of products) and service demand could, in addition to energy efficiency, help reduce GHG emissions below the baseline level in the industry sector (*medium evidence, high agreement*). Many emission-reducing options are cost-effective, profitable and associated with multiple co-benefits (better environmental compliance, health benefits etc.). In the long term, a shift to low-carbon electricity, new industrial processes, radical product innovations (e.g., alternatives to cement), or CCS (e.g., to mitigate process emissions) could contribute to significant GHG emission reductions. Lack of policy and experiences in material and product service efficiency are major barriers. [10.4, 10.7, 10.8, 10.11]

CO_2 emissions dominate GHG emissions from industry, but there are also substantial mitigation opportunities for non-CO_2 gases (*robust evidence, high agreement*). CH_4, N_2O and fluorinated gases from industry accounted for emissions of 0.9 $GtCO_2eq$ in 2010. Key mitigation opportunities include, e.g., the reduction of hydrofluorocarbon emissions by process optimization and refrigerant recovery, recycling and substitution, although there are barriers. [Tables 10.2, 10.7]

Systemic approaches and collaborative activities across companies and sectors can reduce energy and material consumption and thus GHG emissions (*robust evidence, high agreement*). The application of cross-cutting technologies (e.g., efficient motors) and measures (e.g., reducing air or steam leaks) in both large energy intensive industries and small and medium enterprises can improve process performance and plant efficiency cost-effectively. Cooperation across companies (e.g., in industrial parks) and sectors could include the sharing of infrastructure, information, and waste heat utilization. [10.4, 10.5]

Important options for mitigation in waste management are waste reduction, followed by re-use, recycling and energy recovery (*robust evidence, high agreement*). Waste and wastewater accounted for 1.5 $GtCO_2eq$ in 2010. As the share of recycled or reused material is still low (e.g., globally, around 20% of municipal solid waste is recycled), waste treatment technologies and recovering energy to reduce demand for fossil fuels can result in significant direct emission reductions from waste disposal. [10.4, 10.14]

SPM.4.2.4 Agriculture, Forestry and Other Land Use (AFOLU)

The AFOLU sector accounts for about a quarter (~10–12 $GtCO_2eq/yr$) of net anthropogenic GHG emissions mainly from deforestation, agricultural emissions from soil and nutrient management and livestock (*medium evidence, high agreement*). Most recent estimates indicate a decline in AFOLU CO_2 fluxes, largely due to decreasing deforestation rates and increased afforestation. However, the uncertainty in historical net AFOLU emissions is larger than for other sectors, and additional uncertainties in projected baseline net AFOLU emissions exist. Nonetheless, in the future, net annual baseline CO_2 emissions from AFOLU are projected to decline, with net emissions potentially less than half the 2010 level by 2050 and the possibility of the AFOLU sectors becoming a net CO_2 sink before the end of century (*medium evidence, high agreement*). (Figure SPM. 7) [6.3.1.4, 11.2, Figure 6.5]

AFOLU plays a central role for food security and sustainable development. The most cost-effective mitigation options in forestry are afforestation, sustainable forest management and reducing deforestation, with large differences in their relative importance across regions. In agriculture, the most cost-effective mitigation options are cropland management, grazing land management, and restoration of organic soils (*medium evidence, high agreement*). The economic mitigation potential of supply-side measures is estimated to be 7.2 to 11 $GtCO_2eq/year$[25] in 2030 for mitigation efforts consistent with carbon prices[26] up to 100 USD/tCO_2eq, about a third of which can be achieved at a <20 USD/tCO_2eq (*medium evidence, medium agreement*). There are potential barriers to

---

[25] Full range of all studies: 0.49–11 $GtCO_2eq/year$

[26] In many models that are used to assess the economic costs of mitigation, carbon price is used as a proxy to represent the level of effort in mitigation policies (see WGIII AR5 Glossary).

implementation of available mitigation options [11.7, 11.8]. Demand-side measures, such as changes in diet and reductions of losses in the food supply chain, have a significant, but uncertain, potential to reduce GHG emissions from food production (*medium evidence, medium agreement*). Estimates vary from roughly 0.76–8.6 GtCO$_2$eq/yr by 2050 (*limited evidence, medium agreement*). [11.4, 11.6, Figure 11.14]

Policies governing agricultural practices and forest conservation and management are more effective when involving both mitigation and adaptation. Some mitigation options in the AFOLU sector (such as soil and forest carbon stocks) may be vulnerable to climate change (*medium evidence, high agreement*). When implemented sustainably, activities to reduce emissions from deforestation and forest degradation (REDD+[27] is an example designed to be sustainable) are cost-effective policy options for mitigating climate change, with potential economic, social and other environmental and adaptation co-benefits (e.g., conservation of biodiversity and water resources, and reducing soil erosion) (*limited evidence, medium agreement*). [11.3.2, 11.10]

Bioenergy can play a critical role for mitigation, but there are issues to consider, such as the sustainability of practices and the efficiency of bioenergy systems (*robust evidence, medium agreement*) [11.4.4, Box 11.5, 11.13.6, 11.13.7]. Barriers to large-scale deployment of bioenergy include concerns about GHG emissions from land, food security, water resources, biodiversity conservation and livelihoods. The scientific debate about the overall climate impact related to land-use competition effects of specific bioenergy pathways remains unresolved (*robust evidence, high agreement*). [11.4.4, 11.13] Bioenergy technologies are diverse and span a wide range of options and technology pathways. Evidence suggests that options with low lifecycle emissions (e.g., sugar cane, Miscanthus, fast growing tree species, and sustainable use of biomass residues), some already available, can reduce GHG emissions; outcomes are site-specific and rely on efficient integrated 'biomass-to-bioenergy systems', and sustainable land-use management and governance. In some regions, specific bioenergy options, such as improved cookstoves, and small-scale biogas and biopower production, could reduce GHG emissions and improve livelihoods and health in the context of sustainable development (*medium evidence, medium agreement*). [11.13]

SPM.4.2.5   Human settlements, infrastructure and spatial planning

Urbanization is a global trend and is associated with increases in income, and higher urban incomes are correlated with higher consumption of energy and GHG emissions (*medium evidence, high agreement*). As of 2011, more than 52 % of the global population lives in urban areas. In 2006, urban areas accounted for 67–76 % of energy use and 71–76 % of energy-related CO$_2$ emissions. By 2050, the urban population is expected to increase to 5.6–7.1 billion, or 64–69 % of world population. Cities in non-Annex I countries generally have higher levels of energy use compared to the national average, whereas cities in Annex I countries generally have lower energy use per capita than national averages (*medium evidence, medium agreement*). [12.2, 12.3]

The next two decades present a window of opportunity for mitigation in urban areas, as a large portion of the world's urban areas will be developed during this period (*limited evidence, high agreement*). Accounting for trends in declining population densities, and continued economic and population growth, urban land cover is projected to expand by 56–310 % between 2000 and 2030. [12.2, 12.3, 12.4, 12.8]

Mitigation options in urban areas vary by urbanization trajectories and are expected to be most effective when policy instruments are bundled (*robust evidence, high agreement*). Infrastructure and urban form are strongly interlinked, and lock-in patterns of land use, transport choice, housing, and behaviour. Effective mitigation strategies involve packages of mutually reinforcing policies, including co-locating high residential with high employment densities,

[27]   See WGIII AR5 Glossary.

SPM

achieving high diversity and integration of land uses, increasing accessibility and investing in public transport and other demand management measures. [8.4, 12.3, 12.4, 12.5, 12.6]

The largest mitigation opportunities with respect to human settlements are in rapidly urbanizing areas where urban form and infrastructure are not locked in, but where there are often limited governance, technical, financial, and institutional capacities (*robust evidence, high agreement*). The bulk of urban growth is expected in small- to medium-size cities in developing countries. The feasibility of spatial planning instruments for climate change mitigation is highly dependent on a city's financial and governance capability. [12.6, 12.7]

Thousands of cities are undertaking climate action plans, but their aggregate impact on urban emissions is uncertain (*robust evidence, high agreement*). There has been little systematic assessment on their implementation, the extent to which emission reduction targets are being achieved, or emissions reduced. Current climate action plans focus largely on energy efficiency. Fewer climate action plans consider land-use planning strategies and cross-sectoral measures to reduce sprawl and promote transit-oriented development[28]. [12.6, 12.7, 12.9]

Successful implementation of urban-scale climate change mitigation strategies can provide co-benefits (*robust evidence, high agreement*). Urban areas throughout the world continue to struggle with challenges, including ensuring access to energy, limiting air and water pollution, and maintaining employment opportunities and competitiveness. Action on urban-scale mitigation often depends on the ability to relate climate change mitigation efforts to local co-benefits (*robust evidence, high agreement*). [12.5, 12.6, 12.7, 12.8]

SPM.5   Mitigation policies and institutions

SPM.5.1   Sectoral and national policies

Substantial reductions in emissions would require large changes in investment patterns. Mitigation scenarios in which policies stabilize atmospheric concentrations (without overshoot) in the range from 430 to 530 ppm CO_2eq by 2100 lead to substantial shifts in annual investment flows during the period 2010–2029 compared to baseline scenarios (Figure SPM.9). Over the next two decades (2010 to 2029), annual investment in conventional fossil fuel technologies associated with the electricity supply sector is projected to decline by about 30 (2–166) billion USD (median: −20% compared to 2010) while annual investment in low-carbon electricity supply (i.e., renewables, nuclear and electricity generation with CCS) is projected to rise by about 147 (31–360) billion USD (median: +100% compared to 2010) (*limited evidence, medium agreement*). For comparison, global total annual investment in the energy system is presently about 1200 billion USD. In addition, annual incremental energy efficiency investments in transport, buildings and industry is projected to increase by about 336 (1–641) billion USD (*limited evidence, medium agreement*), frequently involving modernization of existing equipment. [13.11, 16.2.2]

There is no widely agreed definition of what constitutes climate finance, but estimates of the financial flows associated with climate change mitigation and adaptation are available. Published assessments of all current annual financial flows whose expected effect is to reduce net GHG emissions and/or to enhance resilience to climate change and climate variability show 343 to 385 billion USD per year globally (*medium confidence*) [Box TS.14]. Most of this goes to mitigation. Out of this, total public climate finance that flowed to developing countries is estimated to be between 35 and 49 billion USD/yr in 2011 and 2012 (*medium confidence*). Estimates of international private climate

---

[28]   See WGIII AR5 Glossary.

SPM

Change in Annual Investment Flows from Baseline Levels

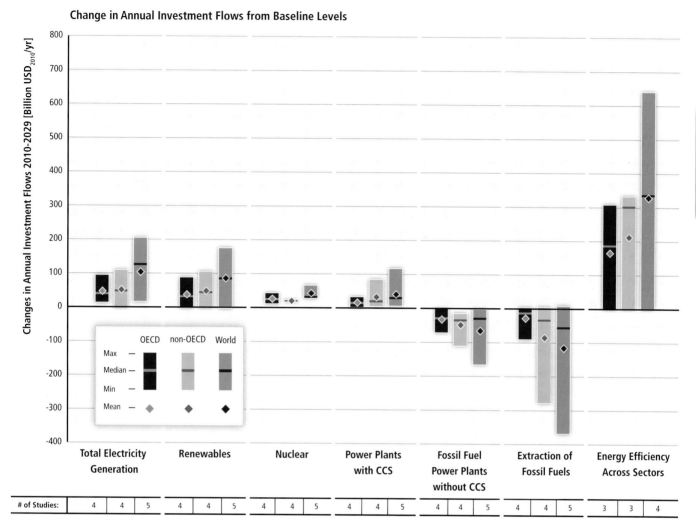

# of Studies:	4	4	5	4	4	5	4	4	5	4	4	5	4	4	5	4	4	5	3	3	4

Figure SPM.9 | Change in annual investment flows from the average baseline level over the next two decades (2010–2029) for mitigation scenarios that stabilize concentrations within the range of approximately 430–530 ppm CO$_2$eq by 2100. Investment changes are based on a limited number of model studies and model comparisons. Total electricity generation (leftmost column) is the sum of renewables, nuclear, power plants with CCS and fossil fuel power plants without CCS. The vertical bars indicate the range between minimum and maximum estimate; the horizontal bar indicates the median. Proximity to this median value does not imply higher likelihood because of the different degree of aggregation of model results, the low number of studies available and different assumptions in the different studies considered. The numbers in the bottom row show the total number of studies in the literature used for the assessment. This underscores that investment needs are still an evolving area of research that relatively few studies have examined. [Figure 16.3]

finance flowing to developing countries range from 10 to 72 billion USD/yr including foreign direct investment as equity and loans in the range of 10 to 37 billion USD/yr over the period of 2008–2011 (*medium confidence*). [16.2.2]

There has been a considerable increase in national and sub-national mitigation plans and strategies since AR4. In 2012, 67 % of global GHG emissions were subject to national legislation or strategies versus 45 % in 2007. However, there has not yet been a substantial deviation in global emissions from the past trend [Figure 1.3c]. These plans and strategies are in their early stages of development and implementation in many countries, making it difficult to assess their aggregate impact on future global emissions (*medium evidence, high agreement*). [14.3.4, 14.3.5, 15.1, 15.2]

Since AR4, there has been an increased focus on policies designed to integrate multiple objectives, increase co-benefits and reduce adverse side-effects (*high confidence*). Governments often explicitly reference co-benefits in climate and sectoral plans and strategies. The scientific literature has sought to assess the size of co-benefits (see Section SPM.4.1) and the greater political feasibility and durability of policies that have large co-benefits and small adverse side-effects. [4.8, 5.7, 6.6, 13.2, 15.2] Despite the growing attention in policymaking and the scientific literature since AR4, the analytical and empirical underpinnings for understanding many of the interactive effects are under-developed [1.2, 3.6.3, 4.2, 4.8, 5.7, 6.6].

Sector-specific policies have been more widely used than economy-wide policies (*medium evidence, high agreement*). Although most economic theory suggests that economy-wide policies for the singular objective of mitigation would be more cost-effective than sector-specific policies, since AR4 a growing number of studies has demonstrated that administrative and political barriers may make economy-wide policies harder to design and implement than sector-specific policies. The latter may be better suited to address barriers or market failures specific to certain sectors, and may be bundled in packages of complementary policies. [6.3.6.5, 8.10, 9.10, 10.10, 15.2, 15.5, 15.8, 15.9]

Regulatory approaches and information measures are widely used, and are often environmentally effective (*medium evidence, medium agreement*). Examples of regulatory approaches include energy efficiency standards; examples of information programmes include labelling programmes that can help consumers make better-informed decisions. While such approaches have often been found to have a net social benefit, the scientific literature is divided on the extent to which such policies can be implemented with negative private costs to firms and individuals. [Box 3.10, 15.5.5, 15.5.6] There is general agreement that rebound effects exist, whereby higher efficiency can lead to lower energy prices and greater consumption, but there is *low agreement* in the literature on the magnitude [3.9.5, 5.7.2, 14.4.2, 15.5.4].

Since AR4, cap and trade systems for GHGs have been established in a number of countries and regions. Their short-run environmental effect has been limited as a result of loose caps or caps that have not proved to be constraining (*limited evidence, medium agreement*). This was related to factors such as the financial and economic crisis that reduced energy demand, new energy sources, interactions with other policies, and regulatory uncertainty. In principle, a cap and trade system can achieve mitigation in a cost-effective way; its implementation depends on national circumstances. Though earlier programmes relied almost exclusively on grandfathering (free allocation of permits), auctioning permits is increasingly applied. If allowances are auctioned, revenues can be used to address other investments with a high social return, and/or reduce the tax and debt burden. [14.4.2, 15.5.3]

In some countries, tax-based policies specifically aimed at reducing GHG emissions—alongside technology and other policies—have helped to weaken the link between GHG emissions and GDP (*high confidence*). In a large group of countries, fuel taxes (although not necessarily designed for the purpose of mitigation) have effects that are akin to sectoral carbon taxes [Table 15.2]. The demand reduction in transport fuel associated with a 1 % price increase is 0.6 % to 0.8 % in the long run, although the short-run response is much smaller [15.5.2]. In some countries revenues are used to reduce other taxes and/or to provide transfers to low-income groups. This illustrates the general principle that mitigation policies that raise government revenue generally have lower social costs than approaches which do not. While it has previously been assumed that fuel taxes in the transport sector are regressive, there have been a number of other studies since AR4 that have shown them to be progressive, particularly in developing countries (*medium evidence, medium agreement*). [3.6.3, 14.4.2, 15.5.2]

The reduction of subsidies for GHG-related activities in various sectors can achieve emission reductions, depending on the social and economic context (*high confidence*). While subsidies can affect emissions in many sectors, most of the recent literature has focused on subsidies for fossil fuels. Since AR4 a small but growing literature based on economy-wide models has projected that complete removal of subsidies for fossil fuels in all countries could result in reductions in global aggregate emissions by mid-century (*medium evidence, medium agreement*) [7.12, 13.13, 14.3.2, 15.5.2]. Studies vary in methodology, the type and definition of subsidies and the time frame for phase out considered. In particular, the studies assess the impacts of complete removal of all fossil fuel subsidies without seeking to assess which subsidies are wasteful and inefficient, keeping in mind national circumstances. Although political economy barriers are substantial, some countries have reformed their tax and budget systems to reduce fuel subsidies. To help reduce possible adverse effects on lower-income groups who often spend a large fraction of their income on energy services, many governments have utilized lump-sum cash transfers or other mechanisms targeted on the poor. [15.5.2]

Interactions between or among mitigation policies may be synergistic or may have no additive effect on reducing emissions (*medium evidence, high agreement*). For instance, a carbon tax can have an additive environmental effect to policies such as subsidies for the supply of RE. By contrast, if a cap and trade system has a binding cap (sufficiently stringent to affect emission-related decisions), then other policies such as RE subsidies have no further impact on reducing emissions within the time period that the cap applies (although they may affect costs and possibly the viability of more stringent future targets) (*medium evidence, high agreement*). In either case, additional policies may be needed to address market failures relating to innovation and technology diffusion. [15.7]

Some mitigation policies raise the prices for some energy services and could hamper the ability of societies to expand access to modern energy services to underserved populations (*low confidence*). **These potential adverse side-effects can be avoided with the adoption of complementary policies** (*medium confidence*). Most notably, about 1.3 billion people worldwide do not have access to electricity and about 3 billion are dependent on traditional solid fuels for cooking and heating with severe adverse effects on health, ecosystems and development. Providing access to modern energy services is an important sustainable development objective. The costs of achieving nearly universal access to electricity and clean fuels for cooking and heating are projected to be between 72 and 95 billion USD per year until 2030 with minimal effects on GHG emissions (*limited evidence, medium agreement*). A transition away from the use of traditional biomass[29] and the more efficient combustion of solid fuels reduce air pollutant emissions, such as sulfur dioxide (SO_2), nitrogen oxides (NO_x), carbon monoxide (CO), and black carbon (BC), and thus yield large health benefits (*high confidence*). [4.3, 6.6, 7.9, 9.3, 9.7, 11.13.6, 16.8]

Technology policy complements other mitigation policies (*high confidence*). Technology policy includes technology-push (e.g., publicly funded R&D) and demand-pull (e.g., governmental procurement programmes). Such policies address market failures related to innovation and technology diffusion. [3.11, 15.6] Technology support policies have promoted substantial innovation and diffusion of new technologies, but the cost-effectiveness of such policies is often difficult to assess [2.6.5, 7.12, 9.10]. Nevertheless, program evaluation data can provide empirical evidence on the relative effectiveness of different policies and can assist with policy design [15.6.5].

In many countries, the private sector plays central roles in the processes that lead to emissions as well as to mitigation. Within appropriate enabling environments, the private sector, along with the public sector, can play an important role in financing mitigation (*medium evidence, high agreement*). The share of total mitigation finance from the private sector, acknowledging data limitations, is estimated to be on average between two-thirds and three-fourths on the global level (2010–2012) (*limited evidence, medium agreement*). In many countries, public finance interventions by governments and national and international development banks encourage climate investments by the private sector [16.2.1] and provide finance where private sector investment is limited. The quality of a country's enabling environment includes the effectiveness of its institutions, regulations and guidelines regarding the private sector, security of property rights, credibility of policies and other factors that have a substantial impact on whether private firms invest in new technologies and infrastructures [16.3]. Dedicated policy instruments, for example, credit insurance, power purchase agreements and feed-in tariffs, concessional finance or rebates, provide an incentive for investment by lowering risks for private actors [16.4].

---

[29]   See WGIII AR5 Glossary.

SPM.5.2

International cooperation

The United Nations Framework Convention on Climate Change (UNFCCC) is the main multilateral forum focused on addressing climate change, with nearly universal participation. Other institutions organized at different levels of governance have resulted in diversifying international climate change cooperation. [13.3.1, 13.4.1.4, 13.5]

Existing and proposed international climate change cooperation arrangements vary in their focus and degree of centralization and coordination. They span: multilateral agreements, harmonized national policies and decentralized but coordinated national policies, as well as regional and regionally-coordinated policies. [Figure TS.38, 13.4.1, 13.13.2, 14.4]

The Kyoto Protocol offers lessons towards achieving the ultimate objective of the UNFCCC, particularly with respect to participation, implementation, flexibility mechanisms, and environmental effectiveness (*medium evidence, low agreement*). [5.3.3, 13.3.4, 13.7.2, 13.13.1.1, 13.13.1.2, 14.3.7.1, Table TS.9]

UNFCCC activities since 2007 have led to an increasing number of institutions and other arrangements for international climate change cooperation. [13.5.1.1, 13.13.1.3, 16.2.1]

Policy linkages among regional, national, and sub-national climate policies offer potential climate change mitigation and adaptation benefits (*medium evidence, medium agreement*). Linkages can be established between national policies, various instruments, and through regional cooperation. [13.3.1, 13.5.3, 13.6, 13.7, 13.13.2.3, 14.4, Figure 13.4]

Various regional initiatives between the national and global scales are either being developed or implemented, but their impact on global mitigation has been limited to date (*medium confidence*). Many climate policies can be more effective if implemented across geographical regions. [13.13, 13.6, 14.4, 14.5]

Technical Summary

Technical Summary

Coordinating Lead Authors:
Ottmar Edenhofer (Germany), Ramón Pichs-Madruga (Cuba), Youba Sokona (Mali/Switzerland), Susanne Kadner (Germany), Jan C. Minx (Germany), Steffen Brunner (Germany)

Lead Authors:
Shardul Agrawala (France), Giovanni Baiocchi (UK/Italy), Igor Alexeyevich Bashmakov (Russian Federation), Gabriel Blanco (Argentina), John Broome (UK), Thomas Bruckner (Germany), Mercedes Bustamante (Brazil), Leon Clarke (USA), Mariana Conte Grand (Argentina), Felix Creutzig (Germany), Xochitl Cruz-Núñez (Mexico), Shobhakar Dhakal (Nepal/Thailand), Navroz K. Dubash (India), Patrick Eickemeier (Germany), Ellie Farahani (Canada/Switzerland/Germany), Manfred Fischedick (Germany), Marc Fleurbaey (France/USA), Reyer Gerlagh (Netherlands), Luis Gómez-Echeverri (Austria/Colombia), Sujata Gupta (India/Philippines), Jochen Harnisch (Germany), Kejun Jiang (China), Frank Jotzo (Germany/Australia), Sivan Kartha (USA), Stephan Klasen (Germany), Charles Kolstad (USA), Volker Krey (Austria/Germany), Howard Kunreuther (USA), Oswaldo Lucon (Brazil), Omar Masera (Mexico), Yacob Mulugetta (Ethiopia/UK), Richard Norgaard (USA), Anthony Patt (Austria/Switzerland), Nijavalli H. Ravindranath (India), Keywan Riahi (IIASA/Austria), Joyashree Roy (India), Ambuj Sagar (USA/India), Roberto Schaeffer (Brazil), Steffen Schlömer (Germany), Karen Seto (USA), Kristin Seyboth (USA), Ralph Sims (New Zealand), Pete Smith (UK), Eswaran Somanathan (India), Robert Stavins (USA), Christoph von Stechow (Germany), Thomas Sterner (Sweden), Taishi Sugiyama (Japan), Sangwon Suh (Republic of Korea/USA), Kevin Urama (Nigeria/UK/Kenya), Diana Ürge-Vorsatz (Hungary), Anthony Venables (UK), David G. Victor (USA), Elke Weber (USA), Dadi Zhou (China), Ji Zou (China), Timm Zwickel (Germany)

Contributing Authors:
Adolf Acquaye (Ghana/UK), Kornelis Blok (Netherlands), Gabriel Chan (USA), Jan Fuglestvedt (Norway), Edgar Hertwich (Austria/Norway), Elmar Kriegler (Germany), Oliver Lah (Germany), Sevastianos Mirasgedis (Greece), Carmenza Robledo Abad (Switzerland/Colombia), Claudia Sheinbaum (Mexico), Steven J. Smith (USA), Detlef van Vuuren (Netherlands)

Review Editors:
Tomás Hernández-Tejeda (Mexico), Roberta Quadrelli (IEA/Italy)

This summary should be cited as:

Edenhofer O., R. Pichs-Madruga, Y. Sokona, S. Kadner, J.C. Minx, S. Brunner, S. Agrawala, G. Baiocchi, I.A. Bashmakov, G. Blanco, J. Broome, T. Bruckner, M. Bustamante, L. Clarke, M. Conte Grand, F. Creutzig, X. Cruz-Núñez, S. Dhakal, N.K. Dubash, P. Eickemeier, E. Farahani, M. Fischedick, M. Fleurbaey, R. Gerlagh, L. Gómez-Echeverri, S. Gupta, J. Harnisch, K. Jiang, F. Jotzo, S. Kartha, S. Klasen, C. Kolstad, V. Krey, H. Kunreuther, O. Lucon, O. Masera, Y. Mulugetta, R.B. Norgaard, A. Patt, N.H. Ravindranath, K. Riahi, J. Roy, A. Sagar, R. Schaeffer, S. Schlömer, K.C. Seto, K. Seyboth, R. Sims, P. Smith, E. Somanathan, R. Stavins, C. von Stechow, T. Sterner, T. Sugiyama, S. Suh, D. Ürge-Vorsatz, K. Urama, A. Venables, D.G. Victor, E. Weber, D. Zhou, J. Zou, and T. Zwickel, 2014: Technical Summary. In: *Climate Change 2014: Mitigation of Climate Change. Contribution of Working Group III to the Fifth Assessment Report of the Intergovernmental Panel on Climate Change* [Edenhofer, O., R. Pichs-Madruga, Y. Sokona, E. Farahani, S. Kadner, K. Seyboth, A. Adler, I. Baum, S. Brunner, P. Eickemeier, B. Kriemann, J. Savolainen, S. Schlömer, C. von Stechow, T. Zwickel and J.C. Minx (eds.)]. Cambridge University Press, Cambridge, United Kingdom and New York, NY, USA.

Contents

TS

TS

TS.1   Introduction and framing

'Mitigation', in the context of climate change, is a human intervention to reduce the sources or enhance the sinks of greenhouse gases (GHGs). One of the central messages from Working Groups I and II of the Intergovernmental Panel on Climate Change (IPCC) is that the consequences of unchecked climate change for humans and natural ecosystems are already apparent and increasing. The most vulnerable systems are already experiencing adverse effects. Past GHG emissions have already put the planet on a track for substantial further changes in climate, and while there are many uncertainties in factors such as the sensitivity of the climate system many scenarios lead to substantial climate impacts, including direct harms to human and ecological well-being that exceed the ability of those systems to adapt fully.

Because mitigation is intended to reduce the harmful effects of climate change, it is part of a broader policy framework that also includes adaptation to climate impacts. Mitigation, together with adaptation to climate change, contributes to the objective expressed in Article 2 of the United Nations Framework Convention on Climate Change (UNFCCC) to stabilize "greenhouse gas concentrations in the atmosphere at a level to prevent dangerous anthropogenic interference with the climate system […] within a time frame sufficient to allow ecosystems to adapt […] to ensure that food production is not threatened and to enable economic development to proceed in a sustainable manner". However, Article 2 is hard to interpret, as concepts such as 'dangerous' and 'sustainable' have different meanings in different decision contexts (see Box TS.1).[1] Moreover, natural science is unable to predict precisely the response of the climate system to rising GHG

---

[1]   Boxes throughout this summary provide background information on main research concepts and methods that were used to generate insight.

Box TS.1 | Many disciplines aid decision making on climate change

Something is dangerous if it leads to a significant risk of considerable harm. Judging whether human interference in the climate system is dangerous therefore divides into two tasks. One is to estimate the risk in material terms: what the material consequences of human interference might be and how likely they are. The other is to set a value on the risk: to judge how harmful it will be.

The first is a task for natural science, but the second is not [Section 3.1]. As the Synthesis Report of AR4 states, "Determining what constitutes 'dangerous anthropogenic interference with the climate system' in relation to Article 2 of the UNFCCC involves value judgements". Judgements of value (valuations) are called for, not just here, but at almost every turn in decision making about climate change [3.2]. For example, setting a target for mitigation involves judging the value of losses to people's well-being in the future, and comparing it with the value of benefits enjoyed now. Choosing whether to site wind turbines on land or at sea requires a judgement of the value of landscape in comparison with the extra cost of marine turbines. To estimate the social cost of carbon is to value the harm that GHG emissions do [3.9.4].

Different values often conflict, and they are often hard to weigh against each other. Moreover, they often involve the conflicting interests of different people, and are subject to much debate and disagreement. Decision makers must therefore find ways to mediate among different interests and values, and also among differing viewpoints about values. [3.4, 3.5]

Social sciences and humanities can contribute to this process by improving our understanding of values in ways that are illustrated in the boxes contained in this summary. The sciences of human and social behaviour—among them psychology, political science, sociology, and non-normative branches of economics—investigate the values people have, how they change through time, how they can be influenced by political processes, and how the process of making decisions affects their acceptability. Other disciplines, including ethics (moral philosophy), decision theory, risk analysis, and the normative branch of economics, investigate, analyze, and clarify values themselves [2.5, 3.4, 3.5, 3.6]. These disciplines offer practical ways of measuring some values and trading off conflicting interests. For example, the discipline of public health often measures health by means of 'disability-adjusted life years' [3.4.5]. Economics uses measures of social value that are generally based on monetary valuation but can take account of principles of distributive justice [3.6, 4.2, 4.7, 4.8]. These normative disciplines also offer practical decision-making tools, such as expected utility theory, decision analysis, cost-benefit and cost-effectiveness analysis, and the structured use of expert judgment [2.5, 3.6, 3.7, 3.9].

There is a further element to decision making. People and countries have rights and owe duties towards each other. These are matters of justice, equity, or fairness. They fall within the subject matter of moral and political philosophy, jurisprudence, and economics. For example, some have argued that countries owe restitution for the harms that result from their past GHG emissions, and it has been debated, on jurisprudential and other grounds, whether restitution is owed only for harms that result from negligent or blameworthy GHG emissions. [3.3, 4.6]

concentrations nor fully understand the harm it will impose on individuals, societies, and ecosystems. Article 2 requires that societies balance a variety of considerations—some rooted in the impacts of climate change itself and others in the potential costs of mitigation and adaptation. The difficulty of that task is compounded by the need to develop a consensus on fundamental issues such as the level of risk that societies are willing to accept and impose on others, strategies for sharing costs, and how to balance the numerous tradeoffs that arise because mitigation intersects with many other goals of societies. Such issues are inherently value-laden and involve different actors who have varied interests and disparate decision-making power.

The Working Group III (WGIII) contribution to the IPCC's Fifth Assessment Report (AR5) assesses literature on the scientific, technological, environmental, economic and social aspects of mitigation of climate change. It builds upon the WGIII contribution to the IPCC's Fourth Assessment Report (AR4), the Special Report on Renewable Energy Sources and Climate Change Mitigation (SRREN) and previous reports and incorporates subsequent new findings and research. Throughout, the focus is on the implications of its findings for policy, without being prescriptive about the particular policies that governments and other important participants in the policy process should adopt. In light of the IPCC's mandate, authors in WGIII were guided by several principles when assembling this assessment: (1) to be explicit about mitigation options, (2) to be explicit about their costs and about their risks and opportunities vis-à-vis other development priorities, (3) and to be explicit about the underlying criteria, concepts, and methods for evaluating alternative policies.

The remainder of this summary offers the main findings of this report. The degree of certainty in findings, as in the reports of all three IPCC Working Groups, is based on the author teams' evaluations of underlying scientific understanding and is expressed as a qualitative level of confidence (from very low to very high) and, when possible, probabilistically with a quantified likelihood (from exceptionally unlikely to virtually certain). Confidence in the validity of a finding is based on the type, amount, quality, and consistency of evidence (e.g., data, mechanistic understanding, theory, models, expert judgment) and the degree of agreement. Probabilistic estimates of quantified measures of uncertainty in a finding are based on statistical analysis of observations or model results, or both, and expert judgment.[2] Where appropriate, findings are also formulated as statements of fact without using uncertainty qualifiers. Within paragraphs of this summary, the confidence, evidence, and agreement terms given for a bolded finding apply to subsequent statements in the paragraph, unless additional terms are provided. References in [square brackets] indicate chapters, sections, figures, tables, and boxes where supporting evidence in the underlying report can be found.

This section continues with providing a framing of important concepts and methods that help to contextualize the findings presented in subsequent sections. Section TS.2 presents evidence on past trends in stocks and flows of GHGs and the factors that drive emissions at the global, regional, and sectoral scales including economic growth, technology, or population changes. Section TS.3.1 provides findings from studies that analyze the technological, economic, and institutional requirements of long-term mitigation scenarios. Section TS.3.2 provides details on mitigation measures and policies that are used within and across different economic sectors and human settlements. Section TS.4 summarizes insights on the interactions of mitigation policies between governance levels, economic sectors, and instrument types.

Climate change is a global commons problem that implies the need for international cooperation in tandem with local, national, and regional policies on many distinct matters. Because the GHG emissions of any agent (individual, company, country) affect every other agent, an effective outcome will not be achieved if individual agents advance their interests independently of others. International cooperation can contribute by defining and allocating rights and responsibilities with respect to the atmosphere [Sections 1.2.4, 3.1, 4.2, 13.2.1]. Moreover, research and development (R&D) in support of mitigation is a public good, which means that international cooperation can play a constructive role in the coordinated development and diffusion of technologies [1.4.4, 3.11, 13.9, 14.4.3]. This gives rise to separate needs for cooperation on R&D, opening up of markets, and the creation of incentives to encourage private firms to develop and deploy new technologies and households to adopt them.

International cooperation on climate change involves ethical considerations, including equitable effort-sharing. Countries have contributed differently to the build-up of GHG in the atmosphere, have varying capacities to contribute to mitigation and adaptation, and have different levels of vulnerability to climate impacts. Many less developed countries are exposed to the greatest impacts but have contributed least to the problem. Engaging countries in effective international cooperation may require strategies for sharing the costs and benefits of mitigation in ways that are perceived to be equitable [4.2]. Evidence suggests that perceived fairness can influence the level of cooperation among individuals, and that finding may suggest that processes and outcomes seen as fair will lead to more international cooperation as well [3.10, 13.2.2.4]. Analysis contained in the literature of moral and political philosophy can contribute to resolving ethical questions raised by climate change [3.2, 3.3, 3.4]. These questions include how much overall mitigation is needed to avoid 'dangerous interference with the climate system' (Box

---

[2] The following summary terms are used to describe the available evidence: limited, medium, or robust; and for the degree of agreement: low, medium, or high. A level of confidence is expressed using five qualifiers: very low, low, medium, high, and very high, and typeset in italics, e.g., *medium confidence*. For a given evidence and agreement statement, different confidence levels can be assigned, but increasing levels of evidence and degrees of agreement are correlated with increasing confidence. The following terms have been used to indicate the assessed likelihood of an outcome or a result: virtually certain 99–100 % probability, very likely 90–100 %, likely 66–100 %, about as likely as not 33–66 %, unlikely 0–33 %, very unlikely 0–10 %, exceptionally unlikely 0–1 %. Additional terms (more likely than not > 50–100 %, and more unlikely than likely 0 –< 50 %) may also be used when appropriate. Assessed likelihood is typeset in italics, e.g., *very likely*. For more details, please refer to the Guidance Note for Lead Authors of the IPCC Fifth Assessment Report on Consistent Treatment of Uncertainties, available at http://www.ipcc.ch/pdf/supporting-material/uncertainty-guidance-note.pdf.

Box TS.2 | Mitigation brings both market and non-market benefits to humanity

The impacts of mitigation consist in the reduction or elimination of some of the effects of climate change. Mitigation may improve people's livelihood, their health, their access to food or clean water, the amenities of their lives, or the natural environment around them.

Mitigation can improve human well-being through both market and non-market effects. Market effects result from changes in market prices, in people's revenues or net income, or in the quality or availability of market commodities. Non-market effects result from changes in the quality or availability of non-marketed goods such as health, quality of life, culture, environmental quality, natural ecosystems, wildlife, and aesthetic values. Each impact of climate change can generate both market and non-market damages. For example, a heat wave in a rural area may cause heat stress for exposed farm labourers, dry up a wetland that serves as a refuge for migratory birds, or kill some crops and damage others. Avoiding these damages is a benefit of mitigation. [3.9]

Economists often use monetary units to value the damage done by climate change and the benefits of mitigation. The monetized value of a benefit to a person is the amount of income the person would be willing to sacrifice in order to get it, or alternatively the amount she would be willing to accept as adequate compensation for not getting it. The monetized value of a harm is the amount of income she would be willing to sacrifice in order to avoid it, or alternatively the amount she would be willing to accept as adequate compensation for suffering it. Economic measures seek to capture how strongly individuals care about one good or service relative to another, depending on their individual interests, outlook, and economic circumstances. [3.9]

Monetary units can be used in this way to measure costs and benefits that come at different times and to different people. But it cannot be presumed that a dollar to one person at one time can be treated as equivalent to a dollar to a different person or at a different time. Distributional weights may need to be applied between people [3.6.1], and discounting (see Box TS.10) may be appropriate between times. [3.6.2]

TS.1) [3.1], how the effort or cost of mitigating climate change should be shared among countries and between the present and future [3.3, 3.6, 4.6], how to account for such factors as historical responsibility for GHG emissions [3.3, 4.6], and how to choose among alternative policies for mitigation and adaptation [3.4, 3.5, 3.6, 3.7]. Ethical issues of well-being, justice, fairness, and rights are all involved. Ethical analysis can identify the different ethical principles that underlie different viewpoints, and distinguish correct from incorrect ethical reasoning [3.3, 3.4].

Evaluation of mitigation options requires taking into account many different interests, perspectives, and challenges between and within societies. Mitigation engages many different agents, such as governments at different levels—regionally [14.1], nationally and locally [15.1], and through international agreements [13.1]—as well as households, firms, and other non-governmental actors. The interconnections between different levels of decision making and among different actors affect the many goals that become linked with climate policy. Indeed, in many countries the policies that have (or could have) the largest impact on emissions are motivated not solely by concerns surrounding climate change. Of particular importance are the interactions and perceived tensions between mitigation and development [4.1, 14.1]. Development involves many activities, such as enhancing access to modern energy services [7.9.1, 14.3.2, 16.8], the building of infrastructures [12.1], ensuring food security [11.1], and eradicating poverty [4.1]. Many of these activities can lead to higher emissions, if achieved by conventional means. Thus, the relationships between development and mitigation can lead to political and ethical conun-

drums, especially for developing countries, when mitigation is seen as exacerbating urgent development challenges and adversely affecting the current well-being of their populations [4.1]. These conundrums are examined throughout this report, including in special boxes highlighting the concerns of developing countries.

Economic evaluation can be useful for policy design and be given a foundation in ethics, provided appropriate distributional weights are applied. While the limitations of economics are widely documented [2.4, 3.5], economics nevertheless provides useful tools for assessing the pros and cons of mitigation and adaptation options. Practical tools that can contribute to decision making include cost-benefit analysis, cost-effectiveness analysis, multi-criteria analysis, expected utility theory, and methods of decision analysis [2.5, 3.7.2]. Economic valuation (see Box TS.2) can be given a foundation in ethics, provided distributional weights are applied that take proper account of the difference in the value of money to rich and poor people [3.6]. Few empirical applications of economic valuation to climate change have been well-founded in this respect [3.6.1]. The literature provides significant guidance on the social discount rate for consumption (see Box TS.10), which is in effect inter-temporal distributional weighting. It suggests that the social discount rate depends in a well-defined way primarily on the anticipated growth in per capita income and inequality aversion [3.6.2].

Most climate policies intersect with other societal goals, either positively or negatively, creating the possibility of 'co-benefits'

Box TS.3 | Deliberative and intuitive thinking are inputs to effective risk management

When people—from individual voters to key decision makers in firms to senior government policymakers—make choices that involve risk and uncertainty, they rely on deliberative as well intuitive thought processes. Deliberative thinking is characterized by the use of a wide range of formal methods to evaluate alternative choices when probabilities are difficult to specify and/or outcomes are uncertain. They can enable decision makers to compare choices in a systematic manner by taking into account both short and long-term consequences. A strength of these methods is that they help avoid some of the well-known pitfalls of intuitive thinking, such as the tendency of decision makers to favour the status quo. A weakness of these deliberative decision aids is that they are often highly complex and require considerable time and attention.

Most analytically based literature, including reports such as this one, is based on the assumption that individuals undertake deliberative and systematic analyses in comparing options. However, when making mitigation and adaptation choices, people are also likely to engage in intuitive thinking. This kind of thinking has the advantage of requiring less extensive analysis than deliberative

thinking. However, relying on one's intuition may not lead one to characterize problems accurately when there is limited past experience. Climate change is a policy challenge in this regard since it involves large numbers of complex actions by many diverse actors, each with their own values, goals, and objectives. Individuals are likely to exhibit well-known patterns of intuitive thinking such as making choices related to risk and uncertainty on the basis of emotional reactions and the use of simplified rules that have been acquired by personal experience. Other tendencies include misjudging probabilities, focusing on short time horizons, and utilizing rules of thumb that selectively attend to subsets of goals and objectives. [2.4]

By recognizing that both deliberative and intuitive modes of decision making are prevalent in the real world, risk management programmes can be developed that achieve their desired impacts. For example, alternative frameworks that do not depend on precise specification of probabilities and outcomes can be considered in designing mitigation and adaptation strategies for climate change. [2.4, 2.5, 2.6]

or 'adverse side-effects'. Since the publication of AR4, a substantial body of literature has emerged looking at how countries that engage in mitigation also address other goals, such as local environmental protection or energy security, as a 'co-benefit' and conversely [1.2.1, 6.6.1, 4.8]. This multi-objective perspective is important because it helps to identify areas where political, administrative, stakeholder, and other support for policies that advance multiple goals will be robust. Moreover, in many societies the presence of multiple objectives may make it easier for governments to sustain the political support needed for mitigation [15.2.3]. Measuring the net effect on social welfare (see Box TS.11) requires examining the interaction between climate policies and pre-existing other policies [3.6.3, 6.3.6.5].

Mitigation efforts generate tradeoffs and synergies with other societal goals that can be evaluated in a sustainable development framework. The many diverse goals that societies value are often called 'sustainable development'. A comprehensive assessment of climate policy therefore involves going beyond a narrow focus on distinct mitigation and adaptation options and their specific co-benefits and adverse side-effects. Instead it entails incorporating climate issues into the design of comprehensive strategies for equitable and sustainable development at regional, national, and local levels [4.2, 4.5]. Maintaining and advancing human well-being, in particular overcoming poverty and reducing inequalities in living standards, while avoiding unsustainable patterns of consumption and production, are fundamental aspects of equitable and sustainable development [4.4, 4.6, 4.8]. Because these aspects are deeply rooted in how societies for-

mulate and implement economic and social policies generally, they are critical to the adoption of effective climate policy.

Variations in goals reflect, in part, the fact that humans perceive risks and opportunities differently. Individuals make their decisions based on different goals and objectives and use a variety of different methods in making choices between alternative options. These choices and their outcomes affect the ability of different societies to cooperate and coordinate. Some groups put greater emphasis on near-term economic development and mitigation costs, while others focus more on the longer-term ramifications of climate change for prosperity. Some are highly risk averse while others are more tolerant of dangers. Some have more resources to adapt to climate change and others have fewer. Some focus on possible catastrophic events while others ignore extreme events as implausible. Some will be relative winners, and some relative losers from particular climate changes. Some have more political power to articulate their preferences and secure their interests and others have less. Since AR4, awareness has grown that such considerations—long the domain of psychology, behavioural economics, political economy, and other disciplines—need to be taken into account in assessing climate policy (see Box TS.3). In addition to the different perceptions of climate change and its risks, a variety of norms can also affect what humans view as acceptable behaviour. Awareness has grown about how such norms spread through social networks and ultimately affect activities, behaviours and lifestyles, and thus development pathways, which can have profound impacts on GHG emissions and mitigation policy. [1.4.2, 2.4, 3.8, 3.10, 4.3]

Box TS.4 | 'Fat tails': unlikely vs. likely outcomes in understanding the value of mitigation

What has become known as the 'fat-tails' problem relates to uncertainty in the climate system and its implications for mitigation and adaptation policies. By assessing the chain of structural uncertainties that affect the climate system, the resulting compound probability distribution of possible economic damage may have a fat right tail. That means that the probability of damage does not decline with increasing temperature as quickly as the consequences rise.

The significance of fat tails can be illustrated for the distribution of temperature that will result from a doubling of atmospheric carbon dioxide (CO_2) (climate sensitivity). IPCC Working Group I (WGI) estimates may be used to calibrate two possible distributions, one fat-tailed and one thin-tailed, that each have a median temperature change of 3 °C and a 15 % probability of a temperature change in excess of 4.5 °C. Although the probability of exceeding 4.5 °C is the same for both distributions, likelihood drops off much more slowly with increasing temperature for the

fat-tailed compared to the thin-tailed distribution. For example, the probability of temperatures in excess of 8 °C is nearly ten times greater with the chosen fat-tailed distribution than with the thin-tailed distribution. If temperature changes are characterized by a fat tailed distribution, and events with large impact may occur at higher temperatures, then tail events can dominate the computation of expected damages from climate change.

In developing mitigation and adaptation policies, there is value in recognizing the higher likelihood of tail events and their consequences. In fact, the nature of the probability distribution of temperature change can profoundly change how climate policy is framed and structured. Specifically, fatter tails increase the importance of tail events (such as 8 °C warming). While research attention and much policy discussion have focused on the most likely outcomes, it may be that those in the tail of the probability distribution are more important to consider. [2.5, 3.9.2]

Effective climate policy involves building institutions and capacity for governance. While there is strong evidence that a transition to a sustainable and equitable path is technically feasible, charting an effective and viable course for climate change mitigation is not merely a technical exercise. It will involve myriad and sequential decisions among states and civil society actors. Such a process benefits from the education and empowerment of diverse actors to participate in systems of decision making that are designed and implemented with procedural equity as a deliberate objective. This applies at the national as well as international levels, where effective governance relating to global common resources, in particular, is not yet mature. Any given approach has potential winners and losers. The political feasibility of that approach will depend strongly on the distribution of power, resources, and decision-making authority among the potential winners and losers. In a world characterized by profound disparities, procedurally equitable systems of engagement, decision making and governance may help enable a polity to come to equitable solutions to the sustainable development challenge. [4.3]

Effective risk management of climate change involves considering uncertainties in possible physical impacts as well as human and social responses. Climate change mitigation and adaptation is a risk management challenge that involves many different decision-making levels and policy choices that interact in complex and often unpredictable ways. Risks and uncertainties arise in natural, social, and technological systems. As Box TS.3 explains, effective risk management strategies not only consider people's values, and their intuitive decision processes but utilize formal models and decision aids for systematically addressing issues of risk and uncertainty [2.4, 2.5]. Research on other such complex and uncertainty-laden policy domains suggest the

importance of adopting policies and measures that are robust across a variety of criteria and possible outcomes [2.5]. As detailed in Box TS.4, a special challenge arises with the growing evidence that climate change may result in extreme impacts whose trigger points and outcomes are shrouded in high levels of uncertainty [2.5, 3.9.2]. A risk management strategy for climate change will require integrating responses in mitigation with different time horizons, adaptation to an array of climate impacts, and even possible emergency responses such as 'geoengineering' in the face of extreme climate impacts [1.4.2, 3.3.7, 6.9, 13.4.4]. In the face of potential extreme impacts, the ability to quickly offset warming could help limit some of the most extreme climate impacts although deploying these geoengineering systems could create many other risks (see Section TS.3.1.3). One of the central challenges in developing a risk management strategy is to have it adaptive to new information and different governing institutions [2.5].

TS.2 Trends in stocks and flows of greenhouse gases and their drivers

This section summarizes historical GHG emissions trends and their underlying drivers. As in most of the underlying literature, all aggregate GHG emissions estimates are converted to CO_2-equivalents based on Global Warming Potentials with a 100-year time horizon (GWP_{100}) (Box TS.5). The majority of changes in GHG emissions trends that are observed in this section are related to changes in drivers such as eco-

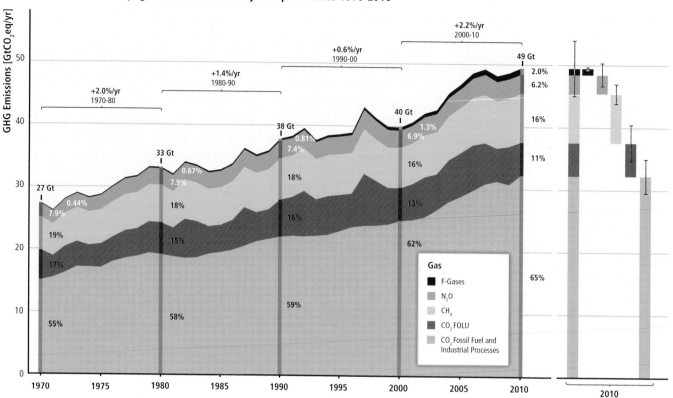

Figure TS.1 | Total annual anthropogenic GHG emissions (GtCO$_2$eq/yr) by groups of gases 1970–2010: carbon dioxide (CO$_2$) from fossil fuel combustion and industrial processes; CO$_2$ from Forestry and Other Land Use[4] (FOLU); methane (CH$_4$); nitrous oxide (N$_2$O); fluorinated gases[5] covered under the Kyoto Protocol (F-gases). At the right side of the figure, GHG emissions in 2010 are shown again broken down into these components with the associated uncertainties (90 % confidence interval) indicated by the error bars. Total anthropogenic GHG emissions uncertainties are derived from the individual gas estimates as described in Chapter 5 [5.2.3.6]. Emissions are converted into CO$_2$-equivalents based on Global Warming Potentials with a 100-year time horizon (GWP$_{100}$) from the IPCC Second Assessment Report (SAR). The emissions data from FOLU represents land-based CO$_2$ emissions from forest and peat fires and decay that approximate to the net CO$_2$ flux from FOLU as described in Chapter 11 of this report. Average annual GHG emissions growth rates for the four decades are highlighted with the brackets. The average annual growth rate from 1970 to 2000 is 1.3 %. [Figure 1.3]

nomic growth, technological change, human behaviour, or population growth. But there are also some smaller changes in GHG emissions estimates that are due to refinements in measurement concepts and methods that have happened since AR4. There is a growing body of literature on uncertainties in global GHG emissions data sets. This section tries to make these uncertainties explicit and reports variations in estimates across global data sets wherever possible.

TS.2.1 Greenhouse gas emission trends

Total anthropogenic GHG emissions have risen more rapidly from 2000 to 2010 than in the previous three decades (*high confidence*). Total anthropogenic GHG emissions were the highest in human history from 2000 to 2010 and reached 49 (±4.5) gigatonnes CO$_2$-equivalents per year (GtCO$_2$eq/yr) in 2010.[3] Current trends are at the high end of levels that had been projected for this last decade. GHG emissions growth has occurred despite the presence of a wide

array of multilateral institutions as well as national policies aimed at mitigation. From 2000 to 2010, GHG emissions grew on average by 1.0 GtCO$_2$eq (2.2 %) per year compared to 0.4 GtCO$_2$eq (1.3 %) per year over the entire period from 1970 to 2000 (Figure TS.1). The global economic crisis 2007/2008 has only temporarily reduced GHG emissions. [1.3, 5.2, 13.3, 15.2.2, Figure 15.1]

---

[3] In this summary, uncertainty in historic GHG emissions data is reported using 90 % uncertainty intervals unless otherwise stated. GHG emissions levels are rounded to two significant digits throughout this document; as a consequence, small differences in sums due to rounding may occur.

[4] FOLU (Forestry and Other Land Use)—also referred to as LULUCF (Land Use, Land-Use Change, and Forestry)—is the subset of Agriculture, Forestry, and Other Land Use (AFOLU) emissions and removals of GHGs related to direct human-induced land use, land-use change and forestry activities excluding agricultural emissions (see WGIII AR5 Glossary).

[5] In this report, data on non-CO$_2$ GHGs, including fluorinated gases, are taken from the EDGAR database (see Annex II.9), which covers substances included in the Kyoto Protocol in its first commitment period.

Total Anthropogenic CO$_2$ Emissions from Fossil Fuel Combustion, Flaring, Cement, as well as Forestry and Other Land Use (FOLU) by Region between 1750 and 2010

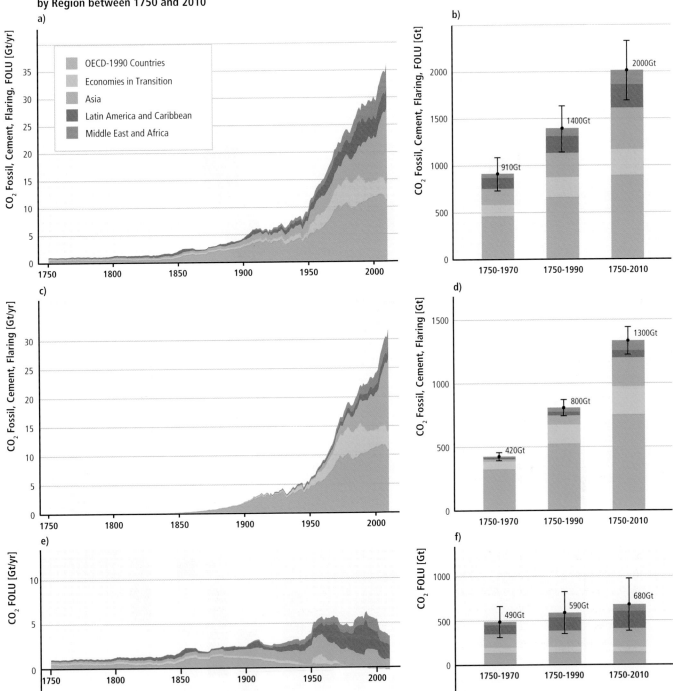

Figure TS.2 | Historical anthropogenic CO$_2$ emissions from fossil fuel combustion, flaring, cement, and Forestry and Other Land Use (FOLU)[4] in five major world regions: OECD-1990 (blue); Economies in Transition (yellow); Asia (green); Latin America and Caribbean (red); Middle East and Africa (brown). Emissions are reported in gigatonnes of CO$_2$ per year (GtCO$_2$/yr). Left panels show regional CO$_2$ emissions 1750–2010 from: (a) the sum of all CO$_2$ sources (c+e); (c) fossil fuel combustion, flaring, and cement; and (e) FOLU. The right panels report regional contributions to cumulative CO$_2$ emissions over selected time periods from: (b) the sum of all CO$_2$ sources (d+f); (d) fossil fuel combustion, flaring and cement; and (f) FOLU. Error bars on panels (b), (d) and (f) give an indication of the uncertainty range (90 % confidence interval). See Annex II.2.2 for definitions of regions. [Figure 5.3]

Greenhouse Gas Emissions by Economic Sectors

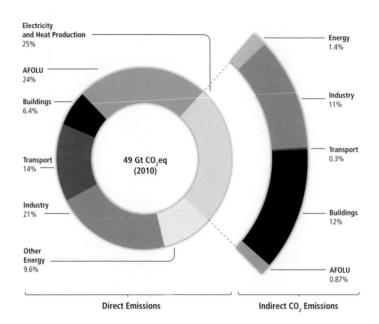

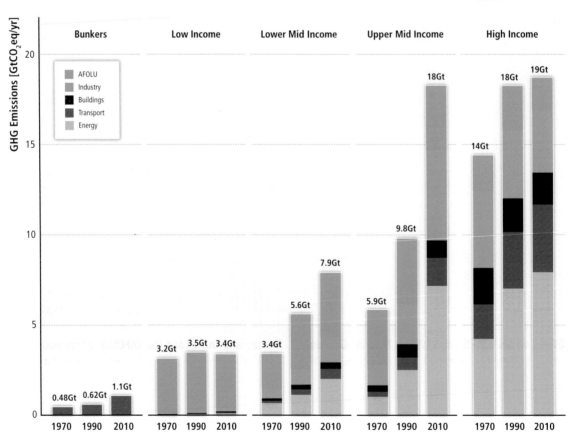

Figure TS.3 | Total anthropogenic GHG emissions (GtCO₂ eq/yr) by economic sectors and country income groups. Upper panel: Circle shows direct GHG emission shares (in % of total anthropogenic GHG emissions) of five major economic sectors in 2010. Pull-out shows how indirect CO₂ emission shares (in % of total anthropogenic GHG emissions) from electricity and heat production are attributed to sectors of final energy use. 'Other Energy' refers to all GHG emission sources in the energy sector other than electricity and heat production. Lower panel: Total anthropogenic GHG emissions in 1970, 1990 and 2010 by five major economic sectors and country income groups. 'Bunkers' refer to GHG emissions from international transportation and thus are not, under current accounting systems, allocated to any particular nation's territory. The emissions data from Agriculture, Forestry and Other Land Use (AFOLU) includes land-based CO₂ emissions from forest and peat fires and decay that approximate to the net CO₂ flux from the Forestry and Other Land Use (FOLU) sub-sector as described in Chapter 11 of this report. Emissions are converted into CO₂-equivalents based on Global Warming Potentials with a 100-year time horizon (GWP₁₀₀) from the IPCC Second Assessment Report (SAR). Assignment of countries to income groups is based on the World Bank income classification in 2013. For details see Annex II.2.3. Sector definitions are provided in Annex II.9.1. [Figure 1.3, Figure 1.6]

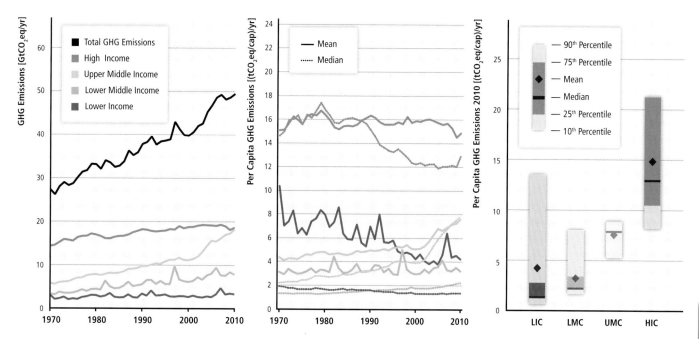

Figure TS.4 | Trends in GHG emissions by country income groups. Left panel: Total annual anthropogenic GHG emissions from 1970 to 2010 (GtCO₂eq/yr). Middle panel: Trends in annual per capita mean and median GHG emissions from 1970 to 2010 (tCO₂eq/cap/yr). Right panel: Distribution of annual per capita GHG emissions in 2010 of countries within each country income group (tCO₂/cap/yr). Mean values show the GHG emissions levels weighed by population. Median values describe GHG emissions levels per capita of the country at the 50th percentile of the distribution within each country income group. Emissions are converted into CO₂-equivalents based on Global Warming Potentials with a 100-year time horizon (GWP₁₀₀) from the IPCC Second Assessment Report (SAR). Assignment of countries to country income groups is based on the World Bank income classification in 2013. For details see Annex II.2.3. [Figures 1.4, 1.8]

CO₂ emissions from fossil fuel combustion and industrial processes contributed about 78 % to the total GHG emissions increase from 1970 to 2010, with similar percentage contribution for the period 2000–2010 (*high confidence*). Fossil fuel-related CO₂ emissions reached 32 (±2.7) GtCO₂/yr in 2010 and grew further by about 3 % between 2010 and 2011 and by about 1–2 % between 2011 and 2012. Since AR4, the shares of the major groups of GHG emissions have remained stable. Of the 49 (±4.5) GtCO₂eq/yr in total anthropogenic GHG emissions in 2010, CO₂ remains the major GHG accounting for 76 % (38±3.8 GtCO₂eq/yr) of total anthropogenic GHG emissions. 16 % (7.8±1.6 GtCO₂eq/yr) come from methane (CH₄), 6.2 % (3.1±1.9 GtCO₂eq/yr) from nitrous oxide (N₂O), and 2.0 % (1.0±0.2 GtCO₂eq/yr) from fluorinated gases (Figure TS.1).[5] Using the most recent GWP₁₀₀ values from the AR5 [WGI 8.7] global GHG emissions totals would be slightly higher (52 GtCO₂eq/yr) and non-CO₂ emission shares would be 20 % for CH₄, 5.0 % for N₂O and 2.2 % for F-gases. Emission shares are sensitive to the choice of emission metric and time horizon, but this has a small influence on global, long-term trends. If a shorter, 20-year time horizon were used, then the share of CO₂ would decline to just over 50 % of total anthropogenic GHG emissions and short-lived gases would rise in relative importance. As detailed in Box TS.5, the choice of emission metric and time horizon involves explicit or implicit value judgements and depends on the purpose of the analysis. [1.2, 3.9, 5.2]

Over the last four decades total cumulative CO₂ emissions have increased by a factor of 2 from about 910 GtCO₂ for the period 1750–1970 to about 2000 GtCO₂ for 1750–2010 (*high confidence*). In 1970, the cumulative CO₂ emissions from fossil fuel combustion, cement production and flaring since 1750 was 420 (±35) GtCO₂; in 2010 that cumulative total had tripled to 1300 (±110) GtCO₂ (Figure TS.2). Cumulative CO₂ emissions associated with FOLU[4] since 1750 increased from about 490 (±180) GtCO₂ in 1970 to approximately 680 (±300) GtCO₂ in 2010. [5.2]

Regional patterns of GHG emissions are shifting along with changes in the world economy (*high confidence*). Since 2000, GHG emissions have been growing in all sectors, except Agriculture, Forestry and Other Land Use (AFOLU)[4] where positive and negative emission changes are reported across different databases and uncertainties in the data are high. More than 75 % of the 10 Gt increase in annual GHG emissions between 2000 and 2010 was emitted in the energy supply (47 %) and industry (30 %) sectors (see Annex II.9.I for sector definitions). 5.9 GtCO₂eq of this sectoral increase occurred in upper-middle income countries,[6] where the most rapid economic development and infrastructure expansion has taken place. GHG emissions growth in the other sectors has been more modest in absolute (0.3–1.1 Gt CO₂eq) as well as in relative terms (3 %–11 %). [1.3, 5.3, Figure 5.18]

---

[6]   When countries are assigned to income groups in this summary, the World Bank income classification for 2013 is used. For details see Annex II.2.3.

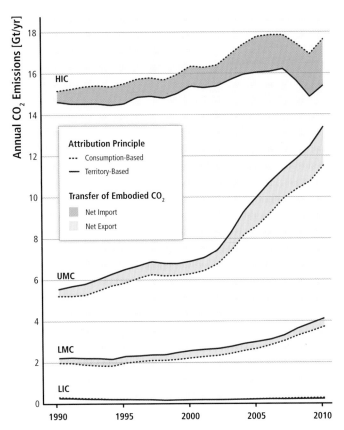

Figure TS.5 | Total annual CO$_2$ emissions (GtCO$_2$/yr) from fossil fuel combustion for country income groups attributed on the basis of territory (solid line) and final consumption (dotted line). The shaded areas are the net CO$_2$ trade balances (differences) between each of the four country income groups and the rest of the world. Blue shading indicates that the country income group is a net importer of embodied CO$_2$ emissions, leading to consumption-based emission estimates that are higher than traditional territorial emission estimates. Orange indicates the reverse situation—the country income group is a net exporter of embodied CO$_2$ emissions. Assignment of countries to country income groups is based on the World Bank income classification in 2013. For details see Annex II.2.3. [Figure 1.5]

Current GHG emission levels are dominated by contributions from the energy supply, AFOLU, and industry sectors; industry and buildings gain considerably in importance if indirect emissions are accounted for (*robust evidence, high agreement*). Of the 49 (±4.5) GtCO$_2$eq emissions in 2010, 35% (17 GtCO$_2$eq) of GHG emissions were released in the energy supply sector, 24% (12 GtCO$_2$eq, net emissions) in AFOLU, 21% (10 GtCO$_2$eq) in industry, 14% (7.0 GtCO$_2$eq) in transport, and 6.4% (3.2 GtCO$_2$eq) in buildings. When indirect emissions from electricity and heat production are assigned to sectors of final energy use, the shares of the industry and buildings sectors in global GHG emissions grow to 31% and 19%,[3] respectively (Figure TS.3 upper panel). [1.3, 7.3, 8.2, 9.2, 10.3, 11.2]

Per capita GHG emissions in 2010 are highly unequal (*high confidence*). In 2010, median per capita GHG emissions (1.4 tCO$_2$eq/cap/yr)

for the group of low-income countries are around nine times lower than median per capita GHG emissions (13 tCO$_2$eq/cap/yr) of high-income countries (Figure TS.4).[6] For low-income countries, the largest part of GHG emissions comes from AFOLU; for high-income countries, GHG emissions are dominated by sources related to energy supply and industry (Figure TS.3 lower panel). There are substantial variations in per capita GHG emissions within country income groups with emissions at the 90th percentile level more than double those at the 10th percentile level. Median per capita emissions better represent the typical country within a country income group comprised of heterogeneous members than mean per capita emissions. Mean per capita GHG emissions are different from median mainly in low-income countries as individual low-income countries have high per capita emissions due to large CO$_2$ emissions from land-use change (Figure TS.4, right panel). [1.3, 5.2, 5.3]

A growing share of total anthropogenic CO$_2$ emissions is released in the manufacture of products that are traded across international borders (*medium evidence, high agreement*). Since AR4, several data sets have quantified the difference between traditional 'territorial' and 'consumption-based' emission estimates that assign all emission released in the global production of goods and services to the country of final consumption (Figure TS.5). A growing share of CO$_2$ emissions from fossil fuel combustion in middle income countries is released in the production of goods and services exported, notably from upper middle income countries to high income countries. Total annual industrial CO$_2$ emissions from the non-Annex I group now exceed those of the Annex I group using territorial and consumption-based accounting methods, but per-capita emissions are still markedly higher in the Annex I group. [1.3, 5.3]

Regardless of the perspective taken, the largest share of anthropogenic CO$_2$ emissions is emitted by a small number of countries (*high confidence*). In 2010, 10 countries accounted for about 70% of CO$_2$ emissions from fossil fuel combustion and industrial processes. A similarly small number of countries emit the largest share of consumption-based CO$_2$ emissions as well as cumulative CO$_2$ emissions going back to 1750. [1.3]

The upward trend in global fossil fuel related CO$_2$ emissions is robust across databases and despite uncertainties (*high confidence*). Global CO$_2$ emissions from fossil fuel combustion are known within 8% uncertainty. CO$_2$ emissions related to FOLU have very large uncertainties attached in the order of 50%. Uncertainty for global emissions of methane (CH$_4$), nitrous oxide (N$_2$O), and the fluorinated gases has been estimated as 20%, 60%, and 20%. Combining these values yields an illustrative total global GHG uncertainty estimate of about 10% (Figure TS.1). Uncertainties can increase at finer spatial scales and for specific sectors. Attributing GHG emissions to the country of final consumption increases uncertainties, but literature on this topic is just emerging. GHG emissions estimates in the AR4 were 5–10% higher than the estimates reported here, but lie within the estimated uncertainty range.[3] [5.2]

Box TS.5 | Emissions metrics depend on value judgements and contain wide uncertainties

Emission metrics provide 'exchange rates' for measuring the contributions of different GHGs to climate change. Such exchange rates serve a variety of purposes, including apportioning mitigation efforts among several gases and aggregating emissions of a variety of GHGs. However, there is no metric that is both conceptually correct and practical to implement. Because of this, the choice of the appropriate metric depends on the application or policy at issue. [3.9.6]

GHGs differ in their physical characteristics. For example, per unit mass in the atmosphere, methane (CH_4) causes a stronger instantaneous radiative forcing than CO_2, but it remains in the atmosphere for a much shorter time. Thus, the time profiles of climate change brought about by different GHGs are different and consequential. Determining how emissions of different GHGs are compared for mitigation purposes involves comparing the resulting temporal profiles of climate change from each gas and making value judgments about the relative significance to humans of these profiles, which is a process fraught with uncertainty. [3.9.6; WGI 8.7]

A commonly used metric is the Global Warming Potential (GWP). It is defined as the accumulated radiative forcing within a specific time horizon (e.g., 100 years—GWP_{100}), caused by emitting one kilogram of the gas, relative to that of the reference gas CO_2. This metric is used to transform the effects of different GHG emissions to a common scale (CO_2-equivalents).[1] One strength of the GWP is

that it can be calculated in a relatively transparent and straightforward manner. However, there are also limitations, including the requirement to use a specific time horizon, the focus on cumulative forcing, and the insensitivity of the metric to the temporal profile of climate effects and its significance to humans. The choice of time horizon is particularly important for short-lived gases, notably methane: when computed with a shorter time horizon for GWP, their share in calculated total warming effect is larger and the mitigation strategy might change as a consequence. [1.2.5]

Many alternative metrics have been proposed in the scientific literature. All of them have advantages and disadvantages, and the choice of metric can make a large difference for the weights given to emissions from particular gases. For instance, methane's GWP_{100} is 28 while its Global Temperature Change Potential (GTP), one alternative metric, is 4 for the same time horizon (AR5 values, see WGI Section 8.7). In terms of aggregate mitigation costs alone, GWP_{100} may perform similarly to other metrics (such as the time-dependent Global Temperature Change Potential or the Global Cost Potential) of reaching a prescribed climate target; however, there may be significant differences in terms of the implied distribution of costs across sectors, regions, and over time. [3.9.6, 6.3.2.5]

An alternative to a single metric for all gases is to adopt a 'multi-basket' approach in which gases are grouped according to their contributions to short and long term climate change. This may solve some problems associated with using a single metric, but the question remains of what relative importance to attach to reducing GHG emissions in the different groups. [3.9.6; WGI 8.7]

[1] In this summary, all quantities of GHG emissions are expressed in CO_2-equivalent (CO_2eq) emissions that are calculated based on GWP_{100}. Unless otherwise stated, GWP values for different gases are taken from IPCC Second Assessment Report (SAR). Although GWP values have been updated several times since, the SAR values are widely used in policy settings, including the Kyoto Protocol, as well as in many national and international emission accounting systems. Modelling studies show that the changes in GWP_{100} values from SAR to AR4 have little impact on the optimal mitigation strategy at the global level. [6.3.2.5, Annex II.9.1]

TS.2.2   Greenhouse gas emission drivers

This section examines the factors that have, historically, been associated with changes in GHG emissions levels. Typically, such analysis is based on a decomposition of total GHG emissions into various components such as growth in the economy (Gross Domestic Product (GDP)/capita), growth in the population (capita), the energy intensity needed per unit of economic output (energy/GDP) and the GHG emissions intensity of that energy (GHGs/energy). As a practical matter, due to data limitations and the fact that most GHG emissions take the form of CO_2 from industry and energy, almost all this research focuses on CO_2 from those sectors.

Globally, economic and population growth continue to be the most important drivers of increases in CO_2 emissions from fossil fuel combustion. The contribution of population growth between 2000 and 2010 remained roughly identical to the previous three decades, while the contribution of economic growth has risen sharply (*high confidence*). Worldwide population increased by 86% between 1970 and 2010, from 3.7 to 6.9 billion. Over the same period, income as measured through production and/ or consumption per capita has grown by a factor of about two. The exact measurement of global economic growth is difficult because countries use different currencies and converting individual national economic figures into global totals can be done in various ways. With rising population and economic output, emissions of CO_2 from fossil fuel combustion have risen as well. Over the last decade, the importance of economic growth as a driver of global CO_2 emissions has risen sharply while population growth has remained roughly steady. Due to changes in technology, changes in the economic structure and the mix of energy sources as well as changes in other inputs such as capital and labour, the energy intensity of economic output has steadily declined worldwide. This decline has had an offsetting effect on global CO_2 emissions that is nearly of the same magnitude as growth in population (Figure TS.6). There are only a few countries that combine economic growth and decreasing territorial CO_2 emissions over longer periods of time. Such decoupling remains largely atypical, especially when considering consumption-based CO_2 emissions. [1.3, 5.3]

Between 2000 and 2010, increased use of coal relative to other energy sources has reversed a long-standing pattern of gradual decarbonization of the world's energy supply (*high confidence*). Increased use of coal, especially in developing Asia, is exacerbating the burden of energy-related GHG emissions (Figure TS.6). Estimates

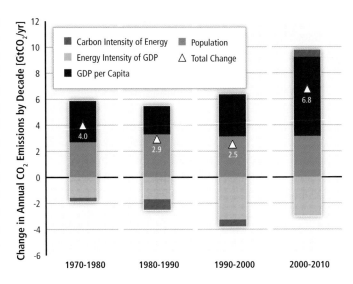

Figure TS.6 | Decomposition of the change in total annual CO_2 emissions from fossil fuel combustion by decade and four driving factors: population, income (GDP) per capita, energy intensity of GDP and carbon intensity of energy. Total emissions changes are indicated by a triangle. The change in emissions over each decade is measured in gigatonnes of CO_2 per year (GtCO$_2$/yr); income is converted into common units using purchasing power parities. [Figure 1.7]

Box TS.6 | The use of scenarios in this report

Scenarios of how the future might evolve capture key factors of human development that influence GHG emissions and our ability to respond to climate change. Scenarios cover a range of plausible futures, because human development is determined by a myriad of factors including human decision making. Scenarios can be used to integrate knowledge about the drivers of GHG emissions, mitigation options, climate change, and climate impacts.

One important element of scenarios is the projection of the level of human interference with the climate system. To this end, a set of four 'representative concentration pathways' (RCPs) has been developed. These RCPs reach radiative forcing levels of 2.6, 4.5, 6.0, and 8.5 Watts per square meter (W/m²) (corresponding to concentrations of 450, 650, 850, and 1370 ppm CO_2eq), respectively, in 2100, covering the range of anthropogenic climate forcing in the 21st century as reported in the literature. The four RCPs are the basis of a new set of climate change projections that have been assessed by WGI AR5. [WGI 6.4, WGI 12.4]

Scenarios of how the future develops without additional and explicit efforts to mitigate climate change ('baseline scenarios') and with the introduction of efforts to limit GHG emissions ('mitigation scenarios'), respectively, generally include socio-economic projections in addition to emission, concentration, and climate change information. WGIII AR5 has assessed the full breadth of

baseline and mitigation scenarios in the literature. To this end, it has collected a database of more than 1200 published mitigation and baseline scenarios. In most cases, the underlying socio-economic projections reflect the modelling teams' individual choices about how to conceptualize the future in the absence of climate policy. The baseline scenarios show a wide range of assumptions about economic growth (ranging from threefold to more than eightfold growth in per capita income by 2100), demand for energy (ranging from a 40 % to more than 80 % decline in energy intensity by 2100) and other factors, in particular the carbon intensity of energy. Assumptions about population are an exception: the vast majority of scenarios focus on the low to medium population range of nine to 10 billion people by 2100. Although the range of emissions pathways across baseline scenarios in the literature is broad, it may not represent the full potential range of possibilities (Figure TS.7). [6.3.1]

The concentration outcomes of the baseline and mitigation scenarios assessed by WGIII AR5 cover the full range of RCPs. However, they provide much more detail at the lower end, with many scenarios aiming at concentration levels in the range of 450, 500, and 550 ppm CO_2eq in 2100. The climate change projections of WGI based on RCPs, and the mitigation scenarios assessed by WGIII AR5 can be related to each other through the climate outcomes they imply. [6.2.1]

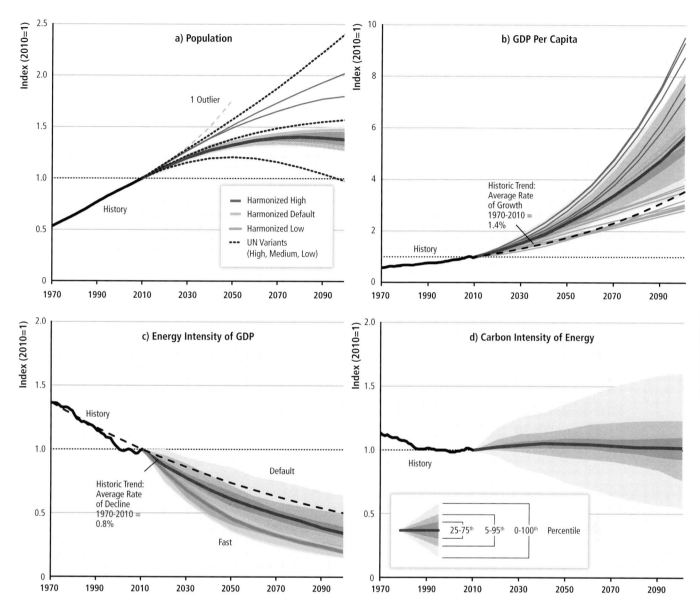

Figure TS.7 | Global baseline projection ranges for four emissions driving factors. Scenarios harmonized with respect to a particular factor are depicted with individual lines. Other scenarios are depicted as a range with median emboldened; shading reflects interquartile range (darkest), 5th–95th percentile range (lighter), and full range (lightest), excluding one indicated outlier in panel a). Scenarios are filtered by model and study for each indicator to include only unique projections. Model projections and historic data are normalized to 1 in 2010. GDP is aggregated using base-year market exchange rates. Energy and carbon intensity are measured with respect to total primary energy. [Figure 6.1]

indicate that coal and unconventional gas and oil resources are large; therefore reducing the carbon intensity of energy may not be primarily driven by fossil resource scarcity, but rather by other driving forces such as changes in technology, values, and socio-political choices. [5.3, 7.2, 7.3, 7.4; SRREN Figure 1.7]

Technological innovations, infrastructural choices, and behaviour affect GHG emissions through productivity growth, energy- and carbon-intensity and consumption patterns (*medium confidence*). Technological innovation improves labour and resource productivity; it can support economic growth both with increasing and with decreasing GHG emissions. The direction and speed of technological change depends on policies. Technology is also central to

the choices of infrastructure and spatial organization, such as in cities, which can have long-lasting effects on GHG emissions. In addition, a wide array of attitudes, values, and norms can inform different lifestyles, consumption preferences, and technological choices all of which, in turn, affect patterns of GHG emissions. [5.3, 5.5, 5.6, 12.3]

Without additional efforts to reduce GHG emissions beyond those in place today, emissions growth is expected to persist, driven by growth in global population and economic activities despite improvements in energy supply and end-use technologies (*high confidence*). Atmospheric concentrations in baseline scenarios collected for this assessment (scenarios without explicit additional efforts to reduce GHG emissions) exceed 450 parts per million (ppm) CO_2eq by 2030.[7] They reach CO_2eq concentration levels from 750 to more than 1300 ppm CO_2eq by 2100 and result in projected global mean surface temperature increases in 2100 from 3.7 to 4.8 °C compared to pre-industrial levels[8] (range based on median climate response; the range is 2.5 °C to 7.8 °C when including climate uncertainty, see Table TS.1).[9] The range of 2100 concentrations corresponds roughly to the range of CO_2eq concentrations in the Representative Concentration Pathways (RCP) 6.0 and RCP8.5 pathways (see Box TS.6), with the majority of scenarios falling below the latter. For comparison, the CO_2eq concentration in 2011 has been estimated to be 430 ppm (uncertainty range 340–520 ppm).[10] The literature does not systematically explore the full range of uncertainty surrounding development pathways and possible evolution of key drivers such as population, technology, and resources. Nonetheless, the scenarios strongly suggest that absent any explicit mitigation efforts, cumulative CO_2 emissions since 2010 will exceed 700 $GtCO_2$ by 2030, 1,500 $GtCO_2$ by 2050, and potentially well over 4,000 $GtCO_2$ by 2100. [6.3.1; WGI Figure SPM.5, WGI 8.5, WGI 12.3]

TS.3 Mitigation pathways and measures in the context of sustainable development

This section assesses the literature on mitigation pathways and measures in the context of sustainable development. Section TS 3.1 first examines the anthropogenic GHG emissions trajectories and potential temperature implications of mitigation pathways leading to a range of future atmospheric CO_2eq concentrations. It then explores the technological, economic, and institutional requirements of these pathways along with their potential co-benefits and adverse side-effects. Section TS 3.2 examines mitigation options by sector and how they may interact across sectors.

TS.3.1 Mitigation pathways

TS.3.1.1 Understanding mitigation pathways in the context of multiple objectives

The world's societies will need to both mitigate and adapt to climate change if it is to effectively avoid harmful climate impacts (*robust evidence, high agreement*). There are demonstrated examples of synergies between mitigation and adaptation [11.5.4, 12.8.1] in which the two strategies are complementary. More generally, the two strategies are related because increasing levels of mitigation imply less future need for adaptation. Although major efforts are now underway to incorporate impacts and adaptation into mitigation scenarios, inherent difficulties associated with quantifying their interdependencies have limited their representation in models used to generate mitigation scenarios assessed in WGIII AR5 (Box TS.7). [2.6.3, 3.7.2.1, 6.3.3]

There is no single pathway to stabilize CO_2eq concentrations at any level; instead, the literature points to a wide range of mitigation pathways that might meet any concentration level (*high confidence*). Choices, whether deliberated or not, will determine which of these pathways is followed. These choices include, among other things, the emissions pathway to bring atmospheric CO_2eq concentrations to a particular level, the degree to which concentrations temporarily exceed (overshoot) the long-term level, the technologies that are deployed to reduce emissions, the degree to which mitigation is coordinated across countries, the policy approaches used to achieve mitigation within and across countries, the treatment of land use, and the manner in which mitigation is meshed with other policy objectives such as sustainable development. A society's development pathway—with its particular socioeconomic, institutional, political, cultural and technological features—enables and constrains the prospects for mitigation. At the national level, change is considered most effective when it reflects country and local visions and approaches to achieving sustainable development according to national circumstances and priorities. [4.2, 6.3–6.8, 11.8]

Mitigation pathways can be distinguished from one another by a range of outcomes or requirements (*high confidence*). Decisions about mitigation pathways can be made by weighing the requirements of different pathways against each other. Although measures of aggregate economic costs and benefits have often been put forward as key decision-making factors, they are far from the only outcomes that matter. Mitigation pathways inherently involve a range of synergies and tradeoffs connected with other policy objectives such as energy and food security, energy access, the distribution of economic impacts, local air quality, other environmental factors associated with different technological solutions, and economic competitiveness (Box TS.11). Many of these fall under the umbrella of sustainable development. In addition, requirements such as the rates of up-scaling of energy technologies or the rates of reductions in GHG emissions may provide important insights into the degree of challenge associated with meeting a particular long-term goal. [4.5, 4.8, 6.3, 6.4, 6.6]

---

[7] These CO_2eq concentrations represent full radiative forcing, including GHGs, halogenated gases, tropospheric ozone, aerosols, mineral dust and albedo change.

[8] Based on the longest global surface temperature dataset available, the observed change between the average of the period 1850–1900 and of the AR5 reference period (1986–2005) is 0.61 °C (5–95 % confidence interval: 0.55 to 0.67 °C) [WGI SPM.E], which is used here as an approximation of the change in global mean surface temperature since pre-industrial times, referred to as the period before 1750.

[9] Provided estimates reflect the 10th to the 90th percentile of baseline scenarios collected for this assessment. The climate uncertainty reflects the 5th to 95th percentile of climate model calculations described in Table TS.1 for each scenario.

[10] This is based on the assessment of total anthropogenic radiative forcing for 2011 relative to 1750 in WGI AR5, i.e., 2.3 W m^{-2}, uncertainty range 1.1 to 3.3 W m^{-2}. [WGI Figure SPM.5, WGI 8.5, WGI 12.3]

Box TS.7 | Scenarios from integrated models can help to understand how actions affect outcomes in complex systems

The long-term scenarios assessed in this report were generated primarily by large-scale computer models, referred to here as 'integrated models', because they attempt to represent many of the most important interactions among technologies, relevant human systems (e.g., energy, agriculture, the economic system), and associated GHG emissions in a single integrated framework. A subset of these models is referred to as 'integrated assessment models', or IAMs. IAMs include not only an integrated representation of human systems, but also of important physical processes associated with climate change, such as the carbon cycle, and sometimes representations of impacts from climate change. Some IAMs have the capability of endogenously balancing impacts with mitigation costs, though these models tend to be highly aggregated. Although aggregate models with representations of mitigation and damage costs can be very useful, the focus in this assessment is on integrated models with sufficient sectoral and geographic resolution to understand the evolution of key processes such as energy systems or land systems.

Scenarios from integrated models are invaluable to help understand how possible actions or choices might lead to different future outcomes in these complex systems. They provide quantitative, long-term projections (conditional on our current state of knowledge) of many of the most important characteristics of mitigation pathways while accounting for many of the most important interactions between the various relevant human and natural systems. For example, they provide both regional and global information about emissions pathways, energy and land-use transitions, and aggregate economic costs of mitigation.

At the same time, these integrated models have particular characteristics and limitations that should be considered when interpreting their results. Many integrated models are based on the rational choice paradigm for decision making, excluding the consideration of some behavioural factors. The models approximate cost-effective solutions that minimize the aggregate economic costs of achieving mitigation outcomes, unless they are specifically constrained to behave otherwise. Scenarios from these models capture only some of the dimensions of development pathways that are relevant to mitigation options, often only minimally treating issues such as distributional impacts of mitigation actions and consistency with broader development goals. In addition, the models in this assessment do not effectively account for the interactions between mitigation, adaptation, and climate impacts. For these reasons, mitigation has been assessed independently from climate impacts. Finally, and most fundamentally, integrated models are simplified, stylized, numerical approaches for representing enormously complex physical and social systems, and scenarios from these models are based on uncertain projections about key events and drivers over often century-long timescales. Simplifications and differences in assumptions are the reason why output generated from different models—or versions of the same model—can differ, and projections from all models can differ considerably from the reality that unfolds. [3.7, 6.2]

TS.3.1.2 Short- and long-term requirements of mitigation pathways

Mitigation scenarios point to a range of technological and behavioral measures that could allow the world's societies to follow GHG emissions pathways consistent with a range of different levels of mitigation (*high confidence*). As part of this assessment, about 900 mitigation and 300 baseline scenarios have been collected from integrated modelling research groups around the world (Box TS.7). The mitigation scenarios span atmospheric concentration levels in 2100 from 430 ppm CO_2eq to above 720 ppm CO_2eq, which is roughly comparable to the 2100 forcing levels between the RCP2.6 and RCP6.0 scenarios (Figure TS.8, left panel). Scenarios have been constructed to reach mitigation goals under very different assumptions about energy demands, international cooperation, technologies, the contributions of CO_2 and other forcing agents to atmospheric CO_2eq concentrations, and the degree to which concentrations temporarily exceed the long-term goal (concentration overshoot, see Box TS.8). Other scenarios were also assessed, including some scenarios

with concentrations in 2100 below 430 ppm CO_2eq (for a discussion of these scenarios see below). [6.3]

Limiting atmospheric peak concentrations over the course of the century—not only reaching long-term concentration levels—is critical for limiting transient temperature change (*high confidence*). Scenarios reaching concentration levels of about 500 ppm CO_2eq by 2100 are *more likely than not* to limit temperature change to less than 2 °C relative to pre-industrial levels, unless they temporarily 'overshoot' concentration levels of roughly 530 ppm CO_2eq before 2100. In this case, they are *about as likely as not* to achieve that goal. The majority of scenarios reaching long-term concentrations of about 450 ppm CO_2eq in 2100 are *likely* to keep temperature change below 2 °C over the course of the century relative to pre-industrial levels (Table TS.1, Box TS.8). Scenarios that reach 530 to 650 ppm CO_2eq concentrations by 2100 are *more unlikely than likely* to keep temperature change below 2 °C relative to pre-industrial levels. Scenarios that exceed about 650 ppm CO_2eq by 2100 are *unlikely* to limit temperature change to below 2 °C relative to pre-industrial levels. Mitigation

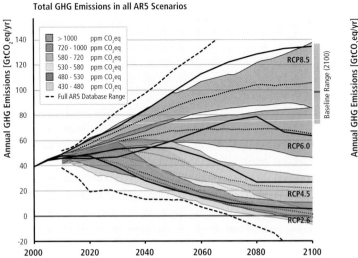

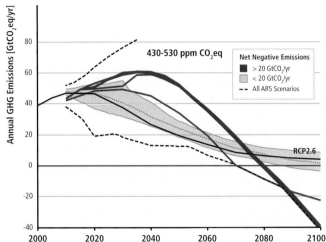

Figure TS.8 | Development of total GHG emissions for different long-term concentration levels (left panel) and for scenarios reaching about 450 to about 500 (430–530) ppm CO₂eq in 2100 with and without net negative CO₂ emissions larger than 20 GtCO₂/yr (right panel). Ranges are given for the 10th–90th percentile of scenarios. [Figure 6.7]

scenarios in which temperature increase is *more likely than not* to be less than 1.5 °C relative to pre-industrial levels by 2100 are characterized by concentrations in 2100 of below 430 ppm CO₂eq. Temperature peaks during the century and then declines in these scenarios. [6.3]

Mitigation scenarios reaching about 450 ppm CO₂eq in 2100 typically involve temporary overshoot of atmospheric concentrations, as do many scenarios reaching about 500 ppm or about 550 ppm CO₂eq in 2100 (*high confidence*). Concentration overshoot means that concentrations peak during the century before descending toward their 2100 levels. Overshoot involves less mitigation in the near term, but it also involves more rapid and deeper emissions reductions in the long run. The vast majority of scenarios reaching about 450 ppm CO₂eq in 2100 involve concentration overshoot, since most models cannot reach the immediate, near-term emissions reductions that would be necessary to avoid overshoot of these concentration levels. Many scenarios have been constructed to reach about 550 ppm CO₂eq by 2100 without overshoot.

Depending on the level of overshoot, many overshoot scenarios rely on the availability and widespread deployment of bioenergy with carbon dioxide capture and storage (BECCS) and/or afforestation in the second half of the century (*high confidence*). These and other carbon dioxide removal (CDR) technologies and methods remove CO₂ from the atmosphere (negative emissions). Scenarios with overshoot of greater than 0.4 W/m² (> 35–50 ppm CO₂eq concentration) typically deploy CDR technologies to an extent that net global CO₂ emissions become negative in the second-half of the century (Figure TS.8, right panel). CDR is also prevalent in many scenarios without concentration overshoot to compensate for residual emissions from sectors where mitigation is more expensive. The availability and potential of BECCS, afforestation, and other CDR technolo-

gies and methods are uncertain and CDR technologies and methods are, to varying degrees, associated with challenges and risks. There is uncertainty about the potential for large-scale deployment of BECCS, large-scale afforestation, and other CDR technologies and methods. [6.3, 6.9]

Reaching atmospheric concentration levels of about 450 to about 500 ppm CO₂eq by 2100 will require substantial cuts in anthropogenic GHG emissions by mid-century (*high confidence*). Scenarios reaching about 450 ppm CO₂eq by 2100 are associated with GHG emissions reductions of about 40 % to 70 % by 2050 compared to 2010 and emissions levels near zero GtCO₂eq or below in 2100.[11] Scenarios with GHG emissions reductions in 2050 at the lower end of this range are characterized by a greater reliance on CDR technologies beyond mid-century. The majority of scenarios that reach about 500 ppm CO₂eq in 2100 without overshooting roughly 530 ppm CO₂eq at any point during the century are associated with GHG emissions reductions of 40 % to 55 % by 2050 compared to 2010 (Figure TS.8, left panel; Table TS.1). In contrast, in some scenarios in which concentrations rise to well above 530 ppm CO₂eq during the century before descending to concentrations below this level by 2100, emissions rise to as high as 20 % above 2010 levels in 2050. However, these high-overshoot scenarios are characterized by negative global emissions of well over 20 GtCO₂ per year in the second half of the century (Figure TS.8, right panel). Cumulative CO₂

---

[11] This range differs from the range provided for a similar concentration category in AR4 (50 % to 85 % lower than 2000 for CO₂ only). Reasons for this difference include that this report has assessed a substantially larger number of scenarios than in AR4 and looks at all GHGs. In addition, a large proportion of the new scenarios include Carbon Dioxide Removal (CDR) technologies and associated increases in concentration overshoot. Other factors include the use of 2100 concentration levels instead of stabilization levels and the shift in reference year from 2000 to 2010.

Box TS.8 | Assessment of temperature change in the context of mitigation scenarios

Long-term climate goals have been expressed both in terms of concentrations and temperature. Article 2 of the UNFCCC calls for the need to 'stabilize' concentrations of GHGs. Stabilization of concentrations is generally understood to mean that the CO_2eq concentration reaches a specific level and then remains at that level indefinitely until the global carbon and other cycles come into a new equilibrium. The notion of stabilization does not necessarily preclude the possibility that concentrations might exceed, or 'overshoot' the long-term goal before eventually stabilizing at that goal. The possibility of 'overshoot' has important implications for the required GHG emissions reductions to reach a long-term concentration level. Concentration overshoot involves less mitigation in the near term with more rapid and deeper emissions reductions in the long run.

The temperature response of the concentration pathways assessed in this report focuses on transient temperature change over the course of the century. This is an important difference with WGIII AR4, which focused on the long-term equilibrium temperature response, a state that is reached millennia after the stabilization of concentrations. The temperature outcomes in this report are thus not directly comparable to those presented in the WGIII AR4 assessment. One reason that this assessment focuses on transient temperature response is that it is less uncertain than the equilibrium response and correlates more strongly with GHG emissions in the near and medium term. An additional reason is that the mitigation pathways assessed in WGIII AR5 do not extend beyond 2100 and are primarily designed to reach specific concentration goals for the year 2100. The majority of these pathways do not stabilize concentrations in 2100, which makes the assessment of the equilibrium temperature response ambiguous and dependent on assumptions about post-2100 emissions and concentrations.

Transient temperature goals might be defined in terms of the temperature in a specific year (e.g., 2100), or based on never exceeding a particular level. This report explores the implications of both types of goals. The assessment of temperature goals are complicated by the uncertainty that surrounds our understanding of key physical relationships in the earth system, most notably the relationship between concentrations and temperature. It is not possible to state definitively whether any long-term concentration pathway will limit either transient or equilibrium temperature change to below a specified level. It is only possible to express the temperature implications of particular concentration pathways in probabilistic terms, and such estimates will be dependent on the source of the probability distribution of different climate parameters and the climate model used for analysis. This report employs the MAGICC model and a distribution of climate parameters that results in temperature outcomes with dynamics similar to those from the Earth System Models assessed in WGI AR5. For each emissions scenario, a median transient temperature response is calculated to illustrate the variation of temperature due to different emissions pathways. In addition, a transient temperature range for each scenario is provided, reflecting the climate system uncertainties. Information regarding the full distribution of climate parameters was utilized for estimating the likelihood that the scenarios would limit transient temperature change to below specific levels (Table TS.1). Providing the combination of information about the plausible range of temperature outcomes as well as the likelihood of meeting different targets is of critical importance for policymaking, since it facilitates the assessment of different climate objectives from a risk management perspective. [2.5.7.2, 6.3.2]

emissions between 2011 and 2100 are 630–1180 $GtCO_2$ in scenarios reaching about 450 ppm CO_2eq in 2100; they are 960–1550 $GtCO_2$ in scenarios reaching about 500 ppm CO_2eq in 2100. The variation in cumulative CO_2 emissions across scenarios is due to differences in the contribution of non-CO_2 GHGs and other radiatively active substances as well as the timing of mitigation (Table TS.1). [6.3]

In order to reach atmospheric concentration levels of about 450 to about 500 ppm CO_2eq by 2100, the majority of mitigation relative to baseline emissions over the course of century will occur in the non-Organisation for Economic Co-operation and Development (OECD) countries (*high confidence*). In scenarios that attempt to cost-effectively allocate emissions reductions across countries and over time, the total CO_2eq emissions reductions from baseline emissions in non-OECD countries are greater than in OECD countries. This is, in large part, because baseline emissions from the non-OECD

countries are projected to be larger than those from the OECD countries, but it also derives from higher carbon intensities in non-OECD countries and different terms of trade structures. In these scenarios, GHG emissions peak earlier in the OECD countries than in the non-OECD countries. [6.3]

Reaching atmospheric concentration levels of about 450 to about 650 ppm CO_2eq by 2100 will require large-scale changes to global and national energy systems over the coming decades (*high confidence*). Scenarios reaching atmospheric concentrations levels of about 450 to about 500 ppm CO_2eq by 2100 are characterized by a tripling to nearly a quadrupling of the global share of zero- and low-carbon energy supply from renewables, nuclear energy, fossil energy with carbon dioxide capture and storage (CCS), and bioenergy with CCS (BECCS), by the year 2050 relative to 2010 (about 17 %) (Figure TS.10, left panel). The increase in total global low-carbon energy sup-

Table TS.1 | Key characteristics of the scenarios collected and assessed for WGIII AR5. For all parameters, the 10th to 90th percentile of the scenarios is shown.[1,2] [Table 6.3]

CO$_2$eq Concentrations in 2100 [ppm CO$_2$eq] Category label (concentration range)[9]	Subcategories	Relative position of the RCPs[5]	Cumulative CO$_2$ emissions[3] [GtCO$_2$]		Change in CO$_2$eq emissions compared to 2010 in [%][4]		Temperature change (relative to 1850–1900)[5,6]				
							2100 Temperature change [°C][7]	Likelihood of staying below temperature level over the 21st century[8]			
			2011–2050	2011–2100	2050	2100		1.5 °C	2.0 °C	3.0 °C	4.0 °C
< 430	Only a limited number of individual model studies have explored levels below 430 ppm CO$_2$eq										
450 (430–480)	Total range[1,10]	RCP2.6	550–1300	630–1180	−72 to −41	−118 to −78	1.5–1.7 (1.0–2.8)	More unlikely than likely	Likely		
500 (480–530)	No overshoot of 530 ppm CO$_2$eq		860–1180	960–1430	−57 to −42	−107 to −73	1.7–1.9 (1.2–2.9)		More likely than not	Likely	Likely
	Overshoot of 530 ppm CO$_2$eq		1130–1530	990–1550	−55 to −25	−114 to −90	1.8–2.0 (1.2–3.3)	Unlikely	About as likely as not		
550 (530–580)	No overshoot of 580 ppm CO$_2$eq		1070–1460	1240–2240	−47 to −19	−81 to −59	2.0–2.2 (1.4–3.6)				
	Overshoot of 580 ppm CO$_2$eq		1420–1750	1170–2100	−16 to 7	−183 to −86	2.1–2.3 (1.4–3.6)		More unlikely than likely[12]		
(580–650)	Total range	RCP4.5	1260–1640	1870–2440	−38 to 24	−134 to −50	2.3–2.6 (1.5–4.2)				
(650–720)	Total range		1310–1750	2570–3340	−11 to 17	−54 to −21	2.6–2.9 (1.8–4.5)		Unlikely	More likely than not	
(720–1000)[2]	Total range	RCP6.0	1570–1940	3620–4990	18 to 54	−7 to 72	3.1–3.7 (2.1–5.8)	Unlikely[11]		More unlikely than likely	
>1000[2]	Total range	RCP8.5	1840–2310	5350–7010	52 to 95	74 to 178	4.1–4.8 (2.8–7.8)		Unlikely[11]	Unlikely	More unlikely than likely

Notes:

1. The 'total range' for the 430–480 ppm CO$_2$eq scenarios corresponds to the range of the 10th–90th percentile of the subcategory of these scenarios shown in Table 6.3.
2. Baseline scenarios (see TS.2.2) fall into the >1000 and 720–1000 ppm CO$_2$eq categories. The latter category also includes mitigation scenarios. The baseline scenarios in the latter category reach a temperature change of 2.5–5.8 °C above preindustrial in 2100. Together with the baseline scenarios in the >1000 ppm CO$_2$eq category, this leads to an overall 2100 temperature range of 2.5–7.8 °C (range based on median climate response: 3.7–4.8 °C) for baseline scenarios across both concentration categories.
3. For comparison of the cumulative CO$_2$ emissions estimates assessed here with those presented in WGI AR5, an amount of 515 [445–585] GtC (1890 [1630–2150] GtCO$_2$), was already emitted by 2011 since 1870 [WGI 12.5]. Note that cumulative CO$_2$ emissions are presented here for different periods of time (2011–2050 and 2011–2100) while cumulative CO$_2$ emissions in WGI AR5 are presented as total compatible emissions for the RCPs (2012–2100) or for total compatible emissions for remaining below a given temperature target with a given likelihood [WGI Table SPM.3, WGI SPM.E.8].
4. The global 2010 emissions are 31 % above the 1990 emissions (consistent with the historic GHG emissions estimates presented in this report). CO$_2$eq emissions include the basket of Kyoto gases (CO$_2$, CH$_4$, N$_2$O as well as F-gases).
5. The assessment in WGIII AR5 involves a large number of scenarios published in the scientific literature and is thus not limited to the RCPs. To evaluate the CO$_2$eq concentration and climate implications of these scenarios, the MAGICC model was used in a probabilistic mode (see Annex II). For a comparison between MAGICC model results and the outcomes of the models used in WGI, see Sections WGI 12.4.1.2, WGI 12.4.8 and 6.3.2.6. Reasons for differences with WGI SPM Table.2 include the difference in reference year (1986–2005 vs. 1850–1900 here), difference in reporting year (2081–2100 vs 2100 here), set-up of simulation (CMIP5 concentration-driven versus MAGICC emission-driven here), and the wider set of scenarios (RCPs versus the full set of scenarios in the WGIII AR5 scenario database here).
6. Temperature change is reported for the year 2100, which is not directly comparable to the equilibrium warming reported in WGIII AR4 [Table 3.5, Chapter 3; see also WGIII AR5 6.3.2]. For the 2100 temperature estimates, the transient climate response (TCR) is the most relevant system property. The assumed 90 % range of the TCR for MAGICC is 1.2–2.6 °C (median 1.8 °C). This compares to the 90 % range of TCR between 1.2–2.4 °C for CMIP5 [WGI 9.7] and an assessed *likely* range of 1–2.5 °C from multiple lines of evidence reported in the WGI AR5 [Box 12.2 in Section 12.5].
7. Temperature change in 2100 is provided for a median estimate of the MAGICC calculations, which illustrates differences between the emissions pathways of the scenarios in each category. The range of temperature change in the parentheses includes in addition the carbon cycle and climate system uncertainties as represented by the MAGICC model [see 6.3.2.6 for further details]. The temperature data compared to the 1850–1900 reference year was calculated by taking all projected warming relative to 1986–2005, and adding 0.61 °C for 1986–2005 compared to 1850–1900, based on HadCRUT4 [see WGI Table SPM.2].
8. The assessment in this table is based on the probabilities calculated for the full ensemble of scenarios in WGIII AR5 using MAGICC and the assessment in WGI AR5 of the uncertainty of the temperature projections not covered by climate models. The statements are therefore consistent with the statements in WGI AR5, which are based on the CMIP5 runs of the RCPs and the assessed uncertainties. Hence, the likelihood statements reflect different lines of evidence from both WGs. This WGI method was also applied for scenarios with intermediate concentration levels where no CMIP5 runs are available. The likelihood statements are indicative only [6.3], and follow broadly the terms used by the WGI AR5 SPM for temperature projections: *likely* 66–100 %, *more likely than not* >50–100 %, *about as likely as not* 33–66 %, and *unlikely* 0–33 %. In addition the term *more unlikely than likely* 0–<50 % is used.
9. The CO$_2$-equivalent concentration includes the forcing of all GHGs including halogenated gases and tropospheric ozone, as well as aerosols and albedo change (calculated on the basis of the total forcing from a simple carbon cycle/climate model, MAGICC).
10. The vast majority of scenarios in this category overshoot the category boundary of 480 ppm CO$_2$eq concentrations.
11. For scenarios in this category no CMIP5 run [WGI Chapter 12, Table 12.3] as well as no MAGICC realization [6.3] stays below the respective temperature level. Still, an *unlikely* assignment is given to reflect uncertainties that might not be reflected by the current climate models.
12. Scenarios in the 580–650 ppm CO$_2$eq category include both overshoot scenarios and scenarios that do not exceed the concentration level at the high end of the category (like RCP4.5). The latter type of scenarios, in general, have an assessed probability of *more unlikely than likely* to stay below the 2 °C temperature level, while the former are mostly assessed to have an *unlikely* probability of staying below this level.

TS

ply is from three-fold to seven-fold over this same period. Many models could not reach 2100 concentration levels of about 450 ppm CO_2eq if the full suite of low-carbon technologies is not available. Studies indicate a large potential for energy demand reductions, but also indicate that demand reductions on their own would not be sufficient to bring about the reductions needed to reach levels of about 650 ppm CO_2eq or below by 2100. [6.3, 7.11]

Mitigation scenarios indicate a potentially critical role for land-related mitigation measures and that a wide range of alternative land transformations may be consistent with similar concentration levels (*medium confidence*). Land-use dynamics in mitigation scenarios are heavily influenced by the production of bioenergy and the degree to which afforestation is deployed as a negative-emissions, or CDR option. They are, in addition, influenced by forces independent of mitigation such as agricultural productivity improvements and increased demand for food. The range of land-use transformations depicted in mitigation scenarios reflects a wide range of

differing assumptions about the evolution of all of these forces. Many scenarios reflect strong increases in the degree of competition for land between food, feed, and energy uses. [6.3, 6.8, 11.4.2]

Delaying mitigation efforts beyond those in place today through 2030 will increase the challenges of, and reduce the options for, limiting atmospheric concentration levels from about 450 to about 500 ppm CO_2eq by the end of the century (*high confidence*). Cost-effective mitigation scenarios leading to atmospheric concentration levels of about 450 to about 500 ppm CO_2eq at the end of the 21st century are typically characterized by annual GHG emissions in 2030 of roughly between 30 GtCO_2eq and 50 GtCO_2eq. Scenarios with emissions above 55 GtCO_2eq in 2030 are characterized by substantially higher rates of emissions reductions from 2030 to 2050 (median emissions reductions of about 6 %/yr as compared to just over 3 %/yr) (Figure TS.9, right panel); much more rapid scale-up of low-carbon energy over this period (more than a tripling compared to a doubling of the low-carbon energy share) (Figure TS.10, right panel);

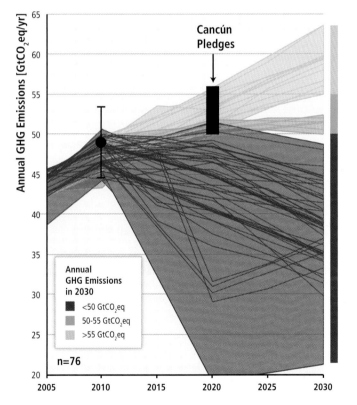

GHG Emissions Pathways to 2030 of Mitigation Scenarios Reaching 430-530 ppm CO_2eq in 2100

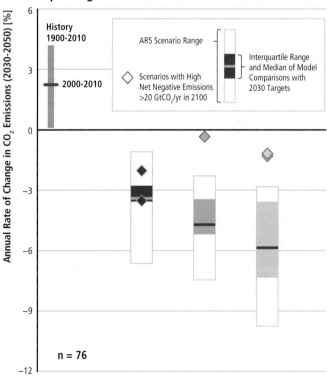

Implications for the Pace of Annual Average CO_2 Emissions Reductions from 2030 to 2050 Depending on Different 2030 GHG Emissions Levels

Figure TS.9 | The implications of different 2030 GHG emissions levels for the rate of CO_2 emissions reductions from 2030 to 2050 in mitigation scenarios reaching about 450 to about 500 (430–530) ppm CO_2eq concentrations by 2100. The scenarios are grouped according to different emissions levels by 2030 (coloured in different shades of green). The left panel shows the pathways of GHG emissions (GtCO_2eq/yr) leading to these 2030 levels. The black bar shows the estimated uncertainty range of GHG emissions implied by the Cancún Pledges. Black dot with whiskers gives historic GHG emission levels and associated uncertainties in 2010 as reported in Figure TS.1. The right panel denotes the average annual CO_2 emissions reduction rates for the period 2030–2050. It compares the median and interquartile range across scenarios from recent intermodel comparisons with explicit 2030 interim goals to the range of scenarios in the Scenario Database for WGIII AR5. Annual rates of historical emissions change between 1900–2010 (sustained over a period of 20 years) and the average annual emissions change between 2000–2010 are shown in grey. Note: Scenarios with large net negative global emissions (> 20 GtCO_2/yr) are not included in the WGII AR5 scenario range, but rather shown as independent points. Only scenarios that apply the full, unconstrained mitigation technology portfolio of the underlying models (default technology assumption) are shown. Scenarios with exogenous carbon price assumptions or other policies affecting the timing of mitigation (other than 2030 interim targets) as well as scenarios with 2010 emissions significantly outside the historical range are excluded. [Figure 6.32, 13.13.1.3]

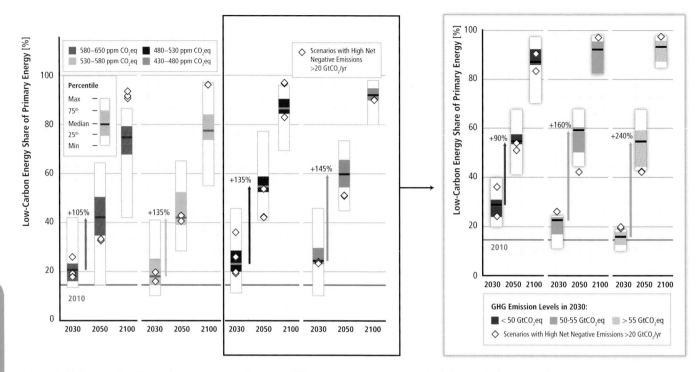

Figure TS.10 | The up-scaling of low-carbon energy in scenarios meeting different 2100 CO₂eq concentration levels (left panel). The right panel shows the rate of up-scaling subject to different 2030 GHG emissions levels in mitigation scenarios reaching about 450 to about 500 (430–530) ppm CO₂eq concentrations by 2100. Colored bars show the inter-quartile range and white bars indicate the full range across the scenarios, excluding those with large, global net negative CO₂ emissions (> 20 GtCO₂/yr). Scenarios with large net negative global emissions are shown as individual points. The arrows indicate the magnitude of zero- and low-carbon energy supply up-scaling from 2030 to 2050. Zero- and low-carbon energy supply includes renewables, nuclear energy, fossil energy with carbon dioxide capture and storage (CCS), and bioenergy with CCS (BECCS). Note: Only scenarios that apply the full, unconstrained mitigation technology portfolio of the underlying models (default technology assumption) are shown. Scenarios with exogenous carbon price assumptions are excluded in both panels. In the right panel, scenarios with policies affecting the timing of mitigation other than 2030 interim targets are also excluded. [Figure 7.16]

a larger reliance on CDR technologies in the long-term (Figure TS.8, right panel); and higher transitional and long term economic impacts (Table TS.2, orange segments, Figure TS.13, right panel). Due to these increased challenges, many models with 2030 GHG emissions in this range could not produce scenarios reaching atmospheric concentrations levels of about 450 to about 500 ppm CO₂eq in 2100. [6.4, 7.11]

Estimated global GHG emissions levels in 2020 based on the Cancún Pledges are not consistent with cost-effective long-term mitigation trajectories that reach atmospheric concentrations levels of about 450 to about 500 ppm CO₂eq by 2100, but they do not preclude the option to meet that goal (*robust evidence, high agreement*). The Cancún Pledges are broadly consistent with cost-effective scenarios reaching about 550 ppm CO₂eq to 650 ppm CO₂eq by 2100. Studies confirm that delaying mitigation through 2030 has a substantially larger influence on the subsequent challenges of mitigation than do delays through 2020 (Figures TS.9, TS.11). [6.4]

Only a limited number of studies have explored scenarios that are *more likely than not* to bring temperature change back to below 1.5 °C by 2100 relative to pre-industrial levels; these scenarios bring atmospheric concentrations to below 430 ppm CO₂eq by 2100 (*high confidence*). Assessing this goal is currently difficult because no multi-model study has explored these scenarios. The

limited number of published studies exploring this goal have produced associated scenarios that are characterized by (1) immediate mitigation; (2) the rapid up-scaling of the full portfolio of mitigation technologies; and (3) development along a low-energy demand trajectory.[12] [6.3, 7.11]

TS.3.1.3 Costs, investments and burden sharing

Globally comprehensive and harmonized mitigation actions would result in significant economic benefits compared to fragmented approaches, but would require establishing effective institutions (*high confidence*). Economic analysis of mitigation scenarios demonstrates that globally comprehensive and harmonized mitigation actions achieve mitigation at least aggregate economic cost, since they allow mitigation to be undertaken where and when it is least expensive (see Box TS.7, Box TS.9). Most of these mitigation scenarios assume a global carbon price, which reaches all sectors of the economy. Instruments with limited coverage of GHG emissions reductions among sectors and climate policy regimes with fragmented regional

---

[12] In these scenarios, the cumulative CO₂ emissions range between 680–800 GtCO₂ for the period 2011–2050 and between 90–310 GtCO₂ for the period 2011–2100. Global CO₂eq emissions in 2050 are between 70–95 % below 2010 emissions, and they are between 110–120 % below 2010 emissions in 2100.

430-530 ppm CO$_2$eq in 2100

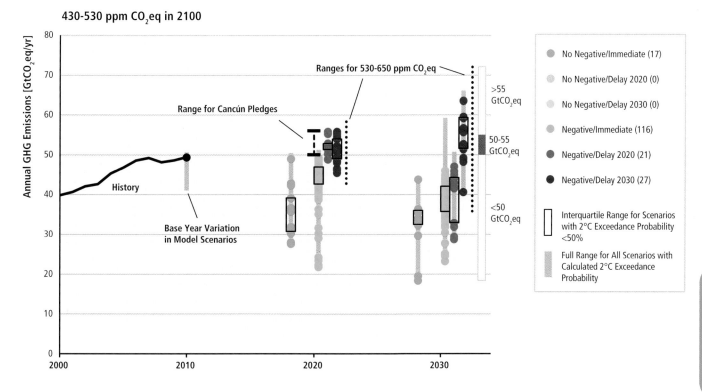

Figure TS.11 | Near-term GHG emissions from mitigation scenarios reaching about 450 to about 500 (430–530) ppm CO$_2$eq concentrations by 2100. The Figure includes only scenarios for which temperature exceedance probabilities were calculated. Individual model results are indicated with a data point when 2 °C exceedance probability is below 50 % as assessed by a simple carbon cycle/climate model (MAGICC). Colours refer to scenario classification in terms of whether net CO$_2$ emissions become negative before 2100 (negative vs. no negative) and the timing of international participation in climate mitigation (immediate vs. delay until 2020 vs. delay until 2030). Number of reported individual results is shown in legend. The range of global GHG emissions in 2020 implied by the Cancún Pledges is based on analysis of alternative interpretations of national pledges. Note: In the WGIII AR5 scenario database, only four reported scenarios were produced based on delayed mitigation without net negative emissions while still lying below 530 ppm CO$_2$eq by 2100. They do not appear in the figure, because the model had insufficient coverage of non-gas species to enable a temperature calculation. Delay in these scenarios extended only to 2020, and their emissions fell in the same range as the 'No Negative/Immediate' category. Delay scenarios include both delayed global mitigation and fragmented action scenarios. [Figure 6.31, 13.13.1.3]

action increase aggregate economic costs. These cost increases are higher at more ambitious levels of mitigation. [6.3.6]

Estimates of the aggregate economic costs of mitigation vary widely, but increase with stringency of mitigation (*high confidence*). Most cost-effective scenarios collected for this assessment that are based on the assumptions that all countries of the world begin mitigation immediately, there is a single global carbon price applied to well-functioning markets, and key technologies are available, estimate that reaching about 450 ppm CO$_2$eq by 2100 would entail global consumption losses of 1 % to 4 % in 2030 (median: 1.7 %), 2 % to 6 % in 2050 (median: 3.4 %), and 3 % to 11 % in 2100 (median: 4.8 %) relative to consumption in baseline scenarios (those without additional mitigation efforts) that grows anywhere from 300 % to more than 900 % between 2010 and 2100 (baseline consumption growth represents the full range of corresponding baseline scenarios; Figure TS.12; Table TS.2 yellow segments). The consumption losses correspond to an annual average reduction of consumption growth by 0.06 to 0.2 percentage points from 2010 through 2030 (median: 0.09), 0.06 to 0.17 percentage points through 2050 (median: 0.09), and 0.04 to 0.14 percentage points over the century (median: 0.06). These numbers are relative to annual

average consumption growth rates in baseline scenarios between 1.9 % and 3.8 % per year through 2050 and between 1.6 % and 3 % per year over the century (Table TS.2, yellow segments). These mitigation cost estimates do not consider the benefits of reduced climate change or co-benefits and adverse side-effects of mitigation (Box TS.9). Costs for maintaining concentrations in the range of 530–650 ppm CO$_2$eq are estimated to be roughly one-third to two-thirds lower than for associated 430–530 ppm CO$_2$eq scenarios. Cost estimates from scenarios can vary substantially across regions. Substantially higher cost estimates have been obtained based on assumptions about less idealized policy implementations and limits on technology availability as discussed below. Both higher and lower estimates have been obtained based on interactions with pre-existing distortions, non-climate market failures, or complementary policies. [6.3.6.2]

Delaying mitigation efforts beyond those in place today through 2030 or beyond could substantially increase mitigation costs in the decades that follow and the second half of the century (*high confidence*). Although delays in mitigation by any major emitter will reduce near-term mitigation costs, they will also result in more investment in carbon-intensive infrastructure and then rely on future

Table TS.2 | Global mitigation costs in cost-effective scenarios[1] and estimated cost increases due to assumed limited availability of specific technologies and delayed additional mitigation. Cost estimates shown in this table do not consider the benefits of reduced climate change as well as co-benefits and adverse side-effects of mitigation. The yellow columns show consumption losses (Figure TS.12, right panel) and annualized consumption growth reductions in cost-effective scenarios relative to a baseline development without climate policy. The grey columns show the percentage increase in discounted costs[2] over the century, relative to cost-effective scenarios, in scenarios in which technology is constrained relative to default technology assumptions (Figure TS.13, left panel).[3] The orange columns show the increase in mitigation costs over the periods 2030–2050 and 2050–2100, relative to scenarios with immediate mitigation, due to delayed additional mitigation through 2030 (see Figure TS.13, right panel).[4] These scenarios with delayed additional mitigation are grouped by emission levels of less or more than 55 GtCO$_2$eq in 2030, and two concentration ranges in 2100 (430–530 ppm CO$_2$eq and 530–650 ppm CO$_2$eq). In all figures, the median of the scenario set is shown without parentheses, the range between the 16th and 84th percentile of the scenario set is shown in the parentheses, and the number of scenarios in the set is shown in square brackets.[5] [Figures TS.12, TS.13, 6.21, 6.24, 6.25, Annex II.10]

Concentration in 2100 [ppm CO$_2$eq]	Consumption losses in cost-effective scenarios[1]						Increase in total discounted mitigation costs in scenarios with limited availability of technologies				Increase in medium- and long-term mitigation costs due to delayed additional mitigation until 2030			
	[% reduction in consumption relative to baseline]			[percentage point reduction in annualized consumption growth rate]			[% increase in total discounted mitigation costs (2015–2100) relative to default technology assumptions]				[% increase in mitigation costs relative to immediate mitigation]			
											≤55 GtCO$_2$eq		>55 GtCO$_2$eq	
	2030	2050	2100	2010–2030	2010–2050	2010–2100	No CCS	Nuclear phase out	Limited Solar/Wind	Limited Bioenergy	2030–2050	2050–2100	2030–2050	2050–2100
450 (430–480)	1.7 (1.0–3.7) [N: 14]	3.4 (2.1–6.2)	4.8 (2.9–11.4)	0.09 (0.06–0.2)	0.09 (0.06–0.17)	0.06 (0.04–0.14)	138 (29–297) [N: 4]	7 (4–18) [N: 8]	6 (2–29) [N: 8]	64 (44–78) [N: 8]	28 (14–50) [N: 34]	15 (5–59)	44 (2–78) [N: 29]	37 (16–82)
500 (480–530)	1.7 (0.6–2.1) [N: 32]	2.7 (1.5–4.2)	4.7 (2.4–10.6)	0.09 (0.03–0.12)	0.07 (0.04–0.12)	0.06 (0.03–0.13)	N/A	N/A	N/A	N/A				
550 (530–580)	0.6 (0.2–1.3) [N: 46]	1.7 (1.2–3.3)	3.8 (1.2–7.3)	0.03 (0.01–0.08)	0.05 (0.03–0.08)	0.04 (0.01–0.09)	39 (18–78) [N: 11]	13 (2–23) [N: 10]	8 (5–15) [N: 10]	18 (4–66) [N: 12]	3 (−5–16) [N: 14]	4 (−4–11)	15 (3–32) [N: 10]	16 (5–24)
580–650	0.3 (0–0.9) [N: 16]	1.3 (0.5–2.0)	2.3 (1.2–4.4)	0.02 (0–0.04)	0.03 (0.01–0.05)	0.03 (0.01–0.05)	N/A	N/A	N/A	N/A				

Notes:

[1] Cost-effective scenarios assume immediate mitigation in all countries and a single global carbon price. In this analysis, they also impose no additional limitations on technology relative to the models' default technology assumptions.

[2] Percentage increase of net present value of consumption losses in percent of baseline consumption (for scenarios from general equilibrium models) and abatement costs in percent of baseline GDP (for scenarios from partial equilibrium models) for the period 2015–2100, discounted (see Box TS.10) at 5 % per year.

[3] No CCS: CCS is not included in these scenarios. Nuclear phase out: No addition of nuclear power plants beyond those under construction, and operation of existing plants until the end of their lifetime. Limited Solar/Wind: a maximum of 20 % global electricity generation from solar and wind power in any year of these scenarios. Limited Bioenergy: a maximum of 100 EJ/yr modern bioenergy supply globally (modern bioenergy used for heat, power, combinations, and industry was around 18 EJ/yr in 2008 [11.13.5]).

[4] Percentage increase of total undiscounted mitigation costs for the periods 2030–2050 and 2050–2100.

[5] The range is determined by the central scenarios encompassing the 16th and 84th percentile of the scenario set. Only scenarios with a time horizon until 2100 are included. Some models that are included in the cost ranges for concentration levels above 530 ppm CO$_2$eq in 2100 could not produce associated scenarios for concentration levels below 530 ppm CO$_2$eq in 2100 with assumptions about limited availability of technologies and/or delayed additional mitigation (see caption of Figure TS.13 for more details).

decision makers to undertake a more rapid, deeper, and costlier future transformation of this infrastructure. Studies have found that aggregate costs, and associated carbon prices, rise more rapidly to higher levels in scenarios with delayed mitigation compared to scenarios where mitigation is undertaken immediately. Recent modelling studies have found that delayed mitigation through 2030 can substantially increase the aggregate costs of meeting 2100 concentrations of about 450 to about 500 ppm CO$_2$eq, particularly in scenarios with emissions greater than 55 GtCO$_2$eq in 2030. (Figure TS.13, right panel; Table TS.2, orange segments) [6.3.6.4]

The technological options available for mitigation greatly influence mitigation costs and the challenges of reaching atmospheric concentration levels of about 450 to about 550 ppm CO$_2$eq by 2100 (*high confidence*). Many models in recent model inter-comparisons could not produce scenarios reaching atmospheric concentrations of about 450 ppm CO$_2$eq by 2100 with broadly pessimistic assumptions about key mitigation technologies. In these studies, the

character and availability of CCS and bioenergy were found to have a particularly important influence on the mitigation costs and the challenges of reaching concentration levels in this range. For those models that could produce such scenarios, pessimistic assumptions about these increased discounted global mitigation costs of reaching concentration levels of about 450 and about 550 ppm CO$_2$eq by the end of the century significantly, with the effect being larger for more stringent mitigation scenarios (Figure TS.13, left panel; Table TS.2, grey segments). The studies also showed that reducing energy demand could potentially decrease mitigation costs significantly. [6.3.6.3]

The distribution of mitigation costs among different countries depends in part on the nature of effort-sharing frameworks and thus need not be the same as the distribution of mitigation efforts. Different effort-sharing frameworks draw upon different ethical principles (*medium confidence*). In cost-effective scenarios reaching concentrations of about 450 to about 550 ppm CO$_2$eq in 2100, the majority of mitigation investments over the course

TS

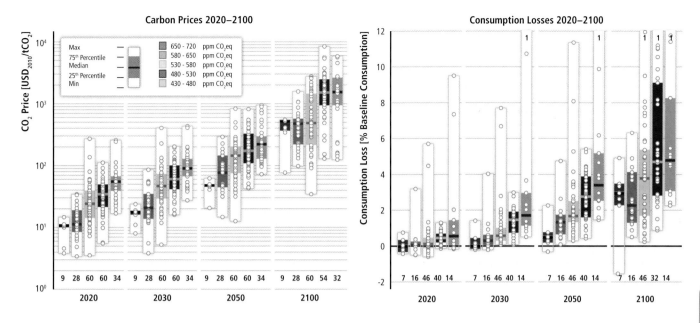

Figure TS.12 | Global carbon prices (left panel) and consumption losses (right panel) over time in cost-effective, idealized implementation scenarios. Consumption losses are expressed as the percentage reduction from consumption in the baseline. The number of scenarios included in the boxplots is indicated at the bottom of the panels. The 2030 numbers also apply to 2020 and 2050. The number of scenarios outside the figure range is noted at the top. Note: The figure shows only scenarios that reported consumption losses (a subset of models with full coverage of the economy) or carbon prices, respectively, to 2050 or 2100. Multiple scenarios from the same model with similar characteristics are only represented by a single scenario in the sample. [Figure 6.21]

Box TS.9 | The meaning of 'mitigation cost' in the context of mitigation scenarios

Mitigation costs represent one component of the change in human welfare from climate change mitigation. Mitigation costs are expressed in monetary terms and generally are estimated against baseline scenarios, which typically involve continued, and sometimes substantial, economic growth and no additional and explicit mitigation efforts [3.9.3, 6.3.6]. Because mitigation cost estimates focus only on direct market effects, they do not take into account the welfare value (if any) of co-benefits or adverse side-effects of mitigation actions (Box TS.11) [3.6.3]. Further, these costs do not capture the benefits of reducing climate impacts through mitigation (Box TS.2).

There are a wide variety of metrics of aggregate mitigation costs used by economists, measured in different ways or at different places in the economy, including changes in GDP, consumption losses, equivalent variation and compensating variation, and loss in consumer and producer surplus. Consumption losses are often used as a metric because they emerge from many integrated models and they directly impact welfare. They can be expressed as a reduction in overall consumption relative to consumption in the corresponding baseline scenario in a given year or as a reduction of the average rate of consumption growth in the corresponding baseline scenario over a given time period.

Mitigation costs need to be distinguished from emissions prices. Emissions prices measure the cost of an additional unit of emissions reduction; that is, the marginal cost. In contrast, mitigation costs usually represent the total costs of all mitigation. In addition, emissions prices can interact with other policies and measures, such as regulatory policies directed at GHG reduction. If mitigation is achieved partly by these other measures, emissions prices may not reflect the actual costs of an additional unit of emissions reductions (depending on how additional emissions reductions are induced).

In general, estimates of global aggregate mitigation costs over the coming century from integrated models are based on largely stylized assumptions about both policy approaches and existing markets and policies, and these assumptions have an important influence on cost estimates. For example, cost-effective idealized implementation scenarios assume a uniform price on CO_2 and other GHGs in every country and sector across the globe, and constitute the least cost approach in the idealized case of largely efficient markets without market failures other than the climate change externality. Most long-term, global scenarios do not account for the interactions between mitigation and pre-existing or new policies, market failures, and distortions. Climate policies can interact with existing policies to increase or reduce the actual cost of climate policies. [3.6.3.3, 6.3.6.5]

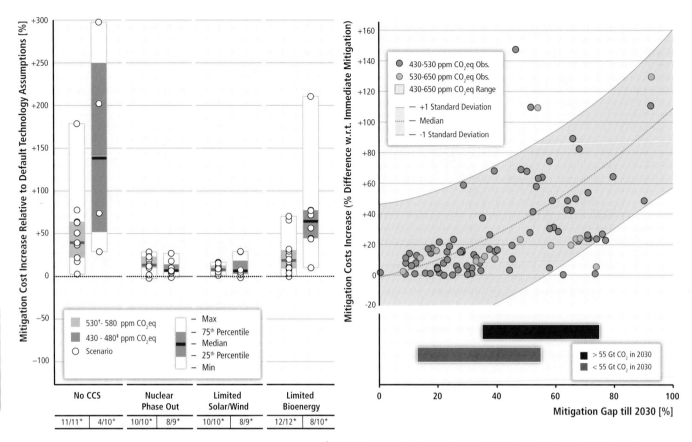

† Scenarios from one model reach concentration levels in 2100 that are slightly below the 530-580 ppm CO₂eq category
‡ Scenarios from two models reach concentration levels in 2100 that are slightly above the 430-480 ppm CO₂eq category
* Number of models successfully vs. number of models attempting running the respective technology variation scenario

Figure TS.13 | Left panel shows the relative increase in net present value mitigation costs (2015–2100, discounted at 5% per year) from technology portfolio variations relative to a scenario with default technology assumptions. Scenario names on the horizontal axis indicate the technology variation relative to the default assumptions: No CCS = unavailability of carbon dioxide capture and storage (CCS); Nuclear phase out = No addition of nuclear power plants beyond those under construction; existing plants operated until the end of their lifetime; Limited Solar/Wind = a maximum of 20% global electricity generation from solar and wind power in any year of these scenarios; Limited Bioenergy = a maximum of 100 exajoules per year (EJ/yr) modern bioenergy supply globally. [Figure 6.24] Right panel shows increase in long-term mitigation costs for the period 2050–2100 (sum over undiscounted costs) as a function of reduced near-term mitigation effort, expressed as the relative change between scenarios implementing mitigation immediately and those that correspond to delayed additional mitigation through 2020 or 2030 (referred to here as 'mitigation gap'). The mitigation gap is defined as the difference in cumulative CO₂ emissions reductions until 2030 between the immediate and delayed additional mitigation scenarios. The bars in the lower right panel indicate the mitigation gap range where 75% of scenarios with 2030 emissions above (dark blue) and below (red) 55 GtCO₂, respectively, are found. Not all model simulations of delayed additional mitigation until 2030 could reach the lower concentration goals of about 450 or 500 (430–530) ppm CO₂eq (for 2030 emissions above 55 GtCO₂eq, 29 of 48 attempted simulations could reach the goal; for 2030 emissions below 55 GtCO₂eq, 34 of 51 attempted simulations could reach the goal). [Figure 6.25]

of century occur in the non-OECD countries. Some studies exploring particular effort-sharing frameworks, under the assumption of a global carbon market, estimate that the associated financial flows could be in the order of hundred billions of USD per year before mid-century to bring concentrations to between about 450 and about 500 ppm CO₂eq in 2100. Most studies assume efficient mechanisms for international carbon markets, in which case economic theory and empirical research suggest that the choice of effort sharing allocations will not meaningfully affect the globally efficient levels of regional abatement or aggregate global costs. Actual approaches to effort-sharing can deviate from this assumption. [3.3, 6.3.6.6, 13.4.2.4]

Geoengineering denotes two clusters of technologies that are quite distinct: carbon dioxide removal (CDR) and solar radiation management (SRM). Mitigation scenarios assessed in AR5 **do not assume any geoengineering options beyond large-scale CDR due to afforestation and BECCS.** CDR techniques include afforestation, using bioenergy along with CCS (BECCS), and enhancing uptake of CO₂ by the oceans through iron fertilization or increasing alkalinity. Most terrestrial CDR techniques would require large-scale land-use changes and could involve local and regional risks, while maritime CDR may involve significant transboundary risks for ocean ecosystems, so that its deployment could pose additional challenges for cooperation between countries. With currently known technologies, CDR could not be deployed quickly on a large scale. SRM includes various technologies to offset crudely some of the climatic effects of the build-up of GHGs in the atmosphere. It works by adjusting the planet's heat balance through a small increase in the reflection of incoming sunlight such as by injecting particles or aerosol precursors in the upper atmosphere. SRM has attracted considerable attention, mainly

Box TS.10 | Future goods should be discounted at an appropriate rate

Investments aimed at mitigating climate change will bear fruit far in the future, much of it more than 100 years from now. To decide whether a particular investment is worthwhile, its future benefits need to be weighed against its present costs. In doing this, economists do not normally take a quantity of commodities at one time as equal in value to the same quantity of the same commodities at a different time. They normally give less value to later commodities than to earlier ones. They 'discount' later commodities, that is to say. The rate at which the weight given to future goods diminishes through time is known as the 'discount rate' on commodities.

There are two types of discount rates used for different purposes. The market discount rate reflects the preferences of presently living people between present and future commodities. The social discount rate is used by society to compare benefits of present members of society with those not yet born. Because living people may be impatient, and because future people do not trade in the market, the market may not accurately reflect the value of commodities that will come to future people relative to those that come to present people. So the social discount rate may differ from the market rate.

The chief reason for social discounting (favouring present people over future people) is that commodities have 'diminishing marginal benefit' and per capita income is expected to increase over time. Diminishing marginal benefit means that the value of

extra commodities to society declines as people become better off. If economies continue to grow, people who live later in time will on average be better off—possess more commodities—than people who live earlier. The faster the growth and the greater the degree of diminishing marginal benefit, the greater should be the discount rate on commodities. If per capita growth is expected to be negative (as it is in some countries), the social discount rate may be negative.

Some authors have argued, in addition, that the present generation of people should give less weight to later people's well-being just because they are more remote in time. This factor would add to the social discount rate on commodities.

The social discount rate is appropriate for evaluating mitigation projects that are financed by reducing current consumption. If a project is financed partly by 'crowding out' other investments, the benefits of those other investments are lost, and their loss must be counted as an opportunity cost of the mitigation project. If a mitigation project crowds out an exactly equal amount of other investment, then the only issue is whether or not the mitigation investment produces a greater return than the crowded-out investment. This can be tested by evaluating the mitigation investment using a discount rate equal to the return that would have been expected from the crowded out investment. If the market functions well, this will be the market discount rate. [3.6.2]

---

because of the potential for rapid deployment in case of climate emergency. The suggestion that deployment costs for individual technologies could potentially be low could result in new challenges for international cooperation because nations may be tempted to prematurely deploy unilaterally systems that are perceived to be inexpensive. Consequently, SRM technologies raise questions about costs, risks, governance, and ethical implications of developing and deploying SRM, with special challenges emerging for international institutions, norms and other mechanisms that could coordinate research and restrain testing and deployment. [1.4, 3.3.7, 6.9, 13.4.4]

Knowledge about the possible beneficial or harmful effects of SRM is highly preliminary. SRM would have varying impacts on regional climate variables such as temperature and precipitation, and might result in substantial changes in the global hydrological cycle with uncertain regional effects, for example on monsoon precipitation. Non-climate effects could include possible depletion of stratospheric ozone by stratospheric aerosol injections. A few studies have begun to examine climate and non-climate impacts of SRM, but there is very little agreement in the scientific community on the results or

on whether the lack of knowledge requires additional research or eventually field testing of SRM-related technologies. [1.4, 3.3.7, 6.9, 13.4.4]

TS.3.1.4   Implications of mitigation pathways for other objectives

Mitigation scenarios reaching about 450 to about 500 ppm CO$_2$eq by 2100 show reduced costs for achieving energy security and air quality objectives (*medium confidence*) (Figure TS.14, lower panel). The mitigation costs of most of the scenarios in this assessment do not consider the economic implications of the cost reductions for these other objectives (Box TS.9). There is a wide range of co-benefits and adverse side-effects other than air quality and energy security (Tables TS.4–8). The impact of mitigation on the overall costs for achieving many of these other objectives as well as the associated welfare implications are less well understood and have not been assessed thoroughly in the literature (Box TS.11). [3.6.3, 4.8, 6.6]

Co-Benefits of Climate Change Mitigation for Energy Security and Air Quality

LIMITS Model Inter-Comparison
Impact of Climate Policy on Energy Security

IPCC AR5 Scenario Ensemble
Impact of Climate Policy on Air Pollutant Emissions (Global, 2005-2050)

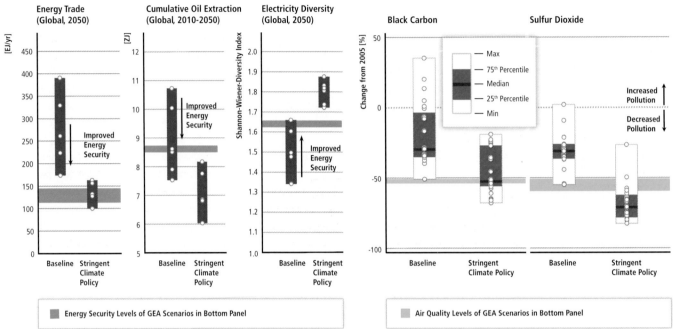

Policy Costs of Achieving Different Objectives

Global Energy Assessment Scenario Ensemble (n=624)

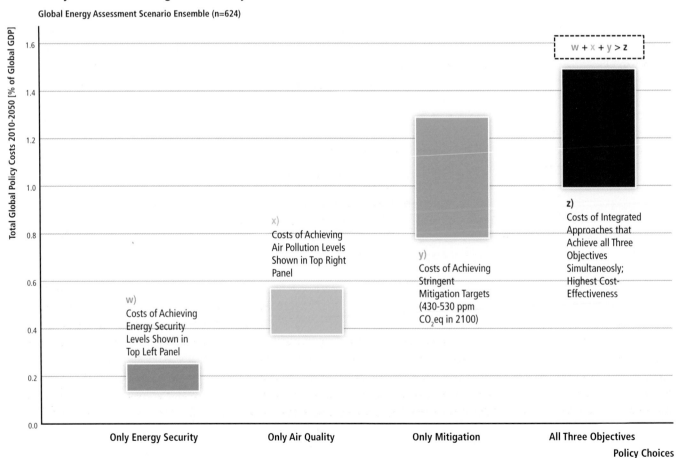

Figure TS.14 | Co-benefits of mitigation for energy security and air quality in scenarios with stringent climate policies reaching about 450 to about 500 (430–530) ppm CO₂eq concentrations in 2100. Upper panels show co-benefits for different security indicators and air pollutant emissions. Lower panel shows related global policy costs of achieving the energy security, air quality, and mitigation objectives, either alone (w, x, y) or simultaneously (z). Integrated approaches that achieve these objectives simultaneously show the highest cost-effectiveness due to synergies (w + x + y > z). Policy costs are given as the increase in total energy system costs relative to a baseline scenario without additional efforts to reduce GHG emissions beyond those in place today. Costs are indicative and do not represent full uncertainty ranges. [Figure 6.33]

Mitigation scenarios reaching about 450 to about 500 ppm CO₂eq by 2100 show co-benefits for energy security objectives, enhancing the sufficiency of resources to meet national energy demand as well as the resilience of the energy system (*medium confidence*). These mitigation scenarios show improvements in terms of the diversity of energy sources and reduction of energy imports, resulting in energy systems that are less vulnerable to price volatility and supply disruptions (Figure TS.14, upper left panel). [6.3.6, 6.6, 7.9, 8.7, 9.7, 10.8, 11.13.6, 12.8]

Mitigation policy could devalue fossil fuel assets and reduce revenues for fossil fuel exporters, but differences between regions and fuels exist (*high confidence*). Most mitigation scenarios are associated with reduced revenues from coal and oil trade for major exporters (*high confidence*). However, a limited number of studies find that mitigation policies could increase the relative competitiveness of conventional oil vis-à-vis more carbon-intensive unconventional oil and 'coal-to-liquids'. The effect of mitigation on natural gas export revenues is more uncertain, with some studies showing possible benefits for export revenues in the medium term until about 2050 (*medium confidence*). The availability of CCS would reduce the adverse effect of mitigation on the value of fossil fuel assets (*medium confidence*). [6.3.6, 6.6, 14.4.2]

Fragmented mitigation policy can provide incentives for emission-intensive economic activity to migrate away from a region that undertakes mitigation (*medium confidence*). Scenario studies have shown that such 'carbon leakage' rates of energy-related emissions are relatively contained, often below 20 % of the emissions reductions. Leakage in land-use emissions could be substantial, though fewer studies have quantified it. While border tax adjustments are seen as enhancing the competitiveness of GHG- and trade-intensive industries within a climate policy regime, they can also entail welfare losses for non-participating, and particularly developing, countries. [5.4, 6.3, 13.8, 14.4]

Mitigation scenarios leading to atmospheric concentration levels of about 450 to about 500 ppm CO₂eq in 2100 are associated with significant co-benefits for air quality and related human health and ecosystem impacts. The benefits from major cuts in air pollutant emissions are particularly high where currently legislated and planned air pollution controls are weak (*high confidence*). Stringent mitigation policies result in co-controls with major cuts in air pollutant emissions significantly below baseline scenarios (Figure TS.14, upper right panel). Co-benefits for health are particularly high in today's developing world. The extent to which air pollution

policies, targeting for example black carbon (BC), can mitigate climate change is uncertain. [5.7, 6.3, 6.6, 7.9, 8.7, 9.7, 10.8, 11.7, 11.13.6, 12.8; WGII 11.9]

There is a wide range of possible adverse side-effects as well as co-benefits and spillovers from climate policy that have not been well-quantified (*high confidence*). Whether or not side-effects materialize, and to what extent side-effects materialize, will be case- and site-specific, as they will depend on local circumstances and the scale, scope, and pace of implementation. Important examples include biodiversity conservation, water availability, food security, income distribution, efficiency of the taxation system, labour supply and employment, urban sprawl, and the sustainability of the growth of developing countries. (Box TS.11)

Some mitigation policies raise the prices for some energy services and could hamper the ability of societies to expand access to modern energy services to underserved populations (*low confidence*). **These potential adverse side-effects can be avoided with the adoption of complementary policies** (*medium confidence*). Most notably, about 1.3 billion people worldwide do not have access to electricity and about 3 billion are dependent on traditional solid fuels for cooking and heating with severe adverse effects on health, ecosystems and development. Providing access to modern energy services is an important sustainable development objective. The costs of achieving nearly universal access to electricity and clean fuels for cooking and heating are projected to be between 72 to 95 billion USD per year until 2030 with minimal effects on GHG emissions (*limited evidence, medium agreement*). A transition away from the use of traditional biomass[13] and the more efficient combustion of solid fuels reduce air pollutant emissions, such as sulfur dioxide (SO₂), nitrogen oxides (NOₓ), carbon monoxide (CO), and black carbon (BC), and thus yield large health benefits (*high confidence*). [4.3, 6.6, 7.9, 9.3, 9.7, 11.13.6, 16.8]

The effect of mitigation on water use depends on technological choices and the portfolio of mitigation measures (*high confidence*). While the switch from fossil energy to renewable energy like photovoltaic (PV) or wind can help reducing water use of the energy system, deployment of other renewables, such as some forms of hydropower, concentrated solar power (CSP), and bioenergy may have adverse effects on water use. [6.6, 7.9, 9.7, 10.8, 11.7, 11.13.6]

---

[13] Traditional biomass refers to the biomass — fuelwood, charcoal, agricultural residues, and animal dung — used with the so-called traditional technologies such as open fires for cooking, rustic kilns and ovens for small industries (see Glossary).

Box TS.11 | Accounting for the co-benefits and adverse side-effects of mitigation

A government policy or a measure intended to achieve one objective (such as mitigation) will also affect other objectives (such as local air quality). To the extent these side-effects are positive, they can be deemed 'co-benefits'; otherwise they are termed 'adverse side-effects'. In this report, co-benefits and adverse side-effects are measured in non-monetary units. Determining the value of these effects to society is a separate issue. The effects of co-benefits on social welfare are not evaluated in most studies, and one reason is that the value of a co-benefit depends on local circumstances and can be positive, zero, or even negative. For example, the value of the extra tonne of sulfur dioxide (SO_2) reduction that occurs with mitigation depends greatly on the stringency of existing SO_2 control policies: in the case of weak existing SO_2 policy, the value of SO_2 reductions may be large, but in the case of stringent existing SO_2 policy it may be near zero. If SO_2 policy is too stringent, the value of the co-benefit may be negative (assuming SO_2 policy is not adjusted). While climate policy affects non-climate objectives (Tables TS.4–8) other policies also affect climate change outcomes. [3.6.3, 4.8, 6.6, Glossary]

Mitigation can have many potential co-benefits and adverse side-effects, which makes comprehensive analysis difficult. The direct benefits of climate policy include, for example, intended effects on global mean surface temperature, sea level rise, agricultural productivity, biodiversity, and health effects of global warming [WGII TS]. The co-benefits and adverse side-effects of climate policy could include effects on a partly overlapping set of objectives such as local air pollutant emissions reductions and related health and ecosystem impacts, biodiversity conservation, water availability, energy and food security, energy access, income distribution, efficiency of the taxation system, labour supply and employment, urban sprawl, and the sustainability of the growth of developing countries [3.6, 4.8, 6.6, 15.2].

All these side-effects are important, because a comprehensive evaluation of climate policy needs to account for benefits and costs related to other objectives. If overall social welfare is to be determined and quantified, this would require valuation methods and a consideration of pre-existing efforts to attain the many objectives. Valuation is made difficult by factors such as interaction between climate policies and pre-existing non-climate policies, externalities, and non-competitive behaviour. [3.6.3]

Mitigation scenarios and sectoral studies show that overall the potential for co-benefits of energy end-use measures outweigh the potential adverse side-effects, whereas the evidence suggests this may not be the case for all energy supply and AFOLU measures *(high confidence).* (Tables TS.4–8) [4.8, 5.7, 6.6, 7.9, 8.7, 9.7, 10.8, 11.7, 11.13.6, 12.8]

TS.3.2 Sectoral and cross-sectoral mitigation measures

Anthropogenic GHG emissions result from a broad set of human activities, most notably those associated with energy supply and consumption and with the use of land for food production and other purposes. A large proportion of emissions arise in urban areas. Mitigation options can be grouped into three broad sectors: (1) energy supply, (2) energy end-use sectors including transport, buildings, industry, and (3) AFOLU. Emissions from human settlements and infrastructures cut across these different sectors. Many mitigation options are linked. The precise set of mitigation actions taken in any sector will depend on a wide range of factors, including their relative economics, policy structures, normative values, and linkages to other policy objectives. The first section examines issues that cut across the sectors and the following subsections examine the sectors themselves.

TS.3.2.1 Cross-sectoral mitigation pathways and measures

Without new mitigation policies GHG emissions are projected to grow in all sectors, except for net CO_2 emissions in the AFOLU[14] sector *(robust evidence, medium agreement).* Energy supply sector emissions are expected to continue to be the major source of GHG emissions in baseline scenarios, ultimately accounting for the significant increases in indirect emissions from electricity use in the buildings and the industry sectors. Deforestation decreases in most of the baseline scenarios, which leads to a decline in net CO_2 emissions from the AFOLU sector. In some scenarios the AFOLU sector changes from an emission source to a net emission sink towards the end of the century. (Figure TS.15) [6.3.1.4, 6.8]

Infrastructure developments and long-lived products that lock societies into GHG-intensive emissions pathways may be difficult or very costly to change, reinforcing the importance of early action for ambitious mitigation *(robust evidence, high agreement).* This lock-in risk is compounded by the lifetime of the infrastructure, by the difference in emissions associated with alternatives, and

---

14   Net AFOLU CO_2 emissions include emissions and removals of CO_2 from the AFOLU sector, including land under forestry and, in some assessments, CO_2 sinks in agricultural soils.

the magnitude of the investment cost. As a result, lock-in related to infrastructure and spatial planning is the most difficult to eliminate, and thus avoiding options that lock high emission patterns in permanently is an important part of mitigation strategies in regions with rapidly developing infrastructure. In mature or established cities, options are constrained by existing urban forms and infrastructure, and limits on the potential for refurbishing or altering them. However, materials, products and infrastructure with long lifetimes and low lifecycle emissions can ensure positive lock-in as well as avoid emissions through dematerialization (i.e., through reducing the total material inputs required to deliver a final service). [5.6.3, 6.3.6.4, 9.4, 10.4, 12.3, 12.4]

Systemic and cross-sectoral approaches to mitigation are expected to be more cost-effective and more effective in cutting emissions than sector-by-sector policies (*medium confidence*). Cost-effective mitigation policies need to employ a system perspective in order to account for inter-dependencies among different economic sectors and to maximize synergistic effects. Stabilizing atmospheric CO_2eq concentrations at any level will ultimately require deep reductions in emissions and fundamental changes to both the end-use and supply-side of the energy system as well as changes in land-use practices and industrial processes. In addition, many low-carbon energy supply technologies (including CCS) and

their infrastructural requirements face public acceptance issues limiting their deployment. This applies also to the adoption of new technologies, and structural and behavioural change, in the energy end-use sectors (*robust evidence, high agreement*) [7.9.4, 8.7, 9.3.10, 9.8, 10.8, 11.3, 11.13]. Lack of acceptance may have implications not only for mitigation in that particular sector, but also for wider mitigation efforts.

Integrated models identify three categories of energy system related mitigation measures: the decarbonization of the energy supply sector, final energy demand reductions, and the switch to low-carbon energy carriers, including electricity, in the energy end-use sectors (*robust evidence, high agreement*) [6.3.4, 6.8, 7.11]. The broad range of sectoral mitigation options available mainly relate to achieving reductions in GHG emissions intensity, energy intensity and changes in activity (Table TS.3) [7.5, 8.3, 8.4, 9.3, 10.4, 12.4]. Direct options in AFOLU involve storing carbon in terrestrial systems (for example, through afforestation) and providing bioenergy feedstocks [11.3, 11.13]. Options to reduce non-CO_2 GHG emissions exist across all sectors, but most notably in agriculture, energy supply, and industry.

Demand reductions in the energy end-use sectors, due to, e.g., efficiency enhancement and behavioural change, are a key miti-

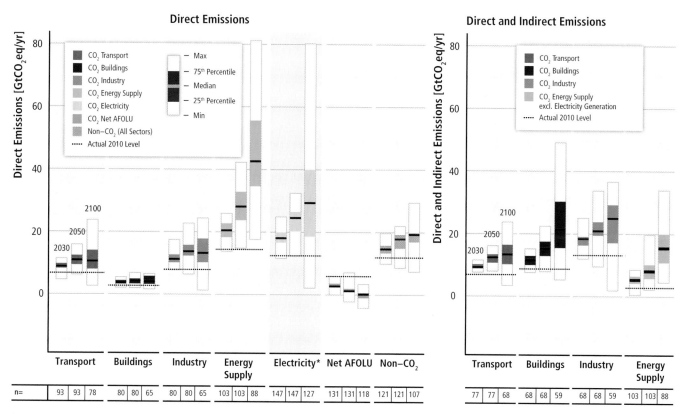

Figure TS.15 | Direct (left panel) and direct and indirect emissions (right panel) of CO_2 and non-CO_2 GHGs across sectors in baseline scenarios. Non-CO_2 GHGs are converted to CO_2-equivalents based on Global Warming Potentials with a 100-year time horizon from the IPCC Second Assessment Report (SAR) (see Box TS.5). Note that in the case of indirect emissions, only electricity generation emissions are allocated from energy supply to end-use sectors. In the left panel electricity sector emissions are shown (Electricity*) in addition to energy supply sector emissions which they are part of, to illustrate their large role on the energy supply side. The numbers at the bottom refer to the number of scenarios included in the ranges that differ across sectors and time due to different sectoral resolutions and time horizons of models. [Figure 6.34]

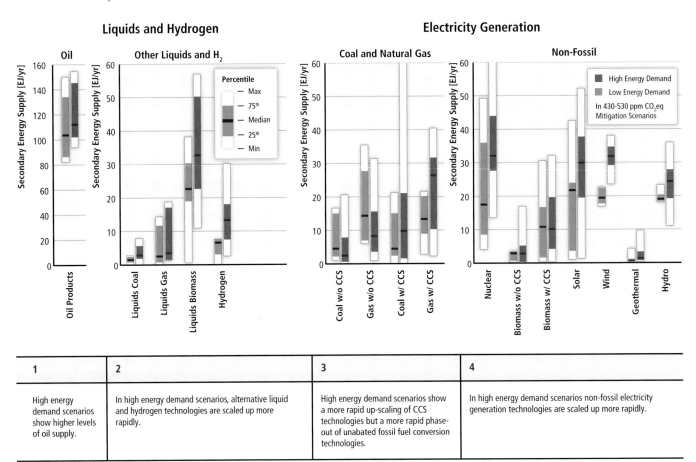

Liquids and Hydrogen | **Electricity Generation**

1	2	3	4
High energy demand scenarios show higher levels of oil supply.	In high energy demand scenarios, alternative liquid and hydrogen technologies are scaled up more rapidly.	High energy demand scenarios show a more rapid up-scaling of CCS technologies but a more rapid phase-out of unabated fossil fuel conversion technologies.	In high energy demand scenarios non-fossil electricity generation technologies are scaled up more rapidly.

Figure TS.16 | Influence of energy demand on the deployment of energy supply technologies in 2050 in mitigation scenarios reaching about 450 to about 500 (430–530) ppm CO₂eq concentrations by 2100. Blue bars for 'low energy demand' show the deployment range of scenarios with limited growth of final energy of < 20 % in 2050 compared to 2010. Red bars show the deployment range of technologies in case of 'high energy demand' (> 20 % growth in 2050 compared to 2010). For each technology, the median, interquartile, and full deployment range is displayed. Notes: Scenarios assuming technology restrictions and scenarios with final energy in the base-year outside ± 5 % of 2010 inventories are excluded. Ranges include results from many different integrated models. Multiple scenario results from the same model were averaged to avoid sampling biases; see Chapter 6 for further details. [Figure 7.11]

gation strategy and affect the scale of the mitigation challenge for the energy supply side (*high confidence*). Limiting energy demand: (1) increases policy choices by maintaining flexibility in the technology portfolio; (2) reduces the required pace for up-scaling low-carbon energy supply technologies and hedges against related supply-side risks (Figure TS.16); (3) avoids lock-in to new, or potentially premature retirement of, carbon-intensive infrastructures; (4) maximizes co-benefits for other policy objectives, since the potential for co-benefits of energy end-use measures outweighs the potential for adverse side-effects which may not be the case for all supply-side measures (see Tables TS.4–8); and (5) increases the cost-effectiveness of the transformation (as compared to mitigation strategies with higher levels of energy demand) (*medium confidence*). However, energy service demand reductions are unlikely in developing countries or for poorer population segments whose energy service levels are low or partially unmet. [6.3.4, 6.6, 7.11, 10.4]

Behaviour, lifestyle, and culture have a considerable influence on energy use and associated emissions, with a high mitigation potential in some sectors, in particular when complementing technological and structural change (*medium evidence, medium agreement*). Emissions can be substantially lowered through: changes in consumption patterns (e.g., mobility demand and mode, energy use in households, choice of longer-lasting products); dietary change and reduction in food wastes; and change of lifestyle (e.g., stabilizing/lowering consumption in some of the most developed countries, sharing economy and other behavioural changes affecting activity) (Table TS.3). [8.1, 8.9, 9.2, 9.3, Box 10.2, 10.4, 11.4, 12.4, 12.6, 12.7]

Evidence from mitigation scenarios indicates that the decarbonization of energy supply is a key requirement for stabilizing atmospheric CO₂eq concentrations below 580 ppm (*robust evidence, high agreement*). In most long-term mitigation scenarios not exceeding 580 ppm CO₂eq by 2100, global energy supply is fully decarbonized at the end of the 21st century with many scenarios relying on a net removal of CO₂ from the atmosphere. However, because existing supply systems are largely reliant on carbon-intensive fossil fuels, energy intensity reductions can equal or outweigh decarbonization of energy supply in the near term. In the buildings and industry sector, for example, efficiency improvements are an important strategy for reducing indirect emissions from electricity generation (Figure TS.15). In the long term, the reduction in electricity generation emissions is accompanied by an increase in the share of electricity in end uses (e.g., for

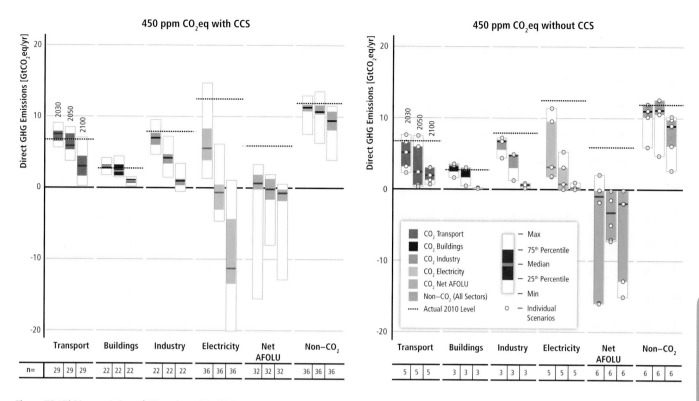

Figure TS.17 | Direct emissions of CO_2 and non-CO_2 GHGs across sectors in mitigation scenarios that reach about 450 (430–480) ppm CO_2eq concentrations in 2100 with using carbon dioxide capture and storage (CCS) (left panel) and without using CCS (right panel). The numbers at the bottom of the graphs refer to the number of scenarios included in the ranges that differ across sectors and time due to different sectoral resolutions and time horizons of models. White dots in the right panel refer to emissions of individual scenarios to give a sense of the spread within the ranges shown due to the small number of scenarios. [Figures 6.35]

space and process heating, and potentially for some modes of transport). Deep emissions reductions in transport are generally the last to emerge in integrated modelling studies because of the limited options to switch to low-carbon energy carriers compared to buildings and industry (Figure TS.17). [6.3.4, 6.8, 8.9, 9.8, 10.10, 7.11, Figure 6.17]

The availability of CDR technologies affects the size of the mitigation challenge for the energy end-use sectors (*robust evidence, high agreement*) [6.8, 7.11]. There are strong interdependencies in mitigation scenarios between the required pace of decarbonization of energy supply and end-use sectors. The more rapid decarbonization of supply generally provides more flexibility for the end-use sectors. However, barriers to decarbonizing the supply side, resulting for example from a limited availability of CCS to achieve negative emissions when combined with bioenergy, require a more rapid and pervasive decarbonisation of the energy end-use sectors in scenarios achieving low-CO_2eq concentration levels (Figure TS.17). The availability of mature large-scale biomass supply for energy, or carbon sequestration technologies in the AFOLU sector also provides flexibility for the development of mitigation technologies in the energy supply and energy end-use sectors [11.3] (*limited evidence, medium agreement*), though there may be adverse impacts on sustainable development.

Spatial planning can contribute to managing the development of new infrastructure and increasing system-wide efficiencies across sectors (*robust evidence, high agreement*). Land use, transport

choice, housing, and behaviour are strongly interlinked and shaped by infrastructure and urban form. Spatial and land-use planning, such as mixed-zoning, transport-oriented development, increasing density, and co-locating jobs and homes can contribute to mitigation across sectors by (1) reducing emissions from travel demand for both work and leisure, and enabling non-motorized transport, (2) reducing floor space for housing, and hence (3) reducing overall direct and indirect energy use through efficient infrastructure supply. Compact and in-fill development of urban spaces and intelligent densification can save land for agriculture and bioenergy and preserve land carbon stocks. [8.4, 9.10, 10.5, 11.10, 12.2, 12.3]

Interdependencies exist between adaptation and mitigation at the sectoral level and there are benefits from considering adaptation and mitigation in concert (*medium evidence, high agreement*). Particular mitigation actions can affect sectoral climate vulnerability, both by influencing exposure to impacts and by altering the capacity to adapt to them [8.5, 11.5]. Other interdependencies include climate impacts on mitigation options, such as forest conservation or hydropower production [11.5.5, 7.7], as well as the effects of particular adaptation options, such as heating or cooling of buildings or establishing more diversified cropping systems in agriculture, on GHG emissions and radiative forcing [11.5.4, 9.5]. There is a growing evidence base for such interdependencies in each sector, but there are substantial knowledge gaps that prevent the generation of integrated results at the cross-sectoral level.

Table TS.3 | Main sectoral mitigation measures categorized by key mitigation strategies (in bold) and associated sectoral indicators (highlighted in yellow) as discussed in Chapters 7–12.

	GHG emissions intensity reduction	Energy intensity reduction by improving technical efficiency	Production and resource efficiency improvement	Structural and systems efficiency improvement	Activity indicator change
Energy [Section 7.5]	*Emissions/ secondary energy output*	*Energy input/ energy output*	*Embodied energy/ energy output*	—	*Final energy use*
	Greater deployment of renewable energy (RE), nuclear energy, and (BE)CCS; fuel switching within the group of fossil fuels; reduction of fugitive (methane) emissions in the fossil fuel chain	Extraction, transport and conversion of fossil fuels; electricity/ heat/ fuel transmission, distribution, and storage; Combined Heat and Power (CHP) or cogeneration (*see Buildings and Human Settlements*)	Energy embodied in manufacturing of energy extraction, conversion, transmission and distribution technologies	Addressing integration needs	Demand from end-use sectors for different energy carriers (*see Transport, Buildings and Industry*)
Transport [8.3]	*Emissions/ final energy*	*Final energy/ transport service*	—	*Shares for each mode*	*Total distance per year*
	Fuel carbon intensity (CO₂eq/megajoule (MJ)): Fuel switching to low-carbon fuels e.g., electricity/hydrogen from low-carbon sources (*see Energy*); specific biofuels in various modes (*see AFOLU*)	**Energy intensity (MJ/passenger-km, tonne-km):** Fuel-efficient engines and vehicle designs; more advanced propulsion systems and designs; use of lighter materials in vehicles	Embodied emissions during vehicle manufacture; material efficiency; and recycling of materials (*see Industry*); infrastructure lifecycle emissions (*see Human Settlements*)	Modal shifts from light-duty vehicles (LDVs) to public transit, cycling/walking, and from aviation and heavy-duty vehicles (HDVs) to rail; eco-driving; improved freight logistics; transport (infrastructure) planning	Journey avoidance; higher occupancy/loading rates; reduced transport demand; urban planning (*see Human Settlements*)
Buildings [9.3]	*Emissions/ final energy*	*Final energy/ useful energy*	*Embodied energy/ operating energy*	*Useful energy/ energy service*	*Energy service demand*
	Fuel carbon intensity (CO₂eq/MJ): Building-integrated RE technologies; fuel switching to low-carbon fuels, e.g., electricity (*see Energy*)	**Device efficiency:** heating/ cooling (high-performance boilers, ventilation, air-conditioning, heat pumps); water heating; cooking (advanced biomass stoves); lighting; appliances	Building lifetime; component, equipment, and appliance durability; low(er) energy and emission material choice for construction (*see Industry*)	**Systemic efficiency:** integrated design process; low/zero energy buildings; building automation and controls; urban planning; district heating/cooling and CHP; smart meters/grids; commissioning	Behavioural change (e.g., thermostat setting, appliance use); lifestyle change (e.g., per capita dwelling size, adaptive comfort)
Industry [10.4]	*Emissions/ final energy*	*Final energy/ material production*	*Material input/ product output*	*Product demand/ service demand*	*Service demand*
	Emissions intensity: Process emissions reductions; use of waste (e.g., municipal solid waste (MSW)/sewage sludge in cement kilns) and CCS in industry; HFCs replacement and leak repair; fuel switching among fossil fuels to low-carbon electricity (*see Energy*) or biomass (*see AFOLU*)	**Energy efficiency/ best available technologies:** Efficient steam systems; furnace and boiler systems; electric motor (pumps, fans, air compressor, refrigerators, and material handling) and electronic control systems; (waste) heat exchanges; recycling	**Material efficiency:** Reducing yield losses; manufacturing/construction: process innovations, new design approaches, re-using old material (e.g., structural steel); product design (e.g., light weight car design); fly ash substituting clinker	**Product-service efficiency:** More intensive use of products (e.g., car sharing, using products such as clothing for longer, new and more durable products)	Reduced demand for, e.g., products such as clothing; alternative forms of travel leading to reduced demand for car manufacturing
Human Settlements [12.4]	*Emissions/ final energy*	*Final energy/ useful energy*	*Material input in infrastructure*	*Useful energy/ energy service*	*Service demand per capita*
	Integration of urban renewables; urban-scale fuel switching programmes	Cogeneration, heat cascading, waste to energy	Managed infrastructure supply; reduced primary material input for infrastructure	Compact urban form; increased accessibility; mixed land use	Increasing accessibility: shorter travel time, and more transport mode options

Agriculture, Forestry and Other Land Use (AFOLU) [11.3]	Supply-side improvements			Demand-side measures
	Emissions/ area or unit product (conserved, restored)			*Animal/crop product consumption per capita*
	Emissions reduction: of methane (e.g., livestock management) and nitrous oxide (fertilizer and manure management) and prevention of emissions to the atmosphere by conserving existing carbon pools in soils or vegetation (reducing deforestation and forest degradation, fire prevention/control, agroforestry); reduced emissions intensity (GHG/unit product).	**Sequestration:** Increasing the size of existing carbon pools, thereby extracting CO₂ from the atmosphere (e.g., afforestation, reforestation, integrated systems, carbon sequestration in soils)	**Substitution:** of biological products for fossil fuels or energy-intensive products, thereby reducing CO₂ emissions, e.g., biomass co-firing/CHP (*see Energy*), biofuels (*see Transport*), biomass-based stoves, and insulation products (*see Buildings*)	**Demand-side measures:** Reducing losses and wastes of food; changes in human diets towards less emission-intensive products; use of long-lived wood products

TS

TS.3.2.2 Energy supply

The energy supply sector is the largest contributor to global GHG emissions (*robust evidence, high agreement*). Annual GHG emissions from the global energy supply sector grew more rapidly between 2000 and 2010 than in the previous decade; their growth accelerated from 1.7 %/yr from 1990–2000 to 3.1 %/yr from 2000–2010. The main contributors to this trend are an increasing demand for energy services and a growing share of coal in the global fuel mix. The energy supply sector, as defined in this report, comprises all energy extraction, conversion, storage, transmission, and distribution processes that deliver final energy to the end-use sectors (industry, transport, buildings, agriculture and forestry). [7.2, 7.3]

In the baseline scenarios assessed in AR5, direct CO_2 emissions from the energy supply sector increase from 14.4 $GtCO_2$/yr in 2010 to 24–33 $GtCO_2$/yr in 2050 (25–75th percentile; full range 15–42 $GtCO_2$/yr), with most of the baseline scenarios assessed in WGIII AR5 showing a significant increase (*medium evidence, medium agreement*) (Figure TS.15). The lower end of the full range is dominated by scenarios with a focus on energy intensity improvements that go well beyond the observed improvements over the past 40 years. The availability of fossil fuels alone will not be sufficient to limit CO_2eq concentration to levels such as 450 ppm, 550 ppm, or 650 ppm. [6.3.4, 6.8, 7.11, Figure 6.15]

The energy supply sector offers a multitude of options to reduce GHG emissions (*robust evidence, high agreement*). These options include: energy efficiency improvements and fugitive emission reductions in fuel extraction as well as in energy conversion, transmission, and distribution systems; fossil fuel switching; and low-GHG energy supply technologies such as renewable energy (RE), nuclear power, and CCS (Table TS.3). [7.5, 7.8.1, 7.11]

The stabilization of GHG concentrations at low levels requires a fundamental transformation of the energy supply system, including the long-term phase-out of unabated fossil fuel conversion technologies and their substitution by low-GHG alternatives (*robust evidence, high agreement*). Concentrations of CO_2 in the atmosphere can only be stabilized if global (net) CO_2 emissions peak and decline toward zero in the long term. Improving the energy efficiencies of fossil fuel power plants and/or the shift from coal to gas will not by themselves be sufficient to achieve this. Low-GHG energy supply technologies would be necessary if this goal were to be achieved (Figure TS.19). [7.5.1, 7.8.1, 7.11]

Decarbonizing (i.e., reducing the carbon intensity of) electricity generation is a key component of cost-effective mitigation strategies in achieving low-stabilization levels (430–530 ppm CO_2eq); in most integrated modelling scenarios, decarbonization happens more rapidly in electricity generation than in the buildings, transport, and industry sectors (*medium evidence, high agreement*) (Figure TS.17). In the majority of mitigation scenar-

ios reaching about 450 ppm CO_2eq concentrations by 2100, the share of low-carbon electricity supply (comprising RE, nuclear, fossil fuels with CCS, and BECCS) increases from the current share of around 30 % to more than 80 % by 2050, and fossil fuel power generation without CCS is phased out almost entirely by 2100 (Figures TS.17 and TS.18) [7.14].

Since AR4, many RE technologies have demonstrated substantial performance improvements and cost reductions, and a growing number of RE technologies have achieved a level of maturity to enable deployment at significant scale (*robust evidence, high agreement*). Some technologies are already economically competitive in various settings. Levelized costs of PV systems fell most substantially between 2009 and 2012, and a less extreme trend has been observed for many others RE technologies. Regarding electricity generation alone, RE accounted for just over half of the new electricity-generating capacity added globally in 2012, led by growth in wind, hydro, and solar power. Decentralized RE to meet rural energy needs has also increased, including various modern and advanced traditional biomass options as well as small hydropower, PV, and wind. Nevertheless, many RE technologies still need direct support (e.g., feed-in tariffs (FITs), RE quota obligations, and tendering/bidding) and/or indirect support (e.g., sufficiently high carbon prices and the internalization of other externalities), if their market shares are to be significantly increased. RE technology policies have been successful in driving the recent growth of RE. Additional enabling policies are needed to address their integration into future energy systems. (*medium evidence, medium agreement*) (Figure TS.19) [7.5.3, 7.6.1, 7.8.2, 7.12, 11.13]

The use of RE is often associated with co-benefits, including the reduction of air pollution, local employment opportunities, few severe accidents compared to some other energy supply technologies, as well as improved energy access and security (*medium evidence, medium agreement*) (Table TS.4). At the same time, however, some RE technologies can have technology and location-specific adverse side-effects, which can be reduced to a degree through appropriate technology selection, operational adjustments, and siting of facilities. [7.9]

Infrastructure and integration challenges vary by RE technology and the characteristics of the existing energy system (*medium evidence, medium agreement*). Operating experience and studies of medium to high penetrations of RE indicate that integration issues can be managed with various technical and institutional tools. As RE penetrations increase, such issues are more challenging, must be carefully considered in energy supply planning and operations to ensure reliable energy supply, and may result in higher costs. [7.6, 7.8.2]

Nuclear energy is a mature low-GHG emission source of baseload power, but its share of global electricity generation has been declining (since 1993). Nuclear energy could make an increasing contribution to low-carbon energy supply, but a variety of barriers and risks exist (*robust evidence, high agree-*

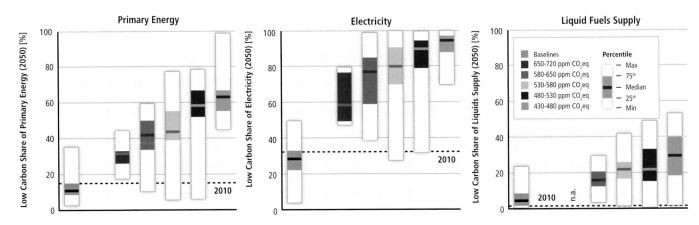

Figure TS.18 | Share of low-carbon energy in total primary energy, electricity and liquid fuels supply sectors for the year 2050. Dashed horizontal lines show the low-carbon share for the year 2010. Low-carbon energy includes nuclear, renewables, fossil fuels with carbon dioxide capture and storage (CCS) and bioenergy with CCS. [Figure 7.14]

ment) (Figure TS.19). Nuclear electricity accounted for 11% of the world's electricity generation in 2012, down from a high of 17% in 1993. Pricing the externalities of GHG emissions (carbon pricing) could improve the competitiveness of nuclear power plants. [7.2, 7.5.4, 7.8.1, 7.12]

Barriers and risks associated with an increasing use of nuclear energy include operational risks and the associated safety concerns, uranium mining risks, financial and regulatory risks, unresolved waste management issues, nuclear weapon proliferation concerns, and adverse public opinion (robust evidence, high agreement) (Table TS.4). New fuel cycles and reactor technologies addressing some of these issues are under development and progress has been made concerning safety and waste disposal. Investigation of mitigation scenarios not exceeding 580 ppm CO₂eq has shown that excluding nuclear power from the available portfolio of technologies would result in only a slight increase in mitigation costs compared to the full technology portfolio (Figure TS.13). If other technologies, such as CCS, are constrained the role of nuclear power expands. [6.3.6, 7.5.4, 7.8.2, 7.9, 7.11]

GHG emissions from energy supply can be reduced significantly by replacing current world average coal-fired power plants with modern, highly efficient natural gas combined cycle power plants or combined heat and power (CHP) plants, provided that natural gas is available and the fugitive emissions associated with its extraction and supply are low or mitigated (robust evidence, high agreement). In mitigation scenarios reaching about 450 ppm CO₂eq concentrations by 2100, natural gas power generation without CCS typically acts as a bridge technology, with deployment increasing before peaking and falling to below current levels by 2050 and declining further in the second half of the century (robust evidence, high agreement). [7.5.1, 7.8, 7.9, 7.11, 7.12]

Carbon dioxide capture and storage (CCS) technologies could reduce the lifecycle GHG emissions of fossil fuel power plants

(medium evidence, medium agreement). While all components of integrated CCS systems exist and are in use today by the fossil fuel extraction and refining industry, CCS has not yet been applied at scale to a large, commercial fossil fuel power plant. CCS power plants could be seen in the market if they are required for fossil fuel facilities by regulation or if they become competitive with their unabated counterparts, for instance, if the additional investment and operational costs faced by CCS plants, caused in part by efficiency reductions, are compensated by sufficiently high carbon prices (or direct financial support). Beyond economic incentives, well-defined regulations concerning short- and long-term responsibilities for storage are essential for a large-scale future deployment of CCS. [7.5.5]

Barriers to large-scale deployment of CCS technologies include concerns about the operational safety and long-term integrity of CO₂ storage, as well as risks related to transport and the required up-scaling of infrastructure (limited evidence, medium agreement) (Table TS.4). There is, however, a growing body of literature on how to ensure the integrity of CO₂ wells, on the potential consequences of a CO₂ pressure build-up within a geologic formation (such as induced seismicity), and on the potential human health and environmental impacts from CO₂ that migrates out of the primary injection zone (limited evidence, medium agreement). [7.5.5, 7.9, 7.11]

Combining bioenergy with CCS (BECCS) offers the prospect of energy supply with large-scale net negative emissions, which plays an important role in many low-stabilization scenarios, while it entails challenges and risks (limited evidence, medium agreement). Until 2050, bottom-up studies estimate the economic potential to be between 2–10 GtCO₂ per year [11.13]. Some mitigation scenarios show higher deployment of BECCS towards the end of the century. Technological challenges and risks include those associated with the upstream provision of the biomass that is used in the CCS facility, as well as those associated with the CCS technology itself. Currently, no large-scale projects have been financed. [6.9, 7.5.5, 7.9, 11.13]

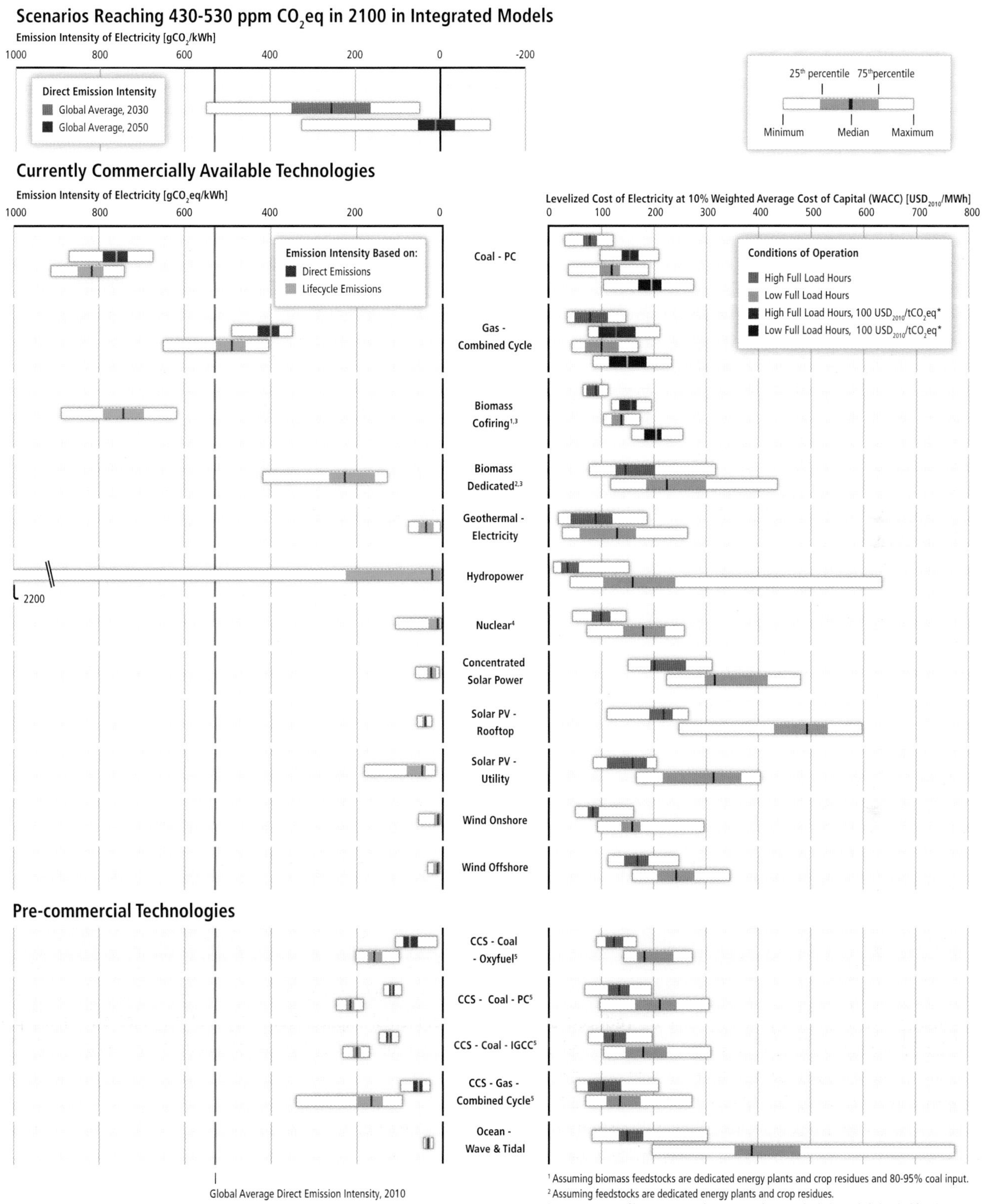

Scenarios Reaching 430-530 ppm CO_2eq in 2100 in Integrated Models

Emission Intensity of Electricity [gCO_2/kWh]

Currently Commercially Available Technologies

Emission Intensity of Electricity [gCO_2eq/kWh]

Levelized Cost of Electricity at 10% Weighted Average Cost of Capital (WACC) [USD_{2010}/MWh]

Global Average Direct Emission Intensity, 2010

Pre-commercial Technologies

[1] Assuming biomass feedstocks are dedicated energy plants and crop residues and 80-95% coal input.
[2] Assuming feedstocks are dedicated energy plants and crop residues.
[3] Direct emissions of biomass power plants are not shown explicitly, but included in the lifecycle emissions. Lifecycle emissions include albedo effect.
[4] LCOE of nuclear include front and back-end fuel costs as well as decommissioning costs.
[5] Transport and storage costs of CCS are set to 10 USD_{2010}/tCO_2.
[*] Carbon price levied on direct emissions. Effects shown where significant.

Figure TS.19| Specific direct and lifecycle emissions (gCO_2eq/ kilowatt hour (kWh)) and levelized cost of electricity (LCOE in USD_{2010}/MWh) for various power-generating technologies (see Annex III.2 for data and assumptions and Annex II.3.1 and II.9.3 for methodological issues). The upper left graph shows global averages of specific direct CO_2 emissions (gCO_2/kWh) of power generation in 2030 and 2050 for the set of about 450 to about 500 (430–530) ppm CO_2eq scenarios that are contained in the WG III AR5 Scenario Database (see Annex II.10). The global average of specific direct CO_2 emissions (gCO_2/kWh) of power generation in 2010 is shown as a vertical line. Note: The inter-comparability of LCOE is limited. For details on general methodological issues and interpretation see Annexes as mentioned above. CCS: CO_2 capture and storage; IGCC: Integrated coal gasification combined cycle; PC: Pulverized hard coal; PV: Photovoltaic; WACC: Weighted average cost of capital. [Figure 7.7]

Table TS.4 | Overview of potential co-benefits (green arrows) and adverse side-effects (orange arrows) of the main mitigation measures in the energy supply sector; arrows pointing up/down denote a positive/negative effect on the respective objective or concern; a question mark (**?**) denotes an uncertain net effect. Co-benefits and adverse side-effects depend on local circumstances as well as on the implementation practice, pace, and scale. For possible upstream effects of biomass supply for bioenergy, see Table TS.8. For an assessment of macroeconomic, cross-sectoral effects associated with mitigation policies (e.g., on energy prices, consumption, growth, and trade), see e.g., Sections 3.9, 6.3.6, 13.2.2.3 and 14.4.2. The uncertainty qualifiers in brackets denote the level of evidence and agreement on the respective effects (see TS.1). Abbreviations for evidence: l=limited, m=medium, r=robust; for agreement: l=low, m=medium, h=high. [Table 7.3]

Energy Supply	Effect on additional objectives/concerns			
	Economic	Social	Environmental	Other
Nuclear replacing coal power	↑ Energy security (reduced exposure to fuel price volatility) (**m/m**) ↑ Local employment impact (but uncertain net effect) (**l/m**) ↑ Legacy cost of waste and abandoned reactors (**m/h**)	Health impact via ↓ Air pollution and coal mining accidents (**m/h**) ↑ Nuclear accidents and waste treatment, uranium mining and milling (**m/l**) ↑ Safety and waste concerns (**r/h**)	Ecosystem impact via ↓ Air pollution (**m/h**) and coal mining (**l/h**) ↑ Nuclear accidents (**m/m**)	Proliferation risk (**m/m**)
RE (wind, PV, concentrated solar power (CSP), hydro, geothermal, bioenergy) replacing coal	↑ Energy security (resource sufficiency, diversity in the near/medium term) (**r/m**) ↑ Local employment impact (but uncertain net effect) (**m/m**) ↑ Irrigation, flood control, navigation, water availability (for multipurpose use of reservoirs and regulated rivers) (**m/h**) ↑ Extra measures to match demand (for PV, wind and some CSP) (**r/h**)	Health impact via ↓ Air pollution (except bioenergy) (**r/h**) ↓ Coal mining accidents (**m/h**) ↑ Contribution to (off-grid) energy access (**m/l**) ? Project-specific public acceptance concerns (e.g., visibility of wind) (**l/m**) ↑ Threat of displacement (for large hydro) (**m/h**)	Ecosystem impact via ↓ Air pollution (except bioenergy) (**m/h**) ↓ Coal mining (**l/h**) ↑ Habitat impact (for some hydro) (**m/m**) ↑ Landscape and wildlife impact (for wind) **m/m** ↓ Water use (for wind and PV) (**m/m**) ↑ Water use (for bioenergy, CSP, geothermal, and reservoir hydro) (**m/h**)	Higher use of critical metals for PV and direct drive wind turbines (**r/m**)
Fossil CCS replacing coal	↑↑ Preservation vs. lock-in of human and physical capital in the fossil industry (**m/m**)	Health impact via ↑ Risk of CO₂ leakage (**m/m**) ↑ Upstream supply-chain activities (**m/h**) ↑ Safety concerns (CO₂ storage and transport) (**m/h**)	↑ Ecosystem impact via upstream supply-chain activities (**m/m**) ↑ Water use (**m/h**)	Long-term monitoring of CO₂ storage (**m/h**)
BECCS replacing coal	*See fossil CCS where applicable. For possible upstream effect of biomass supply, see Table TS.8.*			
Methane leakage prevention, capture or treatment	↑ Energy security (potential to use gas in some cases) (**l/h**)	↓ Health impact via reduced air pollution (**m/m**) ↑ Occupational safety at coal mines (**m/m**)	↓ Ecosystem impact via reduced air pollution (**l/m**)	

TS.3.2.3   Transport

Since AR4, emissions in the global transport sector have grown in spite of more efficient vehicles (road, rail, watercraft, and aircraft) and policies being adopted (*robust evidence, high agreement*). Road transport dominates overall emissions but aviation could play an increasingly important role in total CO₂ emissions in the future. [8.1, 8.3, 8.4]

The global transport sector accounted for 27% of final energy use and 6.7 GtCO₂ direct emissions in 2010, with baseline CO₂ emissions projected to increase to 9.3–12 GtCO₂/yr in 2050 (25–75th percentile; full range 6.2–16 GtCO₂/yr); most of the baseline scenarios assessed in WGIII AR5 foresee a significant increase (*medium evidence/medium agreement*) (Figure TS.15). With-

out aggressive and sustained mitigation policies being implemented, transport sector emissions could increase faster than in the other energy end-use sectors and could lead to more than a doubling of CO₂ emissions by 2050. [6.8, 8.9, 8.10]

While the continuing growth in passenger and freight activity constitutes a challenge for future emission reductions, analyses of both sectoral and integrated studies suggest a higher mitigation potential in the transport sector than reported in the AR4 (*medium evidence, medium agreement*). Transport energy demand per capita in developing and emerging economies is far lower than in OECD countries but is expected to increase at a much faster rate in the next decades due to rising incomes and the development of infrastructure. Baseline scenarios thus show increases in transport energy demand from 2010 out to 2050 and beyond. However, sectoral and integrated mitigation scenarios indicate that energy demand reductions of 10–45 % are possible by 2050 relative to baseline (Figure TS.20, left panel) (*medium evidence, medium agreement*). [6.8.4, 8.9.1, 8.9.4, 8.12, Figure 8.9.4]

A combination of low-carbon fuels, the uptake of improved vehicle and engine performance technologies, behavioural change leading to avoided journeys and modal shifts, investments in related infrastructure and changes in the built environment, together offer a high mitigation potential (*high confidence*) [8.3, 8.8]. Direct (tank-to-wheel) GHG emissions from passenger and freight transport can be reduced by:

- using fuels with lower carbon intensities (CO_2eq/ megajoule (MJ));
- lowering vehicle energy intensities (MJ/passenger-km or MJ/tonne-km);
- encouraging modal shift to lower-carbon passenger and freight transport systems coupled with investment in infrastructure and compact urban form; and
- avoiding journeys where possible (Table TS.3).

Other short-term mitigation strategies include reducing black carbon (BC), aviation contrails, and nitrogen oxides (NO_x) emissions. [8.4]

Strategies to reduce the carbon intensities of fuel and the rate of reducing carbon intensity are constrained by challenges associated with energy storage and the relatively low energy density of low-carbon transport fuels; integrated and sectoral studies broadly agree that opportunities for fuel switching exist in the short term and will grow over time (*medium evidence, medium agreement*) (Figure TS.20, right panel). Electric, hydrogen, and some biofuel technologies could help reduce the carbon intensity of fuels, but their total mitigation potentials are very uncertain (*medium evidence, medium agreement*). Methane-based fuels are already increasing their share for road vehicles and waterborne craft. Electricity produced from low-carbon sources has near-term potential for electric rail and short- to medium-term potential as electric buses, light-duty and 2-wheel road vehicles are deployed. Hydrogen fuels from low-carbon sources constitute longer-term options. Commercially available liquid and gaseous biofuels already provide co-benefits together with mitigation options that can be increased by technology advances, particularly drop-in biofuels for aircraft. Reducing transport emissions of particulate matter (including BC), tropospheric ozone and aerosol precursors (including NO_x) can have human health and mitigation co-benefits in the short term (*medium evidence, medium agreement*). Up to 2030, the majority of integrated studies expect a continued reliance on liquid and gaseous fuels, supported by an increase in the use of biofuels. During the second half of the century, many integrated studies also show substantial shares of electricity and/or hydrogen to fuel electric and fuel-cell light-duty vehicles (LDVs). [8.2, 8.3, 11.13]

Energy efficiency measures through improved vehicle and engine designs have the largest potential for emissions reduc-

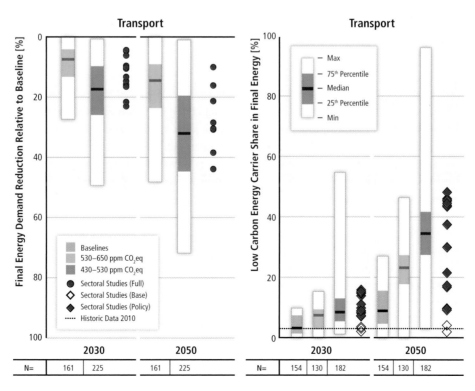

Figure TS.20 | Final energy demand reduction relative to baseline (left panel) and development of final low-carbon energy carrier share in final energy (including electricity, hydrogen, and liquid biofuels; right panel) in transport by 2030 and 2050 in mitigation scenarios from three different CO_2eq concentrations ranges shown in boxplots (see Section 6.3.2) compared to sectoral studies shown in shapes assessed in Chapter 8. Filled circles correspond to sectoral studies with full sectoral coverage. [Figures 6.37 and 6.38]

tions in the short term (*high confidence*). Potential energy efficiency and vehicle performance improvements range from 30–50% relative to 2010 depending on transport mode and vehicle type (Figures TS.21, TS.22). Realizing this efficiency potential will depend on large investments by vehicle manufacturers, which may require strong incentives and regulatory policies in order to achieve GHG emissions reduction goals (*medium evidence, medium agreement*). [8.3, 8.6, 8.9, 8.10]

Shifts in transport mode and behaviour, impacted by new infrastructure and urban (re)development, can contribute to the reduction of transport emissions (*medium evidence, low agreement*). Over the medium term (up to 2030) to long term (to 2050 and beyond), urban redevelopment and investments in new infrastructure, linked with integrated urban planning, transit-oriented development, and more compact urban form that supports cycling and walking can all lead to modal shifts. Such mitigation measures are challenging, have uncertain outcomes, and could reduce transport GHG emissions by 20–50% compared to baseline (*limited evidence, low agreement*). Pricing strategies, when supported by public acceptance initiatives and public and non-motorized transport infrastructures, can reduce travel demand, increase the demand for more efficient vehicles (e.g., where fuel economy standards exist) and induce a shift to low-carbon modes (*medium evidence, medium agreement*). While infrastructure investments may appear expensive at the margin, the case for sustainable urban planning and related policies is reinforced when co-benefits, such as improved health, accessibility, and resilience, are accounted for (Table TS.5). Business initiatives to decarbonize freight transport have begun but will need further support from fiscal, regulatory, and advisory policies to encourage shifting from road to low-carbon modes such as rail or waterborne options where feasible, as well as improving logistics (Figure TS.22). [8.4, 8.5, 8.7, 8.8, 8.9, 8.10]

Sectoral and integrated studies agree that substantial, sustained, and directed policy interventions could limit transport emissions to be consistent with low concentration goals, but the societal mitigation costs (USD/tCO_2eq avoided) remain uncertain (Figures TS.21, TS.22, TS.23). There is good potential to reduce emissions from LDVs and long-haul heavy-duty vehicles (HDVs) from both lower energy intensity vehicles and fuel switching, and the levelized costs of conserved carbon (LCCC) for efficiency improvements can be very low and negative (*limited evidence, low agreement*). Rail, buses, two-wheel motorbikes, and waterborne craft for freight already have relatively low emissions so their emissions reduction potential is limited. The mitigation cost of electric vehicles is currently high, especially if using grid electricity with a high emissions factor, but their LCCC are expected to decline by 2030. The emissions intensity of aviation could decline by around 50% in 2030 but the LCCC, although uncertain, are probably over USD 100/tCO_2eq. While it is expected that mitigation costs will decrease in the future, the magnitude of such reductions is uncertain. (*limited evidence, low agreement*) [8.6, 8.9]

Barriers to decarbonizing transport for all modes differ across regions but can be overcome, in part, through economic incentives (*medium evidence, medium agreement*). Financial, institutional, cultural, and legal barriers constrain low-carbon technology uptake and behavioural change. They include the high investment costs needed to build low-emissions transport systems, the slow turnover of stock and infrastructure, and the limited impact of a carbon price on petroleum fuels that are already heavily taxed. Regional differences are likely due to cost and policy constraints. Oil price trends, price instruments on GHG emissions, and other measures such as road pricing and airport charges can provide strong economic incentives for consumers to adopt mitigation measures. [8.8]

There are regional differences in transport mitigation pathways with major opportunities to shape transport systems and infrastructure around low-carbon options, particularly in developing and emerging countries where most future urban growth will occur (*robust evidence, high agreement*). Possible transformation pathways vary with region and country due to differences in the dynamics of motorization, age and type of vehicle fleets, existing infrastructure, and urban development processes. Prioritizing infrastructure for pedestrians, integrating non-motorized and transit services, and managing excessive road speed for both urban and rural travellers can create economic and social co-benefits in all regions. For all economies, especially those with high rates of urban growth, investments in public transport systems and low-carbon infrastructure can avoid lock-in to carbon-intensive modes. Established infrastructure may limit the options for modal shift and lead to a greater reliance on advanced vehicle technologies; a slowing of growth in LDV demand is already evident in some OECD countries. (*medium evidence, medium agreement*) [8.4, 8.9]

A range of strong and mutually supportive policies will be needed for the transport sector to decarbonize and for the co-benefits to be exploited (*robust evidence, high agreement*). Transport mitigation strategies associated with broader non-climate policies at all government levels can usually target several objectives simultaneously to give lower travel costs, improved access and mobility, better health, greater energy security, improved safety, and increased time savings. Activity reduction measures have the largest potential to realize co-benefits. Realizing the co-benefits depends on the regional context in terms of economic, social, and political feasibility as well as having access to appropriate and cost-effective advanced technologies (Table TS.5). (*medium evidence, high agreement*) Since rebound effects can reduce the CO_2 benefits of efficiency improvements and undermine a particular policy, a balanced package of policies, including pricing initiatives, could help to achieve stable price signals, avoid unintended outcomes, and improve access, mobility, productivity, safety, and health (*medium evidence, medium agreement*). [8.4, 8.7, 8.10]

Passenger Transport

Currently Commercially Available and Future (2030) Expected Technologies

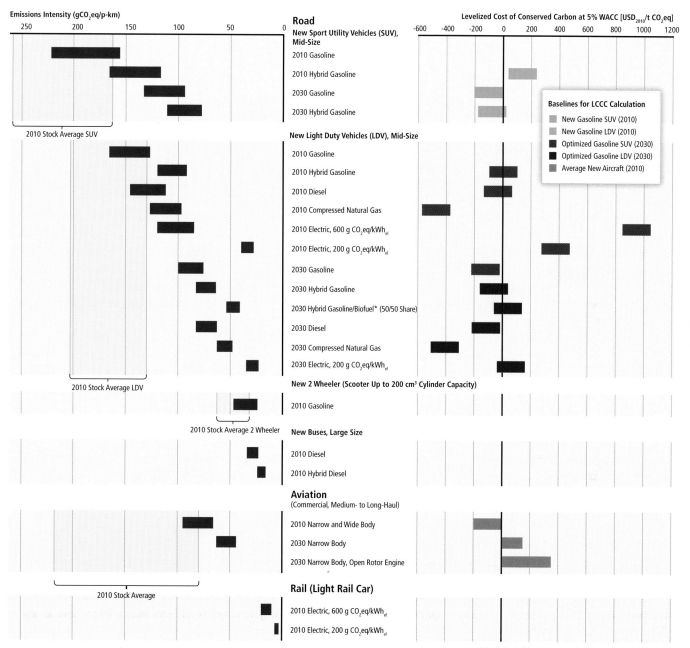

Figure TS.21 | Indicative emissions intensity (tCO$_2$eq/p-km) and levelized costs of conserved carbon (LCCC in USD$_{2010}$/tCO$_2$eq saved) of selected passenger transport technologies. Variations in emissions intensities stem from variation in vehicle efficiencies and occupancy rates. Estimated LCCC for passenger road transport options are point estimates ±100 USD$_{2010}$/tCO$_2$eq based on central estimates of input parameters that are very sensitive to assumptions (e.g., specific improvement in vehicle fuel economy to 2030, specific biofuel CO$_2$eq intensity, vehicle costs, fuel prices). They are derived relative to different baselines (see legend for colour coding) and need to be interpreted accordingly. Estimates for 2030 are based on projections from recent studies, but remain inherently uncertain. LCCC for aviation are taken directly from the literature. Table 8.3 provides additional context (see Annex III.3 for data and assumptions on emissions intensities and cost calculations and Annex II.3.1 for methodological issues on levelized cost metrics). WACC: Weighted average cost of capital. [Table 8.3]

Freight Transport

Currently Commercially Available and Future (2030) Expected Technologies

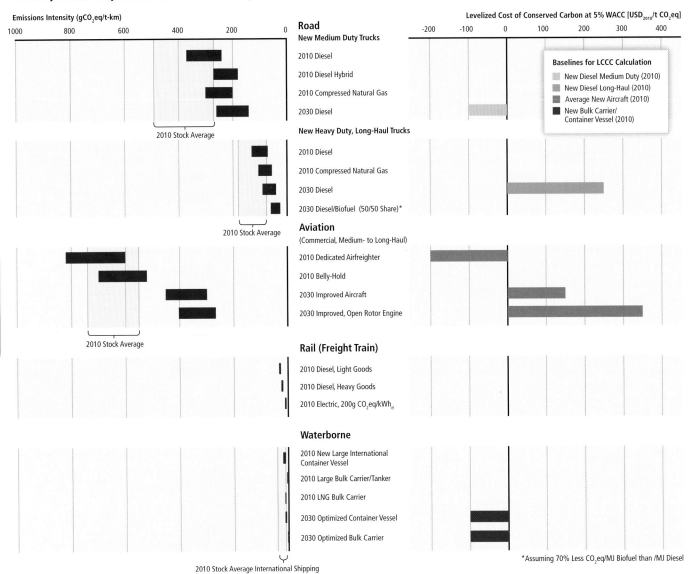

Figure TS.22| Indicative emissions intensity (tCO₂eq/t-km) and levelized costs of conserved carbon (LCCC in USD₂₀₁₀/tCO₂eq saved) of selected freight transport technologies. Variations in emissions intensities largely stem from variation in vehicle efficiencies and load rates. Levelized costs of conserved carbon are taken directly from the literature and are very sensitive to assumptions (e.g., specific improvement in vehicle fuel economy to 2030, specific biofuel CO₂eq intensity, vehicle costs, and fuel prices). They are expressed relative to current baseline technologies (see legend for colour coding) and need to be interpreted accordingly. Estimates for 2030 are based on projections from recent studies but remain inherently uncertain. Table 8.3 provides additional context (see Annex III.3 for data and assumptions on emissions intensities and cost calculations and Annex II.3.1 for methodological issues on levelized cost metrics). LNG: Liquefied natural gas; WACC: Weighted average cost of capital. [Table 8.3]

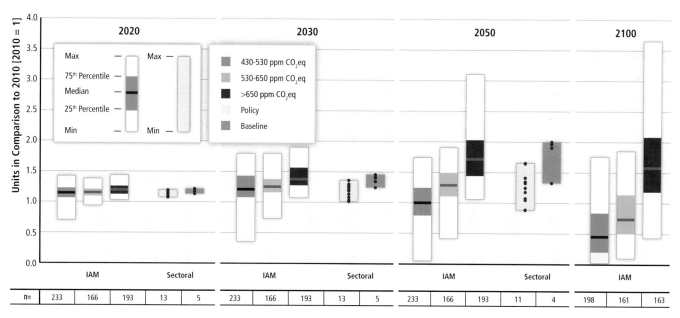

n=	233	166	193	13	5	233	166	193	13	5	233	166	193	11	4	198	161	163

Figure TS.23 | Direct global CO_2 emissions from all passenger and freight transport are indexed relative to 2010 values for each scenario with integrated model studies grouped by CO_2eq concentration levels by 2100, and sectoral studies grouped by baseline and policy categories. [Figure 8.9]

Table TS.5 | Overview of potential co-benefits (green arrows) and adverse side-effects (orange arrows) of the main mitigation measures in the transport sector; arrows pointing up/down denote a positive/negative effect on the respective objective or concern; a question mark (**?**) denotes an uncertain net effect. Co-benefits and adverse side-effects depend on local circumstances as well as on implementation practice, pace and scale. For possible upstream effects of low-carbon electricity, see Table TS.4. For possible upstream effects of biomass supply, see Table TS.8. For an assessment of macroeconomic, cross-sectoral effects associated with mitigation policies (e.g., on energy prices, consumption, growth, and trade), see e.g., Sections 3.9, 6.3.6, 13.2.2.3 and 14.4.2. The uncertainty qualifiers in brackets denote the level of evidence and agreement on the respective effects (see TS.1). Abbreviations for evidence: l = limited, m = medium, r = robust; for agreement: l = low, m = medium, h = high. [Table 8.4]

Transport	Effect on additional objectives/concerns					
	Economic		**Social**		**Environmental**	
Reduction of fuel carbon intensity: electricity, hydrogen (H₂), compressed natural gas (CNG), biofuels, and other fuels	↑	Energy security (diversification, reduced oil dependence and exposure to oil price volatility) (**m/m**)	? ↓ ↑	Health impact via urban air pollution by CNG, biofuels: net effect unclear (**m/l**) Electricity, H₂: reducing most pollutants (**r/h**) Shift to diesel: potentially increasing pollution (**l/m**)	↓ ↑	Ecosystem impact of electricity and hydrogen via Urban air pollution (**m/m**) Material use (unsustainable resource mining) (**l/l**)
	↑	Technological spillovers (e.g., battery technologies for consumer electronics) (**l/l**)	↓	Health impact via reduced noise (electricity and fuel cell LDVs) (**l/m**)	?	Ecosystem impact of biofuels: *see AFOLU*
			↓	Road safety (silent electric LDVs at low speed) (**l/l**)		
Reduction of energy intensity	↑	Energy security (reduced oil dependence and exposure to oil price volatility) (**m/m**)	↓ ↑	Health impact via reduced urban air pollution (**r/h**) Road safety (via increased crash-worthiness) (**m/m**)	↓	Ecosystem and biodiversity impact via reduced urban air pollution (**m/h**)
Compact urban form and improved transport infrastructure **Modal shift**	↑	Energy security (reduced oil dependence and exposure to oil price volatility) (**m/m**)	↓ ↑ ↓	Health impact for non-motorized modes via Increased physical activity (**r/h**) Potentially higher exposure to air pollution (**r/h**) Noise (modal shift and travel reduction) (**r/h**)	↓ ↓	Ecosystem impact via Urban air pollution (**r/h**) Land-use competition (**m/m**)
	↑	Productivity (reduced urban congestion and travel times, affordable and accessible transport) (**m/h**)	↑	Equitable mobility access to employment opportunities, particularly in developing countries (**r/h**)		
	?	Employment opportunities in the public transport sector vs. car manufacturing (**l/m**)	↑	Road safety (via modal shift and/or infrastructure for pedestrians and cyclists) (**r/h**)		
Journey distance reduction and avoidance	↑	Energy security (reduced oil dependence and exposure to oil price volatility) (**r/h**)	↓	Health impact (for non-motorized transport modes) (**r/h**)		Ecosystem impact via
	↑	Productivity (reduced urban congestion, travel times, walking) (**r/h**)			↓ ↑ ↓	Urban air pollution (**r/h**) New/shorter shipping routes (**r/h**) Land-use competition from transport infrastructure (**r/h**)

TS.3.2.4   Buildings

GHG emissions from the buildings sector[15] have more than doubled since 1970, accounting for 19% of global GHG emissions in 2010, including indirect emissions from electricity generation. The share rises to 25% if AFOLU emissions are excluded from the total. The buildings sector also accounted for 32% of total global final energy use, approximately one-third of black carbon emissions, and an eighth to a third of F-gases, with significant uncertainty (*medium evidence, medium agreement*). (Figure TS.3) [9.2]

Direct and indirect CO$_2$ emissions from buildings are projected to increase from 8.8 GtCO$_2$/yr in 2010 to 13–17 GtCO$_2$/yr in 2050 (25–75th percentile; full range 7.9–22 GtCO$_2$/yr) in baseline scenarios; most of the baseline scenarios assessed in WGIII AR5 show a significant increase (*medium evidence, medium agreement*) (Figure TS.15) [6.8]. The lower end of the full range is dominated by scenarios with a focus on energy intensity improvements that go well beyond the observed improvements over the past 40 years. Without further policies, final energy use of the buildings sector may grow from approximately 120 exajoules per year (EJ/yr) in 2010 to 270 EJ/yr in 2050 [9.9].

Significant lock-in risks arise from the long lifespans of buildings and related infrastructure (*robust evidence, high agreement*). If only currently planned policies are implemented, the final energy use in buildings that could be locked-in by 2050, compared to a scenario where today's best practice buildings become the standard in newly built structures and retrofits, is equivalent to approximately 80% of the final energy use of the buildings sector in 2005. [9.4]

Improvements in wealth, lifestyle change, the provision of access to modern energy services and adequate housing, and urbanization will drive the increases in building energy demand (*robust evidence, high agreement*). The manner in which those without access to adequate housing (about 0.8 billion people), modern energy carriers, and sufficient levels of energy services including clean cooking and heating (about 3 billion people) meet these needs will influence the development of building-related emissions. In addition, migration to cities, decreasing household size, increasing levels of wealth, and lifestyle changes, including increasing dwelling size and number and use of appliances, all contribute to considerable increases in building energy services demand. The substantial amount of new construction taking place in developing countries represents both a risk and opportunity from a mitigation perspective. [9.2, 9.4, 9.9]

Recent advances in technologies, know-how, and policies in the buildings sector, however, make it feasible that the global total sector final energy use stabilizes or even declines by mid-century (*robust evidence, medium agreement*). Recent advances in technology,

design practices and know-how, coupled with behavioural changes, can achieve a two to ten-fold reduction in energy requirements of individual new buildings and a two to four-fold reduction for individual existing buildings largely cost-effectively or sometimes even at net negative costs (see Box TS.12) (*robust evidence, high agreement*). [9.6]

Advances since AR4 include the widespread demonstration worldwide of very low, or net zero energy buildings both in new construction and retrofits (*robust evidence, high agreement*). In some jurisdictions, these have already gained important market shares with, for instance, over 25 million m^2 of building floorspace in Europe complying with the 'Passivehouse' standard in 2012. However, zero energy/carbon buildings may not always be the most cost-optimal solution, nor even be feasible in certain building types and locations. [9.3]

High-performance retrofits are key mitigation strategies in countries with existing building stocks, as buildings are very long-lived and a large fraction of 2050 developed country buildings already exists (*robust evidence, high agreement*). Reductions of heating/cooling energy use by 50–90% have been achieved using best practices. Strong evidence shows that very low-energy construction and retrofits can be economically attractive. [9.3]

With ambitious policies it is possible to keep global building energy use constant or significantly reduce it by mid-century compared to baseline scenarios which anticipate an increase of more than two-fold (*medium evidence, medium agreement*) (Figure TS.24). Detailed building sector studies indicate a larger energy savings potential by 2050 than do integrated studies. The former indicate a potential of up to 70% of the baseline for heating and cooling only, and around 35–45% for the whole sector. In general, deeper reductions are possible in thermal energy uses than in other energy services mainly relying on electricity. With respect to additional fuel switching as compared to baseline, both sectoral and integrated studies find modest opportunities. In general, both sectoral and integrated studies indicate that electricity will supply a growing share of building energy demand over the long term, especially if heating demand decreases due to a combination of efficiency gains, better architecture, and climate change. [6.8.4, 9.8.2, Figure 9.19]

The history of energy efficiency programmes in buildings shows that 25–30% efficiency improvements have been available at costs substantially lower than those of marginal energy supply (*robust evidence, high agreement*). Technological progress enables the potential for cost-effective energy efficiency improvements to be maintained, despite continuously improving standards. There has been substantial progress in the adoption of voluntary and mandatory standards since AR4, including ambitious building codes and targets, voluntary construction standards, and appliance standards. At the same time, in both new and retrofitted buildings, as well as in appliances and information, communication and media technology equipment, there have been notable performance and cost improvements. Large

---

[15]   The buildings sector covers the residential, commercial, public and services sectors; emissions from construction are accounted for in the industry sector.

Box TS.12 | Negative private mitigation costs

A persistent issue in the analysis of mitigation options and costs is whether there are mitigation opportunities that are privately beneficial—generating private benefits that more than offset the costs of implementation—but which consumers and firms do not voluntarily undertake. There is some evidence of unrealized mitigation opportunities that would have negative private cost. Possible examples include investments in vehicles [8.1], lighting and heating technology in homes and commercial buildings [9.3], as well as industrial processes [10.1].

Examples of negative private costs imply that firms and individuals do not take opportunities to save money. This might be explained in a number of ways. One is that status-quo bias can inhibit the switch to new technologies or products [2.4, 3.10.1]. Another is that firms and individuals may focus on short-term goals and discount future costs and benefits sharply; consumers

have been shown to do this when choosing energy conservation measures or investing in energy-efficient technologies [2.4.3, 2.6.5.3, 3.10.1]. Risk aversion and ambiguity aversion may also account for this behaviour when outcomes are uncertain [2.4.3, 3.10.1]. Other possible explanations include: insufficient information on opportunities to conserve energy; asymmetric information—for example, landlords may be unable to convey the value of energy efficiency improvements to renters; split incentives, where one party pays for an investment but another party reaps the benefits; and imperfect credit markets, which make it difficult or expensive to obtain finance for energy savings [3.10.1, 16.4].

Some engineering studies show a large potential for negative-cost mitigation. The extent to which such negative-cost opportunities can actually be realized remains a matter of contention in the literature. Empirical evidence is mixed. [Box 3.10]

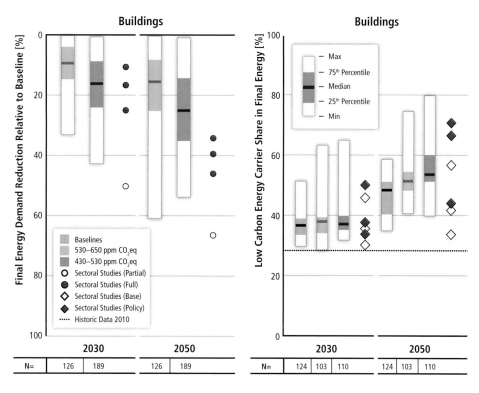

Figure TS.24 | Final energy demand reduction relative to baseline (left panel) and development of final low-carbon energy carrier share in final energy (from electricity; right panel) in buildings by 2030 and 2050 in mitigation scenarios from three different CO₂eq concentrations ranges shown in boxplots (see Section 6.3.2) compared to sectoral studies shown in shapes assessed in Chapter 9. Filled circles correspond to sectoral studies with full sectoral coverage while empty circles correspond to studies with only partial sectoral coverage (e.g., heating and cooling). [Figures 6.37 and 6.38]

reductions in thermal energy use in buildings are possible at costs lower than those of marginal energy supply, with the most cost-effective options including very high-performance new commercial buildings; the same holds for efficiency improvements in some appliances and cooking equipment. [9.5, 9.6, 9.9]

Lifestyle, culture, and other behavioural changes may lead to further large reductions in building and appliance energy requirements beyond those achievable through technologies and architecture. A three- to five-fold difference in energy use has been shown for provision of similar building-related energy

service levels in buildings. (*limited evidence, high agreement*) For developed countries, scenarios indicate that lifestyle and behavioural changes could reduce energy demand by up to 20% in the short term and by up to 50% of present levels by mid-century (*medium evidence, medium agreement*). There is a high risk that emerging countries follow the same path as developed economies in terms of building-related architecture, lifestyle, and behaviour. But the literature suggests that alternative development pathways exist that provide high levels of building services at much lower energy inputs, incorporating strategies such as learning from traditional lifestyles, architecture, and construction techniques. [9.3]

Most mitigation options in the building sector have considerable and diverse co-benefits (*robust evidence, high agreement*). These include, but are not limited to: energy security; less need for energy subsidies; health and environmental benefits (due to reduced indoor and outdoor air pollution); productivity and net employment gains; the alleviation of fuel poverty; reduced energy expenditures; increased value for building infrastructure; and improved comfort and services. (Table TS.6) [9.6, 9.7]

Especially strong barriers in this sector hinder the market-based uptake of cost-effective technologies and practices; as a consequence, programmes and regulation are more effective than pricing instruments alone (*robust evidence, high agreement*). Barriers include imperfect information and lack of awareness, principal/agent problems and other split incentives, transaction costs, lack of access to financing, insufficient training in all construction-related trades, and cognitive/behavioural barriers. In developing countries, the large informal sector, energy subsidies, corruption, high implicit discount rates, and insufficient service levels are further barriers. Therefore, market forces alone are not expected to achieve the necessary transformation without external stimuli. Policy intervention addressing all stages of the building and appliance lifecycle and use, plus new business and financial models, are essential. [9.8, 9.10]

A large portfolio of building-specific energy efficiency policies was already highlighted in AR4, but further considerable advances in available instruments and their implementation have occurred since (*robust evidence, high agreement*). Evidence shows that many building energy efficiency policies worldwide have

Table TS.6 | Overview of potential co-benefits (green arrows) and adverse side-effects (orange arrows) of the main mitigation measures in the buildings sector; arrows pointing up/down denote a positive/negative effect on the respective objective or concern. Co-benefits and adverse side-effects depend on local circumstances as well as on implementation practice, pace and scale. For possible upstream effects of fuel switching and RE, see Tables TS.4 and TS.8. For an assessment of macroeconomic, cross-sectoral effects associated with mitigation policies (e.g., on energy prices, consumption, growth, and trade), see e.g., Sections 3.9, 6.3.6, 13.2.2.3 and 14.4.2. The uncertainty qualifiers in brackets denote the level of evidence and agreement on the respective effects (see TS.1). Abbreviations for evidence: l = limited, m = medium, r = robust; for agreement: l = low, m = medium, h = high. [Table 9.7]

Buildings	Effect on additional objectives/concerns			
	Economic	Social	Environmental	Other
Fuel switching, RES incorporation, green roofs, and other measures reducing GHG emissions intensity	↑ Energy security (**m/h**) ↑ Employment impact (**m/m**) ↑ Lower need for energy subsidies (**l/l**) ↑ Asset values of buildings (**l/m**)	Fuel poverty (residential) via ↓ Energy demand (**m/h**) ↑ Energy cost (**l/m**) ↓ Energy access (for higher energy cost) (**l/m**) ↑ Productive time for women/children (for replaced traditional cookstoves) (**m/h**)	Health impact in residential buildings via ↓ Outdoor air pollution (**r/h**) ↓ Indoor air pollution (in developing countries) (**r/h**) ↓ Fuel poverty (**r/h**) ↓ Ecosystem impact (less outdoor air pollution) (**r/h**) ↑ Urban biodiversity (for green roofs) (**m/m**)	Reduced Urban Heat Island (UHI) effect (**l/m**)
Retrofits of existing buildings (e.g., cool roof, passive solar, etc.) Exemplary new buildings Efficient equipment	↑ Energy security (**m/h**) ↑ Employment impact (**m/m**) ↑ Productivity (for commercial buildings) (**m/h**) ↑ Lower need for energy subsidies (**l/l**) ↑ Asset values of buildings (**l/m**) ↑ Disaster resilience (**l/m**)	↓ Fuel poverty (for retrofits and efficient equipment) (**m/h**) ↓ Energy access (higher cost for housing due to the investments needed) (**l/m**) ↑ Thermal comfort (for retrofits and exemplary new buildings) (**m/h**) ↑ Productive time for women and children (for replaced traditional cookstoves) (**m/h**)	Health impact via ↓ Outdoor air pollution (**r/h**) ↓ Indoor air pollution (for efficient cookstoves) (**r/h**) ↓ Improved indoor environmental conditions (**m/h**) ↓ Fuel poverty (**r/h**) ↓ Insufficient ventilation (**m/m**) ↓ Ecosystem impact (less outdoor air pollution) (**r/h**) ↓ Water consumption and sewage production (**l/l**)	Reduced UHI effect (for retrofits and new exemplary buildings) (**l/m**)
Behavioural changes reducing energy demand	↑ Energy security (**m/h**) ↑ Lower need for energy subsidies (**l/l**)		↓ Health impact via less outdoor air pollution (**r/h**) and improved indoor environmental conditions (**m/h**) ↓ Ecosystem impact (less outdoor air pollution) (**r/h**)	

already been saving GHG emissions at large negative costs. Among the most environmentally and cost-effective policies are regulatory instruments such as building and appliance energy performance standards and labels, as well as public leadership programmes and procurement policies. Progress in building codes and appliance standards in some developed countries over the last decade have contributed to stabilizing or even reducing total building energy use, despite growth in population, wealth, and corresponding energy service level demands. Developing countries have also been adopting different effective policies, most notably appliance standards. However, in order to reach ambitious climate goals, these standards need to be substantially strengthened and adopted in further jurisdictions, and to other building and appliance types. Due to larger capital requirements, financing instruments are essential both in developed and developing countries to achieve deep reductions in energy use. [9.10]

TS.3.2.5   Industry

In 2010, the industry sector accounted for around 28 % of final energy use, and direct and indirect GHG emissions (the latter being associated with electricity consumption) are larger than the emissions from either the buildings or transport end-use sectors and represent just over 30 % of global GHG emissions in 2010 (the share rises to 40 % if AFOLU emissions are excluded

from the total) (*high confidence*). Despite the declining share of industry in global GDP, global industry and waste/wastewater GHG emissions grew from 10 GtCO$_2$eq in 1990 to 13 GtCO$_2$eq in 2005 and to 15 GtCO$_2$eq in 2010 (of which waste/wastewater accounted for 1.4 GtCO$_2$eq). [10.3]

Carbon dioxide emissions from industry, including direct and indirect emissions as well as process emissions, are projected to increase from 13 GtCO$_2$/yr in 2010 to 20–24 GtCO$_2$/yr in 2050 (25–75th percentile; full range 9.5–34 GtCO$_2$/yr) in baseline scenarios; most of the baseline scenarios assessed in WGIII AR5 show a significant increase (*medium evidence, medium agreement*) (Figure TS.15) [6.8]. The lower end of the full range is dominated by scenarios with a focus on energy intensity improvements that go well beyond the observed improvements over the past 40 years.

The wide-scale upgrading, replacement and deployment of best available technologies, particularly in countries where these are not in practice, and in non-energy intensive industries, could directly reduce the energy intensity of the industry sector by about 25 % compared to the current level (*robust evidence, high agreement*). Despite long-standing attention to energy efficiency in industry, many options for improved energy efficiency still remain. Through innovation, additional reductions of about 20 % in energy intensity may potentially be realized (*limited evidence, medium agree-*

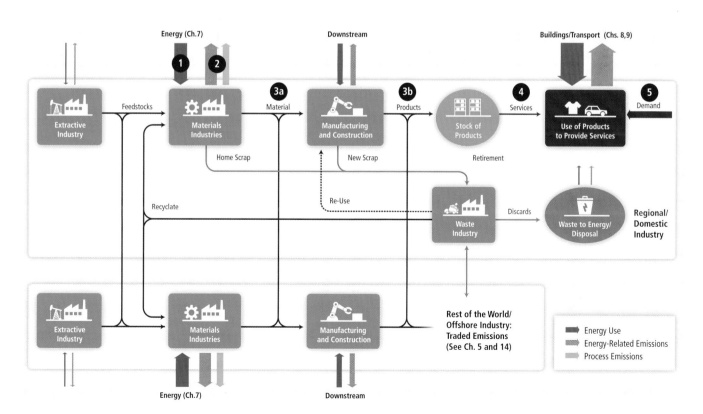

Figure TS.25 | A schematic illustration of industrial activity over the supply chain. Options for mitigation in the industry sector are indicated by the circled numbers: (1) energy efficiency; (2) emissions efficiency; (3a) material efficiency in manufacturing; (3b) material efficiency in product design; (4) product-service efficiency; (5) service demand reduction. [Figure 10.2]

ment). Barriers to implementing energy efficiency relate largely to the initial investment costs and lack of information. Information programmes are a prevalent approach for promoting energy efficiency, followed by economic instruments, regulatory approaches, and voluntary actions. [10.4, 10.7, 10.9, 10.11]

An absolute reduction in emissions from the industry sector will require deployment of a broad set of mitigation options that go beyond energy efficiency measures (*medium evidence, high agreement*) [10.4, 10.7]. In the context of continued overall growth in industrial demand, substantial reductions from the sector will require parallel efforts to increase emissions efficiency (e.g., through fuel and feedstock switching or CCS); material use efficiency (e.g., less scrap, new product design); recycling and re-use of materials and products; product-service efficiency (e.g., more intensive use of products through car sharing, longer life for products); radical product innovations (e.g., alternatives to cement); as well as service demand reductions. Lack of policy and experiences in material and product-service efficiency are major barriers. (Table TS.3, Figure TS.25) [10.4, 10.7, 10.11]

While detailed industry sector studies tend to be more conservative than integrated studies, both identify possible industrial final energy demand savings of around 30% by 2050 in mitigation scenarios not exceeding 650 ppm CO₂eq by 2100 relative to baseline scenarios (*medium evidence, medium agreement*) (Figure TS.26). Integrated models in general treat the industry sector in a more aggregated fashion and mostly do not explicitly provide detailed sub-sectoral material flows, options for reducing material demand, and price-induced inter-input substitution possibilities. Due to the heterogeneous character of the industry sector, a coherent comparison between sectoral and integrated studies remains difficult. [6.8.4, 10.4, 10.7, 10.10.1, Figure 10.14]

Mitigation in the industry sector can also be achieved by reducing material and fossil fuel demand by enhanced waste use, which concomitantly reduces direct GHG emissions from waste disposal (*robust evidence, high agreement*). The hierarchy of waste management places waste reduction at the top, followed by re-use, recycling, and energy recovery. As the share of recycled or reused material is still low, applying waste treatment technologies and recovering energy to reduce demand for fossil fuels can result in direct emission reductions from waste disposal. Globally, only about 20% of municipal solid waste (MSW) is recycled and about 14% is treated with energy recovery while the rest is deposited in open dumpsites or landfills. About 47% of wastewater produced in the domestic and manufacturing sectors is still untreated. The largest cost range is for reducing GHG emissions from landfilling through the treatment of waste by anaerobic digestion. The costs range from negative (see Box TS.12) to very high. Advanced wastewater treatment technologies may enhance GHG emissions reduction in wastewater treatment but they are clustered among the higher cost options (*medium evidence, medium agreement*). (Figure TS.29) [10.4, 10.14]

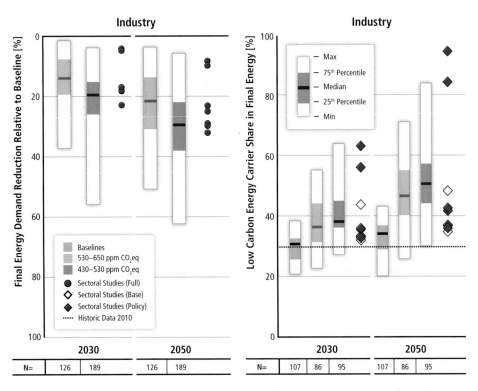

Figure TS.26 | Final energy demand reduction relative to baseline (left panel) and development of final low-carbon energy carrier share in final energy (including electricity, heat, hydrogen, and bioenergy; right panel) in industry by 2030 and 2050 in mitigation scenarios from three different CO₂eq concentration ranges shown in boxplots (see Section 6.3.2) compared to sectoral studies shown in shapes assessed in Chapter 10. Filled circles correspond to sectoral studies with full sectoral coverage. [Figures 6.37 and 6.38]

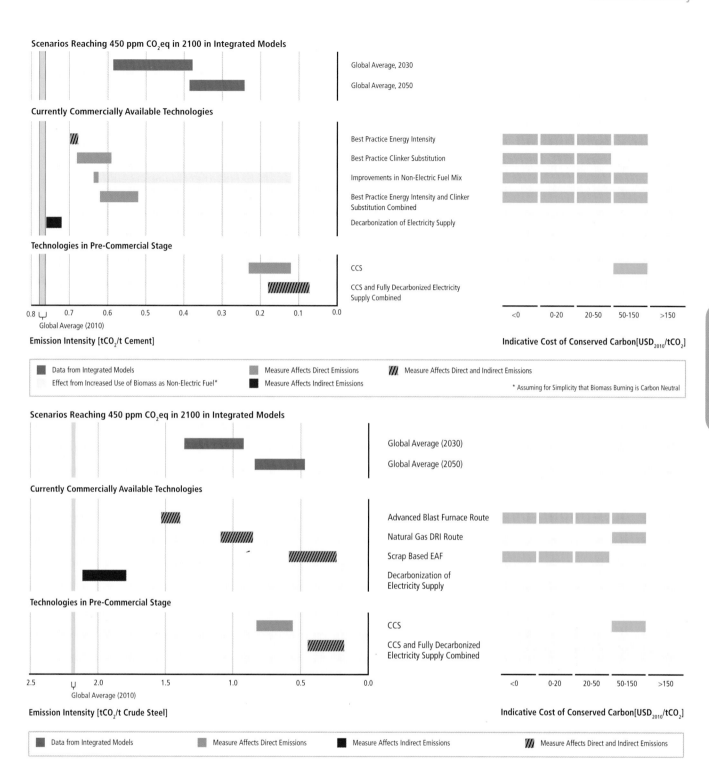

Figure TS.27 | Indicative CO$_2$ emission intensities for cement (upper panel) and steel (lower panel) production, as well as indicative levelized cost of conserved carbon (LCCC) shown for various production practices/technologies and for 450 ppm CO$_2$eq scenarios of a limited selection of integrated models (for data and methodology, see Annex III). DRI: Direct reduced iron; EAF: Electric arc furnace. [Figures 10.7, 10.8]

Waste policy and regulation have largely influenced material consumption, but few policies have specifically pursued material efficiency or product-service efficiency (*robust evidence, high agreement*) [10.11]. Barriers to improving material efficiency include lack of human and institutional capacities to encourage management decisions and public participation. Also, there is a lack of experience

and often there are no clear incentives either for suppliers or consumers to address improvements in material or product-service efficiency, or to reduce product demand. [10.9]

CO$_2$ emissions dominate GHG emissions from industry, but there are also substantial mitigation opportunities for non-CO$_2$ gases

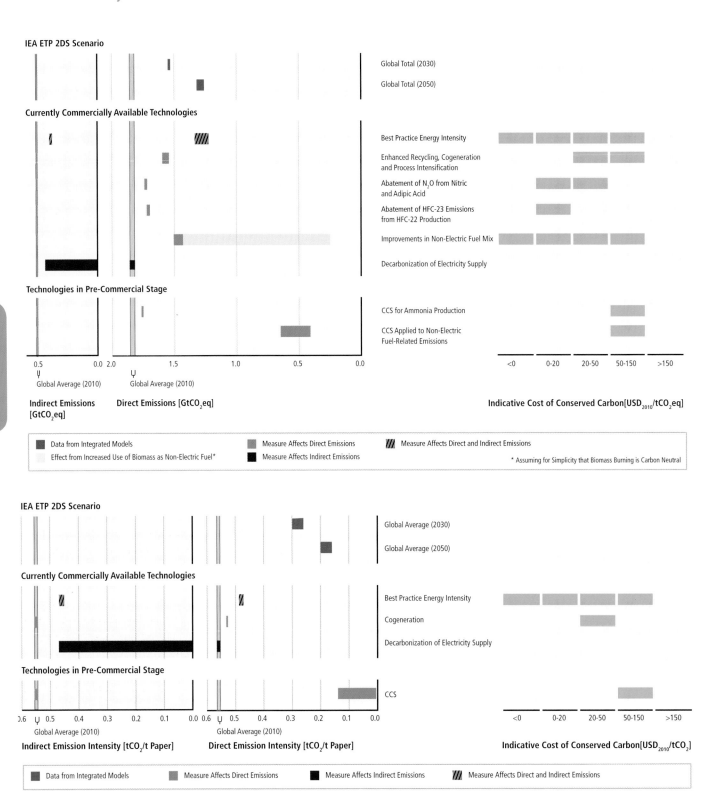

Figure TS.28 | Indicative global CO₂eq emissions for chemicals production (upper panel) and indicative global CO₂ emission intensities for paper production (lower panel) as well as indicative levelized cost of conserved carbon (LCCC) shown for various production practices/technologies and for 450 ppm CO₂eq scenarios of a limited selection of integrated models (for data and methodology, see Annex III). [Figures 10.9, 10.10]

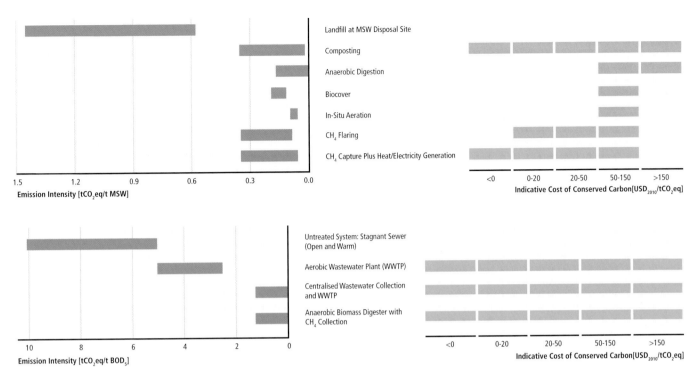

Figure TS.29 | Indicative CO₂eq emission intensities for waste (upper panel) and wastewater (lower panel) of various practices as well as indicative levelized cost of conserved carbon (for data and methodology, see Annex III). MSW: Municipal solid waste. [Figures 10.19 and 10.20]

TS

(*robust evidence, high agreement*). Methane (CH₄), nitrous oxide (N₂O) and fluorinated gases (F-gases) from industry accounted for emissions of 0.9 GtCO₂eq in 2010. Key mitigation opportunities comprise, e.g., reduction of hydrofluorocarbon (HFC) emissions by leak repair, refrigerant recovery and recycling, and proper disposal and replacement by alternative refrigerants (ammonia, HC, CO₂). N₂O emissions from adipic and nitric acid production can be reduced through the implementation of thermal destruction and secondary catalysts. The reduction of non-CO₂ GHGs also faces numerous barriers. Lack of awareness, lack of economic incentives and lack of commercially available technologies (e.g., for HFC recycling and incineration) are typical examples. [Table 10.2, 10.7]

Systemic approaches and collaborative activities across companies (large energy-intensive industries and Small and Medium Enterprises (SMEs)) and sectors can help to reduce GHG emissions (*robust evidence, high agreement*). Cross-cutting technologies such as efficient motors, and cross-cutting measures such as reducing air or steam leaks, help to optimize performance of industrial processes and improve plant efficiency very often cost-effectively with both energy savings and emissions benefits. Industrial clusters also help to realize mitigation, particularly from SMEs. [10.4] Cooperation and cross-sectoral collaboration at different levels—for example, sharing of infrastructure, information, waste heat, cooling, etc.—may provide further mitigation potential in certain regions/industry types [10.5].

Several emission-reducing options in the industrial sector are cost-effective and profitable (*medium evidence, medium agreement*). While options in cost ranges of 0–20 and 20–50 USD/tCO₂eq

and even below 0 USD/tCO₂eq exist, achieving near-zero emissions intensity levels in the industry sector would require the additional realization of long-term step-change options (e.g., CCS), which are associated with higher levelized costs of conserved carbon (LCCC) in the range of 50–150 USD/tCO₂eq. Similar cost estimates for implementing material efficiency, product-service efficiency, and service demand reduction strategies are not available. With regard to long-term options, some sector-specific measures allow for significant reductions in specific GHG emissions but may not be applicable at scale, e.g., scrap-based iron and steel production. Decarbonized electricity can play an important role in some subsectors (e.g., chemicals, pulp and paper, and aluminium), but will have limited impact in others (e.g., cement, iron and steel, waste). In general, mitigation costs vary regionally and depend on site-specific conditions. (Figures TS.27, TS.28, TS.29) [10.7]

Mitigation measures are often associated with co-benefits (*robust evidence, high agreement*). Co-benefits include enhanced competitiveness through cost-reductions, new business opportunities, better environmental compliance, health benefits through better local air and water quality and better work conditions, and reduced waste, all of which provide multiple indirect private and social benefits (Table TS.7). [10.8]

There is no single policy that can address the full range of mitigation measures available for industry and overcome associated barriers. Unless barriers to mitigation in industry are resolved, the pace and extent of mitigation in industry will be limited and even profitable measures will remain untapped (*robust evidence, high agreement*). [10.9, 10.11]

Table TS.7 | Overview of potential co-benefits (green arrows) and adverse side-effects (orange arrows) of the main mitigation measures in the industry sector; arrows pointing up/down denote a positive/negative effect on the respective objective or concern. Co-benefits and adverse side-effects depend on local circumstances as well as on the implementation practice, pace and scale. For possible upstream effects of low-carbon energy supply (includes CCS), see Table TS.4. For possible upstream effects of biomass supply, see Table TS.8. For an assessment of macroeconomic, cross-sectoral, effects associated with mitigation policies (e.g., on energy prices, consumption, growth, and trade), see e.g., Sections 3.9, 6.3.6, 13.2.2.3 and 14.4.2. The uncertainty qualifiers in brackets denote the level of evidence and agreement on the respective effects (see TS.1). Abbreviations for evidence: l = limited, m = medium, r = robust; for agreement: l = low, m = medium, h = high. [Table 10.5]

Industry	Effect on additional objectives/concerns		
	Economic	Social	Environmental
CO$_2$ and non-CO$_2$ GHG emissions intensity reduction	↑ Competitiveness and productivity (**m/h**)	↓ Health impact via reduced local air pollution and better work conditions (for perfluorocarbons from aluminium) (**m/m**)	↓ Ecosystem impact via reduced local air pollution and reduced water pollution (**m/m**) ↑ Water conservation (**l/m**)
Technical energy efficiency improvements via new processes and technologies	↑ Energy security (via lower energy intensity) (**m/m**) ↑ Employment impact (**l/l**) ↑ Competitiveness and productivity (**m/h**) ↑ Technological spillovers in developing countries (due to supply chain linkages) (**l/l**)	↓ Health impact via reduced local pollution (**l/m**) ↑ New business opportunities (**m/m**) ↑ Water availability and quality (**l/l**) ↑ Safety, working conditions and job satisfaction (**m/m**)	Ecosystem impact via: ↓ Fossil fuel extraction (**l/l**) ↓ Local pollution and waste (**m/m**)
Material efficiency of goods, recycling	↓ National sales tax revenue in medium term (**l/l**) ↑ Employment impact in waste recycling market (**l/l**) ↑ Competitiveness in manufacturing (**l/l**) ↑ New infrastructure for industrial clusters (**l/l**)	↓ Health impacts and safety concerns (**l/m**) ↑ New business opportunities (**m/m**) ↓ Local conflicts (reduced resource extraction) (**l/m**)	↓ Ecosystem impact via reduced local air and water pollution and waste material disposal (**m/m**) ↓ Use of raw/virgin materials and natural resources implying reduced unsustainable resource mining (**l/l**)
Product demand reductions	↓ National sales tax revenue in medium term (**l/l**)	↑ Wellbeing via diverse lifestyle choices (**l/l**)	↓ Post-consumption waste (**l/l**)

TS.3.2.6 Agriculture, Forestry and Other Land Use (AFOLU)

Since AR4, GHG emissions from the AFOLU sector have stabilized but the share of total anthropogenic GHG emissions has decreased (*robust evidence, high agreement*). The average annual total GHG flux from the AFOLU sector was 10–12 GtCO$_2$eq in 2000–2010, with global emissions of 5.0–5.8 GtCO$_2$eq/yr from agriculture on average and around 4.3–5.5 GtCO$_2$eq/yr from forestry and other land uses. Non-CO$_2$ emissions derive largely from agriculture, dominated by N$_2$O emissions from agricultural soils and CH$_4$ emissions from livestock enteric fermentation, manure management, and emissions from rice paddies, totalling 5.0–5.8 GtCO$_2$eq/yr in 2010 (*robust evidence, high agreement*). Over recent years, most estimates of FOLU CO$_2$ fluxes indicate a decline in emissions, largely due to decreasing deforestation rates and increased afforestation (*limited evidence, medium agreement*). The absolute levels of emissions from deforestation and degradation have fallen from 1990 to 2010 (*robust evidence, high agreement*). Over the same time period, total emissions for high-income countries decreased while those of low-income countries increased. In general, AFOLU emissions from high-income countries are dominated by agriculture activities while those from low-income countries are dominated by deforestation and degradation. [Figure 1.3, 11.2]

Net annual baseline CO$_2$ emissions from AFOLU are projected to decline over time with net emissions potentially less than half of the 2010 level by 2050, and the possibility of the AFOLU sector becoming a net sink before the end of century. However, the uncertainty in historical net AFOLU emissions is larger than for other sectors, and additional uncertainties in projected baseline net AFOLU emissions exist. (*medium evidence, high agreement*) (Figure TS.15) [6.3.1.4, 6.8, Figure 6.5] As in AR4, most projections suggest declining annual net CO$_2$ emissions in the long run. In part, this is driven by technological change, as well as projected declining rates of agriculture area expansion related to the expected slowing in population growth. However, unlike AR4, none of the more recent scenarios projects growth in the near-term. There is also a somewhat larger range of variation later in the century, with some models projecting a stronger net sink starting in 2050 (*limited evidence, medium agreement*). There are few reported projections of baseline global land-related N$_2$O and CH$_4$ emissions and they indicate an increase over time. Cumulatively, land CH$_4$ emissions are projected to be 44–53 % of total CH$_4$ emissions through 2030, and 41–59 % through 2100, and land N$_2$O emissions 85–89 % and 85–90 %, respectively (*limited evidence, medium agreement*). [11.9]

Opportunities for mitigation in the AFOLU sector include supply- and demand-side mitigation options (*robust evidence, high agreement*). Supply-side measures involve reducing emissions arising from land-use change, in particular reducing deforestation, and land and livestock management, increasing carbon stocks by sequestration in soils and biomass, or the substitution of fossil fuels by biomass for energy production (Table TS.3). Further new supply-side technologies not assessed in AR4, such as biochar or wood products for energy-intensive building materials, could contribute to the mitigation potential of the AFOLU sector, but there are still few studies upon which to make robust estimates. Demand-side measures include dietary change and waste reduction in the food supply chain. Increasing forestry and agricultural production without a commensurate increase in emissions (i.e., one component of sustainable intensification; Figure TS.30) also reduces emissions intensity (i.e., the GHG emissions per unit of product), a mitigation mechanism largely unreported for AFOLU in AR4, which could reduce absolute emissions as long as production volumes do not increase. [11.3, 11.4]

Among supply-side measures, the most cost-effective forestry options are afforestation, sustainable forest management and reducing deforestation, with large differences in their relative importance across regions; in agriculture, low carbon prices[16] **(20 USD/tCO₂eq) favour cropland and grazing land management and high carbon prices (100 USD/tCO₂eq) favour restoration of organic soils** (*medium evidence, medium agreement*). When considering only studies that cover both forestry and agriculture and include agricultural soil carbon sequestration, the economic mitigation potential in the AFOLU sector is estimated to be 7.18 to 10.6 (full range of all studies: 0.49–10.6) GtCO₂eq/yr in 2030 for mitigation efforts consistent with carbon prices up to 100 USD/ tCO₂eq, about a third of which can be achieved at < 20 USD/ tCO₂eq (*medium evidence, medium agreement*). The range of global estimates at a given carbon price partly reflects uncertainty surrounding AFOLU mitigation

potentials in the literature and the land-use assumptions of the scenarios considered. The ranges of estimates also reflect differences in the GHGs and options considered in the studies. A comparison of estimates of economic mitigation potential in the AFOLU sector published since AR4 is shown in Figure TS.31. [11.6]

While demand-side measures are under-researched, changes in diet, reductions of losses in the food supply chain, and other measures have a significant, but uncertain, potential to reduce GHG emissions from food production (0.76–8.55 GtCO₂eq/yr by 2050) (Figure TS.31) (*limited evidence, medium agreement*). Barriers to implementation are substantial, and include concerns about jeopardizing health and well-being, and cultural and societal resistance to behavioural change. However, in countries with a high consumption of animal protein, co-benefits are reflected in positive health impacts resulting from changes in diet (*robust evidence, high agreement*). [11.4.3, 11.6, 11.7, 11.9]

The mitigation potential of AFOLU is highly dependent on broader factors related to land-use policy and patterns (*medium evidence, high agreement*). The many possible uses of land can compete or work in synergy. The main barriers to mitigation are institutional (lack of tenure and poor governance), accessibility to financing mechanisms, availability of land and water, and poverty. On the other hand, AFOLU mitigation options can promote innovation, and many technological supply-side mitigation options also increase agricultural and silvicultural efficiency, and can reduce climate vulnerability by improving resilience. Multifunctional systems that allow the delivery of multiple services from land have the capacity to deliver to many policy goals in addition to mitigation, such as improving land tenure, the governance of natural resources, and equity [11.8] (*limited evidence, high agreement*). Recent frameworks, such as those for assessing environmental or ecosystem services, could provide tools for valuing the multiple synergies and tradeoffs that may arise from mitigation actions (Table TS.8) (*medium evidence, medium agreement*). [11.7, 11.8]

---

[16] In many models that are used to assess the economic costs of mitigation, carbon price is used as a proxy to represent the level of effort in mitigation policies (see Glossary).

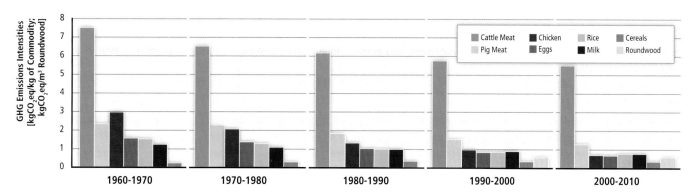

Figure TS.30 | GHG emissions intensities of selected major AFOLU commodities for decades 1960s–2000s. (1) Cattle meat, defined as GHG (enteric fermentation + manure management of cattle, dairy and non-dairy)/meat produced; (2) pig meat, defined as GHG (enteric fermentation + manure management of swine, market and breeding)/meat produced; (3) chicken meat, defined as GHG (manure management of chickens)/meat produced; (4) milk, defined as GHG (enteric fermentation + manure management of cattle, dairy)/milk produced; (5) eggs, defined as GHG (manure management of chickens, layers)/egg produced; (6) rice, defined as GHG (rice cultivation)/rice produced; (7) cereals, defined as GHG (synthetic fertilizers)/cereals produced; (8) wood, defined as GHG (carbon loss from harvest)/roundwood produced. [Figure 11.15]

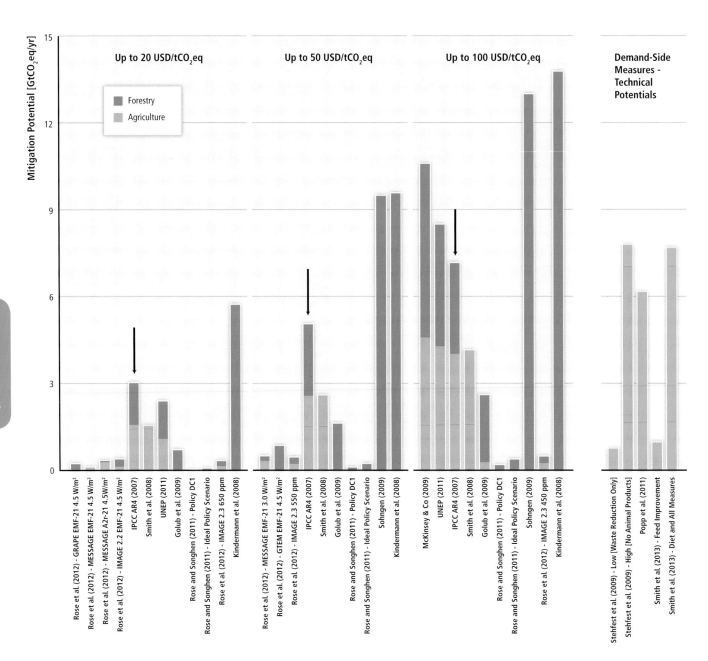

Figure TS.31 | Estimates of economic mitigation potentials in the AFOLU sector published since AR4 (AR4 estimates shown for comparison, denoted by black arrows), including bottom-up, sectoral studies, and top-down, multi-sector studies. Supply-side mitigation potentials are estimated for around 2030, ranging from 2025 to 2035, and are for agriculture, forestry or both sectors combined. Studies are aggregated for potentials up to ~20 USD/tCO$_2$eq (actual range 1.64–21.45), up to ~50 USD/tCO$_2$eq (actual range 31.39–50.00), and up to ~100 USD/tCO$_2$eq (actual range 70.0–120.91). Demand-side measures (shown on the right hand side of the figure) are for ~2050 and are not assessed at a specific carbon price, and should be regarded as technical potentials. Smith et al. (2013) values are the mean of the range. Not all studies consider the same measures or the same GHGs. [11.6.2, Figure 11.14]

Policies governing practices in agriculture as well as forest conservation and management need to account for the needs of both mitigation and adaptation (*medium evidence, high agreement*). Some mitigation options in the AFOLU sector (such as soil and forest carbon stocks) may be vulnerable to climate change. Economic incentives (e.g., special credit lines for low-carbon agriculture, sustainable agriculture and forestry practices, tradable credits, payment for ecosystem services) and regulatory approaches (e.g., enforcement of environmental law to protect forest carbon stocks by reducing defor-estation, set-aside policies, air and water pollution control reducing nitrate load and N$_2$O emissions) have been effective in different cases. Investments in research, development, and diffusion (e.g., increase of resource use-efficiency (fertilizers), livestock improvement, better forestry management practices) could result in synergies between adaptation and mitigation. Successful cases of deforestation reduction in different regions are found to combine different policies such as land planning, regulatory approaches and economic incentives (*limited evidence, high agreement*). [11.3.2, 11.10, 15.11]

Table TS.8 | Overview of potential co-benefits (green arrows) and adverse side-effects (orange arrows) of the main mitigation measures in the AFOLU sector; arrows pointing up/down denote a positive/negative effect on the respective objective or concern. These effects depend on the specific context (including bio-physic, institutional and socio-economic aspects) as well as on the scale of implementation. For an assessment of macroeconomic, cross-sectoral effects associated with mitigation policies (e.g., on energy prices, consumption, growth, and trade), see e.g., Sections 3.9, 6.3.6, 13.2.2.3 and 14.4.2. The uncertainty qualifiers in brackets denote the level of evidence and agreement on the respective effects (see TS.1). Abbreviations for evidence: l = limited, m = medium, r = robust; for agreement: l = low, m = medium, h = high. [Tables 11.9 and 11.12]

AFOLU	Effect on additional objectives/concerns			
	Economic	Social	Environmental	Institutional
Supply side: Forestry, land-based agriculture, livestock, integrated systems, and bioenergy (**marked by ***) **Demand side:** Reduced losses in the food supply chain, changes in human diets, changes in demand for wood and forestry products	* Employment impact via Entrepreneurship development (**m/h**) ↓ Use of less labour-intensive technologies in agriculture (**m/m**) ↑* Diversification of income sources and access to markets (**r/h**) ↑* Additional income to (sustainable) landscape management (**m/h**) ↑* Income concentration (**m/m**) ↑* Energy security (resource sufficiency) (**m/h**) ↑ Innovative financing mechanisms for sustainable resource management (**m/h**) ↑ Technology innovation and transfer (**m/m**)	↑* Food-crops production through integrated systems and sustainable agriculture intensification (**r/m**) ↓* Food production (locally) due to large-scale monocultures of non-food crops (**r/l**) ↑ Cultural habitats and recreational areas via (sustainable) forest management and conservation (**m/m**) ↑* Human health and animal welfare e.g., through less pesticides, reduced burning practices, and practices like agroforestry and silvo-pastoral systems (**m/h**) ↓* Human health when using burning practices (in agriculture or bioenergy) (**m/m**) * Gender, intra- and inter-generational equity via ↑ Participation and fair benefit sharing (**r/h**) ↑ Concentration of benefits (**m/m**)	Provision of ecosystem services via ↑ Ecosystem conservation and sustainable management as well as sustainable agriculture (**r/h**) ↓* Large scale monocultures (**r/h**) ↑* Land-use competition (**r/m**) ↑ Soil quality (**r/h**) ↓ Erosion (**r/h**) ↑ Ecosystem resilience (**m/h**) ↑ Albedo and evaporation (**r/h**)	↑↓* Tenure and use rights at the local level (for indigenous people and local communities) especially when implementing activities in natural forests (**r/h**) ↑↓ Access to participative mechanisms for land management decisions (**r/h**) ↑ Enforcement of existing policies for sustainable resource management (**r/h**)

Reducing Emissions from Deforestation and Forest Degradation (REDD+)[17] can be a very cost-effective policy option for mitigating climate change, if implemented in a sustainable manner (*limited evidence, medium agreement*). REDD+ includes: reducing emissions from deforestation and forest degradation; conservation of forest carbon stocks; sustainable management of forests; and enhancement of forest carbon stocks. It could supply a large share of global abatement of emissions from the AFOLU sector, especially through reducing deforestation in tropical regions, with potential economic, social and other environmental co-benefits. To assure these co-benefits, the implementation of national REDD+ strategies would need to consider financing mechanisms to local stakeholders, safeguards (such as land rights, conservation of biodiversity and other natural resources), and the appropriate scale and institutional capacity for monitoring and verification. [11.10]

Bioenergy can play a critical role for mitigation, but there are issues to consider, such as the sustainability of practices and the efficiency of bioenergy systems (*robust evidence, medium agreement*) [11.4.4, Box 11.5, 11.13.6, 11.13.7]. Barriers to large-scale deployment of bioenergy include concerns about GHG emissions from land, food security, water resources, biodiversity conservation and livelihoods. The scientific debate about the overall climate impact related to land-use competition effects of specific bioenergy pathways remains unresolved (*robust evidence, high agreement*). [11.4.4, 11.13] Bioenergy technologies are diverse and span a wide range of options and technology pathways. Evidence suggests that options with low lifecycle emissions (e.g., sugar cane, Miscanthus, fast growing tree species, and sustainable use of biomass residues), some already available, can reduce GHG emissions; outcomes are site-specific and rely on efficient integrated 'biomass-to-bioenergy systems', and sustainable land-use management and governance. In some regions, specific bioenergy options, such as improved cookstoves, and small-scale biogas and biopower production, could reduce GHG emissions and improve livelihoods and health in the context of sustainable development (*medium evidence, medium agreement*). [11.13]

---

[17] UN Programme on Reducing Emissions from Deforestation and Forest Degradation in developing countries, including conservation, sustainable management of forests and enhancement of forest carbon stocks.

TS.3.2.7 Human settlements, infrastructure, and spatial planning

Urbanization is a global trend transforming human settlements, societies, and energy use (*robust evidence, high agreement*). In 1900, when the global population was 1.6 billion, only 13% of the population, or some 200 million, lived in urban areas. As of 2011, more than 52% of the world's population—roughly 3.6 billion—lives in urban areas. By 2050, the urban population is expected to increase to 5.6–7.1 billion, or 64–69% of the world population. [12.2]

Urban areas account for more than half of global primary energy use and energy-related CO_2 emissions (*medium evidence, high agreement*). The exact share of urban energy and GHG emissions varies with emission accounting frameworks and definitions. Taking account of direct and indirect emissions, urban areas account for 67–76% of global energy use (central estimate) and 71–76% of global energy-related CO_2 emissions. Taking account of direct emissions only, the urban share of emissions is 44% (Figure TS.32). [12.2, 12.3]

No single factor explains variations in per-capita emissions across cities, and there are significant differences in per capita GHG emissions between cities within a single country (*robust evidence, high agreement*). Urban GHG emissions are influenced by a variety of physical, economic and social factors, development levels, and urbanization histories specific to each city. Key influences on urban GHG emissions include income, population dynamics, urban form, locational factors, economic structure, and market failures. Per capita final energy use and CO_2 emissions in cities of Annex I countries tend to be lower than national averages, in cities of non-Annex I countries they tend to be higher. [12.3]

The majority of infrastructure and urban areas have yet to be built (*limited evidence, high agreement*). Accounting for trends in declining population densities, and continued economic and population growth, urban land cover is projected to expand by 56–310% between 2000 and 2030. If the global population increases to 9.3 billion by 2050 and developing countries expand their built environment and infrastructure to current global average levels using available technology of today, the production of infrastructure materials alone would generate about 470 $GtCO_2$ emissions. Currently, average per capita CO_2 emissions embodied in the infrastructure of industrialized countries is five times larger than those in developing countries. [12.2, 12.3]

Infrastructure and urban form are strongly interlinked, and lock in patterns of land use, transport choice, housing, and behaviour (*medium evidence, high agreement*). Urban form and infrastructure shape long-term land-use management, influence individual transport choice, housing, and behaviour, and affect the system-wide efficiency of a city. Once in place, urban form and infrastructure are difficult to change (Figure TS.33). [12.2, 12.3, 12.4]

Mitigation options in urban areas vary by urbanization trajectories and are expected to be most effective when policy instruments are bundled (*robust evidence, high agreement*). For rapidly developing cities, options include shaping their urbanization and infrastructure development towards more sustainable and low-carbon pathways. In mature or established cities, options are constrained by existing urban forms and infrastructure and the potential for refurbishing existing systems and infrastructures. Key mitigation strategies include co-locating high residential with high employment densities,

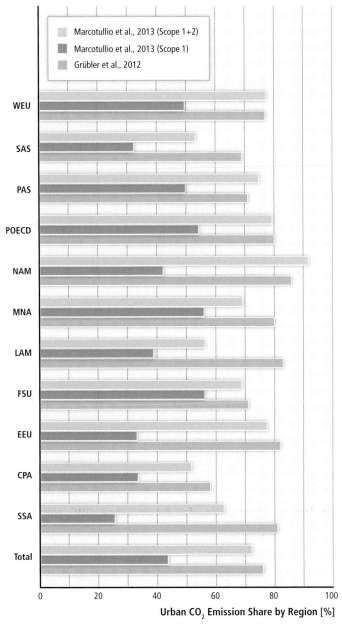

Figure TS.32 | Estimated shares of direct (Scope 1) and indirect urban CO_2 emissions in total emissions across world regions ($GtCO_2$). Indirect emissions (Scope 2) allocate emissions from thermal power plants to urban areas. CPA: Centrally Planned Asia and China; EEU: Central and Eastern Europe; FSU: Former Soviet Union; LAM: Latin America and Caribbean; MNA: Middle East and North Africa; NAM: North America; PAS: South-East Asia and Pacific; POECD: Pacific OECD; SAS: South Asia; SSA: Sub Saharan Africa; WEU: Western Europe. [12.2.2, Figure 12.4]

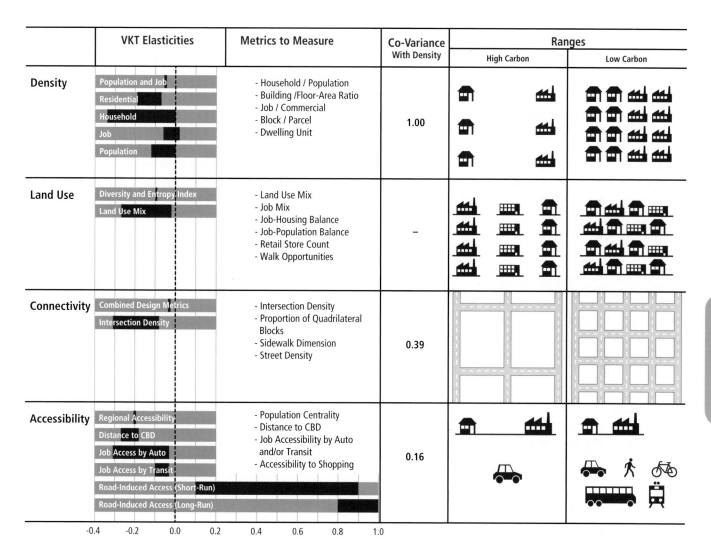

	VKT Elasticities	Metrics to Measure	Co-Variance With Density	Ranges	
				High Carbon	Low Carbon
Density	Population and Job / Residential / Household / Job / Population	- Household / Population - Building /Floor-Area Ratio - Job / Commercial - Block / Parcel - Dwelling Unit	1.00		
Land Use	Diversity and Entropy Index / Land Use Mix	- Land Use Mix - Job Mix - Job-Housing Balance - Job-Population Balance - Retail Store Count - Walk Opportunities	–		
Connectivity	Combined Design Metrics / Intersection Density	- Intersection Density - Proportion of Quadrilateral Blocks - Sidewalk Dimension - Street Density	0.39		
Accessibility	Regional Accessibility / Distance to CBD / Job Access by Auto / Job Access by Transit / Road-Induced Access (Short-Run) / Road-Induced Access (Long-Run)	- Population Centrality - Distance to CBD - Job Accessibility by Auto and/or Transit - Accessibility to Shopping	0.16		

-0.4 -0.2 0.0 0.2 0.4 0.6 0.8 1.0

Figure TS.33 | Four key aspects of urban form and structure (density, land-use mix, connectivity, and accessibility), their vehicle kilometers travelled (VKT) elasticities, commonly used metrics, and stylized graphics. The dark blue row segments under the VKT elasticities column provide the range of elasticities for the studies included. CBD: Central business district. [Figure 12.14]

achieving high diversity and integration of land uses, increasing accessibility and investing in public transit and other supportive demand-management measures (Figure TS.33). Bundling these strategies can reduce emissions in the short term and generate even higher emissions savings in the long term. [12.4, 12.5]

The largest opportunities for future urban GHG emissions reduction might be in rapidly urbanizing countries where urban form and infrastructure are not locked-in but where there are often limited governance, technical, financial, and institutional capacities (*robust evidence, high agreement*). The bulk of future infrastructure and urban growth is expected in small- to medium-size cities in developing countries, where these capacities can be limited or weak. [12.4, 12.5, 12.6, 12.7]

Thousands of cities are undertaking climate action plans, but their aggregate impact on urban emissions is uncertain (*robust evidence, high agreement*). Local governments and institutions possess unique opportunities to engage in urban mitigation activities and

local mitigation efforts have expanded rapidly. However, little systematic assessment exists regarding the overall extent to which cities are implementing mitigation policies and emissions reduction targets are being achieved, or emissions reduced. Climate action plans include a range of measures across sectors, largely focused on energy efficiency rather than broader land-use planning strategies and cross-sectoral measures to reduce sprawl and promote transit-oriented development (Figure TS.34). [12.6, 12.7, 12.9]

The feasibility of spatial planning instruments for climate change mitigation is highly dependent on a city's financial and governance capability (*robust evidence, high agreement*). Drivers of urban GHG emissions are interrelated and can be addressed by a number of regulatory, management, and market-based instruments. Many of these instruments are applicable to cities in both developed and developing countries, but the degree to which they can be implemented varies. In addition, each instrument varies in its potential to generate public revenues or require government expenditures, and the administrative scale at which it can be applied (Figure TS.35). A bun-

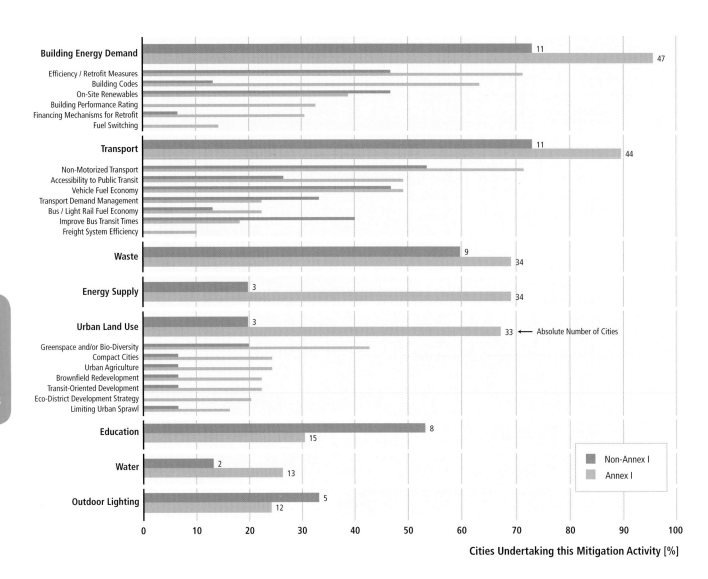

Figure TS.34 | Common mitigation measures in Climate Action Plans. [Figure 12.22]

dling of instruments and a high level of coordination across institutions can increase the likelihood of achieving emissions reductions and avoiding unintended outcomes. [12.6, 12.7]

For designing and implementing climate policies effectively, institutional arrangements, governance mechanisms, and financial resources should be aligned with the goals of reducing urban GHG emissions (*high confidence*). These goals will reflect the specific challenges facing individual cities and local governments. The following have been identified as key factors: (1) institutional arrangements that facilitate the integration of mitigation with other high-priority urban agendas; (2) a multilevel governance context that empowers cities to promote urban transformations; (3) spatial planning competencies and political will to support integrated land-use

and transportation planning; and (4) sufficient financial flows and incentives to adequately support mitigation strategies. [12.6, 12.7]

Successful implementation of urban climate change mitigation strategies can provide co-benefits (*robust evidence, high agreement*). Urban areas throughout the world continue to struggle with challenges, including ensuring access to energy, limiting air and water pollution, and maintaining employment opportunities and competitiveness. Action on urban-scale mitigation often depends on the ability to relate climate change mitigation efforts to local co-benefits. The co-benefits of local climate change mitigation can include public savings, air quality and associated health benefits, and productivity increases in urban centres, providing additional motivation for undertaking mitigation activities. [12.5, 12.6, 12.7, 12.8]

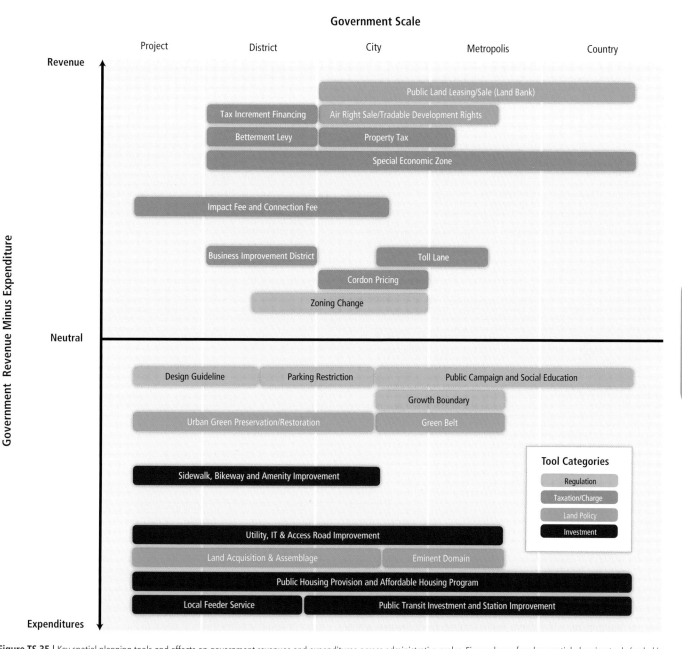

Figure TS.35 | Key spatial planning tools and effects on government revenues and expenditures across administrative scales. Figure shows four key spatial planning tools (coded in colours) and the scale of governance at which they are administered (x-axis) as well as how much public revenue or expenditure the government generates by implementing each instrument (y-axis). [Figure 12.20]

TS.4   Mitigation policies and institutions

The previous section shows that since AR4 the scholarship on mitigation pathways has begun to consider in much more detail how a variety of real-world considerations—such as institutional and political constraints, uncertainty associated with climate change risks, the availability of technologies and other factors—affect the kinds of policies and measures that are adopted. Those factors have important implications for the design, cost, and effectiveness of mitigation action. This sec-

tion focuses on how governments and other actors in the private and public sectors design, implement, and evaluate mitigation policies. It considers the 'normative' scientific research on how policies should be designed to meet particular criteria. It also considers research on how policies are actually designed and implemented a field known as 'positive' analysis. The discussion first characterizes fundamental conceptual issues, and then presents a summary of the main findings from WGIII AR5 on local, national, and sectoral policies. Much of the practical policy effort since AR4 has occurred in these contexts. From there the summary looks at ever-higher levels of aggregation, ultimately ending at the global level and cross-cutting investment and finance issues.

TS.4.1 Policy design, behaviour and political economy

There are multiple criteria for evaluating policies. Policies are frequently assessed according to four criteria [3.7.1, 13.2.2, 15.4.1]:

- Environmental effectiveness—whether policies achieve intended goals in reducing emissions or other pressures on the environment or in improving measured environmental quality.
- Economic effectiveness—the impact of policies on the overall economy. This criterion includes the concept of economic efficiency, the principle of maximizing net economic benefits. Economic welfare also includes the concept of cost-effectiveness, the principle of attaining a given level of environmental performance at lowest aggregate cost.
- Distributional and social impacts—also known as 'distributional equity,' this criterion concerns the allocation of costs and benefits of policies to different groups and sectors within and across economies over time. It includes, often, a special focus on impacts on the least well-off members of societies within countries and around the world.
- Institutional and political feasibility—whether policies can be implemented in light of available institutional capacity, the political constraints that governments face, and other factors that are essential to making a policy viable.

All criteria can be applied with regard to the immediate 'static' impacts of policies and from a long-run 'dynamic' perspective that accounts for the many adjustments in the economic, social and political systems. Criteria may be mutually reinforcing, but there may also be conflicts or tradeoffs among them. Policies designed for maximum environmental effectiveness or economic performance may fare less well on other criteria, for example. Such tradeoffs arise at multiple levels of governing systems. For example, it may be necessary to design international agreements with flexibility so that it is feasible for a large number of diverse countries to accept them, but excessive flexibility may undermine incentives to invest in cost-effective long-term solutions.

Policymakers make use of many different policy instruments at the same time. Theory can provide some guidance on the normative advantages and disadvantages of alternative policy instruments in light of the criteria discussed above. The range of different policy instruments includes [3.8, 15.3]:

- Economic incentives, such as taxes, tradable allowances, fines, and subsidies
- Direct regulatory approaches, such as technology or performance standards
- Information programmes, such as labelling and energy audits
- Government provision, for example of new technologies or in state enterprises
- Voluntary actions, initiated by governments, firms, and non-governmental organizations (NGOs)

Since AR4, the inventory of research on these different instruments has grown, mostly with reference to experiences with policies adopted within particular sectors and countries as well as the many interactions between policies. One implication of that research has been that international agreements that aim to coordinate across countries reflect the practicalities on the particular policy choices of national governments and other jurisdictions.

The diversity in policy goals and instruments highlights differences in how sectors and countries are organized economically and politically as well as the multi-level nature of mitigation. Since AR4, one theme of research in this area has been that the success of mitigation measures depends in part on the presence of institutions capable of designing and implementing regulatory policies and the willingness of respective publics to accept these policies. Many policies have effects, sometimes unanticipated, across multiple jurisdictions—across cities, regions and countries—because the economic effects of policies and the technological options are not contained within a single jurisdiction. [13.2.2.3, 14.1.3, 15.2, 15.9]

Interactions between policy instruments can be welfare-enhancing or welfare-degrading. The chances of welfare-enhancing interactions are particularly high when policy instruments address multiple different market failures—for example, a subsidy or other policy instrument aimed at boosting investment in R&D on less emission-intensive technologies can complement policies aimed at controlling emissions, as can regulatory intervention to support efficient improvement of end-use energy efficiency. By contrast, welfare-degrading interactions are particularly likely when policies are designed to achieve identical goals. Narrowly targeted policies such as support for deployment (rather than R&D) of particular energy technologies that exist in tandem with broader economy-wide policies aimed at reducing emissions (for example, a cap-and-trade emissions scheme) can have the effect of shifting the mitigation effort to particular sectors of the economy in ways that typically result in higher overall costs. [3.8.6, 15.7, 15.8]

There are a growing number of countries devising policies for adaptation, as well as mitigation, and there may be benefits to considering the two within a common policy framework (*medium evidence, low agreement*). However, there are divergent views on whether adding adaptation to mitigation measures in the policy portfolio encourages or discourages participation in international cooperation [1.4.5, 13.3.3]. It is recognized that an integrated approach can be valuable, as there exist both synergies and tradeoffs [16.6].

Traditionally, policy design, implementation, and evaluation has focused on governments as central designers and implementers of policies, but new studies have emerged on government acting in a coordinating role (*medium confidence*). In these cases, governments themselves seek to advance voluntary approaches, especially when traditional forms of regulation are thought to be inadequate or the best choices of policy instruments and goals is not yet apparent. Examples include voluntary schemes that allow individuals and firms to purchase emission credits that offset the emissions associated with their own activities such as flying and driving. Since AR4, a substantial new literature has emerged to examine these schemes from positive and normative perspectives. [13.12, 15.5.7]

The successful implementation of policy depends on many factors associated with human and institutional behaviour (*very high confidence*). One of the challenges in designing effective instruments is that the activities that a policy is intended to affect—such as the choice of energy technologies and carriers and a wide array of agricultural and forestry practices—are also influenced by social norms, decision-making rules, behavioural biases, and institutional processes [2.4, 3.10]. There are examples of policy instruments made more effective by taking these factors into account, such as in the case of financing mechanisms for household investments in energy efficiency and renewable energy that eliminate the need for up-front investment [2.4, 2.6.5.3]. Additionally, the norms that guide acceptable practices could have profound impacts on the baselines against which policy interventions are evaluated, either magnifying or reducing the required level of policy intervention [1.2.4, 4.3, 6.5.2].

Climate policy can encourage investment that may otherwise be suboptimal because of market imperfections (*very high con-*

fidence). Many of the options for energy efficiency as well as low-carbon energy provision require high up-front investment that is often magnified by high-risk premiums associated with investments in new technologies. The relevant risks include those associated with future market conditions, regulatory actions, public acceptance, and technology cost and performance. Dedicated financial instruments exist to lower these risks for private actors—for example, credit insurance, feed-in tariffs (FITs), concessional finance, or rebates [16.4]. The design of other mitigation policies can also incorporate elements to help reduce risks, such as a cap-and-trade regime that includes price floors and ceilings [2.6.5, 15.5, 15.6].

TS.4.2   Sectoral and national policies

There has been a considerable increase in national and subnational mitigation plans and strategies since AR4 (Figure TS.36). These plans and strategies are in their early stages of development and implementation in many countries, making it difficult to assess whether and how they will result in appropriate institutional and policy change, and therefore, their impact on future GHG emissions. However, to date these policies, taken together, have not yet achieved a substantial deviation in GHG emissions from the past trend. Theories of institutional change suggest they might play a role in shaping incentives, political contexts, and policy paradigms in a way that encourages

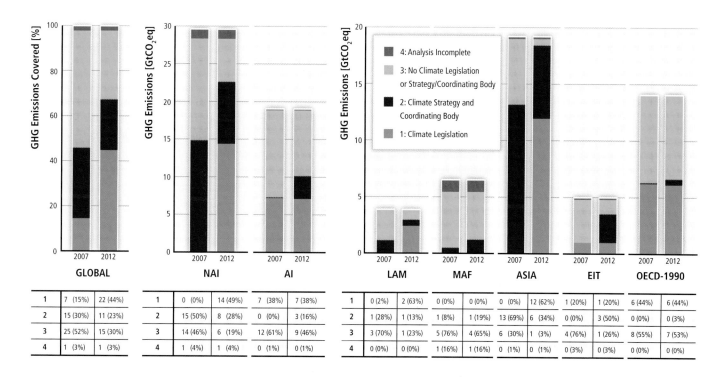

Figure TS.36| National climate legislation and strategies in 2007 and 2012. Regions include NAI (Non Annex I countries—developing countries), AI (Annex I countries—developed countries), LAM (Latin America), MAF (Middle East and Africa), ASIA (Asia), EIT (Economies in Transition), OECD-1990; see Annex II.2 for more details. In this figure, climate legislation is defined as mitigation-focused legislation that goes beyond sectoral action alone. Climate strategy is defined as a non-legislative plan or framework aimed at mitigation that encompasses more than a small number of sectors, and that includes a coordinating body charged with implementation. International pledges are not included, nor are subnational plans and strategies. The panel shows proportion of GHG emissions covered. [Figure 15.1]

GHG emissions reductions in the future [15.1, 15.2]. However, many baseline scenarios (i.e., those without additional mitigation policies) show concentrations that exceed 1000 ppm CO_2eq by 2100, which is far from a concentration with a *likely* probability of maintaining temperature increases below 2 °C this century. Mitigation scenarios suggest that a wide range of environmentally effective policies could be enacted that would be consistent with such goals [6.3]. In practice, climate strategies and the policies that result are influenced by political economy factors, sectoral considerations, and the potential for realizing co-benefits. In many countries, mitigation policies have also been actively pursued at state and local levels. [15.2, 15.5, 15.8]

Since AR4, there is growing political and analytical attention to co-benefits and adverse side-effects of climate policy on other objectives and vice versa that has resulted in an increased focus on policies designed to integrate multiple objectives (*high confidence*). Co-benefits are often explicitly referenced in climate and sectoral plans and strategies and often enable enhanced political support [15.2]. However, the analytical and empirical underpinnings for many of these interactive effects, and particularly for the associated welfare impacts, are under-developed [1.2, 3.6.3, 4.2, 4.8, 6.6]. The scope for co-benefits is greater in low-income countries, where complementary policies for other objectives, such as air quality, are often weak [5.7, 6.6, 15.2].

The design of institutions affects the choice and feasibility of policy options as well as the sustainable financing of mitigation measures. Institutions designed to encourage participation by representatives of new industries and technologies can facilitate transitions to low-GHG emissions pathways [15.2, 15.6]. Policies vary in the extent to which they require new institutional capabilities to be implemented. Carbon taxation, in most settings, can rely mainly on existing tax infrastructure and is administratively easier to implement than many other alternatives such as cap-and-trade systems [15.5]. The extent of institutional innovation required for policies can be a factor in instrument choice, especially in developing countries.

Sector-specific policies have been more widely used than economy-wide, market-based policies (*medium evidence, high agreement*). Although economic theory suggests that market-based, economy-wide policies for the singular objective of mitigation would generally be more cost-effective than sector-specific policies, political economy considerations often make economy-wide policies harder to design and implement than sector-specific policies [15.2.3, 15.2.6, 15.5.1]. In some countries, emission trading and taxes have been enacted to address the market externalities associated with GHG emissions, and have contributed to the fulfilment of sector-specific GHG reduction goals (*medium evidence, medium agreement*) [7.12]. In the longer term, GHG pricing can support the adoption of low-GHG energy technologies. Even if economy-wide policies were implemented, sector-specific policies may be needed to overcome sectoral market failures. For example, building codes can require energy-efficient investments where private investments would otherwise not exist [9.10]. In transport, pricing policies that raise the cost of carbon-intensive forms of private transport are

more effective when backed by public investment in viable alternatives [8.10]. Table TS.9 presents a range of sector-specific policies that have been implemented in practice. [15.1, 15.2, 15.5, 15.8, 15.9]

Carbon taxes have been implemented in some countries and—alongside technology and other policies—have contributed to decoupling of emissions from GDP (*high confidence*). Differentiation by sector, which is quite common, reduces cost-effectiveness that arises from the changes in production methods, consumption patterns, lifestyle shifts, and technology development, but it may increase political feasibility, or be preferred for reasons of competitiveness or distributional equity. In some countries, high carbon and fuel taxes have been made politically feasible by refunding revenues or by lowering other taxes in an environmental fiscal reform. Mitigation policies that raise government revenue (e.g., auctioned emission allowances under a cap-and-trade system or emission taxes) generally have lower social costs than approaches that do not, but this depends on how the revenue is used [3.6.3]. [15.2, 15.5.2, 15.5.3]

Fuel taxes are an example of a sector-specific policy and are often originally put in place for objectives such as revenue—they are not necessarily designed for the purpose of mitigation (*high confidence*). In Europe, where fuel taxes are highest, they have contributed to reductions in carbon emissions from the transport sector of roughly 50 % for this group of countries. The short-run response to higher fuel prices is often small, but long-run price elasticities are quite high, or roughly −0.6 to −0.8. This means that in the long run, 10 % higher fuel prices correlate with 7 % reduction in fuel use and emissions. In the transport sector, taxes have the advantage of being progressive or neutral in most countries and strongly progressive in low-income countries. [15.5.2]

Cap-and-trade systems for GHG emissions are being established in a growing number of countries and regions. Their environmental effect has so far been limited because caps have either been loose or have not yet been binding (*limited evidence, medium agreement*). There appears to have been a tradeoff between the political feasibility and environmental effectiveness of these programmes, as well as between political feasibility and distributional equity in the allocation of permits. Greater environmental effectiveness through a tighter cap may be combined with a price ceiling that improves political feasibility. [14.4.2, 15.5.3]

Different factors reduced the price of European Union Emissions Trading System (EU ETS) allowances below anticipated levels, thereby slowing investment in mitigation (*high confidence*). While the European Union demonstrated that a cross-border cap-and-trade system can work, the low price of EU ETS allowances in recent years provided insufficient incentives for significant additional investment in mitigation. The low price is related to unexpected depth and duration of the economic recession, uncertainty about the long-term reduction targets for GHG emissions, import of credits from the Clean Development Mechanism (CDM), and the interaction with other policy instruments,

Table TS.9 | Sector policy instruments. The table brings together evidence on mitigation policy instruments discussed in Chapters 7 to 12. [Table 15.2]

Policy Instruments	Energy [7.12]	Transport [8.10]	Buildings [9.10]	Industry [10.11]	AFOLU [11.10]	Human Settlements and Infrastructure
Economic Instruments—Taxes (Carbon taxes may be economy-wide)	• Carbon taxes	• Fuel taxes • Congestion charges, vehicle registration fees, road tolls • Vehicle taxes	• Carbon and/or energy taxes (either sectoral or economy wide)	• Carbon tax or energy tax • Waste disposal taxes or charges	• Fertilizer or Nitrogen taxes to reduce nitrous oxide	• Sprawl taxes, Impact fees, exactions, split-rate property taxes, tax increment finance, betterment taxes, congestion charges
Economic Instruments—Tradable Allowances (May be economy-wide)	• Emissions trading (e.g., EU ETS) • Emission credits under CDM • Tradable Green Certificates	• Fuel and vehicle standards	• Tradable certificates for energy efficiency improvements (white certificates)	• Emissions trading • Emission credit under CDM • Tradable Green Certificates	• Emission credits under the Kyoto Protocol's Clean Development Mechanism (CDM) • Compliance schemes outside Kyoto protocol (national schemes) • Voluntary carbon markets	• Urban-scale Cap and Trade
Economic Instruments—Subsidies	• Fossil fuel subsidy removal • Feed-in-tariffs for renewable energy • Capital subsidies and insurance for 1st generation Carbon Dioxide Capture and Storage (CCS)	• Biofuel subsidies • Vehicle purchase subsidies • Feebates	• Subsidies or Tax exemptions for investment in efficient buildings, retrofits and products • Subsidized loans	• Subsidies (e.g., for energy audits) • Fiscal incentives (e.g., for fuel switching)	• Credit lines for low carbon agriculture, sustainable forestry.	• Special Improvement or Redevelopment Districts
Regulatory Approaches	• Efficiency or environmental performance standards • Renewable Portfolio standards for renewable energy • Equitable access to electricity grid • Legal status of long term CO_2 storage	• Fuel economy performance standards • Fuel quality standards • GHG emission performance standards • Regulatory restrictions to encourage modal shifts (road to rail) • Restriction on use of vehicles in certain areas • Environmental capacity constraints on airports • Urban planning and zoning restrictions	• Building codes and standards • Equipment and appliance standards • Mandates for energy retailers to assist customers invest in energy efficiency	• Energy efficiency standards for equipment • Energy management systems (also voluntary) • Voluntary agreements (where bound by regulation) • Labelling and public procurement regulations	• National policies to support REDD+ including monitoring, reporting and verification • Forest law to reduce deforestation • Air and water pollution control GHG precursors • Land-use planning and governance	• Mixed use zoning • Development restrictions • Affordable housing mandates • Site access controls • Transfer development rights • Design codes • Building codes • Street codes • Design standards
Information Programmes		• Fuel labelling • Vehicle efficiency labelling	• Energy audits • Labelling programmes • Energy advice programmes	• Energy audits • Benchmarking • Brokerage for industrial cooperation	• Certification schemes for sustainable forest practices • Information policies to support REDD+ including monitoring, reporting and verification	
Government Provision of Public Goods or Services	• Research and development • Infrastructure expansion (district heating/cooling or common carrier)	• Investment in transit and human powered transport • Investment in alternative fuel infrastructure • Low emission vehicle procurement	• Public procurement of efficient buildings and appliances	• Training and education • Brokerage for industrial cooperation	• Protection of national, state, and local forests. • Investment in improvement and diffusion of innovative technologies in agriculture and forestry	• Provision of utility infrastructure such as electricity distribution, district heating/cooling and wastewater connections, etc. • Park improvements • Trail improvements • Urban rail
Voluntary Actions			• Labelling programmes for efficient buildings • Product eco-labelling	• Voluntary agreements on energy targets or adoption of energy management systems, or resource efficiency	• Promotion of sustainability by developing standards and educational campaigns	

TS

particularly related to the expansion of renewable energy as well as regulation on energy efficiency. It has proven to be politically difficult to address this problem by removing GHG emission permits temporarily, tightening the cap, or providing a long-term mitigation goal. [14.4.2]

Adding a mitigation policy to another may not necessarily enhance mitigation. For instance, if a cap-and-trade system has a sufficiently stringent cap then other policies such as renewable subsidies have no further impact on total GHG emissions (although they may affect costs and possibly the viability of more stringent future targets). If the cap is loose relative to other policies, it becomes ineffective. This is an example of a negative interaction between policy instruments. Since other policies cannot be 'added on' to a cap-and-trade system, if it is to meet any particular target, a sufficiently low cap is necessary. A carbon tax, on the other hand, can have an additive environmental effect to policies such as subsidies to renewables. [15.7]

Reduction of subsidies to fossil energy can achieve significant emission reductions at negative social cost (*very high confidence*). Although political economy barriers are substantial, many countries have reformed their tax and budget systems to reduce fuel subsidies that actually accrue to the relatively wealthy, and utilized lump-sum cash transfers or other mechanisms that are more targeted to the poor. [15.5.3]

Direct regulatory approaches and information measures are widely used, and are often environmentally effective, though debate remains on the extent of their environmental impacts **and cost-effectiveness** (*medium confidence*). Examples of regulatory approaches include energy efficiency standards; examples of information programmes include labelling programmes that can help consumers make better-informed decisions. While such approaches often work at a net social benefit, the scientific literature is divided on whether such policies are implemented with negative private costs (see Box TS.12) to firms and individuals [3.9.3, 15.5.5, 15.5.6]. Since AR4 there has been continued investigation into the 'rebound' effects (see Box TS.13) that arise when higher efficiency leads to lower energy costs and greater consumption. There is general agreement that such rebound effects exist, but there is low agreement in the literature on the magnitude [3.9.5, 5.7.2, 15.5.4].

There is a distinct role for technology policy as a complement to other mitigation policies (*high confidence*). Properly implemented technology policies reduce the cost of achieving a given environmental target. Technology policy will be most effective when technology-push policies (e.g., publicly funded R&D) and demand-pull policies (e.g., governmental procurement programmes or performance regulations) are used in a complementary fashion. While technology-push and demand-pull policies are necessary, they are unlikely to be sufficient without complementary framework conditions. Managing social challenges of technology policy change may require innovations in policy and institutional design, including building integrated policies that make complementary use of market incentives, authority, and norms (*medium confidence*). Since AR4, a large number of countries and subnational jurisdictions have introduced support policies for renewable

---

Box TS.13 | The rebound effect can reduce energy savings from technological improvement

Technological improvements in energy efficiency (EE) have direct effects on energy consumption and thus GHG emissions, but can cause other changes in consumption, production, and prices that will, in turn, affect GHG emissions. These changes are generally called 'rebound' or 'takeback' because in most cases they reduce the net energy or emissions reduction associated with the efficiency improvement. The size of EE rebound is controversial, with some research papers suggesting little or no rebound and others concluding that it offsets most or all reductions from EE policies [3.9.5, 5.7.2].

Total EE rebound can be broken down into three distinct parts: substitution-effect, income-effect, and economy-wide effect [3.9.5]. In end-use consumption, substitution-effect rebound, or 'direct rebound' assumes that a consumer will make more use of a device if it becomes more energy efficient because it will be cheaper to use. Income-effect rebound or 'indirect rebound', arises if the improvement in EE makes the consumer wealthier and leads her to consume additional products that require energy. Economy-wide rebound refers to impacts beyond the behaviour of the entity benefiting directly from the EE improvement, such as the impact of EE on the price of energy.

Analogous rebound effects for EE improvements in production are substitution towards an input with improved energy efficiency, and substitution among products by consumers when an EE improvement changes the relative prices of goods, as well as an income effect when an EE improvement lowers production costs and creates greater wealth.

Rebound is sometimes confused with the concept of carbon leakage, which often describes the incentive for emissions-intensive economic activity to migrate away from a region that restricts GHGs (or other pollutants) towards areas with fewer or no restrictions on such emissions [5.4.1, 14.4]. Energy efficiency rebound can occur regardless of the geographic scope of the adopted policy. As with leakage, however, the potential for significant rebound illustrates the importance of considering the full equilibrium effects of a mitigation policy [3.9.5, 15.5.4].

energy such as feed-in tariffs and renewable portfolio standards. These have promoted substantial diffusion and innovation of new energy technologies such as wind turbines and photovoltaic panels, but have raised questions about their economic efficiency, and introduced challenges for grid and market integration. [2.6.5, 7.12, 15.6.5]

Worldwide investment in research in support of mitigation is small relative to overall public research spending (*medium confidence*). The effectiveness of research support will be greatest if it is increased slowly and steadily rather than dramatically or erratically. It is important that data collection for program evaluation is built into technology policy programmes, because there is limited empirical evidence on the relative effectiveness of different mechanisms for supporting the invention, innovation and diffusion of new technologies. [15.6.2, 15.6.5]

Government planning and provision can facilitate shifts to less energy- and GHG-intensive infrastructure and lifestyles (*high confidence*). This applies particularly when there are indivisibilities in the provision of infrastructure as in the energy sector [7.6] (e.g., for electricity transmission and distribution or district heating networks); in the transport sector [8.4] (e.g., for non-motorized or public transport); and in urban planning [12.5]. The provision of adequate infrastructure is important for behavioural change [15.5.6].

Successful voluntary agreements on mitigation between governments and industries are characterized by a strong institutional framework with capable industrial associations (*medium confidence*). The strengths of voluntary agreements are speed and flexibility in phasing measures, and facilitation of barrier removal activities for energy efficiency and low-emission technologies. Regulatory threats, even though the threats are not always explicit, are also an important factor for firms to be motivated. There are few environmental impacts without a proper institutional framework. [15.5.7]

TS.4.3   Development and regional cooperation

Regional cooperation offers substantial opportunities for mitigation due to geographic proximity, shared infrastructure and policy frameworks, trade, and cross-border investment that would be difficult for countries to implement in isolation (*high confidence*). Examples of possible regional cooperation policies include regionally-linked development of renewable energy power pools, networks of natural gas supply infrastructure, and coordinated policies on forestry. [14.1]

At the same time, there is a mismatch between opportunities and capacities to undertake mitigation (*medium confidence*). The regions with the greatest potential to leapfrog to low-carbon development trajectories are the poorest developing regions where there are few lock-in effects in terms of modern energy systems and urbanization patterns. However, these regions also have the lowest financial, technological, and institutional capacities to embark on such low-carbon development paths (Figure TS.37) and their cost of waiting is high due to unmet energy and development needs. Emerging economies already have more lock-in effects but their rapid build-up of modern energy systems and urban settlements still offers substantial opportunities for low-carbon development. Their capacity to reorient themselves to low-carbon development strategies is higher, but also faces constraints in terms of finance, technology, and the high cost of delaying the installation of new energy capacity. Lastly, industrialized economies have the largest lock-in effects, but the highest capacities to reorient their energy, transport, and urbanizations systems towards low-carbon development. [14.1.3, 14.3.2]

Regional cooperation has, to date, only had a limited (positive) impact on mitigation (*medium evidence, high agreement*). Nonetheless, regional cooperation could play an enhanced role in promoting mitigation in the future, particularly if it explicitly incorporates mitigation objectives in trade, infrastructure and energy policies and promotes direct mitigation action at the regional level. [14.4.2, 14.5]

Most literature suggests that climate-specific regional cooperation agreements in areas of policy have not played an important role in addressing mitigation challenges to date (*medium confidence*). This is largely related to the low level of regional integration and associated willingness to transfer sovereignty to supra-national regional bodies to enforce binding agreements on mitigation. [14.4.2, 14.4.3]

Climate-specific regional cooperation using binding regulation-based approaches in areas of deep integration, such as EU directives on energy efficiency, renewable energy, and biofuels, have had some impact on mitigation objectives (*medium confidence*). Nonetheless, theoretical models and past experience suggest that there is substantial potential to increase the role of climate-specific regional cooperation agreements and associated instruments, including economic instruments and regulatory instruments. In this context it is important to consider carbon leakage of such regional initiatives and ways to address it. [14.4.2, 14.4.1]

In addition, non-climate-related modes of regional cooperation could have significant implications for mitigation, even if mitigation objectives are not a component (*medium confidence*). Regional cooperation with non-climate-related objectives but possible mitigation implications, such as trade agreements, cooperation on technology, and cooperation on infrastructure and energy, has to date also had negligible impacts on mitigation. Modest impacts have been found on the level of GHG emissions of members of regional preferential trade areas if these agreements are accompanied with environmental agreements. Creating synergies between adaptation and mitigation can increase the cost-effectiveness of climate change actions. Linking electricity and gas grids at the regional level has also had a modest impact on mitigation as it facilitated greater use of low-carbon and renewable technologies; there is substantial further mitigation potential in such arrangements. [14.4.2]

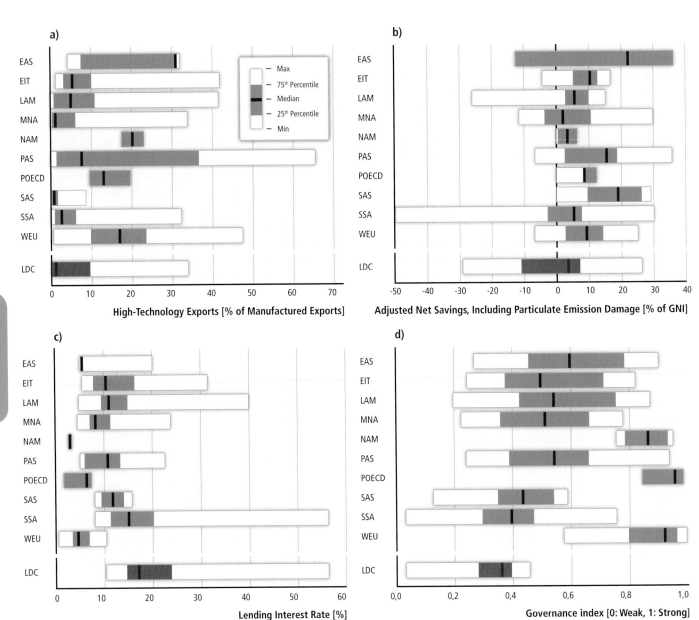

Figure TS.37 | Economic and governance indicators affecting regional capacities to embrace mitigation policies. Regions include EAS (East Asia), EIT (Economies in Transition), LAM (Latin America and Caribbean), MNA (Middle East and North Africa), NAM (North America), POECD (Pacific Organisation for Economic Co-operation and Development (OECD)-1990 members), PAS (South East Asia and Pacific), SAS (South Asia), SSA (sub-Saharan Africa), WEU (Western Europe), LDC (least-developed countries). Statistics refer to the year 2010 or the most recent year available. Note: The lending interest rate refers to the average interest rate charged by banks to private sector clients for short- to medium-term financing needs. The governance index is a composite measure of governance indicators compiled from various sources, rescaled to a scale of 0 to 1, with 0 representing weakest governance and 1 representing strongest governance. [Figure 14.2]

TS.4.4   International cooperation

Climate change mitigation is a global commons problem that requires international cooperation, but since AR4, scholarship has emerged that emphasizes a more complex and multi-faceted view of climate policy (*very high confidence*). Two characteristics of climate change necessitate international cooperation: climate change is a global commons problem, and it is characterized by a high degree of heterogeneity in the origins of GHG emissions, mitigation opportunities, climate impacts, and capacity for mitigation and adapta-

tion [13.2.1.1]. Policymaking efforts to date have primarily focused on international cooperation as a task centrally focused on the coordination of national policies that would be adopted with the goal of mitigation. More recent policy developments suggest that there is a more complicated set of relationships between national, regional, and global policymaking, based on a multiplicity of goals, a recognition of policy co-benefits, and barriers to technological innovation and diffusion [1.2, 6.6, 15.2]. A major challenge is assessing whether decentralized policy action is consistent with and can lead to total mitigation efforts that are effective, equitable, and efficient [6.1.2.1, 13.13].

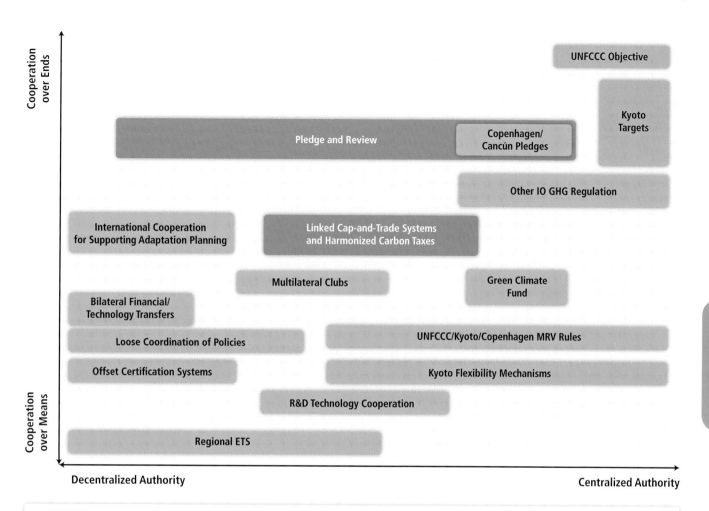

Loose coordination of policies: examples include transnational city networks and Nationally Appropriate Mitigation Actions (NAMAs); R&D technology cooperation: examples include the Major Economies Forum on Energy and Climate (MEF), Global Methane Initiative (GMI), or Renewable Energy and Energy Efficiency Partnership (REEEP); Other international organization (IO) GHG regulation: examples include the Montreal Protocol, International Civil Aviation Organization (ICAO), International Maritime Organization (IMO).

Figure TS.38 | Alternative forms of international cooperation. The figure represents a compilation of existing and possible forms of international cooperation, based upon a survey of published research, but is not intended to be exhaustive of existing or potential policy architectures, nor is it intended to be prescriptive. Examples in orange are existing agreements. Examples in blue are structures for agreements proposed in the literature. The width of individual boxes indicates the range of possible degrees of centralization for a particular agreement. The degree of centralization indicates the authority an agreement confers on an international institution, not the process of negotiating the agreement. [Figure 13.2]

International cooperation on climate change has become more institutionally diverse over the past decade (*very high confidence*). Perceptions of fairness can facilitate cooperation by increasing the legitimacy of an agreement [3.10, 13.2.2.4]. UNFCCC remains a primary international forum for climate negotiations, but other institutions have emerged at multiple scales, namely: global, regional, national, and local [13.3.1, 13.4.1.4, 13.5]. This institutional diversity arises in part from the growing inclusion of climate change issues in other policy arenas (e.g., sustainable development, international trade, and human rights). These and other linkages create opportunities, potential co-benefits, or harms that have not yet been thoroughly examined. Issue linkage also creates the possibility for countries to experiment with different forums of cooperation ('forum shopping'), which may increase negotiation costs and potentially distract from or dilute the performance of international cooperation toward climate goals. [13.3, 13.4, 13.5] Finally, there

has been an emergence of new transnational climate-related institutions not centred on sovereign states (e.g., public-private partnerships, private sector governance initiatives, transnational NGO programmes, and city level initiatives) [13.3.1, 13.12].

Existing and proposed international climate agreements vary in the degree to which their authority is centralized. As illustrated in Figure TS.38, the range of centralized formalization spans strong multilateral agreements (such as the Kyoto Protocol targets), harmonized national policies (such as the Copenhagen/Cancún pledges), and decentralized but coordinated national policies (such as planned linkages of national and sub-national emissions trading schemes) [13.4.1, 13.4.3]. Four other design elements of international agreements have particular relevance: legal bindingness, goals and targets, flexible mechanisms, and equitable methods for effort-shar-

Table TS.10 | Summary of performance assessments of existing and proposed forms of cooperation. Forms of cooperation are evaluated along the four evaluation criteria described in Sections 3.7.1 and 13.2.2. [Table 13.3]

Mode of International Cooperation		Assessment Criteria			
		Environmental Effectiveness	Aggregate Economic Performance	Distributional Impacts	Institutional Feasibility
Existing Cooperation [13.13.1]	**UNFCCC**	Aggregate GHG emissions in Annex I countries declined by 6.0 to 9.2 % below 1990 levels by 2000, a larger reduction than the apparent 'aim' of returning to 1990 levels by 2000.	Authorized joint fulfilment of commitments, multi-gas approach, sources and sinks, and domestic policy choice. Cost and benefit estimates depend on baseline, discount rate, participation, leakage, co-benefits, adverse effects, and other factors.	Commitments distinguish between Annex I (industrialized) and non-Annex I countries. Principle of 'common but differentiated responsibility.' Commitment to 'equitable and appropriate contributions by each [party].'	Ratified (or equivalent) by 195 countries and regional organizations. Compliance depends on national communications.
	The Kyoto Protocol (KP)	Aggregate emissions in Annex I countries were reduced by 8.5 to 13.6 % below 1990 levels by 2011, more than the first commitment period (CP1) collective reduction target of 5.2 %. Reductions occurred mainly in EITs; emissions; increased in some others. Incomplete participation in CP1 (even lower in CP2).	Cost-effectiveness improved by flexible mechanisms (Joint Implementation (JI), CDM, International Emissions Trading (IET)) and domestic policy choice. Cost and benefit estimates depend on baseline, discount rate, participation, leakage, co-benefits, adverse effects, and other factors.	Commitments distinguish between developed and developing countries, but dichotomous distinction correlates only partly (and decreasingly) with historical emissions trends and with changing economic circumstances. Intertemporal equity affected by short-term actions.	Ratified (or equivalent) by 192 countries and regional organizations, but took 7 years to enter into force. Compliance depends on national communications, plus KP compliance system. Later added approaches to enhance measurement, reporting, and verification (MRV).
	The Kyoto Mechanisms	About 1.4 billion tCO_2eq credits under the CDM, 0.8 billion under JI, and 0.2 billion under IET (through July 2013). Additionality of CDM projects remains an issue but regulatory reform underway.	CDM mobilized low cost options, particularly industrial gases, reducing costs. Underperformance of some project types. Some evidence that technology is transferred to non-Annex I countries.	Limited direct investment from Annex I countries. Domestic investment dominates, leading to concentration of CDM projects in few countries. Limited contributions to local sustainable development.	Helped enable political feasibility of Kyoto Protocol. Has multi-layered governance. Largest carbon markets to date. Has built institutional capacity in developing countries.
	Further Agreements under the UNFCCC	Pledges to limit emissions made by all major emitters under Cancun Agreements. Unlikely sufficient to limit temperature change to 2 °C. Depends on treatment of measures beyond current pledges for mitigation and finance. Durban Platform calls for new agreement by 2015, to take effect in 2020, engaging all parties.	Efficiency not assessed. Cost-effectiveness might be improved by market-based policy instruments, inclusion of forestry sector, commitments by more nations than Annex I countries (as envisioned in Durban Platform).	Depends on sources of financing, particularly for actions of developing countries.	Cancún Conference of the Parties (COP) decision; 97 countries made pledges of emission reduction targets or actions for 2020.
	Agreements outside the UNFCCC — **G8, G20, Major Economies Forum on Energy and Climate (MEF)**	G8 and MEF have recommended emission reduction by all major emitters. G20 may spur GHG reductions by phasing out of fossil fuel subsidies.	Action by all major emitters may reduce leakage and improve cost-effectiveness, if implemented using flexible mechanisms. Potential efficiency gains through subsidy removal. Too early to assess economic performance empirically.	Has not mobilized climate finance. Removing fuel subsidies would be progressive but have negative effects on oil-exporting countries and on those with very low incomes unless other help for the poorest is provided.	Lower participation of countries than UNFCCC, yet covers 70 % of global emissions. Opens possibility for forum-shopping, based on issue preferences.
	Montreal Protocol on Ozone-Depleting Substances (ODS)	Spurred emission reductions through ODS phaseouts approximately 5 times the magnitude of Kyoto CP1 targets. Contribution may be negated by high-GWP substitutes, though efforts to phase out HFCs are growing.	Cost-effectiveness supported by multi-gas approach. Some countries used market-based mechanisms to implement domestically.	Later compliance period for phaseouts by developing countries. Montreal Protocol Fund provided finance to developing countries.	Universal participation. but the timing of required actions vary for developed and developing countries
	Voluntary Carbon Market	Covers 0.13 billion tCO_2eq, but certification remains an issue	Credit prices are heterogeneous, indicating market inefficiencies	[No literature cited.]	Fragmented and non-transparent market.

Mode of International Cooperation			Assessment Criteria			
			Environmental Effectiveness	Aggregate Economic Performance	Distributional Impacts	Institutional Feasibility
Proposed Cooperation [13.13.2]	Proposed architectures	Strong multilateralism	Tradeoff between ambition (deep) and participation (broad).	More cost-effective with greater reliance on market mechanisms.	Multilateralism facilitates integrating distributional impacts into negotiations and may apply equity-based criteria as outlined in Ch. 4	Depends on number of parties; degree of ambition
		Harmonized national policies	Depends on net aggregate change in ambition across countries resulting from harmonization.	More cost-effective with greater reliance on market mechanisms.	Depends on specific national policies	Depends on similarity of national policies; more similar may support harmonization but domestic circumstances may vary. National enforcement.
		Decentralized architectures, coordinated national policies	Effectiveness depends on quality of standards and credits across countries	Often (though not necessarily) refers to linkage of national cap-and-trade systems, in which case cost effective.	Depends on specific national policies	Depends on similarity of national policies. National enforcement.
	Effort (burden) sharing arrangements		Refer to Sections 4.6.2 for discussion of the principles on which effort (burden) sharing arrangements may be based, and Section 6.3.6.6 for quantitative evaluation.			

ing [13.4.2]. Existing and proposed modes of international cooperation are assessed in Table TS.10. [13.13]

The UNFCCC is currently the only international climate policy venue with broad legitimacy, due in part to its virtually universal membership (*high confidence*). The UNFCCC continues to evolve institutions and systems for governance of climate change. [13.2.2.4, 13.3.1, 13.4.1.4, 13.5]

Incentives for international cooperation can interact with other policies (*medium confidence*). Interactions between proposed and existing policies, which may be counterproductive, inconsequential, or beneficial, are difficult to predict, and have been understudied in the literature [13.2, 13.13, 15.7.4]. The game-theoretic literature on climate change agreements finds that self-enforcing agreements engage and maintain participation and compliance. Self-enforcement can be derived from national benefits due to direct climate benefits, co-benefits of mitigation on other national objectives, technology transfer, and climate finance. [13.3.2]

Decreasing uncertainty concerning the costs and benefits of mitigation can reduce the willingness of states to make commitments in forums of international cooperation (*medium confidence*). In some cases, the reduction of uncertainty concerning the costs and benefits of mitigation can make international agreements less effective by creating a disincentive for states to participate [13.3.3, 2.6.4.1]. A second dimension of uncertainty, that concerning whether the policies states implement will in fact achieve desired outcomes, can lessen the willingness of states to agree to commitments regarding those outcomes [2.6.3].

International cooperation can stimulate public and private investment and the adoption of economic incentives and direct regulations that promote technological innovation (*medium confidence*). Technology policy can help lower mitigation costs, thereby increasing incentives for participation and compliance with international cooperative efforts, particularly in the long run. Equity issues can be affected by domestic intellectual property rights regimes, which can alter the rate of both technology transfer and the development of new technologies. [13.3, 13.9]

In the absence of—or as a complement to—a binding, international agreement on climate change, policy linkages between and among existing and nascent international, regional, national, and sub-national climate policies offer potential climate change mitigation and adaptation benefits (*medium confidence*). Direct and indirect linkages between and among sub-national, national, and regional carbon markets are being pursued to improve market efficiency. Linkage between carbon markets can be stimulated by competition between and among public and private governance regimes, accountability measures, and the desire to learn from policy experiments. Yet integrating climate policies raises a number of concerns about the performance of a system of linked legal rules and economic activities. [13.3.1, 13.5.3, 13.13.2.3] Prominent examples of linkages are among national and regional climate initiatives (e.g., planned linkage between the EU ETS and the Australian Emission Trading Scheme, international offsets planned for recognition by a number of jurisdictions), and national and regional climate initiatives with the Kyoto Protocol (e.g., the EU ETS is linked to international carbon markets through the project-based Kyoto Mechanisms) [13.6, 13.7, Figure 13.4, 14.4.2].

International trade can promote or discourage international cooperation on climate change (*high confidence*). Developing constructive relationships between international trade and climate agreements involves considering how existing trade policies and rules can be modified to be more climate-friendly; whether border adjustment measures or other trade measures can be effective in meeting the goals of international climate policy, including participation in and compliance with climate agreements; or whether the UNFCCC, World Trade Organization (WTO), a hybrid of the two, or a new institution is the best forum for a trade-and-climate architecture. [13.8]

The Montreal Protocol, aimed at protecting the stratospheric ozone layer, achieved reductions in global GHG emissions (*very high confidence*). The Montreal Protocol set limits on emissions of ozone-depleting gases that are also potent GHGs, such as chlorofluorocarbons (CFCs) and hydrochlorofluorocarbons (HCFCs). Substitutes for those ozone-depleting gases (such as hydrofluorocarbons (HFCs), which are not ozone-depleting) may also be potent GHGs. Lessons learned from the Montreal Protocol, for example about the effect of financial and technological transfers on broadening participation in an international environmental agreement, could be of value to the design of future international climate change agreements (see Table TS.10). [13.3.3, 13.3.4, 13.13.1.4]

The Kyoto Protocol was the first binding step toward implementing the principles and goals provided by the UNFCCC, but it has had limited effects on global GHG emissions because some countries did not ratify the Protocol, some Parties did not meet their commitments, and its commitments applied to only a portion of the global economy (*medium evidence, low agreement*). The Parties collectively surpassed their collective emission reduction target in the first commitment period, but the Protocol credited emissions reductions that would have occurred even in its absence. The Kyoto Protocol does not directly influence the emissions of non-Annex I countries, which have grown rapidly over the past decade. [5.2, 13.13.1.1]

The flexible mechanisms under the Protocol have cost-saving potential, but their environmental effectiveness is less clear (*medium confidence*). The CDM, one of the Protocol's flexible mechanisms, created a market for GHG emissions offsets from developing countries, generating credits equivalent to nearly 1.4 GtCO$_2$eq as of October 2013. The CDM's environmental effectiveness has been mixed due to concerns about the limited additionality of projects, the validity of baselines, the possibility of emissions leakage, and recent credit price decreases. Its distributional impact has been unequal due to the concentration of projects in a limited number of countries. The Protocol's other flexible mechanisms, Joint Implementation (JI) and International Emissions Trading (IET), have been undertaken both by governments and private market participants, but have raised concerns related to government sales of emission units. (Table TS.10) [13.7.2, 13.13.1.2, 14.3.7.1]

Recent UNFCCC negotiations have sought to include more ambitious contributions from the countries with commitments under the Kyoto Protocol, mitigation contributions from a broader set of countries, and new finance and technology mechanisms.

Under the 2010 Cancún Agreement, developed countries formalized voluntary pledges of quantified, economy-wide GHG emission reduction targets and some developing countries formalized voluntary pledges to mitigation actions. The distributional impact of the agreement will depend in part on the magnitude and sources of financing, although the scientific literature on this point is limited, because financing mechanisms are evolving more rapidly than respective scientific assessments (*limited evidence, low agreement*). Under the 2011 Durban Platform for Enhanced Action, delegates agreed to craft a future legal regime that would be 'applicable to all Parties [...] under the Convention' and would include substantial new financial support and technology arrangements to benefit developing countries, but the delegates did not specify means for achieving those ends. [13.5.1.1, 13.13.1.3, 16.2.1]

TS.4.5 Investment and finance

A transformation to a low-carbon economy implies new patterns of investment. A limited number of studies have examined the investment needs for different mitigation scenarios. Information is largely limited to energy use with global total annual investment in the energy sector at about 1200 billion USD. Mitigation scenarios that reach atmospheric CO$_2$eq concentrations in the range from 430 to 530 ppm CO$_2$eq by 2100 (without overshoot) show substantial shifts in annual investment flows during the period 2010–2029 if compared to baseline scenarios (Figure TS.39): annual investment in the existing technologies associated with the energy supply sector (e.g., conventional fossil fuelled power plants and fossil fuel extraction) would decline by 30 (2 to 166) billion USD per year (median: −20 % compared to 2010) (*limited evidence, medium agreement*). Investment in low-emissions generation technologies (renewables, nuclear, and power plants with CCS) would increase by 147 (31 to 360) billion USD per year (median: +100 % compared to 2010) during the same period (*limited evidence, medium agreement*) in combination with an increase by 336 (1 to 641) billion USD in energy efficiency investments in the building, transport and industry sectors (*limited evidence, medium agreement*). Higher energy efficiency and the shift to low-emission generation technologies contribute to a reduction in the demand for fossil fuels, thus causing a decline in investment in fossil fuel extraction, transformation and transportation. Scenarios suggest that average annual reduction of investment in fossil fuel extraction in 2010–2029 would be 116 (−8 to 369) billion USD (*limited evidence, medium agreement*). Such spill-over effects could yield adverse effects on the revenues of countries that export fossil fuels. Mitigation scenarios also reduce deforestation against current deforestation trends by 50 % reduction with an investment of 21 to 35 billion USD per year (*low confidence*). [16.2.2]

Estimates of total climate finance range from 343 to 385 billion USD per year between 2010 and 2012 (*medium confidence*). The range is based on 2010, 2011, and 2012 data. Climate finance was almost evenly invested in developed and developing countries. Around 95 % of the total was invested in mitigation (*medium confidence*). The figures reflect the total financial flow for the underlying investments, *not the incremental investment*, i.e., the portion attributed to the mitigation/adaptation cost increment (see Box TS.14). In general, quantitative data on climate finance are limited, relate to different concepts, and are incomplete. [16.2.1.1]

Depending on definitions and approaches, climate finance flows to developing countries are estimated to range from 39 to 120 billion USD per year during the period 2009 to 2012 (*medium confidence*). The range covers public and private flows for mitigation and adaptation. Public climate finance was 35 to 49 billion USD (2011/2012 USD) (*medium confidence*). Most public climate finance provided to developing countries flows through bilateral and multilateral institutions usually as concessional loans and grants. Under the UNFCCC, climate finance is funding provided to developing countries by Annex II Parties and averaged nearly 10 billion USD per year from

2005 to 2010 (*medium confidence*). Between 2010 and 2012, the 'fast start finance' provided by some developed countries amounted to over 10 billion USD per year (*medium confidence*). Estimates of international private climate finance flowing to developing countries range from 10 to 72 billion USD (2009/2010 USD) per year, including foreign direct investment as equity and loans in the range of 10 to 37 billion USD (2010 USD and 2008 USD) per year over the period of 2008–2011 (*medium confidence*). Figure TS.40 provides an overview of climate finance, outlining sources and managers of capital, financial instruments, project owners, and projects. [16.2.1.1]

Within appropriate enabling environments, the private sector, along with the public sector, can play an important role in financing mitigation. The private sector contribution to total climate finance is estimated at an average of 267 billion USD (74%) per year in the period 2010 to 2011 and at 224 billion USD (62%) per year in the

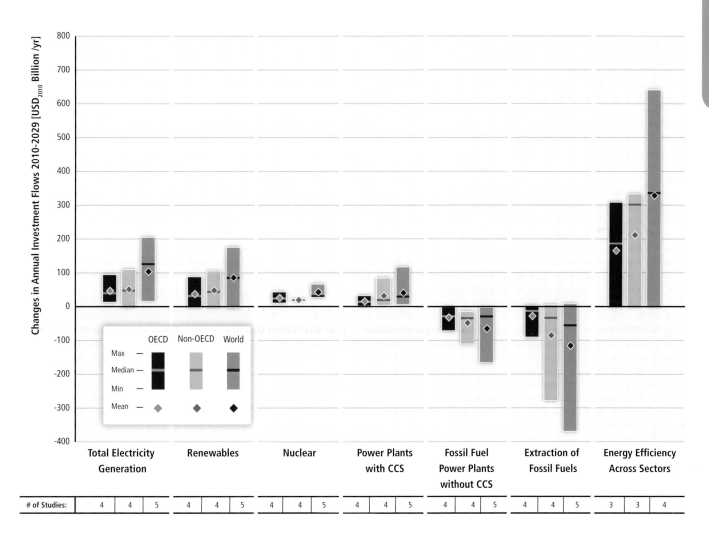

Figure TS.39 | Change of average annual investment flows in mitigation scenarios (2010–2029). Investment changes are calculated by a limited number of model studies and model comparisons for mitigation scenarios that reach concentrations within the range of 430–530 ppm CO_2eq by 2100 compared to respective average baseline investments. The vertical bars indicate the range between minimum and maximum estimate of investment changes; the horizontal bar indicates the median of model results. Proximity to this median value does not imply higher likelihood because of the different degree of aggregation of model results, low number of studies available and different assumptions in the different studies considered. The numbers in the bottom row show the total number of studies assessed. [Figure 16.3]

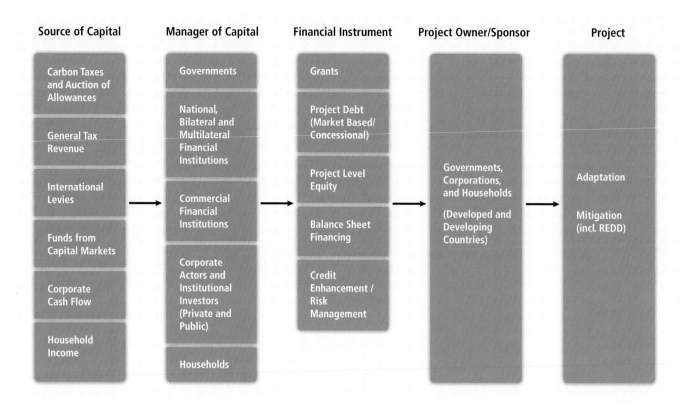

Source of Capital	Manager of Capital	Financial Instrument	Project Owner/Sponsor	Project
Carbon Taxes and Auction of Allowances	Governments	Grants	Governments, Corporations, and Households (Developed and Developing Countries)	Adaptation Mitigation (incl. REDD)
General Tax Revenue	National, Bilateral and Multilateral Financial Institutions	Project Debt (Market Based/Concessional)		
International Levies	Commercial Financial Institutions	Project Level Equity		
Funds from Capital Markets		Balance Sheet Financing		
Corporate Cash Flow	Corporate Actors and Institutional Investors (Private and Public)	Credit Enhancement / Risk Management		
Household Income	Households			

Figure TS.40 | Types of climate finance flows. 'Capital' includes all relevant financial flows. The size of the boxes is not related to the magnitude of the financial flow. [Figure 16.1]

Box TS.14 | There are no agreed definitions of 'climate investment' and 'climate finance'

'*Total climate finance*' includes all financial flows whose expected effect is to reduce net GHG emissions and/or to enhance resilience to the impacts of climate variability and the projected climate change. This covers private and public funds, domestic and international flows, expenditures for mitigation and adaptation, and adaptation to current climate variability as well as future climate change. It covers the full value of the financial flow rather than the share associated with the climate change benefit. The share associated with the climate change benefit is the incremental cost. The '*total climate finance flowing to developing countries*' is the amount of the total climate finance invested in developing countries that comes from developed countries. This covers private and public funds for mitigation and adaptation. '*Public climate finance provided to developing countries*' is the finance provided by developed countries' governments and bilateral institutions as well as multilateral institutions for mitigation and adaptation activities in developing countries. '*Private climate finance flowing to developing countries*' is finance and investment by private actors in/from developed countries for mitigation and adaptation activities in developing countries. Under the UNFCCC, climate finance is not well-defined. Annex II Parties provide and mobilize funding for climate-related activities in developing countries.

The '*incremental investment*' is the extra capital required for the initial investment for a mitigation or adaptation project in comparison to a reference project. Incremental investment for mitigation and adaptation projects is not regularly estimated and reported, but estimates are available from models. The '*incremental cost*' reflects the cost of capital of the incremental investment and the change of operating and maintenance costs for a mitigation or adaptation project in comparison to a reference project. It can be calculated as the difference of the net present values of the two projects. Many mitigation measures have higher investment costs and lower operating and maintenance costs than the measures displaced so incremental cost tends to be lower than the incremental investment. Values depend on the incremental investment as well as projected operating costs, including fossil fuel prices, and the discount rate. The '*macroeconomic cost of mitigation policy*' is the reduction of aggregate consumption or GDP induced by the reallocation of investments and expenditures induced by climate policy (see Box TS.9). These costs do not account for the benefit of reducing anthropogenic climate change and should thus be assessed against the economic benefit of avoided climate change impacts. [16.1]

period 2011 to 2012 (*limited evidence, medium agreement*) [16.2.1]. In a range of countries, a large share of private sector climate investment relies on low-interest and long-term loans as well as risk guarantees provided by public sector institutions to cover the incremental costs and risks of many mitigation investments. The quality of a country's enabling environment—including the effectiveness of its institutions, regulations and guidelines regarding the private sector, security of property rights, credibility of policies, and other factors—has a substantial impact on whether private firms invest in new technologies and infrastructure [16.3]. By the end of 2012, the 20 largest emitting developed and developing countries with lower risk country grades for private sector investments produced 70 % of global energy related CO_2 emissions (*low confidence*). This makes them attractive for international private sector investment in low-carbon technologies. In many other countries, including most least-developed countries, low-carbon investment will often have to rely mainly on domestic sources or international public finance. [16.4.2]

A main barrier to the deployment of low-carbon technologies is a low risk-adjusted rate of return on investment vis-à-vis high-carbon alternatives (*high confidence*). Public policies and support instruments can address this either by altering the average rates of return for different investment options, or by creating mechanisms to lessen the risks that private investors face [15.12, 16.3]. Carbon pricing mechanisms (carbon taxes, cap-and-trade systems), as well as renewable energy premiums, FITs, RPSs, investment grants, soft loans and credit insurance can move risk-return profiles into the required direction [16.4]. For some instruments, the presence of substantial uncertainty about their future levels (e.g., the future size of a carbon tax relative to differences in investment and operating costs) can lead to a lessening of the effectiveness and/or efficiency of the instrument. Instruments that create a fixed or immediate incentive to invest in low-emission technologies, such as investment grants, soft loans, or FITs, do not appear to suffer from this problem. [2.6.5]

Annex

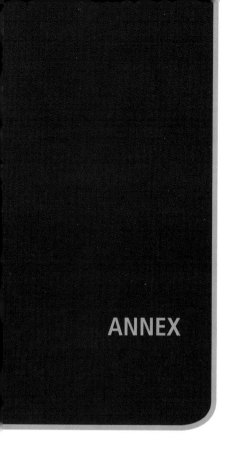

ANNEX

Glossary, Acronyms and Chemical Symbols

Glossary Editors:

Julian M. Allwood (UK), Valentina Bosetti (Italy), Navroz K. Dubash (India), Luis Gómez-Echeverri (Austria/Colombia), Christoph von Stechow (Germany)

Glossary Contributors:

Marcio D'Agosto (Brazil), Giovanno Baiocchi (UK/Italy), John Barrett (UK), John Broome (UK), Steffen Brunner (Germany), Micheline Cariño Olvera (Mexico), Harry Clark (New Zealand), Leon Clarke (USA), Heleen C. de Coninck (Netherlands), Esteve Corbera (Spain), Felix Creutzig (Germany), Gian Carlo Delgado (Mexico), Manfred Fischedick (Germany), Marc Fleurbaey (France/USA), Don Fullerton (USA), Richard Harper (Australia), Edgar Hertwich (Austria/Norway), Damon Honnery (Australia), Michael Jakob (Germany), Charles Kolstad (USA), Elmar Kriegler (Germany), Howard Kunreuther (USA), Andreas Löschel (Germany), Oswaldo Lucon (Brazil), Axel Michaelowa (Germany/Switzerland), Jan C. Minx (Germany), Luis Mundaca (Chile/Sweden), Jin Murakami (Japan/China), Jos G.J. Olivier (Netherlands), Michael Rauscher (Germany), Keywan Riahi (Austria), H.-Holger Rogner (Germany), Steffen Schlömer (Germany), Ralph Sims (New Zealand), Pete Smith (UK), David I. Stern (Australia), Neil Strachan (UK), Kevin Urama (Nigeria/UK/Kenya), Diana Ürge-Vorsatz (Hungary), David G. Victor (USA), Elke Weber (USA), Jonathan Wiener (USA), Mitsutsune Yamaguchi (Japan), Azni Zain Ahmed (Malaysia)

This annex should be cited as:

Allwood J.M., V. Bosetti, N.K. Dubash, L. Gómez-Echeverri, and C. von Stechow, 2014: Glossary. In: *Climate Change 2014: Mitigation of Climate Change. Contribution of Working Group III to the Fifth Assessment Report of the Intergovernmental Panel on Climate Change* [Edenhofer, O., R. Pichs-Madruga, Y. Sokona, E. Farahani, S. Kadner, K. Seyboth, A. Adler, I. Baum, S. Brunner, P. Eickemeier, B. Kriemann, J. Savolainen, S. Schlömer, C. von Stechow, T. Zwickel and J.C. Minx (eds.)]. Cambridge University Press, Cambridge, United Kingdom and New York, NY, USA.

Contents

Glossary

This glossary defines some specific terms as the Lead Authors intend them to be interpreted in the context of this report. Glossary **entries** (highlighted in bold) are by preference subjects; a main entry can contain **_subentries_**, in bold and italic, for example, **_Primary Energy_** is defined under the entry **Energy**. Blue, italicized _words_ indicate that the term is defined in the Glossary. The glossary is followed by a list of acronyms and chemical symbols. Please refer to Annex II for standard units, prefixes, and unit conversion (Section A.II.1) and for regions and country groupings (Section A.II.2).

Abrupt climate change: A large-scale change in the _climate system_ that takes place over a few decades or less, persists (or is anticipated to persist) for at least a few decades, and causes substantial disruptions in human and natural systems. See also _Climate threshold_.

Adaptability: See _Adaptive capacity_.

Adaptation: The process of adjustment to actual or expected _climate_ and its effects. In human systems, adaptation seeks to moderate or avoid harm or exploit beneficial opportunities. In some natural systems, human intervention may facilitate adjustment to expected _climate_ and its effects.[1]

Adaptation Fund: A Fund established under the _Kyoto Protocol_ in 2001 and officially launched in 2007. The Fund finances _adaptation_ projects and programmes in _developing countries_ that are Parties to the _Kyoto Protocol_. Financing comes mainly from sales of _Certified Emissions Reductions (CERs)_ and a share of proceeds amounting to 2% of the value of CERs issued each year for _Clean Development Mechanism (CDM)_ projects. The Adaptation Fund can also receive funds from government, private sector, and individuals.

Adaptive capacity: The ability of systems, _institutions_, humans, and other organisms to adjust to potential damage, to take advantage of opportunities, or to respond to consequences.[2]

Additionality: _Mitigation_ projects (e.g., under the _Kyoto Mechanisms_), _mitigation policies_, or _climate finance_ are additional if they go beyond a _business-as-usual_ level, or _baseline_. Additionality is required to guarantee the environmental integrity of project-based offset mechanisms, but difficult to establish in practice due to the counterfactual nature of the _baseline_.

Adverse side-effects: The negative effects that a _policy_ or _measure_ aimed at one objective might have on other objectives, without yet evaluating the net effect on overall social welfare. Adverse side-effects are often subject to _uncertainty_ and depend on, among others, local circumstances and implementation practices. See also _Co-benefits_, _Risk_, and _Risk tradeoff_.

Aerosol: A suspension of airborne solid or liquid particles, with a typical size between a few nanometres and 10 μm that reside in the _atmosphere_ for at least several hours. For convenience the term aerosol, which includes both the particles and the suspending gas, is often used in this report in its plural form to mean aerosol _particles_. Aerosols may be of either natural or anthropogenic origin. Aerosols may influence _climate_ in several ways: directly through scattering and absorbing radiation, and indirectly by acting as cloud condensation nuclei or ice nuclei, modifying the optical properties and lifetime of clouds. Atmospheric aerosols, whether natural or anthropogenic, originate from two different pathways: emissions of primary _particulate matter (PM)_, and formation of secondary _PM_ from gaseous _precursors_. The bulk of aerosols are of natural origin. Some scientists use group labels that refer to the chemical composition, namely: sea salt, organic carbon, _black carbon (BC)_, mineral species (mainly desert dust), sulphate, nitrate, and ammonium. These labels are, however, imperfect as aerosols combine particles to create complex mixtures. See also _Short-lived climate pollutants (SLCPs)_.

Afforestation: Planting of new _forests_ on lands that historically have not contained _forests_. Afforestation projects are eligible under a number of schemes including, among others, _Joint Implementation (JI)_ and the _Clean Development Mechanism (CDM)_ under the _Kyoto Protocol_ for which particular criteria apply (e.g., proof must be given that the land was not forested for at least 50 years or converted to alternative uses before 31 December 1989).

For a discussion of the term _forest_ and related terms such as afforestation, _reforestation_ and _deforestation_, see the IPCC Special Report on Land Use, Land-Use Change and Forestry (IPCC, 2000). See also the report on Definitions and Methodological Options to Inventory Emissions from Direct Human-induced Degradation of Forests and Devegetation of Other Vegetation Types (IPCC, 2003).

Agreement: In this report, the degree of agreement is the level of concurrence in the literature on a particular finding as assessed by the authors. See also _Evidence_, _Confidence_, _Likelihood_, and _Uncertainty_.

Agricultural emissions: See _Emissions_.

Agriculture, Forestry and Other Land Use (AFOLU): Agriculture, Forestry and Other Land Use plays a central role for _food security_ and _sustainable development (SD)_. The main _mitigation_ options within AFOLU involve one or more of three strategies: _prevention_ of emissions to the _atmosphere_ by conserving existing _carbon pools_ in soils or vegetation or by reducing emissions of _methane (CH₄)_ and _nitrous_

---

[1]   Reflecting progress in science, this glossary entry differs in breadth and focus from the entry used in the Fourth Assessment Report and other IPCC reports.

[2]   This glossary entry builds from definitions used in previous IPCC reports and the Millennium Ecosystem Assessment (MEA, 2005).

oxide (N₂O); *sequestration*—increasing the size of existing *carbon pools*, and thereby extracting *carbon dioxide (CO₂)* from the *atmosphere*; and *substitution*—substituting biological products for *fossil fuels* or energy-intensive products, thereby reducing CO₂ emissions. Demand-side measures (e.g., by reducing losses and wastes of food, changes in human diet, or changes in wood consumption) may also play a role. FOLU (Forestry and Other Land Use)—also referred to as *LULUCF (Land use, land-use change, and forestry)*—is the subset of AFOLU emissions and removals of *greenhouse gases (GHGs)* resulting from direct human-induced land use, land-use change and forestry activities excluding *agricultural emissions*.

Albedo: The fraction of solar radiation reflected by a surface or object, often expressed as a percentage. Snow-covered surfaces have a high albedo, the albedo of soils ranges from high to low, and vegetation-covered surfaces and oceans have a low albedo. The earth's planetary albedo varies mainly through varying cloudiness, snow, ice, leaf area and land cover changes.

Alliance of Small Island States (AOSIS): The Alliance of Small Island States (AOSIS) is a coalition of small islands and low-lying coastal countries with a membership of 44 states and observers that share and are active in global debates and negotiations on the environment, especially those related to their vulnerability to the adverse effects of *climate change*. Established in 1990, AOSIS acts as an ad-hoc lobby and negotiating voice for small island development states (SIDS) within the United Nations including the *United Nations Framework Convention on Climate Change (UNFCCC)* climate change negotiations.

Ancillary benefits: See *Co-benefits*.

Annex I Parties/countries: The group of countries listed in Annex I to the *United Nations Framework Convention on Climate Change (UNFCCC)*. Under Articles 4.2 (a) and 4.2 (b) of the UNFCCC, Annex I Parties were committed to adopting national *policies* and *measures* with the non-legally binding aim to return their *greenhouse gas (GHG)* emissions to 1990 levels by 2000. The group is largely similar to the *Annex B Parties* to the *Kyoto Protocol* that also adopted emissions reduction targets for 2008–2012. By default, the other countries are referred to as *Non-Annex I Parties*.

Annex II Parties/countries: The group of countries listed in Annex II to the *United Nations Framework Convention on Climate Change (UNFCCC)*. Under Article 4 of the UNFCCC, these countries have a special obligation to provide financial resources to meet the agreed full incremental costs of implementing *measures* mentioned under Article 12, paragraph 1. They are also obliged to provide financial resources, including for the transfer of technology, to meet the agreed incremental costs of implementing *measures* covered by Article 12, paragraph 1 and agreed between *developing country* Parties and international entities referred to in Article 11 of the UNFCCC. This group of countries shall also assist countries that are particularly vulnerable to the adverse effects of *climate change*.

Annex B Parties/countries: The subset of *Annex I Parties* that have accepted *greenhouse gas (GHG)* emission reduction targets for the period 2008–2012 under Article 3 of the *Kyoto Protocol*. By default, the other countries are referred to as *Non-Annex I Parties*.

Anthropogenic emissions: See *Emissions*.

Assigned Amount (AA): Under the *Kyoto Protocol*, the AA is the quantity of *greenhouse gas (GHG)* emissions that an *Annex B country* has agreed to as its *cap* on its emissions in the first five-year commitment period (2008–2012). The AA is the country's total GHG emissions in 1990 multiplied by five (for the five-year commitment period) and by the percentage it agreed to as listed in Annex B of the *Kyoto Protocol* (e.g., 92 % for the EU). See also *Assigned Amount Unit (AAU)*.

Assigned Amount Unit (AAU): An AAU equals 1 tonne (metric ton) of *CO₂-equivalent emissions* calculated using the *Global Warming Potential (GWP)*. See also *Assigned Amount (AA)*.

Atmosphere: The gaseous envelope surrounding the earth, divided into five layers—the *troposphere* which contains half of the earth's atmosphere, the *stratosphere*, the mesosphere, the thermosphere, and the exosphere, which is the outer limit of the atmosphere. The dry atmosphere consists almost entirely of nitrogen (78.1 % volume mixing ratio) and oxygen (20.9 % volume mixing ratio), together with a number of *trace gases*, such as argon (0.93 % volume mixing ratio), helium and radiatively active *greenhouse gases (GHGs)* such as *carbon dioxide (CO₂)* (0.035 % volume mixing ratio) and *ozone (O₃)*. In addition, the atmosphere contains the GHG water vapour (H₂O), whose amounts are highly variable but typically around 1 % volume mixing ratio. The atmosphere also contains clouds and *aerosols*.

Backstop technology: *Models* estimating *mitigation* often use an arbitrary carbon-free technology (often for power generation) that might become available in the future in unlimited supply over the horizon of the *model*. This allows modellers to explore the consequences and importance of a generic solution technology without becoming enmeshed in picking the actual technology. This 'backstop' technology might be a nuclear technology, fossil technology with *Carbon Dioxide Capture and Storage (CCS)*, *solar energy*, or something as yet unimagined. The backstop technology is typically assumed either not to currently exist, or to exist only at higher costs relative to conventional alternatives.

Banking (of Assigned Amount Units): Any transfer of *Assigned Amount Units (AAUs)* from an existing period into a future commitment period. According to the *Kyoto Protocol* [Article 3 (13)], Parties included in Annex I to the *United Nations Framework Convention on Climate Change (UNFCCC)* may save excess AAUs from the first commitment period for compliance with their respective *cap* in subsequent commitment periods (post-2012).

Baseline/reference: The state against which change is measured. In the context of *transformation pathways*, the term 'baseline scenarios' refers to *scenarios* that are based on the assumption that no *mitigation policies* or *measures* will be implemented beyond those that are already in force and/or are legislated or planned to be adopted. Baseline scenarios are not intended to be predictions of the future, but rather counterfactual constructions that can serve to highlight the level of emissions that would occur without further *policy* effort. Typically, baseline scenarios are then compared to *mitigation scenarios* that are constructed to meet different goals for *greenhouse gas (GHG)* emissions, atmospheric concentrations, or temperature change. The term 'baseline scenario' is used interchangeably with 'reference scenario' and 'no policy scenario'. In much of the literature the term is also synonymous with the term 'business-as-usual (BAU) scenario,' although the term 'BAU' has fallen out of favour because the idea of 'business-as-usual' in century-long socioeconomic projections is hard to fathom. See also *Climate scenario*, *Emission scenario*, *Representative concentration pathways (RCPs)*, *Shared socio-economic pathways*, *Socio-economic scenarios*, *SRES scenarios*, and *Stabilization*.

Behaviour: In this report, behaviour refers to human decisions and actions (and the perceptions and judgments on which they are based) that directly or indirectly influence *mitigation* or the effects of potential *climate change* impacts (*adaptation*). Human decisions and actions are relevant at different levels, from international, national, and sub-national actors, to NGO, tribe, or firm-level decision makers, to communities, households, and individual citizens and consumers. See also *Behavioural change* and *Drivers of behaviour*.

Behavioural change: In this report, behavioural change refers to alteration of human decisions and actions in ways that mitigate *climate change* and/or reduce negative consequences of *climate change* impacts. See also *Drivers of behaviour*.

Biochar: *Biomass* stabilization can be an alternative or enhancement to *bioenergy* in a land-based *mitigation* strategy. Heating *biomass* with exclusion of air produces a stable carbon-rich co-product (char). When added to soil a system, char creates a system that has greater abatement potential than typical *bioenergy*. The relative benefit of biochar systems is increased if changes in crop yield and soil emissions of *methane (CH_4)* and *nitrous oxide (N_2O)* are taken into account.

Biochemical oxygen demand (BOD): The amount of dissolved oxygen consumed by micro-organisms (bacteria) in the bio-chemical oxidation of organic and inorganic matter in wastewater. See also *Chemical oxygen demand (COD)*.

Biodiversity: The variability among living organisms from terrestrial, marine, and other *ecosystems*. Biodiversity includes variability at the genetic, species, and *ecosystem* levels.[3]

---

[3] This glossary entry builds from definitions used in the Global Biodiversity Assessment (Heywood, 1995) and the Millennium Ecosystem Assessment (MEA, 2005).

Bioenergy: *Energy* derived from any form of *biomass* such as recently living organisms or their metabolic by-products.

Bioenergy and Carbon Dioxide Capture and Storage (BECCS): The application of *Carbon Dioxide Capture and Storage (CCS)* technology to *bioenergy* conversion processes. Depending on the total life-cycle emissions, including total marginal consequential effects (from *indirect land use change (iLUC)* and other processes), BECCS has the potential for net *carbon dioxide (CO_2)* removal from the *atmosphere*. See also *Sequestration*.

Bioethanol: Ethanol produced from *biomass* (e.g., sugar cane or corn). See also *Biofuel*.

Biofuel: A fuel, generally in liquid form, produced from organic matter or combustible oils produced by living or recently living plants. Examples of biofuel include alcohol (*bioethanol*), black liquor from the paper-manufacturing process, and soybean oil.

First-generation manufactured biofuel: First-generation manufactured biofuel is derived from grains, oilseeds, animal fats, and waste vegetable oils with mature conversion technologies.

Second-generation biofuel: Second-generation biofuel uses non-traditional biochemical and thermochemical conversion processes and feedstock mostly derived from the lignocellulosic fractions of, for example, agricultural and forestry residues, municipal solid waste, etc.

Third-generation biofuel: Third-generation biofuel would be derived from feedstocks such as algae and energy crops by advanced processes still under development.

These second- and third-generation biofuels produced through new processes are also referred to as next-generation or advanced biofuels, or advanced biofuel technologies.

Biomass: The total mass of living organisms in a given area or volume; dead plant material can be included as dead biomass. In the context of this report, biomass includes products, by-products, and waste of biological origin (plants or animal matter), excluding material embedded in geological formations and transformed to *fossil fuels* or peat.

Traditional biomass: Traditional biomass refers to the biomass—fuelwood, charcoal, agricultural residues, and animal dung—used with the so-called traditional technologies such as open fires for cooking, rustic kilns and ovens for small industries. Widely used in *developing countries*, where about 2.6 billion people cook with open wood fires, and hundreds of thousands small-industries. The use of these rustic technologies leads to high pollution levels and, in specific circumstances, to *forest* degradation and *deforestation*. There are many successful initiatives around the world to make traditional biomass burned more efficiently

Annex

and cleanly using efficient cookstoves and kilns. This last use of traditional biomass is sustainable and provides large health and economic benefits to local populations in *developing countries*, particularly in rural and peri-urban areas.

Modern biomass: All biomass used in high efficiency conversion systems.

Biomass burning: Biomass burning is the burning of living and dead vegetation.

Biosphere (terrestrial and marine): The part of the earth system comprising all *ecosystems* and living organisms, in the *atmosphere*, on land (terrestrial biosphere) or in the oceans (marine biosphere), including derived dead organic matter, such as litter, soil organic matter and oceanic detritus.

Black carbon (BC): Operationally defined *aerosol* species based on measurement of light absorption and chemical reactivity and/or thermal stability. It is sometimes referred to as soot. BC is mostly formed by the incomplete combustion of *fossil fuels*, *biofuels*, and *biomass* but it also occurs naturally. It stays in the *atmosphere* only for days or weeks. It is the most strongly light-absorbing component of *particulate matter (PM)* and has a warming effect by absorbing heat into the *atmosphere* and reducing the *albedo* when deposited on ice or snow.

Burden sharing (also referred to as Effort sharing): In the context of *mitigation*, burden sharing refers to sharing the effort of reducing the *sources* or enhancing the *sinks* of *greenhouse gases (GHGs)* from historical or projected levels, usually allocated by some criteria, as well as sharing the cost burden across countries.

Business-as-usual (BAU): See *Baseline/reference*.

Cancún Agreements: A set of decisions adopted at the 16th Session of the *Conference of the Parties (COP)* to the *United Nations Framework Convention on Climate Change (UNFCCC)*, including the following, among others: the newly established *Green Climate Fund (GCF)*, a newly established technology mechanism, a process for advancing discussions on *adaptation*, a formal process for reporting *mitigation* commitments, a goal of limiting *global mean surface temperature* increase to 2 °C, and an agreement on MRV—Measuring, Reporting and Verifying for those countries that receive international support for their *mitigation* efforts.

Cancún Pledges: During 2010, many countries submitted their existing plans for controlling *greenhouse gas (GHG)* emissions to the Climate Change Secretariat and these proposals have now been formally acknowledged under the *United Nations Framework Convention on Climate Change (UNFCCC)*. *Developed countries* presented their plans in the shape of economy-wide targets to reduce emissions, mainly up to 2020, while *developing countries* proposed ways to limit their growth of emissions in the shape of plans of action.

Cap, on emissions: Mandated restraint as an upper limit on emissions within a given period. For example, the *Kyoto Protocol* mandates emissions caps in a scheduled timeframe on the anthropogenic *greenhouse gas (GHG)* emissions released by *Annex B countries*.

Carbon budget: The area under a *greenhouse gas (GHG)* emissions trajectory that satisfies assumptions about limits on cumulative emissions estimated to avoid a certain level of *global mean surface temperature* rise. Carbon budgets may be defined at the global level, national, or sub-national levels.

Carbon credit: See *Emission allowance*.

Carbon cycle: The term used to describe the flow of carbon (in various forms, e.g., as *carbon dioxide*) through the *atmosphere*, ocean, terrestrial and marine *biosphere* and lithosphere. In this report, the reference unit for the global carbon cycle is GtC or $GtCO_2$ (1 GtC corresponds to 3.667 $GtCO_2$). Carbon is the major chemical constituent of most organic matter and is stored in the following major *reservoirs*: organic molecules in the *biosphere*, *carbon dioxide (CO_2)* in the *atmosphere*, organic matter in the soils, in the lithosphere, and in the oceans.

Carbon dioxide (CO_2): A naturally occurring gas, also a by-product of burning *fossil fuels* from fossil carbon deposits, such as oil, gas and coal, of burning *biomass*, of *land use changes (LUC)* and of industrial processes (e.g., cement production). It is the principal anthropogenic *greenhouse gas (GHG)* that affects the earth's radiative balance. It is the reference gas against which other GHGs are measured and therefore has a *Global Warming Potential (GWP)* of 1. See Annex II.9.1 for GWP values for other GHGs.

Carbon Dioxide Capture and Storage (CCS): A process in which a relatively pure stream of *carbon dioxide (CO_2)* from industrial and energy-related *sources* is separated (captured), conditioned, compressed, and transported to a storage location for long-term isolation from the *atmosphere*. See also *Bioenergy and carbon capture and storage (BECCS)*, *CCS-ready*, and *Sequestration*.

Carbon dioxide fertilization: The enhancement of the growth of plants as a result of increased atmospheric *carbon dioxide (CO_2)* concentration.

Carbon Dioxide Removal (CDR): Carbon Dioxide Removal methods refer to a set of techniques that aim to remove *carbon dioxide (CO_2)* directly from the *atmosphere* by either (1) increasing natural *sinks* for carbon or (2) using chemical engineering to remove the CO_2, with the intent of reducing the atmospheric CO_2 concentration. CDR methods involve the ocean, land, and technical systems, including such methods as *iron fertilization*, large-scale *afforestation*, and *direct capture* of CO_2 from the *atmosphere* using engineered chemical means. Some CDR methods fall under the category of *geoengineering*, though this may not be the case for others, with the distinction being based on the magnitude, scale, and impact of the particular CDR activities. The boundary between CDR and *mitigation* is not clear and there could be some overlap between the two given current definitions (IPCC, 2012, p. 2). See also *Solar Radiation Management (SRM)*.

Carbon footprint: Measure of the exclusive total amount of emissions of *carbon dioxide (CO₂)* that is directly and indirectly caused by an activity or is accumulated over the life stages of a product (Wiedmann and Minx, 2008).

Carbon intensity: The amount of emissions of *carbon dioxide (CO₂)* released per unit of another variable such as *gross domestic product (GDP)*, output energy use, or transport.

Carbon leakage: See *Leakage*.

Carbon pool: See *Reservoir*.

Carbon price: The price for avoided or released *carbon dioxide (CO₂)* or CO_2-*equivalent* emissions. This may refer to the rate of a *carbon tax*, or the price of *emission permits*. In many *models* that are used to assess the economic costs of *mitigation*, carbon prices are used as a proxy to represent the level of effort in *mitigation policies*.

Carbon sequestration: See *Sequestration*.

Carbon tax: A levy on the carbon content of *fossil fuels*. Because virtually all of the carbon in *fossil fuels* is ultimately emitted as *carbon dioxide (CO₂)*, a carbon tax is equivalent to an emission tax on CO_2 emissions.

CCS-ready: New large-scale, stationary *carbon dioxide (CO₂)* point *sources* intended to be retrofitted with *Carbon Dioxide Capture and Storage (CCS)* could be designed and located to be 'CCS-ready' by reserving space for the capture installation, designing the unit for optimal performance when capture is added, and siting the plant to enable access to storage locations. See also *Bioenergy and Carbon Dioxide Capture and Storage (BECCS)*.

Certified Emission Reduction Unit (CER): Equal to one metric tonne of CO_2-*equivalent emissions* reduced or of *carbon dioxide (CO₂)* removed from the *atmosphere* through the *Clean Development Mechanism (CDM)* (defined in Article 12 of the *Kyoto Protocol*) project, calculated using *Global Warming Potentials (GWP)*. See also *Emissions Reduction Units (ERU)* and *Emissions trading*.

Chemical oxygen demand (COD): The quantity of oxygen required for the complete oxidation of organic chemical compounds in water; used as a measure of the level of organic pollutants in natural and waste waters. See also *Biochemical oxygen demand (BOD)*.

Chlorofluorocarbons (CFCs): A chlorofluorocarbon is an organic compound that contains chlorine, carbon, hydrogen, and fluorine and is used for refrigeration, air conditioning, packaging, plastic foam, insulation, solvents, or *aerosol* propellants. Because they are not destroyed in the lower *atmosphere*, CFCs drift into the upper *atmosphere* where, given suitable conditions, they break down *ozone (O₃)*. It is one of the *greenhouse gases (GHGs)* covered under the 1987 *Montreal Protocol* as a result of which manufacturing of these gases has been phased out and they are being replaced by other compounds, including *hydrofluorocarbons (HFCs)* which are GHGs covered under the *Kyoto Protocol*.

Clean Development Mechanism (CDM): A mechanism defined under Article 12 of the *Kyoto Protocol* through which investors (governments or companies) from developed (*Annex B*) *countries* may finance *greenhouse gas (GHG)* emission reduction or removal projects in developing (*Non-Annex B*) *countries*, and receive *Certified Emission Reduction Units (CERs)* for doing so. The *CERs* can be credited towards the commitments of the respective *developed countries*. The *CDM* is intended to facilitate the two objectives of promoting *sustainable development (SD)* in *developing countries* and of helping *industrialized countries* to reach their emissions commitments in a *cost-effective* way. See also *Kyoto Mechanisms*.

Climate: Climate in a narrow sense is usually defined as the average weather, or more rigorously, as the statistical description in terms of the mean and variability of relevant quantities over a period of time ranging from months to thousands or millions of years. The classical period for averaging these variables is 30 years, as defined by the World Meteorological Organization. The relevant quantities are most often surface variables such as temperature, precipitation and wind. Climate in a wider sense is the state, including a statistical description, of the *climate system*.

Climate change: Climate change refers to a change in the state of the *climate* that can be identified (e.g., by using statistical tests) by changes in the mean and/or the variability of its properties, and that persists for an extended period, typically decades or longer. Climate change may be due to natural internal processes or external forcings such as modulations of the solar cycles, volcanic eruptions and persistent anthropogenic changes in the composition of the *atmosphere* or in *land use*. Note that the *United Nations Framework Convention on Climate Change (UNFCCC)*, in its Article 1, defines climate change as: 'a change of climate which is attributed directly or indirectly to human activity that alters the composition of the global atmosphere and which is in addition to natural climate variability observed over comparable time periods'. The UNFCCC thus makes a distinction between climate change attributable to human activities altering the atmospheric composition, and climate variability attributable to natural causes. See also *Climate change commitment*.

Climate change commitment: Due to the thermal inertia of the ocean and slow processes in the cryosphere and land surfaces, the *climate* would continue to change even if the atmospheric composition were held fixed at today's values. Past change in atmospheric composition leads to a committed *climate change*, which continues for as long as a radiative imbalance persists and until all components of the *climate system* have adjusted to a new state. The further change in temperature after the composition of the *atmosphere* is held constant is referred to as the constant composition temperature commitment or simply committed warming or warming commitment. Climate change commitment includes other future changes, for example in the hydrological cycle, in extreme weather events, in extreme climate events, and in sea level change. The constant emission commitment is the committed climate change that would result from keeping *anthropogenic emissions* constant and the zero emission commitment is the climate change commitment when emissions are set to zero. See also *Climate change*.

Climate (change) feedback: An interaction in which a perturbation in one *climate* quantity causes a change in a second, and the change in the second quantity ultimately leads to an additional change in the first. A negative feedback is one in which the initial perturbation is weakened by the changes it causes; a positive feedback is one in which the initial perturbation is enhanced. In this Assessment Report, a somewhat narrower definition is often used in which the climate quantity that is perturbed is the *global mean surface temperature*, which in turn causes changes in the global radiation budget. In either case, the initial perturbation can either be externally forced or arise as part of internal variability.

Climate engineering: See *Geoengineering*.

Climate finance: There is no agreed definition of climate finance. The term 'climate finance' is applied both to the financial resources devoted to addressing *climate change* globally and to financial flows to *developing countries* to assist them in addressing *climate change*. The literature includes several concepts in these categories, among which the most commonly used include:

Incremental costs: The cost of capital of the *incremental investment* and the change of operating and maintenance costs for a *mitigation* or *adaptation* project in comparison to a reference project. It can be calculated as the difference of the net present values of the two projects. See also *Additionality*.

Incremental investment: The extra capital required for the initial investment for a *mitigation* or *adaptation* project in comparison to a reference project. See also *Additionality*.

Total climate finance: All financial flows whose expected effect is to reduce net *greenhouse gas (GHG)* emissions and/or to enhance *resilience* to the impacts of *climate variability* and the projected *climate change*. This covers private and public funds, domestic and international flows, expenditures for *mitigation* and *adaptation* to current *climate variability* as well as future *climate change*.

Total climate finance flowing to developing countries: The amount of the *total climate finance* invested in *developing countries* that comes from *developed countries*. This covers private and public funds.

Private climate finance flowing to developing countries: Finance and investment by private actors in/from *developed countries* for *mitigation* and *adaptation* activities in *developing countries*.

Public climate finance flowing to developing countries: Finance provided by *developed countries'* governments and bilateral institutions as well as by multilateral institutions for *mitigation* and *adaptation* activities in *developing countries*. Most of the funds provided are concessional loans and grants.

Climate model (spectrum or hierarchy): A numerical representation of the *climate system* based on the physical, chemical and biological properties of its components, their interactions and feedback processes, and accounting for some of its known properties. The climate system can be represented by models of varying complexity, that is, for any one component or combination of components a spectrum or hierarchy of models can be identified, differing in such aspects as the number of spatial dimensions, the extent to which physical, chemical or biological processes are explicitly represented, or the level at which empirical parametrizations are involved. Coupled Atmosphere-Ocean *General Circulation Models* (AOGCMs) provide a representation of the *climate system* that is near or at the most comprehensive end of the spectrum currently available. There is an evolution towards more complex models with interactive chemistry and biology. Climate models are applied as a research tool to study and simulate the *climate*, and for operational purposes, including monthly, seasonal and interannual *climate predictions*.

Climate prediction: A climate prediction or climate forecast is the result of an attempt to produce (starting from a particular state of the *climate system*) an estimate of the actual evolution of the climate in the future, for example, at seasonal, interannual, or decadal time scales. Because the future evolution of the *climate system* may be highly sensitive to initial conditions, such predictions are usually probabilistic in nature. See also *Climate projection*, and *Climate scenario*.

Climate projection: A climate projection is the simulated response of the *climate system* to a scenario of future *emission* or concentration of *greenhouse gases (GHGs)* and *aerosols*, generally derived using *climate models*. Climate projections are distinguished from *climate predictions* by their dependence on the emission/concentration/*radiative forcing* scenario used, which is in turn based on assumptions concerning, for example, future socioeconomic and technological developments that may or may not be realized. See also *Climate scenario*.

Climate scenario: A plausible and often simplified representation of the future *climate*, based on an internally consistent set of climatological relationships that has been constructed for explicit use in investigating the potential consequences of anthropogenic *climate* *change*, often serving as input to impact models. *Climate projections* often serve as the raw material for constructing climate scenarios, but climate scenarios usually require additional information such as the observed current *climate*. See also *Baseline/reference*, *Emission scenario*, *Mitigation scenario*, *Representative concentration pathways (RCPs)*, *Scenario*, *Shared socio-economic pathways*, *Socio-economic scenario*, *SRES scenarios*, *Stabilization*, and *Transformation pathway*.

Climate sensitivity: In IPCC reports, equilibrium climate sensitivity (units: °C) refers to the equilibrium (steady state) change in the annual *global mean surface temperature* following a doubling of the atmospheric CO_2-*equivalent concentration*. Owing to computational constraints, the equilibrium climate sensitivity in a *climate model* is sometimes estimated by running an atmospheric *general circulation model (GCM)* coupled to a mixed-layer ocean model, because equilibrium climate sensitivity is largely determined by atmospheric processes. Efficient models can be run to equilibrium with a dynamic ocean. The climate sensitivity parameter (units: °C (W m^{-2})$^{-1}$) refers to the equilibrium change in the annual *global mean surface temperature* following a unit change in *radiative forcing*.

The effective climate sensitivity (units: °C) is an estimate of the *global mean surface temperature* response to doubled *carbon dioxide (CO_2)* concentration that is evaluated from model output or observations for evolving non-equilibrium conditions. It is a measure of the strengths of the *climate feedbacks* at a particular time and may vary with forcing history and *climate* state, and therefore may differ from equilibrium climate sensitivity.

The transient climate response (units: °C) is the change in the *global mean surface temperature*, averaged over a 20-year period, centred at the time of atmospheric CO_2 doubling, in a *climate model* simulation in which CO_2 increases at 1 % yr^{-1}. It is a measure of the strength and rapidity of the surface temperature response to *greenhouse gas (GHG)* forcing.

Climate system: The climate system is the highly complex system consisting of five major components: the *atmosphere*, the hydrosphere, the cryosphere, the lithosphere and the *biosphere*, and the interactions between them. The climate system evolves in time under the influence of its own internal dynamics and because of external forcings such as volcanic eruptions, solar variations and anthropogenic forcings such as the changing composition of the *atmosphere* and *land use change (LUC)*.

Climate threshold: A limit within the *climate system* that, when crossed, induces a non-linear response to a given forcing. See also *Abrupt climate change*.

Climate variability: Climate variability refers to variations in the mean state and other statistics (such as standard deviations, the occurrence of extremes, etc.) of the *climate* on all spatial and temporal scales beyond that of individual weather events. Variability may be due to natural internal processes within the *climate system* (internal variability), or to variations in natural or anthropogenic external forcing (external variability). See also *Climate change*.

CO_2-equivalent concentration: The concentration of *carbon dioxide (CO_2)* that would cause the same *radiative forcing* as a given mixture of CO_2 and other forcing components. Those values may consider only *greenhouse gases (GHGs)*, or a combination of GHGs, *aerosols*, and surface *albedo* changes. CO_2-equivalent concentration is a metric for comparing *radiative forcing* of a mix of different forcing components at a particular time but does not imply equivalence of the corresponding *climate change* responses nor future forcing. There is generally no connection between *CO_2-equivalent emissions* and resulting CO_2-equivalent concentrations.

CO_2-equivalent emission: The amount of *carbon dioxide (CO_2)* emission that would cause the same integrated *radiative forcing*, over a given time horizon, as an emitted amount of a *greenhouse gas (GHG)* or a mixture of GHGs. The CO_2-equivalent emission is obtained by multiplying the emission of a GHG by its *Global Warming Potential (GWP)* for the given time horizon (see Annex II.9.1 and WGI AR5 Table 8.A.1 for GWP values of the different GHGs). For a mix of GHGs it is obtained by summing the CO_2-equivalent emissions of each gas. CO_2-equivalent emission is a common scale for comparing emissions of different GHGs but does not imply equivalence of the corresponding *climate change* responses. See also *CO_2-equivalent concentration*.

Co-benefits: The positive effects that a *policy* or *measure* aimed at one objective might have on other objectives, without yet evaluating the net effect on overall social welfare. Co-benefits are often subject to *uncertainty* and depend on, among others, local circumstances and implementation practices. Co-benefits are often referred to as ancillary benefits. See also *Adverse side-effect*, *Risk*, and *Risk tradeoff*.

Cogeneration: Cogeneration (also referred to as combined heat and power, or CHP) is the simultaneous generation and useful application of electricity and useful heat.

Combined-cycle gas turbine: A power plant that combines two processes for generating electricity. First, fuel combustion drives a gas turbine. Second, exhaust gases from the turbine are used to heat water to drive a steam turbine.

Combined heat and power (CHP): See *Cogeneration*.

Computable General Equilibrium (CGE) Model: See *Models*.

Conference of the Parties (COP): The supreme body of the *United Nations Framework Convention on Climate Change (UNFCCC)*, comprising countries with a right to vote that have ratified or acceded to the convention. See also *Meeting of the Parties (CMP)*.

Confidence: The validity of a finding based on the type, amount, quality, and consistency of *evidence* (e.g., mechanistic understanding, theory, data, *models*, expert judgment) and on the degree of *agreement*. In this report, confidence is expressed qualitatively (Mastrandrea et al., 2010). See WGI AR5 Figure 1.11 for the levels of confidence and WGI AR5 Table 1.2 for the list of *likelihood* qualifiers. See also *Uncertainty*.

Consumption-based accounting: Consumption-based accounting provides a measure of emissions released to the *atmosphere* in order to generate the goods and services consumed by a certain entity (e.g., person, firm, country, or region). See also *Production-based accounting*.

Contingent valuation method: An approach to quantitatively assess values assigned by people in monetary (willingness to pay) and non-monetary (willingness to contribute with time, resources etc.) terms. It is a direct method to estimate economic values for *ecosystem* and environmental services. In a survey, people are asked their willingness to pay/contribute for access to, or their willingness to accept compensation for removal of, a specific environmental service, based on a hypothetical *scenario* and description of the environmental service.

Conventional fuels: See *Fossil fuels*.

Copenhagen Accord: The political (as opposed to legal) agreement that emerged at the 15th Session of the *Conference of the Parties (COP)* at which delegates 'agreed to take note' due to a lack of consensus that an agreement would require. Some of the key elements include: recognition of the importance of the scientific view on the need to limit the increase in *global mean surface temperature* to 2°C; commitment by *Annex I Parties* to implement economy-wide emissions targets by 2020 and *non-Annex I Parties* to implement *mitigation* actions; agreement to have emission targets of *Annex I Parties* and their delivery of finance for *developing countries* subject to Measurement, Reporting and Verification (MRV) and actions by *developing countries* to be subject to domestic MRV; calls for scaled up financing including a fast track financing of USD 30 billion and USD 100 billion by 2020; the establishment of a new *Green Climate Fund (GCF)*; and the establishment of a new technology mechanism. Some of these elements were later adopted in the *Cancún Agreements*.

Cost-benefit analysis (CBA): Monetary measurement of all negative and positive impacts associated with a given action. Costs and benefits are compared in terms of their difference and/or ratio as an indicator of how a given investment or other *policy* effort pays off seen from the society's point of view.

Cost of conserved energy (CCE): See *Levelized cost of conserved energy (LCCE)*.

Cost-effectiveness: A *policy* is more cost-effective if it achieves a goal, such as a given pollution abatement level, at lower cost. A critical condition for cost-effectiveness is that marginal abatement costs be equal among obliged parties. *Integrated models* approximate cost-effective solutions, unless they are specifically constrained to behave otherwise. Cost-effective *mitigation scenarios* are those based on a stylized implementation approach in which a single price on *carbon dioxide (CO_2)* and other *greenhouse gases (GHGs)* is applied across the globe in every sector of every country and that rises over time in a way that achieves lowest global discounted costs.

Cost-effectiveness analysis (CEA): A tool based on constrained optimization for comparing *policies* designed to meet a prespecified target.

Crediting period, Clean Development Mechanism (CDM): The time during which a project activity is able to generate *Certified Emission Reduction Units (CERs)*. Under certain conditions, the crediting period can be renewed up to two times.

Cropland management: The system of practices on land on which agricultural crops are grown and on land that is set aside or temporarily not being used for crop production (UNFCCC, 2002).

Decarbonization: The process by which countries or other entities aim to achieve a low-carbon economy, or by which individuals aim to reduce their carbon consumption.

Decomposition approach: Decomposition methods disaggregate the total amount of historical changes of a policy variable into contributions made by its various determinants.

Deforestation: Conversion of *forest* to non-forest is one of the major *sources* of *greenhouse gas (GHG)* emissions. Under Article 3.3 of the *Kyoto Protocol*, "the net changes in greenhouse gas emissions by sources and removals by sinks resulting from direct human-induced land-use change and forestry activities, limited to afforestation, reforestation and deforestation since 1990, measured as verifiable changes in carbon stocks in each commitment period, shall be sued to meet the commitments under this Article of each Party included in Annex I". Reducing emissions from deforestation is not eligible for *Joint Implementation (JI)* or *Clean Development Mechanism (CDM)* projects but has been introduced in the program of work under *REDD (Reducing Emissions from Deforestation and Forest Degradation)* under the *United Nations Framework Convention on Climate Change (UNFCCC)*.

For a discussion of the term *forest* and related terms such as *afforestation*, *reforestation*, and deforestation see the IPCC Special Report on Land Use, Land-Use Change and Forestry (IPCC, 2000). See also the report on Definitions and Methodological Options to Inventory Emissions from Direct Human-induced Degradation of Forests and Devegetation of Other Vegetation Types (IPCC, 2003).

Dematerialization: The ambition to reduce the total material inputs required to deliver a final service.

Descriptive analysis: Descriptive (also termed positive) approaches to analysis focus on how the world works or actors behave, not how they should behave in some idealized world. See also *Normative analysis*.

Desertification: Land degradation in arid, semi-arid, and dry sub-humid areas resulting from various factors, including climatic variations and human activities. Land degradation in arid, semi-arid, and dry sub-humid areas is a reduction or loss of the biological or economic productivity and complexity of rainfed cropland, irrigated cropland, or range, pasture, *forest*, and woodlands resulting from *land uses* or from a process or combination of processes, including processes arising from human activities and habitation patterns, such as (1) soil erosion caused by wind and/or water; (2) deterioration of the physical, chemical, biological, or economic properties of soil; and (3) long-term loss of natural vegetation (UNCCD, 1994).

Designated national authority (DNA): A designated national authority is a national *institution* that authorizes and approves *Clean Development Mechansim (CDM)* projects in that country. In CDM host countries, the DNA assesses whether proposed projects assist the host country in achieving its *sustainable development (SD)* goals, certification of which is a prerequisite for registration of the project by the CDM Executive Board.

Developed/developing countries: See *Industrialized/developing countries*.

Development pathway: An evolution based on an array of technological, economic, social, institutional, cultural, and biophysical characteristics that determine the interactions between human and natural systems, including consumption and production patterns in all countries, over time at a particular scale.

Direct Air Capture (DAC): Chemical process by which a pure *carbon dioxide (CO_2)* stream is produced by capturing CO_2 from the ambient air.

Direct emissions: See *Emissions*.

Discounting: A mathematical operation making monetary (or other) amounts received or expended at different times (years) comparable across time. The discounter uses a fixed or possibly time-varying discount rate (> 0) from year to year that makes future value worth less today. See also *Present value*.

Double dividend: The extent to which revenue-generating instruments, such as *carbon taxes* or auctioned (tradable) *emission permits* can (1) contribute to *mitigation* and (2) offset at least part of the potential welfare losses of climate *policies* through recycling the revenue in the economy to reduce other taxes likely to cause distortions.

Drivers of behaviour: Determinants of human decisions and actions, including peoples' values and goals and the factors that constrain action, including economic factors and incentives, information access, regulatory and technological constraints, cognitive and emotional processing capacity, and social norms. See also *Behaviour* and *Behavioural change*.

Drivers of emissions: Drivers of emissions refer to the processes, mechanisms and properties that influence emissions through factors. Factors comprise the terms in a decomposition of emissions. Factors and drivers may in return affect *policies*, *measures* and other drivers.

Economic efficiency: Economic efficiency refers to an economy's allocation of resources (goods, services, inputs, productive activities). An allocation is efficient if it is not possible to reallocate resources so as to make at least one person better off without making someone else worse off. An allocation is inefficient if such a reallocation is possible. This is also known as the Pareto Criterion for efficiency. See also *Pareto optimum*.

Economies in Transition (EITs): Countries with their economies changing from a planned economic system to a market economy. See Annex II.2.1.

Ecosystem: A functional unit consisting of living organisms, their non-living environment, and the interactions within and between them. The components included in a given ecosystem and its spatial boundaries depend on the purpose for which the ecosystem is defined: in some cases they are relatively sharp, while in others they are diffuse. Ecosystem boundaries can change over time. Ecosystems are nested within other ecosystems, and their scale can range from very small to the entire *biosphere*. In the current era, most ecosystems either contain people as key organisms, or are influenced by the effects of human activities in their environment.

Ecosystem services: Ecological processes or functions having monetary or non-monetary value to individuals or society at large. These are frequently classified as (1) supporting services such as productivity or *biodiversity* maintenance, (2) provisioning services such as food, fiber, or fish, (3) regulating services such as *climate* regulation or carbon *sequestration*, and (4) cultural services such as tourism or spiritual and aesthetic appreciation.

Embodied emissions: See *Emissions*.

Embodied energy: See *Energy*.

Emission allowance: See *Emission permit*.

Emission factor/Emissions intensity: The emissions released per unit of activity. See also *Carbon intensity*.

Emission permit: An entitlement allocated by a government to a legal entity (company or other emitter) to emit a specified amount of a substance. Emission permits are often used as part of *emissions trading* schemes.

Emission quota: The portion of total allowable emissions assigned to a country or group of countries within a framework of maximum total emissions.

Emission scenario: A plausible representation of the future development of emissions of substances that are potentially radiatively active (e.g., *greenhouse gases, aerosols*) based on a coherent and internally consistent set of assumptions about driving forces (such as demographic and socioeconomic development, *technological change, energy* and *land use*) and their key relationships. Concentration scenarios, derived from emission scenarios, are used as input to a *climate model* to compute *climate projections*. In IPCC (1992) a set of emission scenarios was presented which were used as a basis for the *climate projections* in IPCC (1996). These emission scenarios are referred to as the IS92 scenarios. In the IPCC Special Report on Emission Scenarios (Nakićenović and Swart, 2000) emission scenarios, the so-called *SRES scenarios*, were published, some of which were used, among others, as a basis for the *climate projections* presented in Chapters 9 to 11 of IPCC (2001) and Chapters 10 and 11 of IPCC (2007). New emission scenarios for *climate change*, the four *Representative Concentration Pathways (RCPs)*, were developed for, but independently of, the present IPCC assessment. See also *Baseline/reference, Climate scenario, Mitigation scenario, Shared socio-economic pathways, Scenario, Socio-economic scenario, Stabilization*, and *Transformation pathway*.

Emission trajectories: A projected development in time of the emission of a *greenhouse gas (GHG)* or group of GHGs, *aerosols*, and GHG *precursors*.

Emissions:

Agricultural emissions: Emissions associated with agricultural systems—predominantly *methane (CH_4)* or *nitrous oxide (N_2O)*. These include emissions from enteric fermentation in domestic livestock, manure management, rice cultivation, prescribed burning of savannas and grassland, and from soils (IPCC, 2006).

Anthropogenic emissions: Emissions of *greenhouse gases (GHGs), aerosols*, and *precursors* of a GHG or *aerosol* caused by human activities. These activities include the burning of *fossil fuels, deforestation, land use changes (LUC)*, livestock production, fertilization, waste management, and industrial processes.

Direct emissions: Emissions that physically arise from activities within well-defined boundaries of, for instance, a region, an economic sector, a company, or a process.

Embodied emissions: Emissions that arise from the production and delivery of a good or service or the build-up of infrastructure. Depending on the chosen system boundaries, upstream emissions are often included (e.g., emissions resulting from the extraction of raw materials). See also *Lifecycle assessment (LCA)*.

Indirect emissions: Emissions that are a consequence of the activities within well-defined boundaries of, for instance, a region, an economic sector, a company or process, but which occur outside the specified boundaries. For example, emissions are described as indirect if they relate to the use of heat but physically arise outside the boundaries of the heat user, or to electricity production but physically arise outside of the boundaries of the power supply sector.

Scope 1, Scope 2, and Scope 3 emissions: Emissions responsibility as defined by the GHG Protocol, a private sector initiative. 'Scope 1' indicates direct *greenhouse gas (GHG)* emissions that are from *sources* owned or controlled by the reporting entity. 'Scope 2' indicates indirect GHG emissions associated with the production of electricity, heat, or steam purchased by the reporting entity. 'Scope 3' indicates all other *indirect emissions*, i.e., emissions associated with the extraction and production of purchased materials, fuels, and services, including transport in vehicles not owned or controlled by the reporting entity, outsourced activities, waste disposal, etc. (WBCSD and WRI, 2004).

Territorial emissions: Emissions that take place within the territories of a particular jurisdiction.

Emissions Reduction Unit (ERU): Equal to one metric tonne of *CO_2-equivalent emissions* reduced or of *carbon dioxide (CO_2)* removed from the *atmosphere* through a *Joint Implementation (JI)* (defined in Article 6 of the *Kyoto Protocol*) project, calculated using *Global Warming Potentials (GWPs)*. See also *Certified Emission Reduction Unit (CER)* and *Emissions trading*.

Emission standard: An emission level that, by law or by *voluntary agreement*, may not be exceeded. Many *standards* use *emission factors* in their prescription and therefore do not impose absolute limits on the emissions.

Emissions trading: A market-based instrument used to limit emissions. The environmental objective or sum of total allowed emissions is expressed as an emissions *cap*. The *cap* is divided in tradable *emission permits* that are allocated—either by auctioning or handing out for free (grandfathering)—to entities within the jurisdiction of the trading scheme. Entities need to surrender *emission permits* equal to the amount of their emissions (e.g., tonnes of *carbon dioxide*). An entity may sell excess permits. Trading schemes may occur at the intra-company, domestic, or international level and may apply to *carbon dioxide (CO_2)*, other *greenhouse gases (GHGs)*, or other substances. Emissions trading is also one of the mechanisms under the *Kyoto Protocol*. See also *Kyoto Mechanisms*.

Energy: The power of 'doing work' possessed at any instant by a body or system of bodies. Energy is classified in a variety of types and becomes available to human ends when it flows from one place to another or is converted from one type into another.

Embodied energy: The *energy* used to produce a material substance or product (such as processed metals or building materials), taking into account *energy* used at the manufacturing facility, *energy* used in producing the materials that are used in the manufacturing facility, and so on.

Final energy: See *Primary energy*.

Primary energy: Primary energy (also referred to as energy *sources*) is the *energy* stored in natural resources (e. g., coal, crude oil, natural gas, uranium, and renewable sources). It is defined in several alternative ways. The International Energy Agency (IEA) utilizes the physical energy content method, which defines primary energy as *energy* that has not undergone any anthropogenic conversion. The method used in this report is the direct equivalent method (see Annex II.4), which counts one unit of secondary energy provided from non-combustible sources as one unit of primary energy, but treats combustion energy as the energy potential contained in fuels prior to treatment or combustion. Primary energy is transformed into secondary energy by cleaning (natural gas), refining (crude oil to oil products) or by conversion into electricity or heat. When the secondary energy is delivered at the end-use facilities it is called final energy (e.g., electricity at the wall outlet), where it becomes usable energy in supplying *energy services* (e.g., light).

Renewable energy (RE): Any form of energy from solar, geophysical, or biological sources that is replenished by natural processes at a rate that equals or exceeds its rate of use. For a more detailed description see *Bioenergy*, *Solar energy*, *Hydropower*, *Ocean*, *Geothermal*, and *Wind energy*.

Secondary energy: See *Primary energy*.

Energy access: Access to clean, reliable and affordable *energy services* for cooking and heating, lighting, communications, and productive uses (AGECC, 2010).

Energy carrier: A substance for delivering mechanical work or transfer of heat. Examples of energy carriers include: solid, liquid, or gaseous fuels (e.g., *biomass*, coal, oil, natural gas, hydrogen); pressurized/heated/cooled fluids (air, water, steam); and electric current.

Energy density: The ratio of stored *energy* to the volume or mass of a fuel or battery.

Energy efficiency (EE): The ratio of useful *energy* output of a system, conversion process, or activity to its *energy* input. In economics, the term may describe the ratio of economic output to *energy* input. See also *Energy intensity*.

Energy intensity: The ratio of *energy* use to economic or physical output.

Energy poverty: A lack of access to modern *energy services*. See also *Energy access*.

Energy security: The goal of a given country, or the global community as a whole, to maintain an adequate, stable, and predictable *energy* supply. Measures encompass safeguarding the sufficiency of *energy* resources to meet national *energy* demand at competitive and stable prices and the resilience of the *energy* supply; enabling development and deployment of technologies; building sufficient infrastructure to generate, store and transmit *energy* supplies; and ensuring enforceable contracts of delivery.

Energy services: An energy service is the benefit received as a result of *energy* use.

Energy system: The energy system comprises all components related to the production, conversion, delivery, and use of *energy*.

Environmental effectiveness: A *policy* is environmentally effective to the extent by which it achieves its expected environmental target (e.g., *greenhouse gas (GHG)* emission reduction).

Environmental input-output analysis: An analytical method used to allocate environmental impacts arising in production to categories of final consumption, by means of the Leontief inverse of a country's economic input-output tables. See also Annex II.6.2.

Environmental Kuznets Curve: The hypothesis that various environmental impacts first increase and then eventually decrease as income per capita increases.

Evidence: Information indicating the degree to which a belief or proposition is true or valid. In this report, the degree of evidence reflects the amount, quality, and consistency of scientific/technical information on which the Lead Authors are basing their findings. See also *Agreement*, *Confidence*, *Likelihood* and *Uncertainty*.

Externality/external cost/external benefit: Externalities arise from a human activity when agents responsible for the activity do not take full account of the activity's impacts on others' production and consumption possibilities, and no compensation exists for such impacts. When the impacts are negative, they are external costs. When the impacts are positive, they are external benefits. See also *Social costs*.

Feed-in tariff (FIT): The price per unit of electricity (heat) that a utility or power (heat) supplier has to pay for distributed or renewable electricity (heat) fed into the power grid (heat supply system) by non-utility generators. A public authority regulates the tariff.

Final energy: See *Primary energy*.

Flaring: Open air burning of waste gases and volatile liquids, through a chimney, at oil wells or rigs, in refineries or chemical plants, and at landfills.

Flexibility Mechanisms: See *Kyoto Mechanisms*.

Food security: A state that prevails when people have secure access to sufficient amounts of safe and nutritious food for normal growth, development, and an active and healthy life.[4]

Forest: A vegetation type dominated by trees. Many definitions of the term forest are in use throughout the world, reflecting wide differences in biogeophysical conditions, social structure and economics. According to the 2005 *United Nations Framework Convention on Climate Change (UNFCCC)* definition a forest is an area of land of at least 0.05–1 hectare, of which more than 10–30 % is covered by tree canopy. Trees must have a potential to reach a minimum of 25 meters at maturity in situ. Parties to the Convention can choose to define a forest from within those ranges. Currently, the definition does not recognize different biomes, nor do they distinguish natural forests from plantations, an anomaly being pointed out by many as in need of rectification.

For a discussion of the term forest and related terms such as *afforestation*, *reforestation* and *deforestation* see the IPCC Report on Land Use, Land-Use Change and Forestry (IPCC, 2000). See also the Report on Definitions and Methodological Options to Inventory Emissions from Direct Human-induced Degradation of Forests and Devegetation of Other Vegetation Types (IPCC, 2003).

Forest management: A system of practices for stewardship and use of *forest* land aimed at fulfilling relevant ecological (including *biological diversity*), economic and social functions of the *forest* in a sustainable manner (UNFCCC, 2002).

Forestry and Other Land Use (FOLU): See *Agriculture, Forestry and Other Land Use (AFOLU)*.

Fossil fuels: Carbon-based fuels from fossil hydrocarbon deposits, including coal, peat, oil, and natural gas.

Free Rider: One who benefits from a common good without contributing to its creation or preservation.

---

[4] This glossary entry builds on definitions used in FAO (2000) and previous IPCC reports.

Fuel cell: A fuel cell generates electricity in a direct and continuous way from the controlled electrochemical reaction of hydrogen or another fuel and oxygen. With hydrogen as fuel the cell emits only water and heat (no *carbon dioxide*) and the heat can be utilized (see also *Cogeneration*).

Fuel poverty: A condition in which a household is unable to guarantee a certain level of consumption of domestic *energy services* (especially heating) or suffers disproportionate expenditure burdens to meet these needs.

Fuel switching: In general, fuel switching refers to substituting fuel A for fuel B. In the context of *mitigation* it is implicit that fuel A has lower carbon content than fuel B, e.g., switching from natural gas to coal.

General circulation (climate) model (GCM): See *Climate model*.

General equilibrium analysis: General equilibrium analysis considers simultaneously all the markets and feedback effects among these markets in an economy leading to market clearance. *(Computable) general equilibrium (CGE) models* are the operational tools used to perform this type of analysis.

Geoengineering: Geoengineering refers to a broad set of methods and technologies that aim to deliberately alter the *climate system* in order to alleviate the impacts of *climate change*. Most, but not all, methods seek to either (1) reduce the amount of absorbed *solar energy* in the *climate system* (*Solar Radiation Management*) or (2) increase net carbon *sinks* from the *atmosphere* at a scale sufficiently large to alter *climate* (*Carbon Dioxide Removal*). Scale and intent are of central importance. Two key characteristics of geoengineering methods of particular concern are that they use or affect the *climate system* (e.g., *atmosphere*, land or ocean) globally or regionally and/or could have substantive unintended effects that cross national boundaries. Geoengineering is different from weather modification and ecological engineering, but the boundary can be fuzzy (IPCC, 2012, p. 2).

Geothermal energy: Accessible thermal *energy* stored in the earth's interior.

Global Environment Facility (GEF): The Global Environment Facility, established in 1991, helps *developing countries* fund projects and programmes that protect the global environment. GEF grants support projects related to *biodiversity*, *climate change*, international waters, land degradation, the *ozone (O_3)* layer, and persistent organic pollutants.

Global mean surface temperature: An estimate of the global mean surface air temperature. However, for changes over time, only anomalies, as departures from a climatology, are used, most commonly based on the area-weighted global average of the sea surface temperature anomaly and land surface air temperature anomaly.

Global warming: Global warming refers to the gradual increase, observed or projected, in global surface temperature, as one of the consequences of *radiative forcing* caused by *anthropogenic emissions*.

Global Warming Potential (GWP): An index, based on radiative properties of *greenhouse gases (GHGs)*, measuring the *radiative forcing* following a pulse emission of a unit mass of a given GHG in the present-day *atmosphere* integrated over a chosen time horizon, relative to that of *carbon dioxide (CO_2)*. The GWP represents the combined effect of the differing times these gases remain in the *atmosphere* and their relative effectiveness in causing *radiative forcing*. The *Kyoto Protocol* is based on GWPs from pulse emissions over a 100-year time frame. Unless stated otherwise, this report uses GWP values calculated with a 100-year time horizon which are often derived from the IPCC Second Assessment Report (see Annex II.9.1 for the GWP values of the different GHGs).

Governance: A comprehensive and inclusive concept of the full range of means for deciding, managing, and implementing *policies* and *measures*. Whereas government is defined strictly in terms of the nation-state, the more inclusive concept of governance recognizes the contributions of various levels of government (global, international, regional, local) and the contributing roles of the private sector, of nongovernmental actors, and of civil society to addressing the many types of issues facing the global community.

Grazing land management: The system of practices on land used for livestock production aimed at manipulating the amount and type of vegetation and livestock produced (UNFCCC, 2002).

Green Climate Fund (GCF): The Green Climate Fund was established by the 16th Session of the *Conference of the Parties (COP)* in 2010 as an operating entity of the financial mechanism of the *United Nations Framework Convention on Climate Change (UNFCCC)*, in accordance with Article 11 of the Convention, to support projects, programmes and *policies* and other activities in *developing country* Parties. The Fund is governed by a Board and will receive guidance of the COP. The Fund is headquartered in Songdo, Republic of Korea.

Greenhouse effect: The infrared radiative effect of all infrared-absorbing constituents in the *atmosphere*. *Greenhouse gases (GHGs)*, clouds, and (to a small extent) *aerosols* absorb terrestrial radiation emitted by the earth's surface and elsewhere in the *atmosphere*. These substances emit infrared radiation in all directions, but, everything else being equal, the net amount emitted to space is normally less than would have been emitted in the absence of these absorbers because of the decline of temperature with altitude in the *troposphere* and the consequent weakening of emission. An increase in the concentration of GHGs increases the magnitude of this effect; the difference is sometimes called the enhanced greenhouse effect. The change in a GHG concentration because of *anthropogenic emissions* contributes to an instantaneous *radiative forcing*. Surface temperature and *troposphere*

warm in response to this forcing, gradually restoring the radiative balance at the top of the *atmosphere*.

Greenhouse gas (GHG): Greenhouse gases are those gaseous constituents of the *atmosphere*, both natural and anthropogenic, that absorb and emit radiation at specific wavelengths within the spectrum of terrestrial radiation emitted by the earth's surface, the *atmosphere* itself, and by clouds. This property causes the *greenhouse effect*. Water vapour (H_2O), *carbon dioxide (CO_2)*, *nitrous oxide (N_2O)*, *methane (CH_4)* and *ozone (O_3)* are the primary GHGs in the earth's *atmosphere*. Moreover, there are a number of entirely human-made GHGs in the *atmosphere*, such as the halocarbons and other chlorine- and bromine-containing substances, dealt with under the *Montreal Protocol*. Beside CO_2, N_2O and CH_4, the *Kyoto Protocol* deals with the GHGs *sulphur hexafluoride (SF_6)*, *hydrofluorocarbons (HFCs)* and *perfluorocarbons (PFCs)*. For a list of well-mixed GHGs, see WGI AR5 Table 2.A.1.

Gross domestic product (GDP): The sum of gross value added, at purchasers' prices, by all resident and non-resident producers in the economy, plus any taxes and minus any subsidies not included in the value of the products in a country or a geographic region for a given period, normally one year. GDP is calculated without deducting for depreciation of fabricated assets or depletion and degradation of natural resources.

Gross national expenditure (GNE): The total amount of public and private consumption and capital expenditures of a nation. In general, national account is balanced such that *gross domestic product (GDP)* + import = GNE + export.

Gross national product: The value added from domestic and foreign sources claimed by residents. GNP comprises *gross domestic product (GDP)* plus net receipts of primary income from non-resident income.

Gross world product: An aggregation of the individual country's *gross domestic products (GDP)* to obtain the world or global *GDP*.

Heat island: The relative warmth of a city compared with surrounding rural areas, associated with changes in runoff, effects on heat retention, and changes in surface *albedo*.

Human Development Index (HDI): The Human Development Index allows the assessment of countries' progress regarding social and economic development as a composite index of three indicators: (1) health measured by life expectancy at birth; (2) knowledge as measured by a combination of the adult literacy rate and the combined primary, secondary and tertiary school enrolment ratio; and (3) standard of living as *gross domestic product (GDP)* per capita (in purchasing power parity). The HDI sets a minimum and a maximum for each dimension, called goalposts, and then shows where each country stands in relation to these goalposts, expressed as a value between 0 and 1. The HDI only acts as a broad proxy for some of the key issues of human development; for instance, it does not reflect issues such as political participation or gender inequalities.

Hybrid vehicle: Any vehicle that employs two sources of propulsion, particularly a vehicle that combines an internal combustion engine with an electric motor.

Hydrofluorocarbons (HFCs): One of the six types of *greenhouse gases (GHGs)* or groups of GHGs to be mitigated under the *Kyoto Protocol*. They are produced commercially as a substitute for *chlorofluorocarbons (CFCs)*. HFCs largely are used in refrigeration and semiconductor manufacturing. See also *Global Warming Potential (GWP)* and Annex II.9.1 for GWP values.

Hydropower: Power harnessed from the flow of water.

Incremental costs: See *Climate finance*.

Incremental investment: See *Climate finance*.

Indigenous peoples: Indigenous peoples and nations are those that, having a historical continuity with pre-invasion and pre-colonial societies that developed on their territories, consider themselves distinct from other sectors of the societies now prevailing on those territories, or parts of them. They form at present principally non-dominant sectors of society and are often determined to preserve, develop, and transmit to future generations their ancestral territories, and their ethnic identity, as the basis of their continued existence as peoples, in accordance with their own cultural patterns, social *institutions*, and common law system.[5]

Indirect emissions: See *Emissions*.

Indirect land use change (iLUC): See *Land use*.

Industrial Revolution: A period of rapid industrial growth with far-reaching social and economic consequences, beginning in Britain during the second half of the 18th century and spreading to Europe and later to other countries including the United States. The invention of the steam engine was an important trigger of this development. The industrial revolution marks the beginning of a strong increase in the use of *fossil fuels* and emission of, in particular, fossil *carbon dioxide*. In this report the terms pre-industrial and industrial refer, somewhat arbitrarily, to the periods before and after 1750, respectively.

Industrialized countries/developing countries: There are a diversity of approaches for categorizing countries on the basis of their level of development, and for defining terms such as industrialized, developed, or developing. Several categorizations are used in this report. (1)

In the United Nations system, there is no established convention for designating of developed and developing countries or areas. (2) The United Nations Statistics Division specifies developed and developing regions based on common practice. In addition, specific countries are designated as *Least Developed Countries (LCD)*, landlocked developing countries, small island developing states, and transition economies. Many countries appear in more than one of these categories. (3) The World Bank uses income as the main criterion for classifying countries as low, lower middle, upper middle, and high income. (4) The UNDP aggregates indicators for life expectancy, educational attainment, and income into a single composite *Human Development Index (HDI)* to classify countries as low, medium, high, or very high human development. See WGII AR5 Box 1–2.

Input-output analysis: See *Environmental input-output analysis*.

Institution: Institutions are rules and norms held in common by social actors that guide, constrain and shape human interaction. Institutions can be formal, such as laws and policies, or informal, such as norms and conventions. Organizations—such as parliaments, regulatory agencies, private firms, and community bodies—develop and act in response to institutional frameworks and the incentives they frame. Institutions can guide, constrain and shape human interaction through direct control, through incentives, and through processes of socialization.

Institutional feasibility: Institutional feasibility has two key parts: (1) the extent of administrative workload, both for public authorities and for regulated entities, and (2) the extent to which the *policy* is viewed as legitimate, gains acceptance, is adopted, and is implemented.

Integrated assessment: A method of analysis that combines results and models from the physical, biological, economic, and social sciences, and the interactions among these components in a consistent framework to evaluate the status and the consequences of environmental change and the *policy* responses to it. See also *Integrated Models*.

Integrated models: See *Models*.

IPAT identity: IPAT is the lettering of a formula put forward to describe the impact of human activity on the environment. Impact (I) is viewed as the product of population size (P), affluence (A=GDP/person) and technology (T= impact per GDP unit). In this conceptualization, population growth by definition leads to greater environmental impact if A and T are constant, and likewise higher income leads to more impact (Ehrlich and Holdren, 1971).

Iron fertilization: Deliberate introduction of iron to the upper ocean intended to enhance biological productivity which can sequester additional atmospheric *carbon dioxide (CO_2)* into the oceans. See also *Geoengineering* and *Carbon Dioxide Removal (CDR)*.

Jevon's paradox: See *Rebound effect*.

---

[5] This glossary entry builds on the definitions used in Cobo (1987) and previous IPCC reports.

Joint Implementation (JI): A mechanism defined in Article 6 of the *Kyoto Protocol*, through which investors (governments or companies) from developed *(Annex B) countries* may implement projects jointly that limit or reduce emissions or enhance *sinks*, and to share the *Emissions Reduction Units (ERU)*. See also *Kyoto Mechanisms*.

Kaya identity: In this identity global emissions are equal to the population size, multiplied by per capita output (*gross world product*), multiplied by the *energy intensity* of production, multiplied by the *carbon intensity* of *energy*.

Kyoto Mechanisms (also referred to as Flexibility Mechanisms): Market-based mechanisms that Parties to the *Kyoto Protocol* can use in an attempt to lessen the potential economic impacts of their commitment to limit or reduce *greenhouse gas (GHG)* emissions. They include *Joint Implementation (JI)* (Article 6), *Clean Development Mechanism (CDM)* (Article 12), and *Emissions trading* (Article 17).

Kyoto Protocol: The Kyoto Protocol to the *United Nations Framework Convention on Climate Change (UNFCCC)* was adopted in 1997 in Kyoto, Japan, at the Third Session of the *Conference of the Parties (COP)* to the UNFCCC. It contains legally binding commitments, in addition to those included in the UNFCCC. Countries included in *Annex B* of the Protocol (most Organisation for Economic Cooperation and Development countries and countries with economies in transition) agreed to reduce their anthropogenic *greenhouse gas (GHG)* emissions (*carbon dioxide (CO$_2$), methane (CH$_4$), nitrous oxide (N$_2$O), hydrofluorocarbons (HFCs), perfluorocarbons (PFCs), and sulphur hexafluoride (SF$_6$)*) by at least 5% below 1990 levels in the commitment period 2008–2012. The Kyoto Protocol entered into force on 16 February 2005.

Land use (change, direct and indirect): Land use refers to the total of arrangements, activities and inputs undertaken in a certain land cover type (a set of human actions). The term land use is also used in the sense of the social and economic purposes for which land is managed (e.g., grazing, timber extraction and conservation). In urban settlements it is related to land uses within cities and their hinterlands. Urban land use has implications on city management, structure, and form and thus on energy demand, *greenhouse gas (GHG)* emissions, and mobility, among other aspects.

    Land use change (LUC): Land use change refers to a change in the use or management of land by humans, which may lead to a change in land cover. Land cover and LUC may have an impact on the surface *albedo*, evapotranspiration, *sources* and *sinks* of GHGs, or other properties of the *climate system* and may thus give rise to *radiative forcing* and/or other impacts on *climate*, locally or globally. See also the IPCC Report on Land Use, Land-Use Change, and Forestry (IPCC, 2000).

    Indirect land use change (iLUC): Indirect land use change refers to shifts in land use induced by a change in the production level of an agricultural product elsewhere, often mediated by markets or driven by *policies*. For example, if agricultural land is diverted to fuel production, *forest* clearance may occur elsewhere to replace the former agricultural production. See also *Afforestation*, *Deforestation* and *Reforestation*.

Land use, land use change and forestry (LULUCF): A *greenhouse gas (GHG)* inventory sector that covers *emissions* and removals of GHGs resulting from direct human-induced *land use*, *land use change* and forestry activities excluding *agricultural emissions*. See also *Agriculture, Forestry and Other Land Use (AFOLU)*.

Land value capture: A financing mechanism usually based around transit systems, or other infrastructure and services, that captures the increased value of land due to improved accessibility.

Leakage: Phenomena whereby the reduction in emissions (relative to a *baseline*) in a jurisdiction/sector associated with the implementation of *mitigation policy* is offset to some degree by an increase outside the jurisdiction/sector through induced changes in consumption, production, prices, land use and/or trade across the jurisdictions/sectors. Leakage can occur at a number of levels, be it a project, state, province, nation, or world region. See also *Rebound effect*.

In the context of *Carbon Dioxide Capture and Storage (CCS)*, 'CO$_2$ leakage' refers to the escape of injected *carbon dioxide (CO$_2$)* from the storage location and eventual release to the atmosphere. In the context of other substances, the term is used more generically, such as for '*methane (CH$_4$)* leakage' (e.g., from *fossil fuel* extraction activities), and '*hydrofluorocarbon (HFC)* leakage' (e.g., from refrigeration and air-conditioning systems).

Learning curve/rate: Decreasing cost-prices of technologies shown as a function of increasing (total or yearly) supplies. The learning rate is the percent decrease of the cost-price for every doubling of the cumulative supplies (also called progress ratio).

Least Developed Countries (LDCs): A list of countries designated by the Economic and Social Council of the United Nations (ECOSOC) as meeting three criteria: (1) a low income criterion below a certain threshold of gross national income per capita of between USD 750 and USD 900, (2) a human resource weakness based on indicators of health, education, adult literacy, and (3) an economic vulnerability weakness based on indicators on instability of agricultural production, instability of export of goods and services, economic importance of non-traditional activities, merchandise export concentration, and the handicap of economic smallness. Countries in this category are eligible for a number of programmes focused on assisting countries most in need. These privileges include certain benefits under the articles of the *United Nations Framework Convention on Climate Change (UNFCCC)*. See also *Industrialized/developing countries*.

Levelized cost of conserved carbon (LCCC): See Annex II.3.1.3 for concepts and definition.

Levelized cost of conserved energy (LCCE): See Annex II.3.1.2 for concepts and definition.

Levelized cost of energy (LCOE): See Annex II.3.1.1 for concepts and definition.

Lifecycle assessment (LCA): A widely used technique defined by ISO 14040 as a "compilation and evaluation of the inputs, outputs and the potential environmental impacts of a product system throughout its life cycle". The results of LCA studies are strongly dependent on the system boundaries within which they are conducted. The technique is intended for relative comparison of two similar means to complete a product. See also Annex II.6.3.

Likelihood: The chance of a specific outcome occurring, where this might be estimated probabilistically. This is expressed in this report using a standard terminology (Mastrandrea et al., 2010): virtually certain 99–100 % probability, very likely 90–100 %, likely 66–100 %, about as likely as not 33–66 %, unlikely 0–33 %, very unlikely 0–10 %, exceptionally unlikely 0–1 %. Additional terms (more likely than not > 50–100 %, and more unlikely than likely 0–< 50 %) may also be used when appropriate. Assessed likelihood is typeset in italics, e. g., *very likely*. See also *Agreement, Confidence, Evidence* and *Uncertainty*.

Lock-in: Lock-in occurs when a market is stuck with a *standard* even though participants would be better off with an alternative.

Marginal abatement cost (MAC): The cost of one unit of additional *mitigation*.

Market barriers: In the context of climate change *mitigation*, market barriers are conditions that prevent or impede the diffusion of *cost-effective* technologies or practices that would mitigate *greenhouse gas (GHG)* emissions.

Market-based mechanisms, GHG emissions: Regulatory approaches using price mechanisms (e. g., taxes and auctioned *emission permits*), among other instruments, to reduce the *sources* or enhance the *sinks* of *greenhouse gases (GHGs)*.

Market exchange rate (MER): The rate at which foreign currencies are exchanged. Most economies post such rates daily and they vary little across all the exchanges. For some developing economies, official rates and black-market rates may differ significantly and the MER is difficult to pin down. See also *Purchasing power parity (PPP)* and Annex II.1.3 for the monetary conversion process applied throughout this report.

Market failure: When private decisions are based on market prices that do not reflect the real scarcity of goods and services but rather reflect market distortions, they do not generate an efficient allocation of resources but cause welfare losses. A market distortion is any event

in which a market reaches a market clearing price that is substantially different from the price that a market would achieve while operating under conditions of perfect competition and state enforcement of legal contracts and the ownership of private property. Examples of factors causing market prices to deviate from real economic scarcity are environmental *externalities*, *public goods*, monopoly power, information asymmetry, *transaction costs*, and non-rational *behaviour*. See also *Economic efficiency*.

Material flow analysis (MFA): A systematic assessment of the flows and stocks of materials within a system defined in space and time (Brunner and Rechberger, 2004). See also Annex II.6.1.

Measures: In climate *policy*, measures are technologies, processes or practices that contribute to *mitigation*, for example *renewable energy (RE)* technologies, waste minimization processes, public transport commuting practices.

Meeting of the Parties (CMP): The *Conference of the Parties (COP)* to the *United Nations Framework Convention on Climate Change (UNFCCC)* serves as the CMP, the supreme body of the *Kyoto Protocol*, since the latter entered into force on 16 February 2005. Only Parties to the *Kyoto Protocol* may participate in deliberations and make decisions.

Methane (CH$_4$): One of the six *greenhouse gases (GHGs)* to be mitigated under the *Kyoto Protocol* and is the major component of natural gas and associated with all hydrocarbon fuels. Significant emissions occur as a result of animal husbandry and agriculture and their management represents a major *mitigation* option. See also *Global Warming Potential (GWP)* and Annex II.9.1 for GWP values.

Methane recovery: Any process by which *methane (CH$_4$)* emissions (e. g., from oil or gas wells, coal beds, peat bogs, gas transmission pipelines, landfills, or anaerobic digesters) are captured and used as a fuel or for some other economic purpose (e. g., chemical feedstock).

Millennium Development Goals (MDGs): A set of eight time-bound and measurable goals for combating poverty, hunger, disease, illiteracy, discrimination against women and environmental degradation. These goals were agreed to at the UN Millennium Summit in 2000 together with an action plan to reach the goals.

Mitigation (of climate change): A human intervention to reduce the *sources* or enhance the *sinks* of *greenhouse gases (GHGs)*. This report also assesses human interventions to reduce the *sources* of other substances which may contribute directly or indirectly to limiting *climate change*, including, for example, the reduction of *particulate matter (PM)* emissions that can directly alter the radiation balance (e. g., *black carbon*) or *measures* that control emissions of carbon monoxide, *nitrogen oxides (NO$_x$)*, *Volatile Organic Compounds (VOCs)* and other

Annex

pollutants that can alter the concentration of tropospheric *ozone (O₃)* which has an indirect effect on the *climate*.

Mitigation capacity: A country's ability to reduce anthropogenic *greenhouse gas (GHG)* emissions or to enhance natural *sinks*, where ability refers to skills, competencies, fitness, and proficiencies that a country has attained and depends on technology, *institutions*, wealth, equity, infrastructure, and information. Mitigative capacity is rooted in a country's *sustainable development (SD)* path.

Mitigation scenario: A plausible description of the future that describes how the (studied) system responds to the implementation of *mitigation policies* and *measures*. See also *Baseline/reference*, *Climate scenario*, *Emission scenario*, *Representative Concentration Pathways (RCPs)*, *Scenario*, *Shared socio-economic pathways*, *Socio-economic scenarios*, *SRES scenarios*, *Stabilization*, and *Transformation pathways*.

Models: Structured imitations of a system's attributes and mechanisms to mimic appearance or functioning of systems, for example, the *climate*, the economy of a country, or a crop. Mathematical models assemble (many) variables and relations (often in a computer code) to simulate system functioning and performance for variations in parameters and inputs.

Computable General Equilibrium (CGE) Model: A class of economic models that use actual economic data (i.e., input/output data), simplify the characterization of economic *behaviour*, and solve the whole system numerically. CGE models specify all economic relationships in mathematical terms and predict the changes in variables such as prices, output and economic welfare resulting from a change in economic policies, given information about technologies and consumer preferences (Hertel, 1997). See also *General equilibrium analysis*.

Integrated Model: Integrated models explore the interactions between multiple sectors of the economy or components of particular systems, such as the *energy system*. In the context of *transformation pathways*, they refer to models that, at a minimum, include full and disaggregated representations of the *energy system* and its linkage to the overall economy that will allow for consideration of interactions among different elements of that system. Integrated models may also include representations of the full economy, *land use* and *land use change (LUC)*, and the *climate system*. See also *Integrated assessment*.

Sectoral Model: In the context of this report, sectoral models address only one of the core sectors that are discussed in this report, such as buildings, industry, transport, energy supply, and *Agriculture, Forestry and Other Land Use (AFOLU)*.

Montreal Protocol: The Montreal Protocol on Substances that Deplete the Ozone Layer was adopted in Montreal in 1987, and subse-

quently adjusted and amended in London (1990), Copenhagen (1992), Vienna (1995), Montreal (1997) and Beijing (1999). It controls the consumption and production of chlorine- and bromine- containing chemicals that destroy stratospheric *ozone (O₃)*, such as *chlorofluorocarbons (CFCs)*, methyl chloroform, carbon tetrachloride and many others.

Multi-criteria analysis (MCA): Integrates different decision parameters and values without assigning monetary values to all parameters. Multi-criteria analysis can combine quantitative and qualitative information. Also referred to as multi-attribute analysis.

Multi-attribute analysis: See *Multi-criteria analysis (MCA)*.

Multi-gas: Next to *carbon dioxide (CO₂)*, there are other forcing components taken into account in, e.g., achieving reduction for a basket of *greenhouse gas (GHG)* emissions (CO₂, *methane (CH₄)*, *nitrous oxide (N₂O)*, and fluorinated gases) or *stabilization* of CO₂-equivalent concentrations (multi-gas *stabilization*, including GHGs and *aerosols*).

Nationally Appropriate Mitigation Action (NAMA): Nationally Appropriate Mitigation Actions are a concept for recognizing and financing emission reductions by *developing countries* in a post-2012 climate regime achieved through action considered appropriate in a given national context. The concept was first introduced in the Bali Action Plan in 2007 and is contained in the *Cancún Agreements*.

Nitrogen oxides (NOₓ): Any of several oxides of nitrogen.

Nitrous oxide (N₂O): One of the six *greenhouse gases (GHGs)* to be mitigated under the *Kyoto Protocol*. The main anthropogenic source of N₂O is agriculture (soil and animal manure management), but important contributions also come from sewage treatment, *fossil fuel* combustion, and chemical industrial processes. N₂O is also produced naturally from a wide variety of biological sources in soil and water, particularly microbial action in wet tropical forests. See also *Global Warming Potential (GWP)* and Annex II.9.1 for GWP values.

Non-Annex I Parties/countries: Non-Annex I Parties are mostly *developing countries*. Certain groups of *developing countries* are recognized by the Convention as being especially vulnerable to the adverse impacts of *climate change*, including countries with low-lying coastal areas and those prone to *desertification* and drought. Others, such as countries that rely heavily on income from *fossil fuel* production and commerce, feel more vulnerable to the potential economic impacts of *climate change* response measures. The Convention emphasizes activities that promise to answer the special needs and concerns of these vulnerable countries, such as investment, insurance, and technology transfer. See also *Annex I Parties/countries*.

Normative analysis: Analysis in which judgments about the desirability of various *policies* are made. The conclusions rest on value judgments as well as on facts and theories. See also *Descriptive analysis*.

Annex

Ocean energy: *Energy* obtained from the ocean via waves, tidal ranges, tidal and ocean currents, and thermal and saline gradients.

Offset (in climate policy): A unit of CO_2-*equivalent emissions* that is reduced, avoided, or sequestered to compensate for emissions occurring elsewhere.

Oil sands and oil shale: Unconsolidated porous sands, sandstone rock, and shales containing bituminous material that can be mined and converted to a liquid fuel. See also *Unconventional fuels*.

Overshoot pathways: Emissions, concentration, or temperature pathways in which the metric of interest temporarily exceeds, or 'overshoots', the long-term goal.

Ozone (O_3): Ozone, the triatomic form of oxygen (O_3), is a gaseous atmospheric constituent. In the *troposphere*, it is created both naturally and by photochemical reactions involving gases resulting from human activities (smog). Tropospheric O_3 acts as a *greenhouse gas (GHG)*. In the *stratosphere*, it is created by the interaction between solar ultraviolet radiation and molecular oxygen (O_2). Stratospheric O_3 plays a dominant role in the stratospheric radiative balance. Its concentration is highest in the O_3 layer.

Paratransit: Denotes flexible passenger transportation, often but not only in areas with low population density, that does not follow fixed routes or schedules. Options include minibuses (matatus, marshrutka), shared taxis and jitneys. Sometimes paratransit is also called community transit.

Pareto optimum: A state in which no one's welfare can be increased without reducing someone else's welfare. See also *Economic efficiency*.

Particulate matter (PM): Very small solid particles emitted during the combustion of *biomass* and *fossil fuels*. PM may consist of a wide variety of substances. Of greatest concern for health are particulates of diameter less than or equal to 10 nanometers, usually designated as PM_{10}. See also *Aerosol*.

Passive design: The word 'passive' in this context implies the ideal target that the only *energy* required to use the designed product or service comes from renewable sources.

Path dependence: The generic situation where decisions, events, or outcomes at one point in time constrain *adaptation*, *mitigation*, or other actions or options at a later point in time.

Payback period: Mostly used in investment appraisal as financial payback, which is the time needed to repay the initial investment by the returns of a project. A payback gap exists when, for example, private investors and micro-financing schemes require higher profitability rates from *renewable energy (RE)* projects than from fossil-fired proj-

ects. Energy payback is the time an *energy* project needs to deliver as much *energy* as had been used for setting the project online. Carbon payback is the time a *renewable energy (RE)* project needs to deliver as much net *greenhouse gas (GHG)* savings (with respect to the fossil reference *energy system*) as its realization has caused GHG emissions from a perspective of *lifecycle assessment (LCA)* (including *land use changes (LUC)* and loss of terrestrial carbon stocks).

Perfluorocarbons (PFCs): One of the six types of *greenhouse gases (GHGs)* or groups of GHGs to be mitigated under the *Kyoto Protocol*. PFCs are by-products of aluminium smelting and uranium enrichment. They also replace *chlorofluorocarbons (CFCs)* in manufacturing semiconductors. See also *Global Warming Potential (GWP)* and Annex II.9.1 for GWP values.

Photovoltaic cells (PV): Electronic devices that generate electricity from light *energy*. See also *Solar energy*.

Policies (for mitigation of or adaptation to climate change): Policies are a course of action taken and/or mandated by a government, e.g., to enhance *mitigation* and *adaptation*. Examples of *policies* aimed at *mitigation* are support mechanisms for *renewable energy (RE)* supplies, carbon or energy taxes, fuel efficiency *standards* for automobiles. See also *Measures*.

Polluter pays principle (PPP): The party causing the pollution is responsible for paying for remediation or for compensating the damage.

Positive analysis: See *Descriptive analysis*.

Potential: The possibility of something happening, or of someone doing something in the future. Different metrics are used throughout this report for the quantification of different types of potentials, including the following:

> **Technical potential:** Technical potential is the amount by which it is possible to pursue a specific objective through an increase in deployment of technologies or implementation of processes and practices that were not previously used or implemented. Quantification of technical potentials may take into account other than technical considerations, including social, economic and/or environmental considerations.

Precautionary principle: A provision under Article 3 of the *United Nations Framework Convention on Climate Change (UNFCCC)*, stipulating that the Parties should take precautionary *measures* to anticipate, prevent, or minimize the causes of *climate change* and mitigate its adverse effects. Where there are threats of serious or irreversible damage, lack of full scientific certainty should not be used as a reason to postpone such *measures*, taking into account that *policies* and *measures* to deal with *climate change* should be *cost-effective* in order to ensure global benefits at the lowest possible cost.

Precursors: Atmospheric compounds that are not *greenhouse gases (GHGs)* or *aerosols*, but that have an effect on GHG or *aerosol* concentrations by taking part in physical or chemical processes regulating their production or destruction rates.

Pre-industrial: See *Industrial Revolution*.

Present value: Amounts of money available at different dates in the future are discounted back to a present value, and summed to get the present value of a series of future cash flows. See also *Discounting*.

Primary production: All forms of production accomplished by plants, also called primary producers.

Primary energy: See *Energy*.

Private costs: Private costs are carried by individuals, companies or other private entities that undertake an action, whereas social costs include additionally the *external costs* on the environment and on society as a whole. Quantitative estimates of both private and social costs may be incomplete, because of difficulties in measuring all relevant effects.

Production-based accounting: Production-based accounting provides a measure of emissions released to the *atmosphere* for the production of goods and services by a certain entity (e.g., person, firm, country, or region). See also *Consumption-based accounting*.

Public good: Public goods are non-rivalrous (goods whose consumption by one consumer does not prevent simultaneous consumption by other consumers) and non-excludable (goods for which it is not possible to prevent people who have not paid for it from having access to it).

Purchasing power parity (PPP): The purchasing power of a currency is expressed using a basket of goods and services that can be bought with a given amount in the home country. International comparison of, for example, *gross domestic products (GDP)* of countries can be based on the purchasing power of currencies rather than on current exchange rates. PPP estimates tend to lower per capita *GDP* in *industrialized countries* and raise per capita *GDP* in *developing countries*. (PPP is also an acronym for *polluter pays principle*). See also *Market exchange rate (MER)* and Annex II.1.3 for the monetary conversion process applied throughout this report.

Radiation management: See *Solar Radiation Management*.

Radiative forcing: Radiative forcing is the change in the net, downward minus upward, radiative flux (expressed in W m^{-2}) at the tropopause or top of *atmosphere* due to a change in an external driver of *climate change*, such as, for example, a change in the concentration of *carbon dioxide (CO$_2$)* or the output of the sun. For the purposes of this report, radiative forcing is further defined as the change relative to the year 1750 and refers to a global and annual average value.

Rebound effect: Phenomena whereby the reduction in *energy* consumption or emissions (relative to a *baseline*) associated with the implementation of *mitigation measures* in a jurisdiction is offset to some degree through induced changes in consumption, production, and prices within the same jurisdiction. The rebound effect is most typically ascribed to technological *energy efficiency (EE)* improvements. See also *Leakage*.

Reducing Emissions from Deforestation and Forest Degradation (REDD): An effort to create financial value for the carbon stored in *forests*, offering incentives for *developing countries* to reduce emissions from forested lands and invest in low-carbon paths to *sustainable development (SD)*. It is therefore a mechanism for *mitigation* that results from avoiding *deforestation*. REDD+ goes beyond *reforestation* and *forest* degradation, and includes the role of conservation, sustainable management of forests and enhancement of forest carbon stocks. The concept was first introduced in 2005 in the 11th Session of the *Conference of the Parties (COP)* in Montreal and later given greater recognition in the 13th Session of the COP in 2007 at Bali and inclusion in the Bali Action Plan which called for "policy approaches and positive incentives on issues relating to reducing emissions to deforestation and forest degradation in developing countries (REDD) and the role of conservation, sustainable management of forests and enhancement of forest carbon stock in developing countries". Since then, support for REDD has increased and has slowly become a framework for action supported by a number of countries.

Reference scenario: See *Baseline/reference*.

Reforestation: Planting of *forests* on lands that have previously sustained *forests* but that have been converted to some other use. Under the *United Nations Framework Convention on Climate Change (UNFCCC)* and the *Kyoto Protocol*, reforestation is the direct human-induced conversion of non-forested land to forested land through planting, seeding, and/or human-induced promotion of natural seed sources, on land that was previously forested but converted to non-forested land. For the first commitment period of the *Kyoto Protocol*, reforestation activities will be limited to reforestation occurring on those lands that did not contain forest on 31 December 1989.

For a discussion of the term *forest* and related terms such as *afforestation*, reforestation and *deforestation*, see the IPCC Report on Land Use, Land-Use Change and Forestry (IPCC, 2000). See also the Report on Definitions and Methodological Options to Inventory Emissions from Direct Human-induced Degradation of Forests and Devegetation of Other Vegetation Types (IPCC, 2003).

Renewable energy (RE): See *Energy*.

Annex

131

Representative Concentration Pathways (RCPs): *Scenarios* that include time series of emissions and concentrations of the full suite of *greenhouse gases (GHGs)* and *aerosols* and chemically active gases, as well as *land use*/land cover (Moss et al., 2008). The word *representative* signifies that each RCP provides only one of many possible *scenarios* that would lead to the specific *radiative forcing* characteristics. The term *pathway* emphasizes that not only the long-term concentration levels are of interest, but also the trajectory taken over time to reach that outcome (Moss et al., 2010).

RCPs usually refer to the portion of the concentration pathway extending up to 2100, for which Integrated Assessment Models produced corresponding *emission scenarios*. Extended Concentration Pathways (ECPs) describe extensions of the RCPs from 2100 to 2500 that were calculated using simple rules generated by stakeholder consultations, and do not represent fully consistent *scenarios*.

Four RCPs produced from Integrated Assessment Models were selected from the published literature and are used in the present IPCC Assessment as a basis for the *climate predictions* and *projections* presented in WGI AR5 Chapters 11 to 14:

RCP2.6 One pathway where *radiative forcing* peaks at approximately 3 W m^{-2} before 2100 and then declines (the corresponding ECP assuming constant emissions after 2100);

RCP4.5 and RCP6.0 Two intermediate *stabilization* pathways in which *radiative forcing* is stabilized at approximately 4.5 W m^{-2} and 6.0 W m^{-2} after 2100 (the corresponding ECPs assuming constant concentrations after 2150);

RCP8.5 One high pathway for which *radiative forcing* reaches greater than 8.5 W m^{-2} by 2100 and continues to rise for some amount of time (the corresponding ECP assuming constant emissions after 2100 and constant concentrations after 2250).

For further description of future *scenarios*, see WGI AR5 Box 1.1. See also *Baseline/reference*, *Climate prediction*, *Climate projection*, *Climate scenario*, *Shared socio-economic pathways*, *Socio-economic scenario*, *SRES scenarios*, and *Transformation pathway*.

Reservoir: A component of the *climate system*, other than the *atmosphere*, which has the capacity to store, accumulate or release a substance of concern, for example, carbon, a *greenhouse gas (GHG)* or a *precursor*. Oceans, soils and *forests* are examples of reservoirs of carbon. Pool is an equivalent term (note that the definition of pool often includes the *atmosphere*). The absolute quantity of the substance of concern held within a reservoir at a specified time is called the stock. In the context of *Carbon Dioxide Capture and Storage (CCS)*, this term is sometimes used to refer to a geological *carbon dioxide (CO$_2$)* storage location. See also *Sequestration*.

Resilience: The capacity of social, economic, and environmental systems to cope with a hazardous event or trend or disturbance, responding or reorganizing in ways that maintain their essential function, identity, and structure, while also maintaining the capacity for *adaptation*, learning, and transformation (Arctic Council, 2013).

Revegetation: A direct human-induced activity to increase carbon stocks on sites through the establishment of vegetation that covers a minimum area of 0.05 hectares and does not meet the definitions of *afforestation* and *reforestation* contained here (UNFCCC, 2002).

Risk: In this report, the term risk is often used to refer to the potential, when the outcome is uncertain, for adverse consequences on lives, livelihoods, health, *ecosystems* and species, economic, social and cultural assets, services (including environmental services), and infrastructure.

Risk assessment: The qualitative and/or quantitative scientific estimation of *risks*.

Risk management: The plans, actions, or policies to reduce the likelihood and/or consequences of a given *risk*.

Risk perception: The subjective judgment that people make about the characteristics and severity of a *risk*.

Risk tradeoff: The change in the portfolio of *risks* that occurs when a countervailing *risk* is generated (knowingly or inadvertently) by an intervention to reduce the target *risk* (Wiener and Graham, 2009). See also *Adverse side-effect*, and *Co-benefit*.

Risk transfer: The practice of formally or informally shifting the *risk* of financial consequences for particular negative events from one party to another.

Scenario: A plausible description of how the future may develop based on a coherent and internally consistent set of assumptions about key driving forces (e.g., rate of *technological change (TC)*, prices) and relationships. Note that scenarios are neither predictions nor forecasts, but are useful to provide a view of the implications of developments and actions. See also *Baseline/reference*, *Climate scenario*, *Emission scenario*, *Mitigation scenario*, *Representative Concentration Pathways (RCPs)*, *Shared socio-economic pathways*, *Socioeconomic scenarios*, *SRES scenarios*, *Stabilization*, and *Transformation pathway*.

Scope 1, Scope 2, and Scope 3 emissions: See *Emissions*.

Secondary energy: See *Primary energy*.

Sectoral Models: See *Models*.

Sensitivity analysis: Sensitivity analysis with respect to quantitative analysis assesses how changing assumptions alters the outcomes. For

Annex

example, one chooses different values for specific parameters and re-runs a given *model* to assess the impact of these changes on model output.

Sequestration: The uptake (i.e., the addition of a substance of concern to a *reservoir*) of carbon containing substances, in particular *carbon dioxide (CO$_2$)*, in terrestrial or marine *reservoirs*. Biological sequestration includes direct removal of CO$_2$ from the *atmosphere* through *land-use change (LUC)*, *afforestation*, *reforestation*, *revegetation*, carbon storage in landfills, and practices that enhance soil carbon in agriculture (*cropland management*, *grazing land management*). In parts of the literature, but not in this report, (carbon) sequestration is used to refer to *Carbon Dioxide Capture and Storage (CCS)*.

Shadow pricing: Setting prices of goods and services that are not, or are incompletely, priced by market forces or by administrative regulation, at the height of their social marginal value. This technique is used in *cost-benefit analysis (CBA)*.

Shared socio-economic pathways (SSPs): Currently, the idea of SSPs is developed as a basis for new emissions and *socio-economic scenarios*. An SSP is one of a collection of pathways that describe alternative futures of socio-economic development in the absence of climate *policy* intervention. The combination of SSP-based *socio-economic scenarios* and *Representative Concentration Pathway (RCP)-based climate projections* should provide a useful integrative frame for climate impact and *policy* analysis. See also *Baseline/reference*, *Climate scenario*, *Emission scenario*, *Mitigation scenario*, *Scenario*, *SRES scenarios*, *Stabilization*, and *Transformation pathway*.

Short-lived climate pollutant (SLCP): Pollutant emissions that have a warming influence on *climate* and have a relatively short lifetime in the *atmosphere* (a few days to a few decades). The main SLCPs are *black carbon (BC)* ('soot'), *methane (CH$_4$)* and some *hydroflurorcarbons (HFCs)* some of which are regulated under the *Kyoto Protocol*. Some pollutants of this type, including CH$_4$, are also *precursors* to the formation of tropospheric *ozone (O$_3$)*, a strong warming agent. These pollutants are of interest for at least two reasons. First, because they are short-lived, efforts to control them will have prompt effects on *global warming*—unlike long-lived pollutants that build up in the *atmosphere* and respond to changes in emissions at a more sluggish pace. Second, many of these pollutants also have adverse local impacts such as on human health.

Sink: Any process, activity or mechanism that removes a *greenhouse gas (GHG)*, an *aerosol*, or a *precursor* of a GHG or *aerosol* from the *atmosphere*.

Smart grids: A smart grid uses information and communications technology to gather data on the *behaviours* of suppliers and consumers in the production, distribution, and use of electricity. Through automated responses or the provision of price signals, this information can then

be used to improve the efficiency, reliability, economics, and *sustainability* of the electricity network.

Smart meter: A meter that communicates consumption of electricity or gas back to the utility provider.

Social cost of carbon (SCC): The net present value of climate damages (with harmful damages expressed as a positive number) from one more tonne of carbon in the form of *carbon dioxide (CO$_2$)*, conditional on a global emissions trajectory over time.

Social costs: See *Private costs*.

Socio-economic scenario: A *scenario* that describes a possible future in terms of population, *gross domestic product (GDP)*, and other socio-economic factors relevant to understanding the implications of *climate change*. See also *Baseline/reference*, *Climate scenario*, *Emission scenario*, *Mitigation scenario*, *Representative Concentration Pathways (RCPs)*, *Scenario*, *Shared socio-economic pathways*, *SRES scenarios*, *Stabilization*, and *Transformation pathway*.

Solar energy: *Energy* from the sun. Often the phrase is used to mean *energy* that is captured from solar radiation either as heat, as light that is converted into chemical energy by natural or artificial photosynthesis, or by photovoltaic panels and converted directly into electricity.

Solar Radiation Management (SRM): Solar Radiation Management refers to the intentional modification of the earth's shortwave radiative budget with the aim to reduce *climate change* according to a given metric (e.g., surface temperature, precipitation, regional impacts, etc.). Artificial injection of stratospheric *aerosols* and cloud brightening are two examples of SRM techniques. Methods to modify some fast-responding elements of the longwave radiative budget (such as cirrus clouds), although not strictly speaking SRM, can be related to SRM. SRM techniques do not fall within the usual definitions of *mitigation* and *adaptation* (IPCC, 2012, p. 2). See also *Carbon Dioxide Removal (CDR)* and *Geoengineering*.

Source: Any process, activity or mechanism that releases a *greenhouse gas (GHG)*, an *aerosol* or a *precursor* of a GHG or *aerosol* into the *atmosphere*. Source can also refer to, e.g., an *energy* source.

Spill-over effect: The effects of domestic or sector *mitigation measures* on other countries or sectors. Spill-over effects can be positive or negative and include effects on trade, (carbon) *leakage*, transfer of innovations, and diffusion of environmentally sound technology and other issues.

SRES scenarios: SRES scenarios are *emission scenarios* developed by Nakićenović and Swart (2000) and used, among others, as a basis for some of the *climate projections* shown in Chapters 9 to 11 of IPCC (2001) and Chapters 10 and 11 of IPCC (2007) as well as WGI AR5. The following terms are relevant for a better understanding of the structure and use of the set of SRES scenarios:

Scenario family: *Scenarios* that have a similar demographic, societal, economic and technical change storyline. Four scenario families comprise the SRES scenario set: A1, A2, B1, and B2.

Illustrative Scenario: A *scenario* that is illustrative for each of the six scenario groups reflected in the Summary for Policymakers of Nakićenović and Swart (2000). They include four revised marker scenarios for the scenario groups A1B, A2, B1, B2, and two additional *scenarios* for the A1FI and A1T groups. All scenario groups are equally sound.

Marker Scenario: A *scenario* that was originally posted in draft form on the SRES website to represent a given scenario family. The choice of markers was based on which of the initial quantifications best reflected the storyline, and the features of specific models. Markers are no more likely than other scenarios, but are considered by the SRES writing team as illustrative of a particular storyline. They are included in revised form in Nakićenović and Swart (2000). These scenarios received the closest scrutiny of the entire writing team and via the SRES open process. *Scenarios* were also selected to illustrate the other two scenario groups.

Storyline: A narrative description of a *scenario* (or family of *scenarios*), highlighting the main *scenario* characteristics, relationships between key driving forces and the dynamics of their evolution.

See also *Baseline/reference*, *Climate scenario*, *Emission scenario*, *Mitigation scenario*, *Representative Concentration Pathways (RCPs)*, *Shared socio-economic pathways*, *Socio-economic scenario*, *Stabilization*, and *Transformation pathway*.

Stabilization (of GHG or CO$_2$-equivalent concentration): A state in which the atmospheric concentrations of one *greenhouse gas (GHG)* (e.g., *carbon dioxide*) or of a *CO$_2$-equivalent* basket of GHGs (or a combination of GHGs and *aerosols*) remains constant over time.

Standards: Set of rules or codes mandating or defining product performance (e.g., grades, dimensions, characteristics, test methods, and rules for use). Product, technology or performance standards establish minimum requirements for affected products or technologies. Standards impose reductions in *greenhouse gas (GHG)* emissions associated with the manufacture or use of the products and/or application of the technology.

Stratosphere: The highly stratified region of the *atmosphere* above the *troposphere* extending from about 10 km (ranging from 9 km at high latitudes to 16 km in the tropics on average) to about 50 km altitude.

Structural change: Changes, for example, in the relative share of *gross domestic product (GDP)* produced by the industrial, agricultural, or services sectors of an economy, or more generally, systems transformations whereby some components are either replaced or potentially substituted by other components.

Subsidiarity: The principle that decisions of government (other things being equal) are best made and implemented, if possible, at the lowest most decentralized level, that is, closest to the citizen. Subsidiarity is designed to strengthen accountability and reduce the dangers of making decisions in places remote from their point of application. The principle does not necessarily limit or constrain the action of higher orders of government, but merely counsels against the unnecessary assumption of responsibilities at a higher level.

Sulphur hexafluoride (SF$_6$): One of the six types of *greenhouse gases (GHGs)* to be mitigated under the *Kyoto Protocol*. SF$_6$ is largely used in heavy industry to insulate high-voltage equipment and to assist in the manufacturing of cable-cooling systems and semi-conductors. See *Global Warming Potential (GWP)* and Annex II.9.1 for GWP values.

Sustainability: A dynamic process that guarantees the persistence of natural and human systems in an equitable manner.

Sustainable development (SD): Development that meets the needs of the present without compromising the ability of future generations to meet their own needs (WCED, 1987).

Technical potential: See *Potential*.

Technological change (TC): Economic models distinguish autonomous (exogenous), endogenous, and induced TC.

Autonomous (exogenous) technological change: Autonomous (exogenous) technological change is imposed from outside the model (i.e., as a parameter), usually in the form of a time trend affecting factor and/or energy productivity and therefore *energy* demand and/or economic growth.

Endogenous technological change: Endogenous technological change is the outcome of economic activity within the model (i.e., as a variable) so that factor productivity or the choice of technologies is included within the model and affects *energy* demand and/or economic growth.

Induced technological change: Induced technological change implies endogenous technological change but adds further changes induced by *policies* and *measures*, such as *carbon taxes* triggering research and development efforts.

Technological learning: See *Learning curve/rate*.

Technological/knowledge spillovers: Any positive *externality* that results from purposeful investment in technological innovation or development (Weyant and Olavson, 1999).

Territorial emissions: See *Emissions*.

Trace gas: A minor constituent of the *atmosphere*, next to nitrogen and oxygen that together make up 99 % of all volume. The most important trace gases contributing to the *greenhouse effect* are *carbon dioxide (CO_2), ozone (O_3), methane (CH_4), nitrous oxide (N_2O), perfluorocarbons (PFCs), chlorofluorocarbons (CFCs), hydrofluorocarbons (HFCs), sulphur hexafluoride (SF_6)* and water vapour (H_2O).

Tradable (green) certificates scheme: A *market-based mechanism* to achieve an environmentally desirable outcome (*renewable energy (RE)* generation, *energy efficiency (EE)* requirements) in a *cost-effective* way by allowing purchase and sale of certificates representing under and over-compliance respectively with a quota.

Tradable (emission) permit: See *Emission permit*.

Tradable quota system: See *Emissions trading*.

Transaction costs: The costs that arise from initiating and completing transactions, such as finding partners, holding negotiations, consulting with lawyers or other experts, monitoring agreements, or opportunity costs, such as lost time or resources (Michaelowa et al., 2003).

Transformation pathway: The trajectory taken over time to meet different goals for *greenhouse gas (GHG)* emissions, atmospheric concentrations, or *global mean surface temperature* change that implies a set of economic, *technological*, and *behavioural changes*. This can encompass changes in the way *energy* and infrastructure is used and produced, natural resources are managed, *institutions* are set up, and in the pace and direction of *technological change (TC)*. See also *Baseline/reference*, *Climate scenario*, *Emission scenario*, *Mitigation scenario*, *Representative Concentration Pathways (RCPs)*, *Scenario*, *Shared socio-economic pathways*, *Socio-economic scenarios*, *SRES scenarios*, and *Stabilization*.

Transient climate response: See *Climate sensitivity*.

Transit oriented development (TOD): Urban development within walking distance of a transit station, usually dense and mixed with the character of a walkable environment.

Troposphere: The lowest part of the *atmosphere*, from the surface to about 10 km in altitude at mid-latitudes (ranging from 9 km at high latitudes to 16 km in the tropics on average), where clouds and weather phenomena occur. In the troposphere, temperatures generally decrease with height. See also *Stratosphere*.

Uncertainty: A cognitive state of incomplete knowledge that can result from a lack of information or from disagreement about what is known or even knowable. It may have many types of sources, from imprecision in the data to ambiguously defined concepts or terminol-

ogy, or uncertain projections of human *behaviour*. Uncertainty can therefore be represented by quantitative measures (e. g., a probability density function) or by qualitative statements (e. g., reflecting the judgment of a team of experts) (see Moss and Schneider, 2000; Manning et al., 2004; Mastrandrea et al., 2010). See also *Agreement*, *Evidence*, *Confidence* and *Likelihood*.

Unconventional resources: A loose term to describe *fossil fuel* reserves that cannot be extracted by the well-established drilling and mining processes that dominated extraction of coal, gas, and oil throughout the 20th century. The boundary between conventional and unconventional resources is not clearly defined. Unconventional oils include *oil shales*, tar sands/bitumen, heavy and extra heavy crude oils, and deep-sea oil occurrences. Unconventional natural gas includes gas in Devonian shales, tight sandstone formations, geopressured aquifers, coal-bed gas, and *methane (CH_4)* in clathrate structures (gas hydrates) (Rogner, 1997).

United Nations Framework Convention on Climate Change (UNFCCC): The Convention was adopted on 9 May 1992 in New York and signed at the 1992 Earth Summit in Rio de Janeiro by more than 150 countries and the European Community. Its ultimate objective is the 'stabilisation of greenhouse gas concentrations in the atmosphere at a level that would prevent dangerous anthropogenic interference with the climate system'. It contains commitments for all Parties under the principle of 'common but differentiated responsibilities'. Under the Convention, Parties included in *Annex I* aimed to return *greenhouse gas (GHG)* emissions not controlled by the *Montreal Protocol* to 1990 levels by the year 2000. The convention entered in force in March 1994. In 1997, the UNFCCC adopted the *Kyoto Protocol*.

Urban heat island: See *Heat island*.

Verified Emissions Reductions: Emission reductions that are verified by an independent third party outside the framework of the *United Nations Framework Convention on Climate Change (UNFCCC)* and its *Kyoto Protocol*. Also called 'Voluntary Emission Reductions'.

Volatile Organic Compounds (VOCs): Important class of organic chemical air pollutants that are volatile at ambient air conditions. Other terms used to represent VOCs are *hydrocarbons* (HCs), *reactive organic gases* (ROGs) and *non-methane volatile organic compounds* (NMVOCs). NMVOCs are major contributors—together with *nitrogen oxides (NO_x)*, and carbon monoxide (CO)—to the formation of photochemical oxidants such as *ozone (O_3)*.

Voluntary action: Informal programmes, self-commitments, and declarations, where the parties (individual companies or groups of companies) entering into the action set their own targets and often do their own monitoring and reporting.

Annex

Voluntary agreement (VA): An agreement between a government authority and one or more private parties to achieve environmental objectives or to improve environmental performance beyond compliance with regulated obligations. Not all voluntary agreements are truly voluntary; some include rewards and/or penalties associated with joining or achieving commitments.

Voluntary Emission Reductions: See *Verified Emissions Reductions*.

Watts per square meter (W m^{-2}): See *Radiative forcing*.

Wind energy: Kinetic *energy* from air currents arising from uneven heating of the earth's surface. A wind turbine is a rotating machine for converting the kinetic energy of the wind to mechanical shaft energy to generate electricity. A windmill has oblique vanes or sails and the mechanical power obtained is mostly used directly, for example, for water pumping. A wind farm, wind project, or wind power plant is a group of wind turbines interconnected to a common utility system through a system of transformers, distribution lines, and (usually) one substation.

Acronyms and chemical symbols

AAU	Assigned Amount Unit
ADB	Asian Development Bank
AfDB	African Development Bank
AFOLU	Agriculture, Forestry and Other Land Use
AME	Asian Modeling Exercise
AMPERE	Assessment of Climate Change Mitigation Pathways and Evaluation of the Robustness of Mitigation Cost Estimates
AOSIS	Alliance of Small Island States
APEC	Asia-Pacific Economic Cooperation
AR4	IPCC Fourth Assessment Report
ASEAN	Association of Southeast Asian Nations
ASIA	Non-OECD Asia
BAMs	Border adjustment measures
BAT	Best available technology
BAU	Business-as-usual
BC	Black carbon
BECCS	Bioenergy with carbon dioxide capture and storage
BEVs	Battery electric vehicles
BNDES	Brazilian Development Bank
BOD	Biochemical Oxygen Demand
BRT	Bus rapid transit
C	Carbon
C40	C40 Cities Climate Leadership Group
CBA	Cost-benefit analysis
CBD	Convention on Biological Diversity
CBD	Central business district
CCA	Climate Change Agreement
CCE	Cost of conserved energy
CCL	Climate Change Levy
CCS	Carbon dioxide capture and storage
CDM	Clean Development Mechanism
CDR	Carbon dioxide removal
CEA	Cost-effectiveness analysis
CERs	Certified Emissions Reductions
CFCs	Chlorofluorocarbons
CGE	Computable general equilibrium
CH_4	Methane
CHP	Combined heat and power
CIFs	Climate Investment Funds
CMIP	Coupled Model Intercomparison Project
CNG	Compressed natural gas
CO	Carbon monoxide
CO_2	Carbon dioxide
CO_2eq	Carbon dioxide-equivalent, CO_2-equivalent
COD	Chemical oxygen demand
COP	Conference of the Parties
CRF	Capital recovery factor
CSP	Concentrated solar power
CTCN	Climate Technology Centre and Network

DAC	Direct air capture
DAC	Development Assistance Committee
DALYs	Disability-adjusted life years
DANN	Designated National Authority
DCs	Developing countries
DRI	Direct reduced iron
DSM	Demand-side management
EAF	Electric arc furnace
EAS	East Asia
ECA	Economic Commission for Africa
ECN	Energy Research Center of the Netherlands
ECOWAS	Economic Community of West African States
EDGAR	Emissions Database for Global Atmospheric Research
EE	Energy efficiency
EIA	U.S. Energy Information Administration
EITs	Economies in Transition
EMF	Energy Modeling Forum
EPA	U.S. Environmental Protection Agency
EPC	Energy performance contracting
ERU	Emissions reduction unit
ESCOs	Energy service companies
ETS	Emissions Trading System
EU	European Union
EU ETS	European Union Emissions Trading Scheme
EVs	Electric vehicles
F-gases	Fluorinated gases
FAO	Food and Agriculture Organization of the United Nations
FAQ	Frequently asked questions
FAR	IPCC First Assessment Report
FCVs	Fuel cell vehicles
FDI	Foreign Direct Investment
FE	Final energy
FEEM	Fondazione Eni Enrico Mattei
FF&I	Fossil fuel and industrial
FIT	Feed-in tariff
FOLU	Forestry and Other Land Use
FSF	Fast-start Finance
G20	Group of Twenty Finance Ministers
G8	Group of Eight Finance Ministers
GATT	General Agreement on Tariffs and Trade
GCAM	Global Change Assessment Model
GCF	Green Climate Fund
GCM	General Circulation Model
GDP	Gross domestic product
GEA	Global Energy Assessment
GEF	Global Environment Facility
GHG	Greenhouse gas
GNE	Gross national expenditure
GSEP	Global Superior Energy Performance Partnership
GTM	Global Timber Model
GTP	Global Temperature Change Potential
GWP	Global Warming Potential

137

H$_2$	Hydrogen
HCFCs	Hydrochlorofluorocarbons
HDI	Human Development Index
HDVs	Heavy-duty vehicles
HFCs	Hydrofluorocarbon
HFC-23	Trifluoromethane
Hg	Mercury
HHV	Higher heating value
HIC	High-income countries
HVAC	Heating, ventilation and air conditioning
IAEA	International Atomic Energy Agency
IAMC	Integrated Assessment Modelling Consortium
ICAO	International Civil Aviation Organization
ICE	Internal combustion engine
ICLEI	International Council for Local Environmental Initiatives
ICT	Information and communication technology
IDB	Inter-American Development Bank
IDP	Integrated Design Process
IEA	International Energy Agency
IET	International Emissions Trading
IGCC	Integrated gasification combined cycle
IIASA	International Institute for Applied Systems Analysis
iLUC	Indirect land-use change
IMF	International Monetary Fund
IMO	International Maritime Organization
INT TRA	International transport
IO	International organization
IP	Intellectual property
IPAT	Income-Population-Affluence-Technology
IPCC	Intergovernmental Panel on Climate Change
IRENA	International Renewable Energy Agency
IRR	Internal rate of return
ISO	International Organization for Standardization
JI	Joint Implementation
JICA	Japan International Cooperation Agency
KfW	Kreditanstalt für Wiederaufbau
LAM	Latin America
LCA	Lifecycle Assessment
LCCC	Levelized costs of conserved carbon
LCD	Liquid crystal display
LCCE	Levelized cost of conserved energy
LCOE	Levelized costs of energy
LDCs	Least Developed Countries
LDCF	Least Developed Countries Fund
LDVs	Light-duty vehicles
LED	Light-emitting diode
LHV	Lower heating value
LIC	Low-income countries
LIMITS	Low Climate Impact Scenarios and Implications of Required Tight Emission Control Strategies
LMC	Lower-middle income countries
LNG	Liquefied natural gas

LPG	Liquefied petroleum gas
LUC	Land-use change
LULUCF	Land Use, Land-Use Change and Forestry
MAC	Marginal abatement cost
MAF	Middle East and Africa
MAGICC	Model for the Assessment of Greenhouse Gas Induced Climate Change
MCA	Multi-criteria analysis
MDB	Multilateral Development Bank
MDGs	Millennium Development Goals
MEF	Major Economies Forum on Energy and Climate
MER	Market exchange rate
MFA	Material flow analysis
MNA	Middle East and North Africa
MRIO	Multi-Regional Input-Output Analysis
MRV	Measurement, reporting, and verification
MSW	Municipal solid waste
N	Nitrogen
N$_2$O	Nitrous oxide
NAM	North America
NAMA	Nationally Appropriate Mitigation Action
NAPA	National Adaptation Programmes of Action
NAS	U.S. National Academy of Science
NF$_3$	Nitrogen trifluoride
NGCC	Natural gas combined cycle
NGO	Non-governmental organization
NH$_3$	Ammonia
NO$_x$	Nitrogen oxides
NPV	Net present value
NRC	U.S. National Research Council
NREL	U.S. National Renewable Energy Laboratory
NZEB	Net zero energy buildings
O$_3$	Ozone
O&M	Operation and maintenance
OC	Organic carbon
ODA	Official development assistance
ODS	Ozone-depleting substances
OECD	Organisation for Economic Co-operation and Development
OPEC	Organization of Petroleum Exporting Countries
PACE	Property Assessed Clean Energy
PAS	South-East Asia and Pacific
PBL	Netherlands Environmental Assessment Agency
PC	Pulverized Coal
PDF	Probability density function
PEVs	Plug-in electric vehicles
PFC	Perfluorocarbons
PHEVs	Plug-in hybrid electric vehicles
PIK	Potsdam Institute for Climate Impact Research
PM	Particulate Matter
PNNL	Pacific Northwest National Laboratories
POEDC	Pacific OECD 1990 members (Japan, Aus, NZ)
PPP	Polluter pays principle

PPP	Purchasing power parity		TCR	Transient climate response
PV	Photovoltaic		Th	Thorium
R&D	Research and development		TNAs	Technology Needs Assessments
RCPs	Representative Concentration Pathways		TOD	Transit-oriented development
RD&D	Research, Development and Demonstration		TPES	Total primary energy supply
RE	Renewable energy		TRIPs	Trade Related Intellectual Property Rights
RECIPE	Report on Energy and Climate Policy in Europe		TT	Technology transfer
REDD	Reducing Emissions From Deforestation and Forest Degradation		U	Uranium
			UHI	Urban heat island
REEEP	Renewable Energy and Energy Efficiency Partnership		UMC	Upper-middle income countries
RES	Renewable energy sources		UN	United Nations
RGGI	Regional Greenhouse Gas Initiative		UN DESA	United Nations Department for Economic and Social Affairs
RoSE	Roadmaps towards Sustainable Energy futures			
ROW	Rest of the World		UNCCD	United Nations Convention to Combat Desertification
RPS	Renewable portfolio standards		UNCSD	United Nations Conference on Sustainable Development
SAR	IPCC Second Assessment Report			
SAS	South Asia		UNDP	United Nations Development Programme
SCC	Social cost of carbon		UNEP	United Nations Environment Programme
SCCF	Special Climate Change Fund		UNESCO	United Nations Educational, Scientific and Cultural Organization
SCP	Sustainable consumption and production			
SD	Sustainable development		UNFCCC	United Nations Framework Convention on Climate Change
SF_6	Sulphur hexafluoride			
SLCP	Short-lived climate pollutant		UNIDO	United Nations Industrial Development Organization
SMEs	Small and Medium Enterprises		USD	U.S. Dollars
SO_2	Sulphur dioxide		VAs	Voluntary agreements
SPM	Summary for Policymakers		VOCs	Volatile Organic Compounds
SRES	IPCC Special Report on Emission Scenarios		VKT	Vehicle kilometers travelled
SREX	IPCC Special Report on Managing the Risks of Extreme Events and Disasters to Advance Climate Change Adaptation		WACC	Weighted costs of capital
			WBCSD	World Business Council on Sustainable Development
			WCED	World Commission on Environment and Development
SRM	Solar radiation management		WCI	Western Climate Initiative
SRREN	IPCC Special Report on Renewable Energy Sources and Climate Change Mitigation		WEU	Western Europe
			WGI	IPCC Working Group I
SRCSS	IPCC Special Report on Carbon dioxide Capture and Storage		WGII	IPCC Working Group II
			WGIII	IPCC Working Group III
SSA	Sub-Saharan Africa		WHO	World Health Organization
SUVs	Sport Utility Vehicles		WTP	Willingness to pay
SWF	Social welfare function		WWTP	Wastewater plant
TAR	IPCC Third Assessment Report		WTO	World Trade Organization
TC	Technological change			

Annex

References

United Nations Secretary General's Advisory Group on Energy and Climate (AGECC) (2010). *Energy for a Sustainable Future.* New York, NY, USA.

Arctic Council (2013). Glossary of terms. In: *Arctic Resilience Interim Report 2013.* Stockholm Environment Institute and Stockholm Resilience Centre, Stockholm, Sweden.

Brunner, P.H. and H. Rechberger (2004). Practical handbook of material flow analysis. *The International Journal of Life Cycle Assessment,* **9**(5), 337–338.

Cobo, J.R.M. (1987). *Study of the problem of discrimination against indigenous populations.* Sub-commission on Prevention of Discrimination and Protection of Minorities. New York: United Nations, 1987.

Ehrlich, P.R. and J.P. Holdren (1971). Impact of population growth. *Science,* **171**(3977), 1212–1217.

Food and Agricultural Organization of the United Nations (FAO) (2000). *State of food insecurity in the world 2000.* Rome, Italy.

Hertel, T.T.W. (1997). *Global trade analysis: modeling and applications.* T.W. Hertel (Ed.). Cambridge University Press, Cambridge, United Kingdom.

Heywood, V.H. (ed.) (1995). *The Global Biodiversity Assessment.* United Nations Environment Programme. Cambridge University Press, Cambridge, United Kingdom.

IPCC (1992). *Climate Change 1992: The Supplementary Report to the IPCC Scientific Assessment* [Houghton, J.T., B.A. Callander, and S.K. Varney (eds.)]. Cambridge University Press, Cambridge, United Kingdom and New York, NY, USA, 116 pp.

IPCC (1996). *Climate Change 1995: The Science of Climate Change. Contribution of Working Group I to the Second Assessment Report of the Intergovernmental Panel on Climate Change* [Houghton, J.T., L.G. Meira Filho, B.A. Callander, N. Harris, A. Kattenberg, and K. Maskell (eds.)]. Cambridge University Press, Cambridge, United Kingdom and New York, NY, USA, 572 pp.

IPCC (2000). *Land Use, Land-Use Change, and Forestry. Special Report of the Intergovernmental Panel on Climate Change* [Watson, R.T., I.R. Noble, B. Bolin, N.H. Ravindranath, D.J. Verardo, and D.J. Dokken (eds.)]. Cambridge University Press, Cambridge, United Kingdom and New York, NY, USA, 377 pp.

IPCC (2001). *Climate Change 2001: The Scientific Basis. Contribution of Working Group I to the Third Assessment Report of the Intergovernmental Panel on Climate Change* [Houghton, J.T., Y. Ding, D.J. Griggs, M. Noguer, P.J. van der Linden, X. Dai, K. Maskell, and C.A. Johnson (eds.)]. Cambridge University Press, Cambridge, United Kingdom and New York, NY, USA, 881 pp.

IPCC (2003). *Definitions and Methodological Options to Inventory Emissions from Direct Human-Induced Degradation of Forests and Devegetation of Other Vegetation Types* [Penman, J., M. Gytarsky, T. Hiraishi, T. Krug, D. Kruger, R. Pipatti, L. Buendia, K. Miwa, T. Ngara, K. Tanabe, and F. Wagner (eds.)]. The Institute for Global Environmental Strategies (IGES), Japan, 32 pp.

IPCC (2006). *2006 IPCC Guidelines for National Greenhouse Gas Inventories,* Prepared by the National Greenhouse Gas Inventories Programme [Eggleston H.S., L. Buendia, K. Miwa, T. Ngara and K. Tanabe K. (eds.)]. The Institute for Global Environmental Strategies (IGES), Japan.

IPCC (2007). *Climate Change 2007: The Physical Science Basis. Contribution of Working Group I to the Fourth Assessment Report of the Intergovernmental Panel on Climate Change* [Solomon, S., D. Qin, M. Manning, Z. Chen, M. Marquis, K.B. Averyt, M. Tignor, and H.L. Miller (eds.)]. Cambridge University Press, Cambridge, United Kingdom and New York, NY, USA, 996 pp.

IPCC (2012). *Meeting Report of the Intergovernmental Panel on Climate Change Expert Meeting on Geoengineering* [O. Edenhofer, R. Pichs-Madruga, Y. Sokona, C. Field, V. Barros, T.F. Stocker, Q. Dahe, J. Minx, K. Mach, G.-K. Plattner, S. Schlömer, G. Hansen, and M. Mastrandrea (eds.)]. IPCC Working Group III Technical Support Unit, Potsdam Institute for Climate Impact Research, Potsdam, Germany, 99 pp.

Manning, M.R., M. Petit, D. Easterling, J. Murphy, A. Patwardhan, H-H. Rogner, R. Swart, and G. Yohe (eds.) (2004). *IPCC Workshop on Describing Scientific Uncertainties in Climate Change to Support Analysis of Risk of Options.* Workshop Report. Intergovernmental Panel on Climate Change, Geneva, Switzerland.

Mastrandrea, M.D., C.B. Field, T.F. Stocker, O. Edenhofer, K.L. Ebi, D.J. Frame, H. Held, E. Kriegler, K.J. Mach, P.R. Matschoss, G.-K. Plattner, G.W. Yohe, and F.W. Zwiers (2010). Guidance Note for Lead Authors of the IPCC Fifth Assessment Report on Consistent Treatment of Uncertainties. Intergovernmental Panel on Climate Change (IPCC). Published online at: http://www.ipcc-wg2.gov/meetings/CGCs/index.html#UR

Michaelowa, A., M. Stronzik., F. Eckermann, and A. Hunt (2003). Transaction costs of the Kyoto Mechanisms. *Climate policy,* **3**(3), 261–278.

Millennium Ecosystem Assessment (MEA) (2005). *Ecosystems and Human Wellbeing: Current States and Trends.* World Resources Institute, Washington, D.C. [Appendix D, p. 893].

Moss, R., and S. Schneider (2000). Uncertainties in the IPCC TAR: Recommendations to Lead Authors for More Consistent Assessment and Reporting. In: *IPCC Supporting Material: Guidance Papers on Cross Cutting Issues in the Third Assessment Report of the IPCC* [Pachauri, R., T. Taniguchi, and K. Tanaka (eds.)]. Intergovernmental Panel on Climate Change, Geneva, Switzerland, pp. 33–51.

Moss, R., M. Babiker, S. Brinkman, E. Calvo, T. Carter, J. Edmonds, I. Elgizouli, S. Emori, L. Erda, K. Hibbard, R. Jones, M. Kainuma, J. Kelleher, J.F. Lamarque, M. Manning, B. Matthews, J. Meehl, L. Meyer, J. Mitchell, N. Nakicenovic, B. O'Neill, R. Pichs, K. Riahi, S. Rose, P. Runci, R. Stouffer, D. van Vuuren, J. Weyant, T. Wilbanks, J.P. van Ypersele, and M. Zurek (2008). *Towards new scenarios for analysis of emissions, climate change, impacts and response strategies.* Intergovernmental Panel on Climate Change, Geneva, Switzerland, 132 pp.

Moss, R., J.A. Edmonds, K.A. Hibbard, M.R. Manning, S.K. Rose, D.P. van Vuuren, T.R. Carter, S. Emori, M. Kainuma, T. Kram, G.A. Meehl, J.F.B. Mitchell, N. Nakicenovic, K. Riahi, S.J. Smith, R.J. Stouffer, A.M. Thomson, J.P. Weyant, and T.J. Wilbanks (2010). The next generation of scenarios for climate change research and assessment. *Nature,* **463**, 747–756.

Nakićenović, N. and R. Swart (eds.) (2000). Special Report on Emissions Scenarios. A Special Report of Working Group III of the Intergovernmental Panel on Climate Change. Cambridge University Press, Cambridge, United Kingdom and New York, NY, USA, 599 pp.

Rogner, H.H. (1997). An assessment of world hydrocarbon resources. *Annual review of energy and the environment,* **22**(1), 217–262.

UNFCCC (2000). Report on the Conference of the Parties on its Seventh Session, held at Marrakesh from 29 October to 10 November 2001. Addendum. Part Two: Action Taken by the Conference of the Parties. (FCCC/CP/2001/13/Add.1).

United Nations Convention to Combat Desertification (UNCCD) (1994). *Article 1: Use of terms*. United Nations Convention to Combat Desertification. 17 June 1994: Paris, France.

Weyant, J.P. and T. Olavson (1999). Issues in modeling induced technological change in energy, environmental, and climate policy. Environmental Modeling & Assessment, 4(2–3), 67–85.

World Business Council on Sustainable Development (WBCSD) and World Resources Institute (WRI). (2004). *The Greenhouse Gas Protocol - A Corporate Accounting and Reporting Standard*. Geneva and Washington, DC.

Wiedmann, T. and J. Minx (2007). A definition of carbon footprint. *Ecological economics research trends*, **1**, 1–11.

Wiener, J.B. and J.D. Graham (2009). *Risk vs. risk: Tradeoffs in protecting health and the environment*. Harvard University Press, Cambridge, MA, USA.

World Commission on Environment and Development (WCED) (1987). *Our Common Future*. Oxford University Press, Oxford, United Kingdom

Annex

Claudia Rock   **Suneeti Phadke**

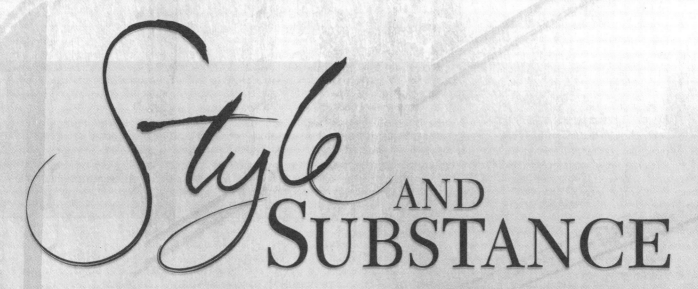

Style AND Substance

Second Edition

PEARSON
Longman

ERPI publishes and distributes PEARSON ELT products in Canada.
1611 Crémazie Boulevard East, 10th Floor, Montréal, Québec H2M 2P2 CANADA
Telephone: 1 800 263-3678   Fax: 514 334-4720
information@pearsonerpi.com   pearsonerpi.com

ACKNOWLEDGEMENTS

Knowledge, skills, patience, effort and perseverance are required to bring a book project to fruition, and the authors extend their heartfelt thanks to all the people who helped make this second edition possible. In particular, we would like to thank Jean-Pierre Albert, Sharnee Chait, Lucie Turcotte, Muriel Normand, Martin Tremblay and Linda Power for their special contributions to this project. Bravo! We think that a second edition is like the proverbial "second time around"—even better than the first!

Managing Editor
Sharnee Chait

Editor
Lucie Turcotte

Copy Editor
Jeremy Lanaway

Proofreaders
Jane Davey
My-Trang Nguyen

Art Director
Hélène Cousineau

Cover design
Martin Tremblay

Book design
Dessine-moi un mouton
Muriel Normand
Martin Tremblay

Page layout
Dessine-moi un mouton

Registration of copyright: Bibliothèque et Archives nationales du Québec, 2007
Registration of copyright: Library and Archives Canada, 2007
Printed in Canada

ISBN 978-2-7613-2090-0   890 MI 21 20 19 18
                            10830 ABCD   0F10

The following authors, publishers and photographers have generously given permission to reprint copyright material.

Chapter 1 p. 18 Photograph of Edgar Allan Poe © Roger Viollet / Topfoto / PONOPRESSE.

Chapter 2 p. 30 Photograph of Frank Stockton © Library of Congress.

Chapter 3 p. 42 Photograph of Oscar Wilde © Roger Viollet / Topfoto / PONOPRESSE.

Chapter 4 p. 54 Photograph of Kate Chopin © Missouri Historical Society.

Chapter 5 p. 63 Photograph of Ambrose Bierce © Bettmann / Corbis.

Chapter 6 p. 76 Photograph of Ernest Hemingway © UPP / Topfoto / PONOPRESSE. p. 77 "Hills Like White Elephants" by Ernest Hemingway reprinted with permission of Scribner, an imprint of Simon & Schuster Adult Publishing Group, from *The Short Stories of Ernest Hemingway*. Copyright 1927 Charles Scribner's Sons. Copyright renewed 1955 by Ernest Hemingway.

Chapter 7 p. 86 Photograph of Roald Dahl © UPP / Topfoto / PONOPRESSE. p. 87 "Beware of the Dog" reprinted from *Over to You: Ten Stories of Flyers and Flying* by Roald Dahl. © Penguin (Non-Classics). New edition January 2, 1980. First published in *Harper's Bazaar*.

Chapter 8 p. 100 "The Lottery" by Shirley Jackson. Copyright 1948, 1949 by Shirley Jackson. Copyright renewed 1976, 1977 by Laurence Hyman, Barry Hyman, Mrs. Sarah Webster and Mrs. Joanne Schnurer. Reprinted by permission of Farrar, Straus and Giroux, L.L.C.

Chapter 9 p. 110 Photograph of Ray Bradbury © Ulf Andersen / Gamma / PONOPRESSE. p. 111 "The Veldt" by Ray Bradbury reprinted by permission of Don Congdon Associates, Inc. Copyright © 1950 by the Curtis Publishing Company, renewed 1977 by Ray Bradbury.

Chapter 10 p. 127 Photograph of Hugh Garner courtesy of Barbara Wong. p. 128 "The Yellow Sweater" reprinted from *Hugh Garner's Best Stories*. © 1952, 1963. Permission granted by McGraw-Hill Ryerson Limited.

Chapter 11 p. 138 Photograph of Jeannette Winterson © CORBIS SYGMA. p. 140 "Newton" by Jeannette Winterson reprinted by permission of International Creative Management, Inc. Copyright © 1988 by Jeannette Winterson.

Chapter 12 p. 152 Photograph of David Bezmozgis © Greg Martin Photography. p. 153 "Tapka" reprinted from *Natasha and Other Stories* by David Bezmozgis. Published by Harper Collins Publishers Ltd. Copyright © 2004 by Nada Films, Inc. All rights reserved.

Chapter 13 p. 167 Photograph of Stephen Leacock © Notman Photographic Archives, McCord Museum, Montreal; photograph of Margaret Atwood © UPP / Topfoto / PONOPRESSE. p. 169 "Bread" reprinted from *Murder in the Dark* by Margaret Atwood. Used by permission of McClelland & Stewart Ltd. p. 171 "Old Habits Die Hard" by Makeda Silvera reprinted from *Her Head a Village*, 1994, Press Gang Publishers. Reprinted by permission of the author.

Chapter 14 p. 196 Photograph of William Blake © Topham Picturepoint / PONOPRESSE. p. 198 Photograph of William Wordsworth © Topham Picturepoint / PONOPRESSE. p. 201 Photograph of George Gordon, Lord Byron © Topham Picturepoint / Topfoto / PONOPRESSE. p. 204 Photograph of Elizabeth Barrett Browning © Fotomas / Topfoto / PONOPRESSE. p. 207 Photograph of Alfred, Lord Tennyson © ARPL – HIP / Topfoto / PONOPRESSE. p. 210 Photograph of Frederick George Scott © Notman Photographic Archives, McCord Museum, Montreal. p. 213 Photograph of Emily Dickinson © Topham Picturepoint / Topfoto / PONOPRESSE. p. 216 Photograph of Stephen Crane © Betmann / CORBIS.

Chapter 15 p. 220 Photograph of John McCrae © Notman Photographic Archives, McCord Museum, Montreal. p. 223 Photograph of Robert Frost © Topham Picturepoint / PONOPRESSE; "Stopping by Woods on a Snowy Evening" reprinted from *The Poetry of Robert Frost*, edited by Edward Connery Lathem. Copyright 1923, © 1969 by Henry Holt and Co., copyright 1951 by Robert Frost. Reprinted by permission of Henry Holt and Company. L.L.C. p. 225 Photograph of Earle Birney courtesy of Wailan Low. p. 226 "David" reprinted from *Poetry of Mid-Century 1940/1960*, edited by Milton Wilson, general editor Malcolm Ross, McClelland & Stewart Ltd. © Wailan Low. p. 232 Photograph of Irving Layton © B. Carrière / PUBLIPHOTO; "The Swimmer" reprinted from *A Wild Peculiar Joy* by Irving Layton. Used by permission of McClelland & Stewart Ltd. p. 234 Photograph of Leonard Cohen © A. Masson / PUBLIPHOTO. p. 235 "A Kite Is a Victim" reprinted from *Stranger Music* by Leonard Cohen. Used by permission of McClellan & Stewart Ltd. p. 237 Photograph of F. R. Scott © Lois Lord. p. 238 "Calamity" by F. R. Scott reprinted with the permission of William Toye, literary executor for the estate of F. R. Scott. p. 240 Photograph courtesy of Phyllis Webb; "Treblinka Gas Chamber" reprinted from *Selected Poems: The Vision Tree* © 1982 Phyllis Webb with the permission of Talon Books Ltd., Vancouver, B.C. p. 242 Photograph courtesy of Marlene Nourbese Phillip; "Meditations on the Declension of Beauty by the Girl with the Flying Cheekbones" reprinted by permission of the author. p. 245 Photograph of Robyn Sarah © Porcupine's Quill; "Levels" reprinted from *A Day's Grace—Poems* by Robyn Sarah, by permission of the Porcupine's Quill. Copyright © Robyn Sarah, 2003.

Chapter 16 p. 254 Photograph of William Shakespeare © Collection Roger Viollet / Topfoto / PONOPRESSE.

Chapter 17 p. 264 Photograph of Susan Glaspell © Topham Picturepoint / PONOPRESSE.

Unit 4 p. 297 "Essay 1" reprinted by permission of Josée Bissonnette; "Essay 2" reprinted by permission of Alexandra Deschamps-Sonsino. p. 303 "A comparison of 'Finishing School' and 'The Veldt'" by Alexandra Deschamps-Sonsino reprinted by permission of the author.

Unit 11 p. 374 "Romantic Elements in Edgar Allan Poe's 'The Tell-Tale-Heart'" by Virginie Lachapelle reprinted by permission of the author. p. 375 "The Role of Women in Kate Chopin's 'The Story of an Hour'" reprinted by permission of the author.

Overview

We have designed *Style and Substance* for the teaching of literary analysis and essay writing at the college level in contexts such as advanced ESL or remedial native language courses where students would benefit from a stimulating but in-depth, step-by-step approach to understanding and writing literary analysis.

For greater ease of use and efficiency, we have divided this second edition of *Style and Substance* into two major parts, "Literature and Analysis" and "Writing and Grammar." In the first part of the book, "Literature and Analysis," we have included a wide variety of texts (stories, poems and plays from different periods) and activities to allow instructors teaching 45- or 60-hour courses to tailor their course to the specific needs of their students. The chapters contain historical and cultural background (where pertinent), analytical exercises aimed at increasing the student's interpretative abilities, writing activities designed to develop the student's writing skills, and oral activities geared at preparing the student to make a finely honed oral analysis by the end of the term.

The students will encounter texts chosen not only for their excellence and style, but also for their embodiment of different epochs, themes, groups of people and points of view. We believe that the texts included in the book will appeal to a wide variety of readers. First and foremost, we hope that students will enjoy their initial direct contact with each piece. Then, as they work through the various exercises and activities, we hope the book will help stimulate them to delve deeper to find other levels of meaning and to appreciate how different authors use words and language in very effective ways.

The second part, "Writing and Grammar," has been designed to teach writing a literary analysis in a step-by-step process and to develop grammatical accuracy. The students gradually develop their writing skills by responding to the literary works in the first part of the book. Many different writing topics are suggested at the end of each chapter. As students become more sensitive to both "style and substance," they in turn will sharpen their own writing skills. We have expanded the explanations and activities for essay writing and reorganized and added to the grammar rules and exercises.

In this second edition, we have kept the elements that instructors and students appreciated in the first edition, such as the in-depth, step-by-step approach to understanding literary analysis, key background information, end-of-the-book glossaries and the original selection of short stories. Moreover, we have added new reading material, and we have integrated My eLab activities into each chapter of the book.

We feel, too, that it is very important for students to experience poetry and drama aurally. Works belonging to both of these genres have much more impact when they are heard and often seem very flat in the written mode. To bring the selected poetic and dramatic pieces to life, we have included audio recordings of the poems and links to an audio version of *Trifles* in My eLab.

In My eLab Documents, instructors will find practical teaching suggestions (including how to integrate new technology to advantage in a literature class), answer keys, assessment rubrics and reproducibles as well as a PowerPoint presentation for teaching literary elements and doing in-class error analysis.

In conclusion, we believe that this second edition of *Style and Substance* offers both teachers and students a very comprehensive and stimulating introduction to literary analysis and essay writing.

In a Nutshell

This second edition of *Style and Substance* features the following:

- fifteen short stories – five **NEW** ones including stories by award-winning authors Roald Dahl and David Bezmozgis, a young Canadian author
- nineteen poems – ten **NEW** ones including poems by four award-winning Canadian poets: Earle Birney, Irving Layton, Leonard Cohen and Robyn Sarah
- two plays – *Trifles* by Susan Glaspell and a **NEW** one, *Othello* by William Shakespeare
- key background information on authors, historical periods, literary movements and the evolution of various genres
- detailed explanations on analyzing literature and on using literary terminology effectively
- written and oral exercises and activities to develop appropriate writing and speaking skills
- a **NEW** writing section that presents a step-by-step process including prewriting, writing paragraphs and essays, revising for style and editing for errors, and which comprises many student writing models of paragraphs and essays
- a **NEW** grammar section that includes most points from the first edition but also the following items: sentence variety, using exact language, avoiding clichés, avoiding slang, avoiding pronoun and verb tense shifts, citing sources using MLA format, editing two student essays
- a completely **NEW** My eLab with interactive activities integrated with each chapter and section of the book and mp3 recordings of all the poems
- glossaries of literary and grammatical terms
- an Instructor's Resource Manual that includes teaching suggestions, answer keys, reproducible material and a **NEW** PowerPoint presentation

---

To the Student

> *"Ye who read are still among the living, but I who write shall have long since gone my way into the region of shadows. For indeed strange things shall happen, and many secret things be known, and many centuries shall pass away, ere these memorials be seen of men. And, when seen, there will be some to disbelieve, and some to doubt, and yet a few who will find much to ponder upon in the characters here graven with a stylus of iron."*
>
> — from "Shadow — a Parable" (1835) by Edgar Allan Poe

You are about to embark upon a journey of discovery in which you will enter different realities that various authors have crafted just for you, the reader. Along the way, you will meet fascinating and strange characters and enter into lives and times that may be very different from your own. As you approach a text for the first time, we invite you to open your mind and imagination to the experience the author is inviting you to explore. Let yourself be carried away by the power of words and storytelling. Let your thoughts and feelings flow where they will.

Next, we will ask you to stand back and re-examine the piece you have just read from various angles in a more formal fashion. You and your classmates will look for meaning on different levels and interpret the work using appropriate terms as well as your own insights and experience. Then you will present your response and analysis first orally in discussions and then in writing. At the end of the term, you will be able to write an analytical essay in which you express your thoughts orally in a logical, concise fashion.

We hope that you will be able to say that not only did you enjoy reading the texts, you also thought and learned about many new things, and subsequently improved your own writing and speaking skills. In the end, we believe that you will surely write, and speak, with more "style and substance."

Highlights

Part 1 and Part 2

Style and Substance is divided into two parts: the first presents literature (short stories, poetry and drama) and the second, writing and grammar.

History of ...

Each new literature section presents a brief historical overview of the genre featured to give students a sense of time and place.

Literary Elements
Poetic Elements

These two handy reference sections, before the short story and poetry sections, introduce students to the basic elements needed for literary analysis.

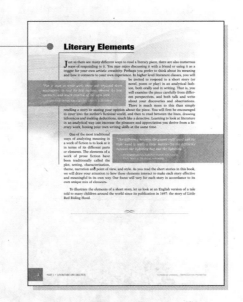

Chapter Contents

The important elements of each chapter are presented in point form so students are immediately aware of the chapter's focus.

About the Author

Students are introduced briefly to the author of the featured work so they can approach the text from a broader perspective.

Literary Trends of the Times

Students read background information about the period so they can situate the work in a particular context. This helps them approach texts from other periods more effectively.

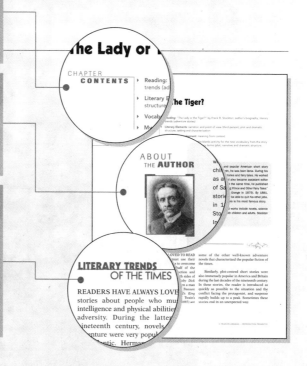

Warm-Up Discussion Topics

This activity appeals to prior knowledge and allows students to prepare for reading.

Initial Reading

Students read the text a first time for enjoyment.

Reader's Response

Students are encouraged to express their spontaneous feelings and thoughts about the text.

Close Reading

Students read the text a second time to check on comprehension and to hone analytic skills.

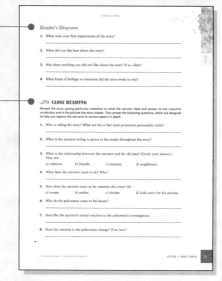

Vocabulary Development

Students work on activities involving vocabulary from each story to increase their command of the language.

Wrap-Up Activities

Students are encouraged to put the skills and knowledge they have acquired into practice both orally and in writing.

My eLab Activity

Extension activities have been designed to extend and increase students' knowledge and interest.

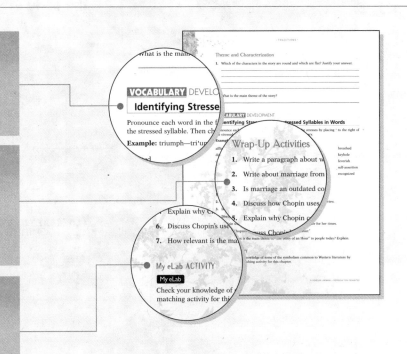

Glossaries

Students can consult glossaries of important literary and grammatical terms to consolidate their knowledge.

Glossary of Literary Terms

Abstractness — not perceptible by the senses, only by the mind
Action — events taking place in a dramatic or narrative work
Acts — divisions of a play that usually last from thirty minutes to an hour
Alliteration — repetition of initial consonant sounds in a series of words
Anapestic foot — a poetic foot of three syllables, the first two unaccented and the last accented
Antagonist — a person or a power who opposes the main character in a narrative or dramatic work
Antithesis — a linking of two contrasting ideas to make each more vivid
Apostrophe — an address to a dead person as if he or she were still alive
Assonance — repetition of the same vowel sound
Atmosphere — the emotional mood of the text
Blank verse — a poem that has meter but no rhyme
Cacophony — unpleasant or harsh combination of sounds
Character — all persons found in a dramatic or narrative work
Climax — the point of maximum conflict of the story
Comparison — looking at similarities and differences
Complication — a conflict in the story
Concreteness — perceptible by the senses
Conflict — the opposition of different characters and forces that may be either internal or external
Connotation — the secondary level of a meaning
Consonance — repetition of consonants in a series
Couplet — two lines that rhyme
Dactylic foot — poetic foot of three syllables, the first accented
Denotation — the literal or dictionary meaning of a word
Denouement — the conclusion of the story
Dialogue — conversation in written form
Dramatic meter — poetic meter of two feet
Dramatic structure — the beginning, middle, and end of a story
Drama — a theatrical production intended for actors to perform
Epic — a long poem about someone heroic or noble
Essay — short written composition on a single subject
Euphony — pleasant or harmonious sounds
Exposition — what happens at the beginning of a story
External action — action that happens outside the dramatic work
Fiction — all written works that are created in the imagination

Glossary of Grammatical Terms

Clause — a group of words containing a subject and a verb, but not necessarily expressing a complete idea
Comma splice — incorrect use of comma between two independent clauses when a period should be used
Complex sentence — one independent clause and at least one dependent clause
Compound sentence — two independent clauses joined by a coordinate such as "and," "but," "or," etc.
Compound-complex sentence — two independent clauses and at least one dependent clause
Co-ordinating conjunction — words such as "and," "but," "or," "so," "yet"
Co-ordination — joining of two or more ideas of equal importance
Dependent or subordinate clause — a clause which expresses an incomplete idea
Direct quotation — repetition of the exact words a person has said or written using double quotation marks
Homographs — words that are spelled the same but pronounced differently and with different meaning
Homonyms — words that are pronounced the same but have different spellings and meanings
Independent clause or principal clause — clause that expresses a complete idea
Indirect quotation — stating indirectly what someone else has said
Modifiers — words, phrases, or clauses that describe or qualify another element in the sentence
Paragraph — a group of sentences with one main idea
Parallel construction — refers to the use of similar grammatical structures
Paraphrasing — restating in your own words someone else's ideas or texts
Phrase — a group of words often beginning with a preposition or participle
Prefix — additions affixed "in front of" root words in order to give them particular meanings
Principal or main clause — see independent clause
Pronouns — words that replace nouns
Quotations — repetitions of exact words a person has said or written
Relative pronouns — words such as "who," "whom," "whose," "which" and "that," which are used to show subordination
Run-on sentence — at two or more independent clauses with no punctuation or too many conjunctions (and, but, or) between them
Sentence — a group of words containing a subject and a verb and expressing a complete idea
Sentence fragment — an incomplete idea, part of a sentence punctuated as if it were a complete sentence
Simple sentence — one independent clause
Subordination — shows a relation between ideas having different degrees of importance; less important idea is subordinated to a more important idea

Table of Contents
Style and Substance, Second Edition

PART 2   WRITING AND GRAMMAR

Part 1

LITERATURE and ANALYSIS

Literary Elements

Just as there are many different ways to read a literary piece, there are also numerous ways of responding to it. You may enjoy discussing it with a friend or using it as a trigger for your own artistic creativity. Perhaps you prefer to think about its meaning and how it connects to your own experience. In higher level literature classes, you will be invited to respond to a short story (or novel, poem or play) in an analytical fashion, both orally and in writing. That is, you will examine the piece carefully from different perspectives, and both talk and write about your discoveries and observations. There is much more to this than simply retelling a story or stating your opinion about the piece. You will first be encouraged to enter into the author's fictional world, and then to read between the lines, drawing inferences and making deductions, much like a detective. Learning to look at literature in an analytical way can increase the pleasure and appreciation you derive from a literary work, honing your own writing skills at the same time.

> *"For a man to write well, there are required three necessaries: to read the best authors, observe the best speakers, and much exercise of his own style."*
>
> —seventeeth-century playwright Ben Johnson in *Discoveries*.

One of the most traditional ways of analyzing meaning in a work of fiction is to look at it in terms of its different parts or elements. The elements of a work of prose fiction have been traditionally called the plot, setting, characterization, theme, narration and point of view, and style. As you read the short stories in this book, we will draw your attention to how these elements interact to make each story effective and meaningful in its own way. Our focus will vary for each story in accordance to its own unique mix of elements.

> *"The difference between the almost right word and the right word is really a large matter—'tis the difference between the lightning bug and the lightning."*
>
> —nineteenth-century humorist, essayist and novelist Mark Twain in *The Art of Authorship*.

To illustrate the elements of a short story, let us look at an English version of a tale told to many children around the world since its publication in 1697: the story of Little Red Riding Hood.

Little Red Riding Hood
Charles Perrault

1 ONCE upon a time there was a little girl, the prettiest one of all the village. Her mother loved her dearly. Her grandmother loved her even more, and made her a little red hood that looked so good on her that everyone called her Little Red Riding Hood.

2 One day, her mother, who had just baked some cakes, called her and said: "Go and see how your grandmother is. I've heard that she is sick. Take her a cake and this little pot of butter."

3 Little Red Riding Hood set off at once for her grandmother's cottage, located in another village. On her way through the forest she met a wolf. He would have very much liked to eat her on the spot, but dared not do so because of some wood-cutters who were in the forest. He asked her where she was going. The poor child, not knowing that it was dangerous to stop and talk to a wolf, said: "I'm on my way to my grandmother's. I'm taking her a cake and a pot of butter from my mother."

4 "Does she live far away?" asked the wolf.

5 "Oh, yes," replied Little Red Riding Hood, pointing; "far over there by the mill, the first house in the village."

6 "Well," said the wolf, "I think I'll go and see her, too. I'll take this path, and you take that one, and we'll see who gets there first." The wolf took off down the shorter road running as fast as he could. The little girl continued happily on her way along the longer one, stopping from time to time to gather nuts, run after butterflies, and pick wild flowers.

7 The wolf soon arrived at the grandmother's house. Knock. Knock.

8 "Who's there?"

9 "It is your granddaughter, Red Riding Hood," said the wolf, in a thin, high voice, "and I've brought you a cake and a little pot of butter as a present from my mother."

10 The grandmother who was sick in bed called out, "Come on in. The door's not locked." The wolf flew in through the open door, sprang upon the poor old lady and gobbled her up in an instant, for he hadn't eaten in three days.

11 Then, he shut the door, lay down under the covers of the grandmother's bed, and waited for Little Red Riding Hood to arrive.

12 Knock. Knock.

13 "Who's there?"

14 Upon hearing the wolf's hoarse voice, Little Red Riding Hood felt frightened at first, but then thinking that her grandmother had a bad cold, she replied: "It's your granddaughter, Little Red Riding Hood. I've brought you a cake and a little pot of butter from my mother."

15 Speaking in a high voice, the Wolf called out to her: "Come on in. The door's not locked." Little Red Riding Hood came in through the open door. The wolf lay hidden under the covers in the bed. "Put the cake and the little pot of butter on the night table," he said, "and come up on the bed nearer to me."

16 Little Red Riding Hood took off her jacket, but when she climbed up on the bed she was astonished to see how her grandmother looked in her nightgown.

17 "Granmama!" she exclaimed, "what big arms you have!"

18 "The better to hug you with, my child!"

19 "Granmama, what big legs you have!"

20 "The better to run with, my child!"

21 "Granmama, what big ears you have!"

22 "The better to hear with, my child!"

23 "Granmama, what big eyes you have!"

24 "The better to see you with, my child!"

25 "Granmama, what big teeth you have!"

26 "The better to eat you with!" And, with these words, the wicked wolf jumped on Little Red Riding Hood and gobbled her up.

(Translated from the original by C. Rock.)

The End

∞ PLOT: DRAMATIC AND NARRATIVE STRUCTURE

One of the most fundamental elements of a traditional short story is its **plot**. The plot is what happens in the story. In our example, a young girl is sent to another village by her mother to deliver a cake to her grandmother, who is ill. On her way, she passes through a forest where she stops to talk to a wolf. He beats her to the grandmother's cottage, eats the old lady, disguises himself and then lies in wait for the girl to arrive. When she does, he eats her too. This is the story as seen on a first level of meaning.

When doing literary analysis, one of the aspects to examine is how the plot is constructed. First, there is the **exposition** or description of the situation at the outset: the mother telling her daughter to take some food to her sick grandmother. In itself, this request would seem to be a simple task, but a **complication** arises: the little girl runs into a wolf in the forest. The **tension** starts to rise at this point because the girl does not realize that she is in danger. The reader knows that wolves are dangerous and wonders what is going to happen; it cannot be good. The tension continues to build as the unsuspecting girl arrives at the cottage and engages in a conversation with the wolf. As the dialogue continues, the reader feels the girl's suspicions about the situation growing as she realizes that it is not her grandmother in the bed. Every reader can relate to the sinking feeling she must be experiencing as she starts to realize the truth about her predicament. The reader wonders if the little girl will react in time. No. Horror of horrors! The **climax** of the story occurs when the wolf gobbles Little Red Riding Hood up, too. Perrault's story ends abruptly with the finality of this awful act, and leaves the reader somewhat suspended in disbelief and hoping that there is more to come, in other words, a happy ending, such as the one the Brothers Grimm added to this tale. However, the story really has ended and the tension has been released by the final outcome, or **dénouement**.

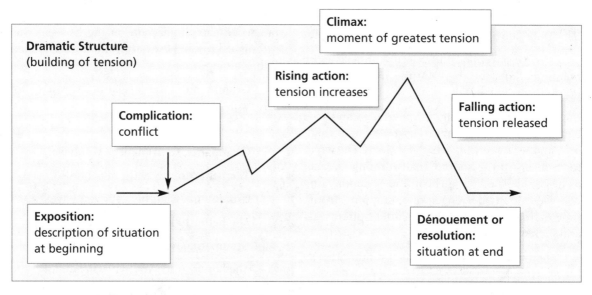

In summary, in its barest form, the **dramatic structure** of a short story often entails the exposition of a situation, followed by a complication of some sort. The main character is faced with a situation of conflict that must be resolved. The tension rises until it reaches a climax, something happens to resolve it, and then the story ends. Of course, in reality, the dramatic structure of short stories is often more complex than this. The tension often rises and falls several times before coming to a climax. It is interesting to note that in many modern stories, the tension is sometimes only partly resolved, and just as in real life, where the reader learns to live with half-solved problems and partial solutions, often there is no final, definite resolution.

The conflict experienced by the main character can be caused by an external agent, such as another character or a difficult situation, or by an internal one, such as his or her conscience or desire. The action in the story can often be seen on two levels at once; in other words, what is going on inside a person's mind as well as the external events taking place around him or her. In the story "Little Red Riding Hood," the **external action**—for example, the meeting of the girl with the wolf in the woods—is described explicitly, whereas the reader is left to imagine the **internal action** going on in the girl's mind as she begins to realize that she is in danger.

The **narrative structure** relates to the order in which the events of the story unfold. The events can be presented in chronological order, where the action starts at one point in time and progresses forward, or they can be presented in non-chronological order, where the time sequence of the story is varied and does not progress forward in a linear way. Writers use literary devices such as **flashback**, a reference to an event in the past, and **foreshadowing**, a hint at what will happen in the future, to create an interesting effect in the plot of the story. "Little Red Riding Hood" is presented mainly in chronological order. She meets the wolf, who hurries to the grandmother's house before the girl and waits for her arrival. The story contains one powerful example of foreshadowing—the moment when the reader learns that the wolf would like to eat her on the spot. This act, of course, is exactly what he does at the end of the story. There is another less direct hint of what is to come in the future when the wolf asks her where she is going. The reader already knows that the wolf wants to eat her, but is unable to do so in the forest. Therefore, the question is very ominous, and the reader guesses that the wolf is going to do something bad.

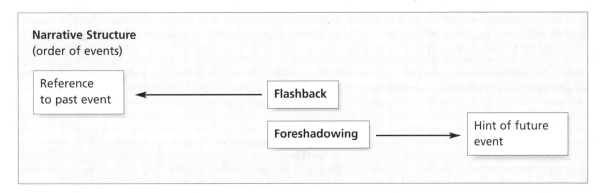

Narrative Structure
(order of events)

Reference to past event ← **Flashback**

Foreshadowing → Hint of future event

One last word about the importance of plot: the plot and dramatic structure are often the elements that move a story along and make it exciting to read. However, these elements are always interwoven with the characters and setting to give an underlying meaning or theme to a text. That is, if the story had a different plot, the characters would react differently, and if it had different characters or a different setting, the plot would not be the same.

USEFUL QUESTIONS FOR DISCUSSING THE PLOT

- What is the dramatic structure of the plot?
  - Is it a traditional or modern structure?
  - Are the events fast moving or slow?
  - Is the reader's interest engaged at the very beginning of the story, or only later on?
  - Can the action be seen on more than one level?
- In what ways is the dramatic structure of the plot effective?
- What is the central conflict in the story?
  - Is it an internal conflict, an external conflict or both?

- How is suspense or tension created by the plot?
- How important is the plot to the story?
  - Does it play a central role or a minor one?
  - How is it interwoven with the characterization and setting?
  - What effect does the plot have on the main theme of the story?
- What is the climax of the story?
- Is the ending effective/suitable?
- In what order do the events unfold in the story? Do they unfold chronologically or non-chronologically?

∽ SETTING

The **setting** of a short story refers to the time and place it occurred. A story can be set in the present, past or future, and the setting usually refers to a physical place, either real or fictional. Sometimes the **general** setting (epoch, historical period, continent, etc.) of a story is more significant than the **specific** setting (particular time and place). Usually a setting is **external** (outside a character), but it can also be **internal** (inside the head of a character), and sometimes the action shifts subtly between the two, for example, in the story "An Occurrence at Owl Creek Bridge."

The setting gives the story a concreteness and helps the reader connect with the characters and the plot. By using vivid imagery to describe the elements of the setting, the author can create a **mood** for each scene and an overall **atmosphere**. In the case of "Little Red Riding Hood," the story is set somewhere in Europe at an undetermined time in the past. The actual date or country of the story is irrelevant because historical significance is unimportant in the context. However, the main character must walk through a forest, which adds atmosphere, tension and meaning to the story.

> **Setting**
> (time and place)
>
> **General** setting: epoch, historical period, continent, etc.
>
> **Specific** setting: exactly where and when the story takes place
>
> **External** setting: outside a person
> **Internal** setting: inside the head of a character
>
> Settings often create a **mood** (a feeling or emotion connected with a scene) and **atmosphere** (a work's general emotional tone).

This particular setting has a direct influence on the plot, characters and meaning (theme) of the story. Not only are forests dark, scary places for many people at night—places where they can lose their way even during the day—but symbolically, forests are also places where evil lurks in waiting to prey on the unsuspecting and innocent. In forests, dangerous animals, such as wolves, lie hidden in wait. Moreover, forests can also be interpreted in Freudian terms as representing the human unconscious, filled with submerged fears of death and violence, and repressed sexuality. Thus, this particular element of the setting in "Little Red

Riding Hood" is absolutely central to the story. It not only determines the action in the plot but also lends an atmosphere of pregnant fear to the tale.

USEFUL QUESTIONS FOR ANALYZING THE SETTING

- Does the setting have a historical or geographical significance?
- How important is the setting of the story (of major or minor importance)?
- How does the setting affect other aspects of the story, such as plot, characterization and theme?
- How does the setting influence the mood and atmosphere of the story?
- Is the setting believable to the reader? Does it establish credibility?
- Does the setting symbolize something?
- Is the setting suggestive about the culture, philosophy or spirit of the times?

∞ CHARACTERIZATION

When reading fiction, the reader makes judgements about the story based on the characters. If a reader cannot identify or sympathize with a character, very often he or she will have a negative opinion of the story. Therefore, characterization is a very important element when it comes to analyzing a work of fiction. Characterization has its root in the word "character," and it refers to the technique in which a writer develops and portrays the characters in the story. There are different terms used for the different characters in a fictional work. The main character of a story is called the **protagonist**, and the person who is in conflict with the protagonist is referred to as the **antagonist**. The terms "protagonist" and "antagonist" do not necessarily have positive and negative connotations respectively. That is, the terms do not connote any value judgements, and therefore, a protagonist and an antagonist may or may not have heroic qualities. While the protagonist is easy to identify because the plot revolves around him or her, the antagonist is sometimes harder to identify. In the story "Little Red Riding Hood," the title of the story identifies the protagonist. The antagonist is not a human being in this case, but a wolf. In other stories, like Hemingway's "The Old Man and the Sea," the antagonist is a marlin with which the old fisherman struggles. Sometimes the antagonist might be a difficult environment or social force, such as a winter storm or a war, against which the protagonist must struggle.

Character development is another important consideration. A **round character** is a well-developed and multi-faceted character. These characters evolve and change as the story progresses. If they evolve, the characters are identified as **dynamic**. The opposite of round characters are **flat** or one-dimensional characters. These types of characters do not grow or change and are therefore called **static** characters. Usually such character types are **stereotypes**. It is important to mention that as with the terms "protagonist" and "antagonist," "flat" and "round" characters do not bear any positive or negative connotations. Each type serves a purpose to the story as a whole. Little Red Riding Hood and the wolf, even though they represent the protagonist and the antagonist, are both flat characters. She is a naïve little girl, and he is the stereotypical villain who wants to eat her. They do not evolve or change throughout the story.

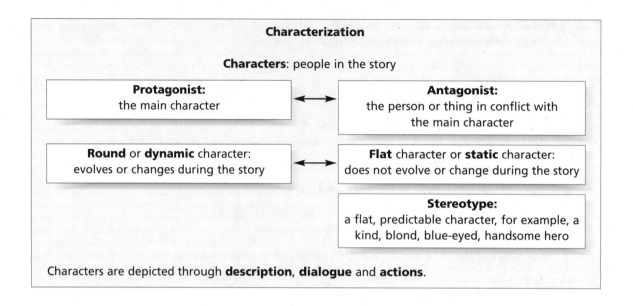

Characterization

Characters: people in the story

| Protagonist: the main character | Antagonist: the person or thing in conflict with the main character |

| Round or dynamic character: evolves or changes during the story | Flat character or static character: does not evolve or change during the story |

Stereotype: a flat, predictable character, for example, a kind, blond, blue-eyed, handsome hero

Characters are depicted through **description**, **dialogue** and **actions**.

Writers reveal their characters in many different ways. A writer can describe a character. For example, a writer can depict a character's physical appearance and personality through detailed **description**. Little Red Riding Hood obviously wears a red cape with a hood. She is also described as being very pretty and loved by her mother and grandmother. The reader can assume that she has a sweet and charming personality. Sometimes a name of a character can be symbolic, revealing something vital about the character. For example, Old Man Warner and Mr. Graves are two characters in the short story "The Lottery" who are meant to be taken seriously.

Furthermore, a writer can establish a character through **dialogue**. The subject and tone of the conversation convey a lot about a character. For example, a reader should analyze the subject of the dialogue in terms of its importance. Who is participating in the conversation is also another important factor. Where and when the conversation is taking place is a further consideration. Also, the quality of the conversation and the attitude of the speakers are very important. In "Little Red Riding Hood," the conversation between her and the wolf in the woods gives clues as to the wolf's motive. The reader knows that the wolf is trying to get as much information as possible from her about her grandmother. The dialogue between her and the wolf at the grandmother's house has an ominous tone to it. Little Red Riding Hood's questions become pointed, showing her growing alarm and suspicions in response to the wolf's answers.

In addition, writers also depict characters through their **actions**. A character's behaviour in different circumstances gives clues to his or her personality. A character can act or react. Very often, as with real life, a character does both, either consciously or unconsciously. Little Red Riding Hood reacts to the threat of the wolf by trying to run away from him. The wolf behaves very predictably since he wants to eat the girl.

Above all, a character must be believable or credible. He or she must behave in a convincing manner, and any changes in his or her personality must be motivated. Characters are crucial in fiction. They make the story a story.

USEFUL QUESTIONS FOR ANALYZING CHARACTERS

- Who is the protagonist in the story?
- Who or what is the antagonist?
- What are the main actions of the protagonist and antagonist?
- What are their character traits or personalities?
- What methods does the writer use to depict characters?
- Are they round or flat characters?
- Who are the secondary characters?
- What is their importance or relevance to the story?
- Are the characters of the story interesting and credible?
- How does the setting and plot influence the characters?

∽ THEME

The **theme** of a piece expresses the underlying meaning a reader derives from it. The main theme, in other words, is the unifying philosophy on life that the author wishes to convey. The reader's philosophy can differ from that of the author, but it is what the author wishes to tell the reader, whether it is a universal human truth or a specific observation, that must be regarded as the theme.

The complexity of a theme can vary from work to work. Readers might interpret the main theme of a work differently because important secondary themes may be present. In certain works, themes are explicitly stated, while in others, they are implied. The different literary elements, such as plot, setting and so on, interact and complement each other in order to develop the theme of a work.

To discover the theme of a work, it is often useful to search out the ideas derived from the observations of the author that the work embodies. It is important not to confuse the **subject** of the work with the ideas or observations of the author. Usually the subject of a work can be expressed in one word, such as "love," "marriage," "hatred" and so on. Ideas and observations are judgements that express the values of the writer. The reader is also able to interpret the ideas of the work and make a value judgement based on his or her life experiences. The theme of a work should be expressed, not only as an idea, but also as a value statement that offers opinion or judgement for discussion. For example, the conflict in this story can be interpreted as the classic duel between good (the child) and evil (the wolf), or between male and female, with the male preying on the female! If a reader rereads the story with an eye open to signs of the latter, he or she will notice quite a lot of sexual imagery, such as the dark forest, the red colour (blood, passion, menstruation) of a young virgin's "hood," the devouring of the female by the wolf in a bed, and so on.

Theme
(meaning)

Theme: the underlying message or meaning

Subject of a work: general topic, such as "love," "growing old" and so on

Theme of a work: value statement, such as "love is wonderful," "growing old is difficult" and so on

---

USEFUL QUESTIONS FOR EVALUATING THE THEME OF A WORK

- What is the work about?

- What are the ideas and values demonstrated in the work?

- Are these implied or stated explicitly?

- Do the other elements convey the same ideas and values? (Look for what characters say and think.)

- Do the other elements, such as characters and setting, symbolize an idea?

- Are the ideas and values universal?

- Does the work contain any secondary themes?

- In what ways do the secondary themes relate to the main theme?

∞ NARRATION AND POINT OF VIEW

The **narrator** is the person or voice who is telling the story. The **point of view** is the perspective from which the story is told. It is important not to confuse the author with the narrator, or the author's opinion about a particular subject with the point of view expressed in the story.

Fiction abounds with different types of narrators. When a story is told by a character who participates in the story, either directly or indirectly, the narrator is referred to as the **first-person narrator**, and the story is told from his or her point of view. The first-person pronoun "I" is used to relate the story. The first-person point of view has advantages and disadvantages. Because the narrator ("I") is a direct participant in the story, the events and actions in the story seem much more immediate and intimate. However, this type of narrator can give only incomplete information about the actions and thoughts of the other characters because he or she can only interpret them from his or her own perspective. That is, the reader receives interpretations of and judgements about other characters from the first-person narrator. The reader then has to decide if the information and the interpretation of the actions are accurate. Writers often use the first-person narrator when writing thrillers or detective novels because they can then give readers only select information.

The **first-person plural narrator** is rare in fiction. This narrator tells the story as part of a group, using the first-person plural pronoun "we."

When a narrator does not directly or indirectly participate in a story, but is an observer, he or she is called the **third-person narrator**. This type of narrator refers to the characters using the third-person pronouns "he" or "she," and can tell the story from two points of view: the **omniscient point of view** and the **limited point of view**. In the first point of view, the narrator knows everything about all the characters and events. Therefore, it is called the omniscient point of view, and the narrator is all-seeing and all-knowing. However, an omniscient narrator can relate the story **objectively**, or he or she can interpret it and offer judgements. In other words, the third-person narrator can be **intrusive**. In the second point of view, the narrator can limit himself or herself to telling the story from the point of view of one character in the story.

In the example, the story is told by a third-person omniscient narrator who knows what is going on in the minds of each character. For instance, the narrator tells the reader what the wolf is thinking in the forest when he meets Little Red Riding Hood (he'd love a snack!), and the reader learns that she is feeling frightened when she first hears the wolf's voice after knocking on the door of her grandmother's house. This type of narration serves to distance the reader from the characters, putting him or her in the position of a spectator who often knows more than the protagonist as the action continues to unfold.

One last point to mention about narration is that many novels written in the latter half of the twentieth century contain **multiple narrators**. That is, parts of a story are told from the point of view of one character, and other parts (or even sometimes the same parts) from the points of view of other characters. This is a very interesting technique because it gives the reader two or more points of view about what is happening in the story.

Narration and Point of View
(how the story is told)

Narrator: the person or voice telling the story

Point of view: the perspective from wich the story is told

First-person: The narrator is **directly involved** in the story; he or she uses "**I**" (first-person singular) or "**we**" (first-person plural) to tell the story.

Third-person: The narrator is **not involved** in the story, only an **observer**; he or she uses "**he**," "**she**" or "**they**" to tell the story.

Omniscient point of view: all-seeing, all-knowing narrator

- **Objective:** The narrator reports only the facts.
- **Intrusive:** The narrator offers opinions and judgements.

Limited point of view: The narrator can only give incomplete information.

Multiple narrators tell the story from several different points of view.

Narration and point of view often affect how close the reader feels to the people and events in the story.

USEFUL QUESTIONS FOR UNDERSTANDING NARRATION AND POINT OF VIEW

- Who is the narrator of the story? Is there only one narrator, or are there multiple narrators?
- Is the narrator credible as an interpreter of the actions in the story?
- How does the narrator gain access to the information in the story?
- Does the narrator have a positive or negative opinion of the characters?
- What is the point of view of the work?
- Is the story told from only one point of view, or from many points of view?
- How does the point of view affect the reader's sense of involvement in the story (closeness to the characters and action)?

∞ STYLE

Style refers to the techniques of language a writer uses in a specific work. When analyzing style, it is common to consider two aspects. The first aspect is called **diction**. Diction refers to a writer's (or a speaker's) choice and use of words, and to the words' characteristics. Usually a writer controls and manipulates language very carefully for effect since it is through language that he or she conveys his or her ideas. Diction consists of several levels of analysis. The reader should look at the relationship between the **denotation**, or literal meaning, and **connotation**, or suggested meaning, of the words. Also, the level of **concreteness** (perceptible by the senses) or **abstractness** (not perceptible by the senses—only by the mind) should be taken into account.

Figurative devices or **figures of speech** are expressions used in a non-literal way in order to create a special effect or to extend meaning beyond the limits of ordinary usage. The most common figures of speech are **similes**, which compare two things using "like" or "as," for example, "eyes like deep pools," and **metaphors**, which compare two things without using "like" or "as," for example, "the deep pools of his eyes." Other common devices include **personification** (giving human qualities to inanimate or abstract objects, for example, "the sighing wind"), **imagery** (using descriptive words to create an image that appeals to one of the senses, for example, "the golden rays of the setting sun"), and **irony** (making a statement or presenting an event that is the opposite of what is intended, expected or expressed, for example, "She died just at the moment she had the most to live for."). All these figurative devices are important elements of style. In addition, the reader should examine the **rhythm** or sound patterns of the sentences in terms of **alliteration** (repeating the initial letter or sound in words), **assonance** (repeating identical vowel sounds), **consonance** (repeating identical consonant sounds), and **repetition** of other sentence elements, such as words and structure. Finally, the reader should consider the **level of language** (language for a specific social setting), such as **formal speech** (hello), **informal speech** (hi), **slang** (howdy), and **dialogue** (conversation between two characters).

The second aspect to look at is called **syntax**. Syntax refers to the way in which the words, phrases and sentences are arranged. Important factors to consider are **sentence length** (long or short), **word order** (subject, verb, complement), and **sentence types** (simple, compound or complex). In conclusion, both diction and syntax contribute to the **tone** or **mood** of the work. Tone refers to the attitude of the writer to the subject of the prose piece. Attitude can reflect irony, anger, sentimentality or humour, or it can be objective or neutral.

In "Little Red Riding Hood," the author uses language associated with fairy tales, for example, "once upon a time" and "one day." This leads the reader to anticipate that something bad is going to happen, as it does in most fairy tales. Moreover, the appearance of the wolf in the forest brings forth connotations of fear and evil in the reader's mind. The instance of foreshadowing (when the wolf states that he would like to eat the girl upon their first meeting) also helps set the tone of growing dread that culminates in the repeated elements of the final dialogue between the girl and the wolf in disguise: "Granmama! What big ... you have!" followed by "The better to ... my child!" The fact that the personified wolf (he can talk) disguises himself as an innocent grandmother reinforces the good-versus-evil theme of the story and reminds the reader that he or she must beware of evil lurking in the most surprising guises.

Throughout the story, the author uses concrete language, for example, "gather nuts, run after butterflies, and pick wild flowers," which enables young readers to visualize the various scenes. The story is told objectively by a third-person omniscient narrator using simple, straightforward vocabulary, but the language strikes the contemporary reader as fairly formal. This formality is mainly due to the fact that the tale was written in Europe in 1697 using the language of the times. All in all, the reader can conclude that the style fits the content and purpose of the piece.

Style
(techniques of language usage)

Diction: a writer's (or a speaker's) choice and use of words and the words' characteristics

Denotation: the literal or dictionary meaning

Connotation: the associated or suggested meaning

Level of concreteness: perceptible by the senses

Level of abstractness: not perceptible by the senses—only by the mind

Figures of speech: expressions used in a non-literal way to create a special effect or to extend meaning
  Simile: a comparison of two things using "like" or "as"
  Metaphor: an implied comparison of two things without using "like" or "as"
  Personification: giving human qualities to inanimate or abstract objects

Imagery: using descriptive words to create an image that appeals to one of the senses

Irony: the opposite of what is intended, expected or expressed

Rhythm: sound patterns
  Alliteration: a repetition of the initial letter or sound in words
  Assonance: a repetition of identical vowel sounds
  Consonance: a repetition of identical consonant sounds
  Repetition: a repetition of other sentence elements, such as words and structure

Level of language: language for a specific social setting
  Formal speech: conventional language that respects the rules
  Informal speech: everyday, unofficial language
  Slang: casual speech

Use of dialogue: conversation between two characters

Syntax: the way in which the words, phrases and sentences are arranged

Sentence length: long, short

Word order: subject, verb, complement

Sentence types: simple, compound, complex

Diction and syntax contribute to a story's **tone** or **mood** (attitude of the writer toward the subject).

USEFUL QUESTIONS FOR ANALYZING A WRITER'S STYLE

- What type of diction is found throughout the work? Consider the meaning of words in terms of their denotations and connotations.

- Are the words concrete or abstract?

- Is the language formal or informal? Is the story written in slang or dialect?

- What kinds of figurative devices, such as similes, metaphors and personification, does the story include?

- Does it include imagery?

- Is there a particular sound to the language with regard to repetition, alliteration, assonance or consonance?

- What types of sentences are found in the work?

- What is their length?

- What is the word order found in the sentences?

- What is the overall tone or atmosphere of the work?

- How is this tone or atmosphere created?

This brings to a close our presentation of the traditional elements comprising literary analysis. As you work through the various sections of the book, you will note that not only do these elements serve in the analysis of short stories, but many of them are also useful for interpreting plays and poems. Although analyzing a work may seem an arduous task at times, you will not be wasting your time. In fact, you will be developing a new awareness of the English language and the many ways it can be used, and your own composition skills will most certainly benefit. Moreover, your new analytical skills and heightened sensitivity will make you a better reader and no doubt increase your appreciation of the excellent texts you will read in the future, be they newspaper articles or novels.

SECTION 1

SHORT STORIES

After a brief summary of the evolution of the short story from a historical perspective, this section presents five early masterpieces that illustrate certain *Traditions* followed by seven more modern stories representing *The Evolving Genre* and three *Short Short Stories*. In each subsection, the works are presented in chronological order.

History *of the* Short Story

The short story developed as a genre in English very gradually over the centuries. It was inspired and influenced by narratives that emanated from other parts of the world. Scholars suggest that the short story had its origins in Middle Eastern narratives. With the dawn of spoken language, early peoples most surely began to tell tales of things that had happened. As early as 2000 BC, the ancient Babylonians recorded one of the oldest narratives in the world, the *Epic of Gilgamesh*. At roughly the same time, Egyptians were writing down tales in prose on papyrus.

The earliest recorded tales from India date from around 700 BC. About a century later, Aesop wrote his famous fables in Greece. Starting in the eighth or ninth century BC, the Hebrews recorded the narratives that now make up part of the Old Testament in the Bible, stories such as those of Adam and Eve, Cain and Abel, David and Goliath, and many more. Later, the New Testament, which contains numerous parables that Jesus was said to have told his followers, completed the Bible as we know it. The Bible had a huge influence on English and European culture and was the first book to have been printed by Johann Gutenberg using movable type in 1455. Before this revolutionary invention, books had been copied painstakingly by hand, and not many people had access to them. With the invention of movable type, books could be printed and put into circulation much more rapidly and widely. As a result, more and more people learned to read.

> *When a book, any sort of book, reaches a certain intensity of artistic performance it becomes literature. That intensity may be a matter of style, situation, character, emotional tone, or idea, or half a dozen other things. It may also be a perfection of control over the movement of a story similar to the control a great pitcher has over the ball.*
>
> —Raymond Chandler (1888-1959), American author. Letter, 29 Jan. 1946, to crime writer Erle Stanley Gardner.

Most of the ancient stories had a didactic purpose; they were aimed at teaching people important values. They taught people how to live a "good" life (and what would happen if they did not!) and how to distinguish right from wrong. In his fables, for example, Aesop ascribed human characteristics to animals and then placed the animals in some sort of conflict designed to teach a moral lesson. It was hoped that people would learn from these stories. In this tradition, tales about the lives of saints (called "exempla") became very popular in the eleventh and twelfth centuries. An exemplum was meant to show and inspire model behaviour.

In late-fourteenth-century England, influenced by ribald, comic short narratives called *fabliaux*, Geoffrey Chaucer (*c.*1343-1400) composed his famous *Canterbury Tales*. Although the tales were primarily written in verse, they are considered to be stories nonetheless. Sir Thomas Malory (*c.*1405-1471) was another early writer of stories, but instead of comedies, he wrote romances. His *Le Morte d'Arthur* represents the first complete account of the legends of King Arthur and the Knights of the Round Table. At roughly the same time, short fiction was taking similar form in other parts of Europe, and it was at this time that the stories of Scheherazade and *The Arabian Nights* became famous.

During the Elizabethan period, from approximately 1550 to 1625, the English were generally far more interested in drama than in short stories. The existing short stories were often copies or translations of flowery Italian novellas and were considered too superficial for popular taste. During the reign of Elizabeth I, Shakespeare came to the height of his powers as a brilliant dramatist who provided audiences with one compelling play after another.

Short fiction became a popular genre in the second half of the eighteenth century because of growing interest in the oriental tale and the Gothic novel. The Gothic novel, with characteristics of the dark, eerie and supernatural, was the precursor to the modern horror tale. For example, Mary Shelley's *Frankenstein* has Gothic elements. Between 1780 and 1840, the modern short story emerged at almost the same time in France, Russia, Germany and the United States. Tales of ghosts, horror and the supernatural, written in German by E. T. A. Hoffman (1776-1822) and Heinrich von Kleist (1777-1811), influenced the development of the genre in English.

In the United States, authors such as Washington Irving (1783-1859), Nathaniel Hawthorne (1804-1864) and Edgar Allan Poe (1809-1849) gave new forms to the short story and greatly accelerated its development by providing models for future writers. In fact, two basic types of short stories developed at this time: one more realistic, focused on the objective reality of events; the other more impressionistic, portraying the impressions left in a person's mind by reality (these impressions were often distortions of reality). Poe is considered by many to be the father of the modern short story in English because he defined the most central characteristic of the genre, the "unity of effect," which implies that the distinct elements of the short story, such as the plot, setting and characterization, must create a united picture. Each of his stories is structured so that an atmosphere of unity is conveyed to the reader.

> *The object of art is to give life shape.*
> —Jean Anouilh (1910-1987), French playwright. *The Rehearsal*

The short story as a literary genre in English-speaking countries continued to appeal to both writers and readers during the latter half of the nineteenth century and all of the twentieth century. Hundreds, if not thousands, of excellent stories were written during those years. In the chapters presenting individual short stories, we will examine how each one fits into the literary trends of its times.

The Tell-Tale Heart

CHAPTER
CONTENTS

▸ **Reading:** "The Tell-Tale Heart" by Edgar Allan Poe; author's biography; literary trends (Gothic novel, Romanticism)

▸ **Literary Elements:** narration and point of view (first person); style and literary devices (repetition, imagery, metaphor and simile, personification, contrast of opposites and irony)

▸ **Vocabulary Development:** Latin and Germanic roots, prefixes and suffixes

▸ **My eLab Activity:** a crossword puzzle on literary terms (narration, point of view, style and literary devices)

CHAPTER 1

ABOUT THE **AUTHOR**

Edgar Allan Poe (1809-1849) is considered to be one of the founders of the modern psychological thriller and short story. He developed theories about the new genre, and was the author of many horror stories. The result of his work ensured the popularity of narrative fiction and the short story in the nineteenth century, a popularity that has continued to the present day.

Poe was born and raised in Virginia. He established his writing career first as a poet and then as a fiction writer, publishing his first set of short stories in 1832. By 1835, Poe had been appointed editor of *The Southern Literary Messenger*, a magazine to which he contributed his own fiction as well as literary criticism. Poe's works show an overall predilection to the grotesque and macabre. His most famous works, "The Raven" (1845), "The Fall of the House of Usher" (1839), "The Pit and the Pendulum" (1842) and "The Tell-Tale Heart" (1843), make use of black elements such as murder, death, madness, melancholy and torture. Controversy still surrounds the circumstances of his death in Baltimore in 1849.

LITERARY TRENDS
OF THE TIMES

WHEN STUDYING LITERATURE, STUDENTS are often confounded about its meaning and utility. It is useful, therefore, to study literature within a context that takes into account its origins, theory or criticism. With this type of approach to literature, we will try to establish a historical context to the fictional works we will be studying.

"The Tell-Tale Heart" was published in 1843 in the magazine *The Pioneer*. Many adjectives come to mind when analyzing this story. For example, "weird," "grotesque," "bizarre" and "gross" are just some of the descriptions that can apply to this tale. All of these impressions are valid, and when put into a historical perspective, they become very understandable. Poe was writing narrative fiction in the first half of the nineteenth century. This period is referred to in European literature as the Romantic Era. Romanticism was, first, a reaction against the earlier period in literature called Classicism, an age that viewed human beings as rational and enlightened creatures. Second, Romanticism had its origins in the English Gothic novel, a popular genre from about the 1760s to the 1820s. Stories in this genre of novel were about mystery, terror or the supernatural. Dark, horrific events, which drew upon the irrational nature of humanity, predominated. Mary Shelley's *Frankenstein* (1817) is an excellent example of a Gothic novel.

During the Romantic Era, the theme of madness was quite often used to show the opposite of rationality. Furthermore, characters who were not part of mainstream society, such as robbers and hermits, dominated the literature of the time. Romantic literature was also saturated with subjects such as mysticism, the battle between good and evil, the spirit world, hallucinations, the dream world and the dark side of nature.

Today, in contemporary horror novels, short stories and films, we are still influenced by the tradition of Gothic novels and the elements of Romanticism. Modern writing is still fascinated by the supernatural and the dark side of nature. Writers such as Dean Koontz, author of *The Husband* (2006), and Clive Barker, author of *The Books of Blood* (1991), are only two of many contemporary writers of horror fiction. Of course, Stephen King is the most famous contemporary writer whose works, such as *Carrie* (1974), *The Shining* (1977) and *Lisey's Story* (2006), contain elements from the Gothic period.

Moreover, many young people worldwide refer to themselves as Goths. This movement started out as a punk subculture in the 1970s and is still popular today; its adherents wear black clothes and pale make-up, and have strange hairstyles and hair colours. Goth culture is represented in such films as *The Crow*, *Nosferatu* and the *Cabinet of Doctor Caligari*, as well as in the music of bands such as Bauhaus, Siouxsie and the Banshees, the Sisters of Mercy and Dead Can Dance.

Warm-Up Discussion Topics

1. What are some stories, novels or films you have read or seen that use Gothic and Romantic elements like the supernatural?

   _____

2. Did you find these literary works or films scary? Why or why not?

   _____

3. What makes scary stories or films so popular?

   _____

4. Have you ever read any works by Edgar Allan Poe? If so, which ones? Did you enjoy them?

   _____

∞ INITIAL READING

Read the story for the first time for personal enjoyment. Do not look up each unfamiliar word in the dictionary. If you were to do that, you would break up the reading process and be distracted from the story. Instead, try to guess the word from clues contained in the paragraph in which the word is found, that is, look for the meaning of unfamiliar words within their context.

The Tell-Tale Heart
Edgar Allan Poe

1 TRUE!—nervous—very, very dreadfully nervous I had been and am; but why will you say that I am **mad**[1]? The disease had sharpened my senses—not destroyed—not dulled them. Above all was the sense of hearing acute. I heard all things in the heaven and in the earth. I heard many things in hell. How, then, am I mad? Hearken! and observe how healthily—how calmly I can tell you the whole story.

2 It is impossible to say how first the idea entered my brain; but once conceived, it haunted me day and night. Object there was none. Passion there was none. I loved the old man. He had never wronged me. He had never given me insult. For his gold I had no desire. I think it was his eye! yes, it was this! He had the eye of a vulture—a pale blue eye, with a film over it. Whenever it fell upon me, my blood ran cold; and so by degrees—very gradually—I made up my mind to take the life of the old man, and thus rid myself of the eye forever.

3 Now this is the point. You fancy me mad. Madmen know nothing. But you should have seen me. You should have seen how wisely I proceeded—with what caution—with what foresight—with what dissimulation I went to work! I was never kinder to the old man than during the whole week before I killed him. And every night, about midnight, I turned the latch of his door and opened it—oh so gently! And then, when I had made an opening sufficient for my head, I put in a dark lantern, all closed, closed, that no light shone out, and then I thrust in my head. Oh, you would have laughed to see how cunningly I thrust it in! I moved it slowly—very, very slowly, so that I might not disturb the old man's sleep. It took me an hour to place my whole head within the opening so far that I could see him as he lay upon his bed. Ha!—would a madman have been so wise as this? And then, when my head was well in the room, I undid the lantern cautiously—oh, so cautiously—cautiously (for the hinges creaked)—I undid it just so much that a single thin ray fell upon the vulture eye. And this I did for seven long nights—every night just at midnight—but I found the eye always closed; and so it was impossible to do the work; for it was not the old man who vexed me, but his Evil Eye. And every morning, when the day broke, I went boldly into the chamber, and spoke courageously to him, calling him by name in a hearty tone, and inquiring how he had passed the night. So you see he would have been a very profound old man, indeed, to suspect that every night, just at twelve, I looked in upon him while he slept.

4 Upon the eighth night I was more than usually cautious in opening the door. A watch's minute hand moves more quickly than did mine. Never before that night had I felt the extent of my own powers—of my sagacity. I could scarcely contain my feelings of triumph. To think that there I was, opening the door, little by little, and he not even to dream of my secret deeds or thoughts. I fairly chuckled at the idea; and perhaps he heard me; for he moved on the bed suddenly, as if startled. Now you may think that I drew back—but no. His room was as black as pitch with the thick darkness (for

---

1. *Mad*: crazy.

the shutters were close fastened, through fear of robbers), and so I knew that he could not see the opening of the door, and I kept pushing it on steadily, steadily.

5   I had my head in, and was about to open the lantern, when my thumb slipped upon the tin fastening, and the old man sprang up in bed, crying out—"Who's there?"

6   I kept quite still and said nothing. For a whole hour I did not move a muscle, and in the meantime I did not hear him lie down. He was still sitting up in the bed listening—just as I have done, night after night, hearkening to the death watches in the wall.

7   Presently I heard a slight groan, and I knew it was the groan of mortal terror. It was not a groan of pain or of grief—oh, no!—it was the low stifled sound that arises from the bottom of the soul when overcharged with awe. I knew the sound well. Many a night, just at midnight, when all the world slept, it has welled up from my own bosom, deepening, with its dreadful echo, the terrors that distracted me. I say I knew it well. I knew what the old man felt, and pitied him, although I chuckled at heart. I knew that he had been lying awake ever since the first slight noise, when he had turned in the bed. His fears had been ever since growing upon him. He had been trying to fancy them causeless, but could not. He had been saying to himself—"It is nothing but the wind in the chimney—it is only a mouse crossing the floor," or "It is merely a cricket which has made a single chirp." Yes, he had been trying to comfort himself with these suppositions: but he had found all in vain. *All in vain*; because Death, in approaching him had stalked with his black shadow before him, and enveloped the victim. And it was the mournful influence of the unperceived shadow that caused him to feel—although he neither saw nor heard—to *feel* the presence of my head within the room.

8   When I had waited a long time, very patiently, without hearing him lie down, I resolved to open a little—a very, very little crevice in the lantern. So I opened it—you cannot imagine how stealthily, stealthily—until, at length a simple dim ray, like the thread of the spider, shot from out the crevice and fell full upon the vulture eye.

9   It was open—wide, wide open—and I grew furious as I gazed upon it. I saw it with perfect distinctness—all a dull blue, with a hideous veil over it that chilled the very marrow in my bones; but I could see nothing else of the old man's face or person: for I had directed the ray as if by instinct, precisely upon the damned spot.

10   And have I not told you that what you mistake for madness is but over-acuteness of the senses?—now, I say, there came to my ears a low, dull, quick sound, such as a watch makes when enveloped in cotton. I knew *that* sound well, too. It was the beating of the old man's heart. It increased my fury, as the beating of a drum stimulates the soldier into courage.

11   But even yet I refrained and kept still. I scarcely breathed. I held the lantern motionless. I tried how steadily I could maintain the ray upon the eye. Meantime the hellish tattoo of the heart increased. It grew quicker and quicker, and louder and louder every instant. The old man's terror must have been extreme! It grew louder, I say, louder every moment!—do you mark me well? I have told you that I am nervous: so I am. And now at the dead hour of the night, amid the dreadful silence of that old house, so strange a noise as this excited me to uncontrollable terror. Yet, for some minutes longer I refrained and stood still. But the beating grew louder, louder! I thought the heart must burst. And now a new anxiety seized me—the sound would be heard by a neighbour! The old man's hour had come! With a loud yell, I threw open the lantern and leaped into the room. He shrieked once—once only. In an instant I dragged him to the floor, and pulled the heavy bed over him. I then smiled gaily, to find the deed so far done. But, for many minutes, the heart beat on with a muffled sound. This, however, did not vex me; it would not be heard through the wall. At length it ceased. The old man was dead. I removed the bed and examined the corpse. Yes, he was stone, stone dead. I placed my hand upon the heart and held it there many minutes. There was no pulsation. He was stone dead. His eye would trouble me no more.

12   If still you think me mad, you will think so no longer when I describe the wise precautions I took for the concealment of the body. The night waned, and I worked hastily, but in silence. First of all I dismembered the corpse. I cut off the head and the arms and the legs.

13   I then took up three planks from the flooring of the chamber, and deposited all between the scantlings. I then replaced the boards so cleverly, so cunningly, that no human eye—not even *his*—could have detected anything wrong. There was nothing to wash out—no stain of any kind—no blood-spot whatever. I had been too wary for that. A tub had caught all—ha! ha!

14   When I had made an end of these labours, it was four o'clock—still dark as midnight. As the bell sounded the hour, there came a knocking at the street door. I went down to open it with a light heart—for what had I *now* to fear? There entered three men, who introduced themselves, with perfect suavity, as officers of the police. A shriek had been heard by a neighbour during the night; suspicion of foul play had been aroused; information had been lodged at the police office, and they (the officers) had been deputed to search the premises.

15   I smiled—for *what* had I to fear? I bade the gentlemen welcome. The shriek, I said, was my own in a dream. The old man, I mentioned, was absent in the country. I took my visitors all over the house. I bade them search—search *well*. I led them, at length, to *his* chamber. I showed them his treasures, secure, undisturbed. In the enthusiasm of my confidence, I brought chairs into the room, and desired them *here* to rest from their fatigues, while I myself, in the wild audacity of my perfect triumph, placed my own seat upon the very spot beneath which reposed the corpse of the victim.

16   The officers were satisfied. My *manner* had convinced them. I was singularly at ease. They sat, and while I answered cheerily, they chatted of familiar things. But, ere long, I felt myself getting pale and wished them gone. My head ached, and I fancied a ringing in my ears: but still they sat and still chatted. The ringing became more distinct:—It continued and became more distinct: I talked more freely to get rid of the feeling: but it continued and gained definiteness—until, at length, I found that the noise was *not* within my ears.

17   No doubt I now grew *very* pale;—but I talked more fluently, and with a heightened voice. Yet the sound increased—and what could I do? It was *a low, dull, quick sound—much such a sound as a watch makes when enveloped in cotton.* I gasped for breath—and yet the officers heard it not. I talked more quickly—more vehemently; but the noise steadily increased. I arose and argued about trifles, in a high key and with violent gesticulations; but the noise steadily increased. Why would they not be gone? I paced the floor to and fro with heavy strides, as if excited to fury by the observation of the men—but the noise steadily increased. Oh God! what *could* I do? I foamed—I raved—I swore! I swung the chair upon which I had been sitting, and grated it upon the boards, but the noise arose over all and continually increased. It grew louder—louder—*louder*! And still the men chatted pleasantly, and smiled. Was it possible they heard not? Almighty God!—no, no! They heard!—they suspected!—they *knew*!—they were making a mockery of my horror!—this I thought, and this I think. But anything was better than this agony! Anything was more tolerable than this derision! I could bear those hypocritical smiles no longer! I felt that I must scream or die! and now—again!—hark! louder! louder! louder! *louder*!

18   "Villains!" I shrieked, "dissemble no more! I admit the deed!—tear up the planks! here, here!—It is the beating of his hideous heart!"

PUBLISHED IN
1843

Reader's Response

1. What were your first impressions of the story?

   _____

2. What did you like best about the story?

   _____

3. Was there anything you did not like about the story? If so, what?

   _____

4. What kinds of feelings or emotions did the story evoke in you?

   _____

∽ CLOSE READING

Reread the story, paying particular attention to what the narrator feels and senses, to the colourful vocabulary and to the pictures the story creates. Then answer the following questions, which are designed to help you explore the text and its various aspects in depth.

1. Who is telling the story? What are his or her most prominent personality traits?

   _____

2. What is the narrator trying to prove to the reader throughout the story?

   _____

3. What is the relationship between the narrator and the old man? (Circle your answer.) They are

   a) relatives.   b) friends.   c) enemies.   d) neighbours.

4. What does the narrator want to do? Why?

   _____

5. How does the narrator react as he commits the crime? He

   a) weeps.   b) smiles.   c) shrieks.   d) feels sorry for his actions.

6. Why do the policemen come to his house?

   _____

7. Describe the narrator's initial reaction to the policemen's investigation.

   _____

8. Does his reaction to the policemen change? If so, how?

   _____

9. What is his emotional state at the end of the story?

_____

10. How does he try to cover up the crime?

_____

11. Which of the following adjectives could be used to describe the mood and atmosphere of this story?

a) serene   b) terrifying   c) nervous   d) sensitive   e) naïve

12. Explain the significance of the story's title.

_____

∞ WRITER'S CRAFT

In preparation for the following questions, read the explanations on narration and point of view and style found in the Literary Elements section at the beginning of this book.

> One of the characteristic features of Poe's works is the atmosphere of fear and terror. His stories horrify and fascinate the reader at the same time. To achieve this overall effect, Poe uses literary devices such as repetition, imagery, metaphor and personification.

Narration and Point of View

1. What point of view is used to tell the story?

_____

2. Does the narrator seem close to or distant from the reader? Justify your answer with proof from the story.

_____

3. How does the narrator engage the reader right at the very beginning of the story?

_____

4. Look carefully at the first sentence. Which points in time (past, present and/or future) are mentioned? (*Hint*: look at the verbs.)

_____

5. Circle one choice in each of the parentheses.

The narrator is telling the reader a story in the (past, present, future) that took place in the (past, present, future) and is concerned about the reader's (past, present, future) judgment.

> By using a first-person narrator who speaks in the present tense, Poe triggers the reader's interest from the start and creates an immediacy to the story.

Style and Literary Devices

■ REPETITION

Poe uses repetition to build up the tension. Even within sentences, he repeats adjectives, adverbs and concrete images. For example, observe the following sentences:

"I moved it slowly—very, very slowly."

"I undid the lantern cautiously—oh, so cautiously—cautiously (for the hinges creaked)—quicker and quicker, louder and louder."

"He had the eye of a vulture—a pale blue eye, with a film over it."

1. Find other examples of repetition in Poe's use of adjectives, adverbs and concrete images.

_____

_____

_____

_____

■ IMAGERY

Imagery is a literary device that allows the writer to make use of one or more of the five senses as a dominant element in descriptive writing.

1. Poe uses sound as a dominant element throughout this story. He creates striking contrasts between sounds that are quiet and sounds that are loud. For example:

Quiet Sounds	versus	Loud Sounds
slight groan		shriek
low stifled sound		ringing in my ears
stealthily		heightened voice

Find and list at least three other expressions that indicate quiet sounds and three that indicate loud sounds.

Quiet sounds:

_____

Loud sounds:

_____

2. How does the transition from quiet to loud sounds echo the narrator's sense of guilt?

_____

3. How does this transition add to the atmosphere of the story?

_____

▓ METAPHOR AND SIMILE

> A **metaphor** is a direct comparison of two things; for example, the heart is a drum.
>
> A **simile** is a slightly less direct comparison of two things; it uses the words "like" or "as"; for example, "the heart is like a drum."

1. Find images from the text that describe the following actions. Write the exact words used and then identify them either as metaphors (M) or similes (S). For a), b) and c), write a phrase from the text. For d), e) and f), fill in the missing word(s) from the text.

   a) The beating of the old man's heart at the beginning of the story

   _____

   b) The beating of the old man's heart toward the end of the story before the sound becomes loud

   _____

   c) The movement of the protagonist's hand as he opens the door on the eighth night

   _____

   d) "… ray fell upon the _____ eye."

   e) as black _____

   f) _____ dead (*Hint*: the same image appears twice.)

   > Vivid descriptions of what the protagonist hears, sees and imagines (through imagery) and of how he moves (through the choice of adverbs) help create mood and atmosphere.

▓ PERSONIFICATION

> **Personification** gives human qualities to non-human objects or animals.

1. In paragraph 7 ("Presently I heard …"), Poe uses personification to describe death. Find this example and explain its meaning.

   _____

   _____

   > Much of Poe's imagery serves to create an overall mood of fear and tension in the story. As well, he effectively creates other imagery to evoke various feelings in the reader, thus adding to the atmosphere of tension. The methods he uses include contrast of opposites and irony.

■ CONTRAST OF OPPOSITES

1. In the first paragraph of the text, find opposites of the following expressions (some of the words are not perfect matches).

    a) dreadfully nervous   how _____

    b) heard all things in heaven   heard many things _____

    c) sharpened my senses   not _____ them

    d) The disease   how _____

2. In paragraph 11 ("But even yet I refrained …"), the dreadful silence of the house is contrasted with "so strange a _____ ."

3. Describe the contrast at the end of the story between the perceptions and feelings of the police officers and those of the narrator.

    _____

    _____

    By juxtaposing (placing side by side or near to each other) opposites throughout the story, Poe creates a great deal of tension.

■ IRONY

Irony refers to the device of writing or saying one thing, but meaning the opposite.

1. How is the end of the story ironic?

    _____

    _____

VOCABULARY DEVELOPMENT

Latin and Germanic Roots, Prefixes and Suffixes

Early in its history, the territory now known as Great Britain was invaded and occupied by several successive invaders: first, Latin-speaking Romans in the early centuries AD, then, German-speaking tribes (Angles, Saxons and Jutes) in the fifth and sixth centuries AD, and finally the French-speaking Normans in 1066. Thus, the English language as we know it is a mix of words from several sources. In fact, about half the words in English come from Germanic languages and half from Latin or French. It is often very useful (and also very interesting!) to see how words are put together in English. This type of exercise can definitely help build a better vocabulary.

1

Linguists call the basic part of a word that carries the essential meaning the **root** (from Germanic *rot*) or **radical** (from Latin *radix, radic-*). We often add a **prefix** (from Latin *prae*, meaning "in front of," and *fixus*, meaning "fastened") before the root or a **suffix** (from Latin *sub*, meaning "under") after it. Most prefixes and suffixes add particular meanings to the roots of words. Here is an example using the root "count":

Root	Prefix	Suffix	Both Prefix and Suffix
count	*dis*count	count*able*	*dis*count*able*
	account		account*able*, *un*account*able*
	*re*count	count*er*	account*ant*
		count*down*	

Using your dictionary, give the meaning of the following prefixes and suffixes and identify their origin as Latin or Germanic (consider Old English, that is, words that were spoken in England before 1066, as Germanic). You will note that a good dictionary will place a hyphen after a prefix and before a suffix. This allows you to distinguish them from other words.

Then find a word from "The Tell-Tale Heart" that uses the particular prefix or suffix and identify its root. One has been done as an example.

Note: Without a prefix or a suffix, the root may not appear to be an English word. For example, in the word "detect," "de" is the prefix and "tect" the root; "tect" does not exist on its own.

Prefix	Meaning	Origin	Word from Text	Root
e.g.: over-	*too much*	*Germanic*	*over-acuteness*	*acute(ness)*
a) dis-	_____	_____	_____	_____
b) im- (in-)	_____	_____	_____	_____
c) fore-	_____	_____	_____	_____
d) de-	_____	_____	_____	_____
e) mid-	_____	_____	_____	_____
f) un-	_____	_____	_____	_____
g) con-	_____	_____	_____	_____
h) pre-	_____	_____	_____	_____
i) intro-	_____	_____	_____	_____
j) ex-	_____	_____	_____	_____
Suffix				
a) –less	_____	_____	_____	_____
b) –able, -ible	_____	_____	_____	_____

Wrap-Up Activities

1. Using vivid imagery, describe an incident when you felt very scared or frightened. Include descriptions of the setting and of your feelings.

2. Choose one example of imagery from the story and explain why you find it effective.

3. Write a paragraph (about eight sentences long) that has a dominant mood. Do not state what the mood is; simply create it by using imagery and a variety of sentence types. Read your paragraph to a classmate and have him or her guess the mood you have created.

4. Discuss whether or not the protagonist of "The Tell-Tale Heart" is insane. Are his arguments and justifications plausible?

5. Compare this story to a horror film that you have seen.

6. Research the evolution of the horror genre and present your findings to the class.

7. Read another story by Poe and compare the two stories.

8. Role play in groups of three. Imagine that you are in a court of law at the end of the narrator's trial for first-degree murder. Two of you are lawyers and the third is the judge. The lawyer defending the narrator gives his or her final arguments; the lawyer for the prosecution presents his or her conclusions; and then the judge, after asking any questions he or she may have, makes a decision and pronounces a sentence followed by his or her reasons.

My eLab ACTIVITY

My eLab 🗁

Check your knowledge of the vocabulary for the terms in the Literary Elements sections relating to narration, point of view, style and literary devices by doing the crossword puzzle for this chapter.

The Lady or The Tiger?

CHAPTER 2

CHAPTER CONTENTS

▸ **Reading:** "The Lady or the Tiger?" by Frank R. Stockton; author's biography; literary trends (adventure stories)

▸ **Literary Elements:** narration and point of view (third person); plot and dramatic structure; setting and characterization

▸ **Vocabulary Development:** meaning from context

▸ **My eLab Activities:** a fill-in-the-blanks activity for the new vocabulary from the story and a crossword puzzle for literary terms (plot, narrative and dramatic structure, setting and characterization)

ABOUT THE AUTHOR

Frank Stockton (1834-1902) was a prolific and popular American short story writer and humorist. The seventh of thirteen children, he was born lame. During his childhood, he regaled his brothers and sisters with stories and fairy tales. He worked as a wood engraver after completing high school and also became assistant editor of *Saint Nicolas Magazine*, a magazine for children. At the same time, he published stories for children ("Ting-a-Ling" in 1870, "The Floating Prince and Other Fairy Tales" in 1881) and a humorous book for adults (*Rudder Grange* in 1879). By 1881, Stockton was earning enough money from his writing to be able to quit his other jobs. In 1882, he wrote "The Lady or the Tiger?" Perhaps this is his most famous story.

The twenty-three volumes comprising his collected works include novels, science fiction and utopian stories, as well as fairy tales for both children and adults. Stockton was one of the most popular writers of his time.

LITERARY TRENDS
OF THE TIMES

READERS HAVE ALWAYS LOVED TO READ stories about people who must use their intelligence and physical abilities to overcome adversity. During the latter half of the nineteenth century, novels of action and adventure were very popular on both sides of the Atlantic. Herman Melville's *Moby Dick* (1851) depicts an epic struggle between a man and a whale. Robert Louis Stevenson's *Treasure Island* (1883), H. Rider Haggard's *King Solomon's Mines* (1885), and Mark Twain's *The Adventures of Huckleberry Finn* (1885) are some of the other well-known adventure novels that characterized the popular fiction of the times.

Similarly, plot-centred short stories were also immensely popular in America and Britain during the last decades of the nineteenth century. In these stories, the reader is introduced as quickly as possible to the situation and the conflict facing the protagonist, and suspense rapidly builds up to a peak. Sometimes these stories end in an unexpected way.

A shorter version of "The Lady or the Tiger?" was originally written to provoke discussion at a party. The story proved so successful that Stockton expanded it for publication in a magazine. Then he and his wife left on a long European vacation, and he did not hear about the intense public debate his story generated until he returned home to the United States. "The Lady or the Tiger?" proved immensely popular, even provoking responses from other well-known writers, and Stockton was inundated with letters.

Warm-Up Discussion Topics

1. How much can you trust your friends? Explain.

2. Have you ever had to make an important decision without having enough information? Describe the situation.

3. What do you know about the punishment faced by early Christians in Roman arenas in the first four centuries AD?

4. Do you think that chance determines what happens to people, or do they control their own destiny? Explain your point of view.

∞ INITIAL READING

In this story, the main character finds himself in a life-threatening situation where he must make a choice. As you read the story for the first time, do not worry about unfamiliar words unless they prevent understanding. In other words, try to guess their meaning from their context in the story.

The Lady or The Tiger?
Frank R. Stockton

1 IN the very olden time there lived a semi-barbaric king, whose ideas, though somewhat polished and sharpened by the progressiveness of distant Latin neighbours, were still large and unrestricted, as became the half of him, which was barbaric. He was a man of exuberant **fancy**[1], and of such irresistible authority that, at his **will**, he turned his varied fancies into facts. He was greatly given to self-communing, and when he and himself agreed upon anything, the thing was done. When every member of his domestic and political systems moved smoothly in its appointed course, his nature was calm and genial; but whenever there was a little hitch,

---

1. *Fancy:* imagination.

and some of his orbs got out of their orbits, he was calmer and more genial still, for nothing pleased him so much as to make the crooked straight and crush down uneven places.

2   Among the borrowed notions by which his barbarism had become semi-civilized was that of the public arena, in which, by exhibitions of manly and beastly **valour**, the minds of his subjects were refined and cultured.

3   But even here the exuberant and barbaric fancy asserted itself. The arena of the king was built, not to give the people an opportunity of hearing the rhapsodies of dying gladiators, nor to enable them to view the inevitable conclusion of a conflict between religious opinions and hungry jaws, but for purposes far better adapted to widen and develop the mental energies of the people. This vast amphitheatre, with its encircling galleries, its mysterious vaults, and its unseen passages, was an agent of **poetic justice**, in which crime was punished, or virtue rewarded, by the decrees of an impartial and incorruptible chance.

4   When a subject was accused of a crime of sufficient importance to interest the king, public notice was given that on an appointed day the fate of the accused person would be decided in the king's arena, a structure which well deserved its name; for, although its form and plan were borrowed from afar, its purpose emanated solely from the brain of this man. Every barley corn a king, he knew no tradition to which he owed more allegiance than pleased his fancy, and he moulded every thought and every institution to please his own barbaric idealism.

5   When all the people had assembled in the galleries, and the king, surrounded by his court, sat high up on his throne of royal state on one side of the arena, he gave a signal, a door beneath him opened, and the accused subject stepped out into the amphitheatre. Directly opposite him, on the other side of the enclosed space, were two doors, exactly alike and side by side. It was the duty and the privilege of the person on trial to walk directly to these doors and open one of them. He could open either door he pleased. He was subject to no guidance or influence but that of the aforementioned impartial and incorruptible chance. If he

opened the one, there came out of it a hungry tiger, the fiercest and most cruel that could be procured, which immediately sprang upon him and tore him to pieces as a punishment for his guilt. The moment that the case of the criminal was thus decided, doleful iron bells were clanged, great **wails** went up from the hired **mourners** posted on the outer rim of the arena, and the vast audience, with bowed heads and downcast hearts, wended slowly their homeward way, mourning greatly that one so young and fair, or so old and respected, should have merited so **dire** a **fate**.

6   But, if the accused person opened the other door, there came forth from it a lady, the most suitable to his years and station that his majesty could select among his **fair** subjects, and to this lady he was immediately married, as a reward of his innocence. It mattered not that he might already possess a wife and family, or that his affections might be engaged upon an object of his own selection; the king allowed no such subordinate arrangements to interfere with his great scheme of retribution and reward. The exercises, as in the other instance, took place immediately, and in the arena. Another door opened beneath the king, and a priest, followed by a band of **choristers**, and dancing **maidens** blowing joyous airs on golden horns and treading a nuptial measure, advanced to where the pair stood, side by side, and the wedding was promptly and cheerily solemnized. Then the gay brass bells rang forth their merry peals, the people shouted glad hurrahs, and the innocent man, preceded by children strewing flowers on his path, led his bride to his home.

7   This was the king's semi-barbaric method of administering justice. Its perfect fairness is obvious. The criminal could not know out of which door would come the lady; he opened either he pleased, without having the slightest idea whether, in the next instant, he was to be devoured or married. On some occasions the tiger came out of one door, and on some out of the other. The decisions of this tribunal were not only fair, they were positively final. The accused person was instantly punished if he found himself guilty, and, if innocent, he was rewarded on the spot, whether he liked it or not. There was no escape from the judgments of the king's arena.

8 The institution was a very popular one. When the people gathered together on one of the great trial days, they never knew whether they were to witness a bloody slaughter or a hilarious wedding. This element of uncertainty lent an interest to the occasion, which it could not otherwise have attained. Thus, the masses were entertained and pleased, and the thinking part of the community could bring no charge of unfairness against this plan, for did not the accused person have the whole matter in his own hands?

9 This semi-barbaric king had a daughter as blooming as his most elaborate fancies, and with a soul as fervent and imperious as his own. As is usual in such cases, she was **the apple of his eye**, and was loved by him above all humanity. Among his courtiers was a young man of that fineness of blood and lowness of station common to the conventional heroes of romance who love royal maidens. This royal maiden was well satisfied with her lover, for he was handsome and brave to a degree **unsurpassed** in all this kingdom, and she loved him with an ardour that had enough barbarism in it to make it exceedingly warm and strong. This love affair moved on happily for many months, until one day the king happened to discover its existence. He did not hesitate nor waver in regard to his duty. The youth was immediately cast into prison, and a day was appointed for his trial in the king's arena. This, of course, was an especially important occasion, and his majesty, as well as all the people, was greatly interested in the workings and development of this trial. Never before had such a case occurred; never before had a subject dared to love the daughter of the king. In after years such things became commonplace enough, but then they were, in no slight degree, novel and **startling**.

10 The tiger cages of the kingdom were searched for the most savage and relentless beasts, from which the fiercest monster might be selected for the arena; and the ranks of maiden youth and beauty throughout the land were carefully surveyed by competent judges in order that the young man might have a fitting bride in case fate did not determine for him a different destiny. Of course, everybody knew that the deed with which the accused was charged had been done. He had loved the princess, and neither he, she, nor anyone else, thought of denying the fact. But the king would not think of allowing any fact of this kind to interfere with the workings of the tribunal, in which he took such great delight and satisfaction. No matter how the affair turned out, the youth would be disposed of, and the king would take an aesthetic pleasure in watching the course of events, which would determine whether or not the young man had done wrong in allowing himself to love the princess.

11 The appointed day arrived. From far and near the people gathered, and thronged the great galleries of the arena, while crowds, unable to gain admittance, massed themselves against its outside walls. The king and his court were in their places, opposite the twin doors, those fearful portals, so terrible in their similarity.

12 All was ready. The signal was given. A door beneath the royal party opened, and the lover of the princess walked into the arena. Tall, beautiful, fair, his appearance was greeted with a low hum of admiration and anxiety. Half the audience had not known so grand a youth had lived among them. No wonder the princess loved him! What a terrible thing for him to be there!

13 As the youth advanced into the arena he turned, as the custom was, to bow to the king, but he did not think at all of that royal personage. His eyes were fixed upon the princess, who sat to the right of her father. Had it not been for her semi-barbaric nature, it is probable that lady would not have been there. But her intense and fervid soul would not allow her to be absent on an occasion in which she was so terribly interested. From the moment that the decree had gone forth that her lover should decide his fate in the king's arena, she had thought of nothing, night or day, but this great event and the various subjects connected with it. Possessed of more power, influence, and force of character than anyone who had ever before been interested in such a case, she had done what no other person had done—she had possessed herself of the secret of the doors. She knew in which of the two rooms behind those doors stood the cage of the tiger, with its open front, and in which waited the lady. Through

2

these thick doors, heavily curtained with skins on the inside, it was impossible that any noise or suggestion should come from within to the person who should approach to raise the latch of one of them. But gold, and the power of a woman's will, had brought the secret to the princess.

14 And not only did she know in which room stood the lady ready to emerge, all blushing and radiant, should her door be opened, but she knew who the lady was. It was one of the fairest and loveliest of the **damsels** of the court who had been selected as the reward of the accused youth, should he be proved innocent of the crime of aspiring to one so far above him; and the princess hated her. Often had she seen, or imagined that she had seen, this fair creature throwing **glances** of admiration upon the person of her lover, and sometimes she thought these glances were perceived, and even returned. Now and then she had seen them talking together. It was but for a moment or two, but much can be said in a brief space. It may have been on most unimportant topics, but how could she know that? The girl was lovely, but she had dared to raise her eyes to the loved one of the princess; and, with all the intensity of the savage blood transmitted to her through long lines of wholly barbaric ancestors, she hated the woman who blushed and trembled behind that silent door.

15 When her lover turned and looked at her, and his eye met hers as she sat there, paler and whiter than anyone in the vast ocean of anxious faces about her, he saw, by that power of quick perception which is given to those whose souls are one, that she knew behind which door crouched the tiger, and behind which stood the lady. He had expected her to know it. He understood her nature, and his soul was assured that she would never rest until she had made plain to herself this thing, hidden to all other lookers-on, even to the king. The only hope for the youth in which there was any element of certainty was based upon the success of the princess in discovering this mystery; and the moment he looked upon her, he saw she had succeeded, as in his soul he knew she would succeed.

16 Then it was that his quick and anxious glance asked the question: "Which?" It was as plain to her as if he shouted it from where he stood. There was not an instant to be lost. The question was asked in a flash; it must be answered in another.

17 Her right arm lay on the cushioned parapet before her. She raised her hand, and made a slight, quick movement toward the right. No one but her lover saw her. Every eye but his was fixed on the man in the arena.

18 He turned, and with a firm and rapid step he walked across the empty space. Every heart stopped beating, every breath was held, every eye was fixed immovably upon that man. Without the slightest hesitation, he went to the door on the right, and opened it.

19 Now, the point of the story is this: Did the tiger come out of that door, or did the lady?

20 The more we reflect upon this question, the harder it is to answer. It involves a study of the human heart, which leads us through devious **mazes** of passion, out of which it is difficult to find our way. Think of it, fair reader, not as if the decision of the question depended upon yourself, but upon that hot-blooded, semi-barbaric princess, her soul at a white heat beneath the combined fires of despair and jealousy. She had lost him, but who should have him?

21 How often, in her waking hours and in her dreams, had she started in wild horror, and covered her face with her hands as she thought of her lover opening the door on the other side of which waited the cruel **fangs** of the tiger!

22 But how much oftener had she seen him at the other door! How in her grievous reveries had she **gnashed** her teeth, and torn her hair, when she saw his start of rapturous delight as he opened the door of the lady! How her soul had burned in agony when she had seen him rush to meet that woman, with her flushing cheek and sparkling eye of triumph; when she had seen him lead her forth, his whole frame kindled with the joy of recovered life; when she had heard the glad shouts from the multitude, and the wild ringing of the happy bells; when she had seen the priest, with his joyous followers, advance to the couple, and make them man and wife before her very eyes; and when she had seen them walk away together upon their path of flowers, followed

by the tremendous shouts of the hilarious multitude, in which her one despairing **shriek** was lost and drowned!

23 Would it not be better for him to die at once, and go to wait for her in the blessed regions of semi-barbaric futurity?

24 And yet, that awful tiger, those shrieks, that blood!

25 Her decision had been indicated in an instant, but it had been made after days and nights of **anguished** deliberation. She had

known she would be asked, she had decided what she would answer, and, without the slightest hesitation, she had moved her hand to the right.

26 The question of her decision is one not to be lightly considered, and it is not for me to presume to set myself up as the one person able to answer it. And so I leave it with all of you: Which came out of the opened door—the lady or the tiger?

Note: The original text has been slightly altered because of archaic language.

PUBLISHED IN
1882

Reader's Response

1. Did you find this story easy or difficult to read?

_____

2. What do you think of the king's system of justice?

_____

3. Did you appreciate the story's ending? Why or why not?

_____

4. In your opinion, which came out of the opened door, the lady or the tiger? Explain.

_____

∽ CLOSE READING

Reread the story, paying particular attention to how the author builds up interest and suspense. Then answer the following questions, which are designed to help you explore the narrative and dramatic structure of the story in depth.

1. Circle the characteristics that describe the king in the story.

a) kind

b) half barbaric, half civilized

c) courageous

d) authoritative

e) interested in others' opinions

f) highly imaginative

g) likes things to go smoothly in his kingdom

2. Why does the king enjoy using the public arena?

a) He wants to refine the minds of his subjects.

b) He wants his people to see gladiators fight tigers.

c) He wants his people to see a crime punished not by a court but by pure chance.

3. How does the king mete out justice for someone accused of a crime?

_____

4. How does the king feel about his daughter?

_____

5. Why does the king have the young man arrested?

_____

6. Circle the characteristics that describe the king's daughter.

a) very beautiful   d) powerful

b) gentle   e) easygoing

c) passionate   f) semi-barbaric

7. Underline the sentence in paragraph 10 that shows that the king is more intellectual than humane and caring.

8. Which words would apply to the type of "aesthetic pleasure" experienced by the king?
a) emotional   b) rational   c) intellectual   d) caring

9. Write the first four words of the sentence in paragraph 11 that shows that the public is very interested in the "trial" of the young man.

_____

10. How does the crowd react when they first see the youth in the arena?

_____

11. What does the princess know and how did she find it out?

_____

12. Why is the princess tormented by the decision she has to make?

_____

13. According to the text, which feelings and emotions does the princess experience before making her decision?

a) despair   d) jealousy   g) anguish

b) sympathy   e) horror   h) joy

c) sadness   f) happiness   i) hatred

14. Why does nobody notice that the princess is communicating with her lover?

_____

15. If you were the princess's lover, which door would you open? Explain.

_____

✆ WRITER'S CRAFT

In preparation for the following questions, read the sections on narration and point of view, plot and dramatic structure, and setting and characterization found in the Literary Elements section at the beginning of this book.

Narration and Point of View

1. What type of narrator tells the story and from what point of view does he tell it?

_____

2. Does the narrator seem close to or distant from the reader at this point? Justify your answer with examples from the story.

_____

3. Read paragraphs 19 and 20 and the last paragraph of the story. How do the narration, style and point of view change in order to make the narrator seem much closer to the reader?

_____

4. Circle one choice in each of the parentheses.

The narrator tells the reader a story that took place in the (past, present, future) but switches to the (past, present, future) to involve the reader directly in the dénouement.

> By suddenly changing the narrator's point of view, Stockton involves the reader directly in the story by making him or her choose the story's outcome.

Plot and Dramatic Structure

1. First, place the **A** items in the order in which they appear in a traditional short story. Then match them with the equivalent **B** items.

A. rising action, dénouement, exposition, falling action, complication, climax

B. situation at the end, moment of greatest tension, tension/suspense increases, description of the situation at the beginning of the story, conflict, tension released

A	B
1. _____	_____
2. _____	_____
3. _____	_____
4. _____	_____
5. _____	_____
6. _____	_____

2. Which element of the dramatic structure (see **A** above) does each of the following represent? (*Hint*: two represent the same element at different points in the narrative.)

a) The king's daughter signals to her lover to choose the door on the right.

   _____

b) A young man falls in love with the daughter of the king, which is against the law. He is imprisoned and then must choose one of the doors in the arena.

   _____

c) The king's daughter finds out what is behind each of the doors and burns with jealousy and hatred toward the beautiful young woman behind one of them. However, choosing the other door means death for her lover.

   _____

d) A king in a far-off kingdom punishes wrongdoers in an arena where they must choose to open one of two doors: behind one is a beautiful lady (life); behind the other is a ferocious tiger (death).

   _____

3. Which three elements are missing from the dramatic structure of the story?

   _____

Short stories have different types of endings. Some have a **linear** ending, where the climax occurs, the conflict is resolved and the story ends. Others have a **circular** ending, where the main character sets out on a quest and ends up coming back to the point where he or she started. Yet others have a **surprise ending**, where the conflict is unexpectedly resolved, or an **open** ending, where the reader is left to decide what happens at the end of the story.

4. What kind of ending does this story have?
   a) linear   b) circular   c) surprise   d) open

5. Would another type of ending have a stronger impact on the reader? Explain.

   _____

6. What is the conflict facing the young man in the arena? Is it internal or external?

   _____

7. What is the conflict facing the princess? Is it internal or external?

   _____

8. Which of the two conflicts is more interesting for the reader? Explain.

   _____

   _____

Setting and Characterization

1. Where and when does the story take place?

_____

2. Is the exact setting of major or minor importance to the story? Explain.

_____

3. Who is the protagonist and who are the antagonists in the story?

Protagonist: _____   Antagonists: _____

4. Are these characters flat or round? _____

5. Which character in the story is the most developed? Explain.

_____

> In a traditional short story, suspense is often created by the conflict faced by the main character. In "The Lady or the Tiger?" the most important conflict takes place in the mind of the princess, a secondary character upon whose decision the life of the protagonist depends. This conflict engages the reader because it remains unresolved.

VOCABULARY DEVELOPMENT

Meaning from Context

The most efficient way to learn new vocabulary is through reading. In other words, the more you read, the bigger your vocabulary will become. Most words are learned through context, that is, through association with other words. As you read, try to figure out the meanings of unfamiliar words from their context.

Look at the context in which the words and expressions in the left-hand column are used in the text, where they appear in bold, in the same order as they are listed below. Then circle the letter corresponding to their closest definition.

1. will:
   a) determination
   b) a legal document distributing a person's goods
   c) the future tense

2. valour:
   a) value
   b) courage
   c) velvet

3. poetic justice:
   a) a legal poem
   b) virtue rewarded; evil punished (sometimes ironically)
   c) a desire for justice

4. wails:

    a) long, sad cries
    b) a large ocean mammal
    c) part of a ship

5. mourners:

    a) people up early in the day
    b) people expressing sadness
    c) people guarding the arena

6. dire:

    a) talking
    b) good
    c) terrible

7. fate:

    a) obese
    b) destiny
    c) reward

8. fair:

    a) the cost of a ticket
    b) good-looking
    c) equal

9. choristers:

    a) singers
    b) birds
    c) angels

10. maidens:

    a) gypsies
    b) young women
    c) slave girls

11. the apple of his eye:

    a) a treasured person
    b) a visual fruit
    c) a lover

12. unsurpassed:

    a) unproven
    b) unexpected
    c) unequalled

13. startling:

    a) surprising
    b) just beginning
    c) usual

14. damsels:

    a) young women
    b) children
    c) dragonflies

15. glances:

    a) arrows
    b) brief looks
    c) flowers

16. mazes:

    a) labyrinths
    b) straight paths
    c) levels

17. fangs:

    a) long claws
    b) strong jaws
    c) sharp teeth

18. gnash(ed):

    a) grind (ground)
    b) lose (lost)
    c) pull (pulled) out

19. shriek: a) a soft sound
b) a shrill cry
c) a loud song

20. anguished: a) difficult, hard
b) unfortunate, unlucky
c) tormented, tortured

Wrap-Up Activities

1. What advice would you give the accused about which door he should open? Be convincing.

2. "The Lady or the Tiger?" had a large public impact in both the U.S.A. and England when it was published in 1882. In your opinion, would the story have the same impact today?

3. Regarding the king's method for administering justice, the narrator states, "Its perfect fairness is obvious" (paragraph 8). Discuss why you agree or disagree with this statement.

4. Stockton (the author) told people that when they chose an ending for the story "The Lady or the Tiger?" they would find out what kind of person they were. Do you agree? Explain.

5. Write a convincing ending to the story.

6. Compare "The Lady or the Tiger?" to a TV program or film that you have seen that has a similar "cliff-hanging" ending.

7. Rewrite the story in a modern context.

8. Evaluate the strengths and weaknesses of "The Lady or the Tiger?" as a short story.

My eLab ACTIVITIES

My eLab

Check your knowledge of the new vocabulary in the story by completing the fill-in-the-blanks activity for this chapter.

My eLab

Complete the crossword puzzle regarding the terms in the Literary Elements sections relating to plot, narrative and dramatic structure, setting and characterization, and theme.

The Nightingale and The Rose

CHAPTER
CONTENTS

▸ **Reading:** "The Nightingale and the Rose" by Oscar Wilde; author's biography; literary trends (Romanticism, fairy tales)

▸ **Literary Elements:** style and literary devices (figures of speech and poetic devices: simile, metaphor, personification, repetition, alliteration, assonance, consonance, imagery and symbolism); theme

▸ **Vocabulary Development:** vivid language

▸ **My eLab Activity:** a fill-in-the-blanks activity regarding the symbolism of colour found in Western traditions and literature

ABOUT THE **AUTHOR**

Oscar Wilde (1854-1900) was born and raised in Dublin, Ireland. He went to Oxford University, where he achieved a distinguished academic record. After finishing university in 1878, he moved to London and began his writing career. He leaned toward the Aesthetic Movement, which believed in the philosophy of "art for art's sake." He was also known for his witty conversational style and flamboyant dress, which made him an appreciated guest in English social circles. As a writer, he was extremely popular and respected, as he wrote not only literary criticism and essays, but also excellent prose, including *The Portrait of Dorian Grey* (1891), plays, including *The Importance of Being Earnest* (1895) and *An Ideal Husband* (1895), and poetry, including "Impression du Matin" (1881) and "The Harlot's House" (1885).

Wilde, who was married and had two sons, met Lord Alfred Douglas in 1891. He and Lord Douglas had a passionate relationship. His popularity in English and American society ceased in 1895 when he was accused by Lord Douglas's father of having a homosexual relationship with his son at a time when homosexuality was considered to be a serious criminal offence. Wilde was found guilty of the crime and served two years in jail, doing hard labour before escaping to Paris, where he died penniless three years later in 1900.

Wilde wrote "The Nightingale and the Rose" for his sons.

LITERARY TRENDS
OF THE TIMES

THE ROMANTIC MOVEMENT CONTINUED into the Victorian Era in England, a period when Oscar Wilde wrote his greatest works. Literature of this time often explored the imaginative, emotional side of human nature, as opposed to the intellectual, rational side. In Victorian times, many readers showed considerable interest in myths and legends, such as those about King Arthur, as well as in literary fairy tales, such as those written by Wilde. Similar to dreams, these narratives often contained archetypal imagery and symbolism. Oscar Wilde and his contemporaries examined the values of their era and wanted to point out the imbalance between feminine (*eros*, the heart) and masculine (*logos*, the mind) principles. They wished to show the danger of placing too much importance on the material side of life (owning things, knowing scientific facts, and so on) and not enough on the emotional side (experiencing emotions, creating art, and so on).

3

Warm-Up Discussion Topics

1. Did you enjoy fairy tales as a child? Why or why not?

_____

2. Do you prefer science and mathematics or the arts (music, dance, painting, and so on)? Explain why.

_____

3. How important is love in a person's life? Are there more important things in life than love?

_____

✄ INITIAL READING

Read the story for the first time as you would read a fairy tale. Visualize the images and listen to the rhythms of the language as you read.

The Nightingale and The Rose
Oscar Wilde

1 "SHE said that she would dance with me if I brought her red roses," cried the young Student; "but in all my garden there is no red rose."

2 From her nest in the holm-oak tree the Nightingale heard him, and she looked out through the leaves, and wondered.

3 "No red rose in all my garden!" he cried, and his beautiful eyes filled with tears. "Ah, on what little things does happiness depend! I have read all that the wise men have written, and all the secrets of philosophy are mine, yet for want of a red rose is my life made wretched."

4 "Here at last is a true lover," said the Nightingale. "Night after night have I sung of him, though I knew him not: night after night have I told his story to the stars, and now I see him. His hair is dark as the hyacinth-blossom, and his lips are red as the rose of his desire; but passion has made his face like pale ivory, and sorrow has set her seal upon his brow."

5 "The Prince gives a ball tomorrow night," murmured the young Student, "and my love will be of the company. If I bring her a red rose, I shall hold her in my arms, and she will lean her head upon my shoulder, and her hand will be clasped in mine. But there is no red rose in

my garden, so I shall sit lonely, and she will pass me by. She will have no heed of me, and my heart will break."

6 "Here indeed is the true lover," said the Nightingale. "What I sing of, he suffers, what is joy to me, to him is pain. Surely Love is a wonderful thing. It is more precious than emeralds, and dearer than fine opals. Pearls and pomegranates cannot buy it, nor is it set forth in the market-place. It may not be purchased of the merchants, nor can it be weighed out in the balance for gold."

7 "The musicians will sit in their gallery," said the young Student, "and play upon their stringed instruments, and my love will dance to the sound of the harp and the violin. She will dance so lightly that her feet will not touch the floor, and the courtiers in their gay dresses will throng round her. But with me she will not dance, for I have no red rose to give her"; and he flung himself down on the grass, and buried his face in his hands, and wept.

8 "Why is he weeping?" asked a little Green Lizard, as he ran past him with his tail in the air.

9 "Why, indeed?" said a Butterfly, who was fluttering about after a sunbeam.

10 "Why, indeed?" whispered a Daisy to his neighbour, in a soft, low voice.

11 "He is weeping for a red rose," said the Nightingale, and the little Lizard, who was something of a cynic, laughed outright.

12 But the Nightingale understood the secret of the Student's sorrow, and she sat silent in the oak-tree, and thought about the mystery of Love.

13 Suddenly she spread her brown wings for flight, and soared into the air. She passed through the grove like a shadow, and like a shadow she sailed across the garden.

14 In the centre of the grass-plot was standing a beautiful Rose-tree, and when she saw it, she flew over to it, and lit upon a spray.

15 "Give me a red rose," she cried, "and I will sing you my sweetest song."

16 But the Tree shook its head.

17 "My roses are white," it answered; "as white as the foam of the sea, and whiter than the snow upon the mountain. But go to my brother who grows round the old sun-dial, and perhaps he will give you what you want."

18 So the Nightingale flew over to the Rose-tree that was growing round the old sun-dial.

19 "Give me a red rose," she cried, "and I will sing you my sweetest song."

20 But the Tree shook its head.

21 "My roses are yellow," it answered; "as yellow as the hair of the mermaiden who sits upon an amber throne, and yellower than the daffodil that blooms in the meadow before the mower comes with his scythe. But go to my brother who grows beneath the Student's window, and perhaps he will give you what you want."

22 So the Nightingale flew over to the Rose-tree that was growing beneath the Student's window.

23 "Give me a red rose," she cried, "and I will sing you my sweetest song."

24 But the Tree shook its head.

25 "My roses are red," it answered, "as red as the feet of the dove, and redder than the great fans of coral that wave in the ocean-cavern. But the winter has chilled my veins, and the frost has nipped my buds, and the storm has broken my branches, and I shall have no roses at all this year."

26 "One red rose is all I want," cried the Nightingale, "only one red rose! Is there no way by which I can get it?"

27 "There is a way," answered the Tree; "but it is so terrible that I dare not tell it to you."

28 "Tell it to me," said the Nightingale, "I am not afraid."

29 "If you want a red rose," said the Tree, "you must build it out of music by moonlight, and stain it with your own heart's-blood. You must sing to me with your breast against a thorn. All night long you must sing to me, and the thorn must pierce your heart, and your life-blood must flow into my veins, and become mine."

30 "Death is a great price to pay for a red rose," cried the Nightingale, "and Life is very dear to all. It is pleasant to sit in the green wood, and to watch the Sun in his chariot of gold, and the Moon in her chariot of pearl. Sweet is the scent of the hawthorn, and sweet are the bluebells that hide in the valley, and the heather that blows on the hill. Yet Love is

better than Life, and what is the heart of a bird compared to the heart of a man?"

31 So she spread her brown wings for flight, and soared into the air. She swept over the garden like a shadow, and like a shadow she sailed through the grove.

32 The young Student was still lying on the grass, where she had left him, and the tears were not yet dry in his beautiful eyes.

33 "Be happy," cried the Nightingale, "be happy; you shall have your red rose. I will build it out of music by moonlight, and stain it with my own heart's-blood. All that I ask of you in return is that you will be a true lover, for Love is wiser than Philosophy, though she is wise, and mightier than Power, though he is mighty. Flame-coloured are his wings, and coloured like flame is his body. His lips are sweet as honey, and his breath is like frankincense."

34 The Student looked up from the grass, and listened, but he could not understand what the Nightingale was saying to him, for he only knew the things that are written down in books.

35 But the Oak-tree understood, and felt sad, for he was very fond of the little Nightingale who had built her nest in his branches.

36 "Sing me one last song," he whispered; "I shall feel very lonely when you are gone."

37 So the Nightingale sang to the Oak-tree, and her voice was like water bubbling from a silver jar.

38 When she had finished her song the Student got up, and pulled a note-book and a lead-pencil out of his pocket.

39 "She has form," he said to himself, as he walked away through the grove—"that cannot be denied to her; but has she got feeling? I am afraid not. In fact, she is like most artists; she is all style, without any sincerity. She would not sacrifice herself for others. She thinks merely of music, and everybody knows that the arts are selfish. Still, it must be admitted that she has some beautiful notes in her voice. What a pity it is that they do not mean anything, or do any practical good." And he went into his room, and lay down on his little pallet-bed, and began to think of his love; and, after a time, he fell asleep.

40 And when the Moon shone in the heavens, the Nightingale flew to the Rose-tree, and set her breast against the thorn. All night long she sang, and the thorn went deeper and deeper into her breast, and her life-blood ebbed away from her.

41 She sang first of the birth of love in the heart of a boy and a girl. And on the top-most spray of the Rose-tree there blossomed a marvellous rose, petal following petal, as song followed song. Pale was it, at first, as the mist that hangs over the river—pale as the feet of the morning, and silver as the wings of the dawn. As the shadow of a rose in a mirror of silver, as the shadow of a rose in a water-pool, so was the rose that blossomed on the topmost spray of the Tree.

42 But the Tree cried to the Nightingale to press closer against the thorn. "Press closer, little Nightingale," cried the Tree, "or the Day will come before the rose is finished."

43 So the Nightingale pressed closer against the thorn, and louder and louder grew her song, for she sang of the birth of passion in the soul of a man and a maid.

44 And a delicate flush of pink came into the leaves of the rose, like the flush in the face of the bridegroom when he kisses the lips of the bride. But the thorn had not yet reached her heart, so the rose's heart remained white, for only a Nightingale's heart's-blood can crimson the heart of a rose.

45 And the Tree cried to the Nightingale to press closer against the thorn. "Press closer, little Nightingale," cried the Tree, "or the Day will come before the rose is finished."

46 So the Nightingale pressed closer against the thorn, and the thorn touched her heart, and a fierce pang of pain shot through her. Bitter, bitter was the pain, and wilder and wilder grew her song, for she sang of the Love that is perfected by Death, of the Love that dies not in the tomb.

47 And the marvellous rose became crimson, like the rose of the eastern sky. Crimson was the girdle of petals, and crimson as a ruby was the heart.

48 But the Nightingale's voice grew fainter, and her little wings began to beat, and a film came over her eyes. Fainter and fainter grew

her song, and she felt something choking in her throat.

49   Then she gave one last burst of music. The white Moon heard it, and she forgot the dawn, and lingered on in the sky. The red rose heard it, and it trembled all over with ecstasy, and opened its petals to the cold morning air. Echo bore it to her purple cavern in the hills, and woke the sleeping shepherds from their dreams. It floated through the reeds of the river, and they carried its message to the sea.

50   "Look, look!" cried the Tree, "the rose is finished now"; but the Nightingale made no answer, for she was lying dead in the long grass, with the thorn in her heart.

51   And at noon the Student opened his window and looked out.

52   "Why, what a wonderful piece of luck!" he cried; "here is a red rose! I have never seen any rose like it in all my life. It is so beautiful that I am sure it has a long Latin name"; and he leaned down and plucked it.

53   Then he put on his hat, and ran up to the Professor's house with the rose in his hand.

54   The daughter of the Professor was sitting in the doorway winding blue silk on a reel, and her little dog was lying at her feet.

55   "You said that you would dance with me if I brought you a red rose," cried the Student.

"Here is the reddest rose in all the world. You will wear it tonight next your heart, and as we dance together, it will tell you how I love you."

56   But the girl frowned.

57   "I am afraid it will not go with my dress," she answered; "and besides, the Chamberlain's nephew has sent me some real jewels, and everybody knows that jewels cost far more than flowers."

58   "Well, upon my word, you are very ungrateful," said the Student angrily; and he threw the rose into the street, where it fell into the gutter, and a cart-wheel went over it.

59   "Ungrateful!" said the girl. "I tell you what, you are very rude; and after all, who are you? Only a Student. Why, I don't believe you have even got silver buckles to your shoes as the Chamberlain's nephew has"; and she got up from her chair and went into the house.

60   "What a silly thing Love is," said the Student as he walked away. "It is not half as useful as Logic, for it does not prove anything, and it is always telling one of things that are not going to happen, and making one believe things that are not true. In fact, it is quite unpractical, and as in this age to be practical is everything, I shall go back to Philosophy and study Metaphysics."

61   So he returned to his room and pulled out a great dusty book, and began to read.

PUBLISHED IN
1888

Reader's Response

1. What was your reaction to the ending of the story?

_____

2. Who do you think this fairy tale is intended for, children or adults? Explain.

_____

3. How is the ending of this story different from that of a traditional fairy tale?

∽ CLOSE READING

Reread the story, paying particular attention to the colourful vocabulary and the pictures the story creates. Think about the underlying meaning and what the major characters and objects might symbolize. Then answer the following questions.

1. Circle the correct answer.

   The student is unhappy because he is in love with a girl

   a) to whom he promised red roses, which he cannot find.

   b) who loves another man.

   c) who wants the nightingale to sing to her.

   d) who is married.

2. Which **two** of the following statements are true?

   a) The lizard believes in true love.

   b) The nightingale believes in true love.

   c) The nightingale thinks the girl knows what true love is.

   d) The nightingale thinks the student knows what true love is.

3. The nightingale identifies herself with the student because

   a) she falls in love with him.

   b) she is in love with the oak tree.

   c) the student feels the same passion as her nightly song.

   d) the student is good-looking.

4. Why does the nightingale decide to sacrifice herself? Give two reasons.

   _____

   _____

5. The nightingale describes love as being wiser than _____, and

   stronger than _____, with wings and a body that are

   _____, and lips as sweet as _____,

   and _____ like frankincense.

6. Why doesn't the student understand what the nightingale is saying?

   _____

3

7. What does the student think about the nightingale's song for the oak tree? He thinks that it

a) is truly a work of art, full of beauty and meaning.

b) is only a meaningless song with some beautiful notes.

c) has a practical use.

d) is full of style and feeling.

8. Why does the student think the beautiful rose has a long Latin name?

_____

9. What three reasons does the professor's daughter give for refusing the red rose?

_____

_____

10. What happens to the rose?

_____

11. At the end of the story, what does the student think about love? (You can choose more than one answer.) He thinks that it is

a) ridiculous. b) intelligent. c) useless. d) truthful. e) practical.

∞ WRITER'S CRAFT

In preparation for the following questions, read the explanations on style and theme in the Literary Elements section at the beginning of the book.

Style and Literary Devices

1. Wilde wrote this story in the form of a fairy tale. Which of the following descriptions apply to the story?

a) It is realistic.

b) It contains imaginative elements.

c) It has fairies in it.

d) It has no plot.

e) It has an underlying message.

2. The sentences mainly found in this story are

a) complex. b) long. c) short. d) simple.

3. The story is written in a

    a) concrete style (non-descriptive with few adjectives).

    b) flowery style (very descriptive with many adjectives).

4. This story contains many examples of

    a) connotation.   b) denotation.   c) neither   d) both

> One of the characteristic features of fairy tales is their use of figurative language or figures of speech—that is, language that is not meant to be taken literally. Fairy tales often use poetic forms of language in order to create atmosphere. Wilde uses many figures of speech and poetic devices to give his story the atmosphere of a fairy tale.

■ FIGURES OF SPEECH AND POETIC DEVICES

1. Read the following excerpts from the story and then identify each one as an example of **a)** simile; **b)** metaphor; **c)** personification; **d)** repetition; **e)** alliteration; **f)** assonance; or **g)** consonance. There may be more than one answer for some excerpts.

1. "She passed through the grove like a shadow, and like a shadow she sailed across the garden."

2. "... but passion has made his face like pale ivory ..."

3. "Bitter, bitter was the pain, and wilder and wilder grew her song."

4. "... for Love is wiser than Philosophy ..."

5. "Night after night have I sung of him, though I knew him not."

6. "... the Sun in his chariot of gold, and the Moon in her chariot of pearl."

7. " ... and the little Lizard, who was something of a cynic, laughed outright."

8. "As the shadow of a rose in a mirror of silver, as the shadow of a rose in a water-pool, so was the rose that blossomed on the topmost spray of the Tree."

9. "It is more precious than emeralds, and dearer than fine opals. Pearls and pomegranates cannot buy it, nor is it set forth in the market-place. It may not be purchased of the merchants ..."

10. "Flame-coloured are his wings ..."

11. "But the Tree shook its head."

2. Wilde capitalizes many words in the middle of his sentences, such as Nightingale, Love, Philosophy, Green Lizard, Butterfly, Tree and so on. These capital letters often point to the use of a particular figure of speech. Which one?

■ **IMAGERY AND SYMBOLISM**

1. Why do you think there are so many variants of the colour red? In other words, what could the colour red symbolize in the context of the story?

   _____

2. Here is a list of all the colours that appear in the story. Find at least one item from the text for each colour.

   yellow: _____   blue: _____

   green: _____   red: _____

   silver: _____   crimson: _____

   gold: _____   pink: _____

   purple: _____   ivory: _____

   pearl: _____   white: _____

   brown: _____

   Imagery is also an example of figurative language. Imagery consists of vivid descriptions that appeal to one of the five senses. These descriptions enable readers to create pictures or images in their mind and add mood and atmosphere to the work.

3. In this story, Wilde uses many vivid images that appeal to different senses. Complete the following chart with an image from the story that you found particularly effective.

Imagery in "The Nightingale and the Rose"	
Senses	**Examples from the text**
Sight	
Hearing	
Touch	
Smell	
Taste	

4. Match each item from the story with the quality you believe it symbolizes.

1. Rose	_____	a) wisdom
2. Nightingale	_____	b) unrequited love
3. Student	_____	c) goodness and virtue
4. Oak tree	_____	d) materialism
5. Professor's daughter	_____	e) cynicism not appreciating true beauty

■ IRONY

Something that is **ironic** is the opposite of what is intended, expected or expressed.

1. Explain the irony in the expectations of both the student and the nightingale.

_____

_____

2. How is what happens to the rose at the end of the story ironic?

_____

3. Find another example of irony in the story.

_____

4. As in all fairy tales, the main character learns a lesson at the end of the tale. What is this lesson and how is it ironic?

_____

An author's use of **irony** often points to the **theme** or meaning of a story.

5. What is the theme of this story?

_____

VOCABULARY DEVELOPMENT

Vivid Language

Some words are more interesting and evocative than others. Compare these phrases.

The girl *told* her friend a secret and *asked* her .../The girl *whispered* a secret to her friend and *begged* her ...

The boy and the girl *looked* into each other's eyes./The lovers *stared* into the *depths* of each other's eyes.

In the second examples, the verbs are much more vivid, adding atmosphere and depth to the sentences. You can make your writing more effective by using colourful language, but be careful not to overdo it. Otherwise, you will be accused of writing what is known as "purple prose." A good thesaurus and dictionary can help you improve your skills.

A. In the following examples, underline the word that is more evocative.

1. The student's life was **a)** bad. **b)** sad. **c)** wretched.

2. The mower cut the grass with his **a)** big knife. **b)** scythe. **c)** cutters.

3. He felt **a)** strongly **b)** greatly **c)** passionately about the matter.

4. She **a)** murmured the answer. **b)** said the answer softly. **c)** answered quietly.

5. The nightingale **a)** flew **b)** rose **c)** soared into the air.

6. She wore a **a)** dark red **b)** crimson **c)** red dress to the ball.

7. Her life-blood **a)** flowed **b)** ran **c)** ebbed away from her.

8. The orchestra finished playing with one last **a)** burst **b)** note **c)** sound of music.

9. She **a)** moved **b)** shook **c)** trembled with **d)** happiness. **e)** ecstasy. **f)** joy.

10. The student read a/an **a)** dusty **b)** old **c)** long-lasting book.

B. Write a more evocative and precise term for each of the words in bold.

John and his friend Jennifer were **going** _____ to the store on a

path beside the frozen river. It was **very dark** _____, and a **cold**

_____ wind blew **strongly** _____ into their faces.

Suddenly they heard a **sound** _____.

"Help me," a girl's voice **said** _____. "I'm in a **very bad**

_____ situation. I've fallen through the ice, and the water is **so cold**

_____.

John **looked** _____ into the dark trying **very hard**

_____ to **see** the girl.

"Please **move fast** _____," the voice **asked** _____.

"I see her," **said** _____ Jenny. "She's over there. Let's use this branch."

"We'll help you," John **said** _____. "Can you **take**

_____ the end of this branch? We'll **get** _____

you out."

Wrap-Up Activities

1. Do you agree with the student's view of love at the end of the story? Why or why not?

2. Using effective imagery, write a paragraph about a place that is special to you.

3. When the student listens to the nightingale's song, he has this reaction:

> "She has form," he said to himself, as he walked away through the grove—"that cannot be denied to her; but has she got feeling? I am afraid not. In fact, she is like most artists; she is all style, without any sincerity. She would not sacrifice herself for others. She thinks merely of music, and everybody knows that the arts are selfish. Still, it must be admitted that she has some beautiful notes in her voice. What a pity it is that they do not mean anything, or do any practical good."

   a) Discuss the student's assessment above and explain why you agree or disagree with it.

   b) Do you think Wilde agreed or disagreed with the student's assessment? Explain.

4. "The Nightingale and the Rose" is written in the genre of a fairy tale. Write whether you agree or disagree with this statement.

5. Research the life and works of Oscar Wilde. Do a short presentation on your findings.

6. Write about the Aesthetic Movement for which Oscar Wilde was a spokesperson.

7. Compare the ending of "The Nightingale and the Rose" to that of another fairy tale.

8. Write about the history and characteristics of the fairy tale genre.

9. Write a fairy tale.

My eLab ACTIVITY

My eLab

Do the fill-in-the-blanks activity for this chapter to check your knowledge of the symbolism of colour found in Western traditions and literature.

The Story of an Hour

CHAPTER 4

ABOUT THE AUTHOR

Kate Chopin (18501904) was born in St. Louis, Missouri, to a wealthy Irish family. In 1870, she married Oscar Chopin. The young couple set up house in New Orleans, Louisiana, and had six children. The Chopin family moved to a plantation in Natchitoches Parish, and it was there that Kate became steeped in the Creole culture that would provide subject matter for many of her works. After the death of her husband and mother, Kate Chopin began to write poems, stories and novels. A kindly doctor had suggested that she express her disappointments with life in writing. Her early success allowed her to support her six children.

Society in the southern United States at that time was oriented around male authority. Women were expected to take subordinate roles and would never question or express concern about the quality of their lives or the way things were. Many of Chopin's stories and novels such as "Désirée's Baby" (1894), "The Story of an Hour" (1894) and *The Awakening* (1899) are concerned with the status of women and the quality of their lives. They portray women searching for their identity and examining their sexuality. The publication of her novel *The Awakening* brought harsh criticism and social ostracism to Chopin. Some critics labelled the work as pornographic and immoral, and it was banned in St. Louis.

After remaining forgotten for over sixty years, Kate Chopin's writings were rediscovered in the 1960s and 1970s. She is now considered to have influenced the modern feminist movement in the United States, and her works are required reading in many college and university literature courses.

LITERARY TRENDS OF THE TIMES

TWO MAJOR LITERARY MOVEMENTS swept through Europe and North America during the nineteenth century. The first, which began in the early 1800s, is called *Romanticism*.

The works of Edgar Allan Poe, with their emphasis on the weird, the supernatural, the dark side of humanity and the Gothic, serve to illustrate this period very well. In the latter half of the century, many writers turned away from the exaggerated feelings of the Romantics and decided instead to turn their attention to the details of everyday life. They took a critical, truthful look at the reality of their times and portrayed it realistically, thus inspiring the name *Realism* for this period.

In Europe, the writers of this Realist period—Stendhal, Balzac and Flaubert of France; Turgenev, Dostoevsky and Tolstoy of Russia; and Dickens of England, among others—wrote about the socio-economic situations of their times. They explored various themes connected with what they perceived as common social ills of the times: child labour and working-class poverty brought on by the Industrial Revolution, the negative aspects of war, the pitfalls of the bourgeois mentality and the social relations between the upper and lower classes, just to name a few.

These writers incorporated vivid imagery in their works to create realistic effects.

In the United States, the Civil War marked an important turning point in American life, with the country rejecting slavery and accepting mechanization and industrialization as the economic basis of society. In theory, black people were now free. However, author Kate Chopin realized that women were not. Her writings, and especially her book *The Awakening*, shocked the literary intelligentsia of the times. Not only was she criticized harshly for questioning the current social and sexual mores—taboo subjects—she was also ostracized. She stopped writing and was relegated into oblivion only to be rediscovered in the latter half of the twentieth century. Her works are fine examples of Realism.

Warm-Up Discussion Topics

1. What does the word "marriage" mean to you? Write at least eight words that you associate with *marriage*.

    _____

    _____

2. Define the word "freedom." How is feeling free important to you?

    _____

    _____

∞ INITIAL READING

This story was written at the end of the nineteenth century, when both men and women were expected to conform to their particular social roles more fully than they do today. As you read this story the first time for personal enjoyment, try to look at life from Mrs. Mallard's point of view and to imagine the feelings and sensations that she is experiencing throughout the narrative. Do not stop to look up each unfamiliar word. Just read the story straight through.

The Story of an Hour
Kate Chopin

1 KNOWING that Mrs. Mallard was afflicted with a heart trouble, great care was taken to break to her as gently as possible the news of her husband's death.

2 It was her sister Josephine who told her, in broken sentences; veiled hints that revealed in half concealing. Her husband's friend Richards was there, too, near her. It was he who had been in the newspaper office when intelligence of the railroad disaster was received, with Brently Mallard's name leading the list of "killed." He had only taken the time to assure himself of its truth by a second telegram, and had hastened to forestall any less careful, less tender friend in bearing the sad message.

3 She did not hear the story as many women have heard the same, with a paralyzed inability to accept its significance. She wept at once, with sudden, wild abandonment, in her sister's arms. When the storm of grief had spent itself, she went away to her room alone. She would have no one follow.

4 There stood, facing the open window, a comfortable, roomy armchair. Into this she sank, pressed down by a physical exhaustion that haunted her body and seemed to reach into her soul.

5 She could see in the open square before her house the tops of trees that were all aquiver with the new spring life. The delicious breath of rain was in the air. In the street below a peddler was crying his wares. The notes of a distant song which someone was singing reached her faintly, and countless sparrows were twittering in the eaves.

6 There were patches of blue sky showing here and there through the clouds that had met and piled one above the other in the west facing her window.

7 She sat with her head thrown back upon the cushion of the chair, quite motionless, except when a sob came up into her throat and shook her, as a child who has cried itself to sleep continues to sob in its dreams.

8 She was young, with a fair, calm face, whose lines bespoke repression and even a certain strength. But now there was a dull stare in her eyes, whose gaze was **fixed**[1] away off yonder on one of those patches of blue sky. It was not a glance of reflection, but rather indicated a suspension of intelligent thought.

9 There was something coming to her and she was waiting for it, fearfully. What was it? She did not know; it was too subtle and elusive to name. But she felt it, creeping out of the sky, reaching toward her through the sounds, the scents, the colour that filled the air.

10 Now her bosom rose and fell tumultuously. She was beginning to recognize this thing that was approaching to possess her, and she was striving to beat it back with her will—as powerless as her two white slender hands would have been.

11 When she abandoned herself, a little whispered word escaped her slightly parted lips. She said it over and over under her breath: "free, free, free!" The vacant stare and the look of terror that had followed it went from her eyes. They stayed keen and bright. Her pulses beat fast, and the coursing blood warmed and relaxed every inch of her body.

12 She did not stop to ask if it were or were not a monstrous joy that held her. A clear and exalted perception enabled her to dismiss the suggestion as trivial.

13 She knew that she would weep again when she saw the kind, tender hands folded in death; the face that had never looked save with love upon her, fixed and grey and dead. But she saw beyond that bitter moment a long procession of years to come that would belong to her absolutely. And she opened and spread her arms out to them in welcome.

14 There would be no one to live for during those coming years; she would live for herself. There would be no powerful will bending hers in that blind persistence with which men and

---

1. *Fixed:* unmoving.

women believe they have a right to impose a private will upon a fellow-creature. A kind intention or a cruel intention made the act seem no less a crime as she looked upon it in that brief moment of illumination.

15   And yet she had loved him—sometimes. Often she had not. What did it matter! What could love, the unsolved mystery, count for in face of this possession of self-assertion which she suddenly recognized as the strongest impulse of her being!

16   "Free! Body and soul free!" she kept whispering.

17   Josephine was kneeling before the closed door with her lips to the keyhole, imploring for admission. "Louise, open the door! I beg; open the door; you will make yourself ill. What are you doing, Louise? For heaven's sake open the door."

18   "Go away. I am not making myself ill." No; she was drinking in a very elixir of life through that open window.

19   Her **fancy**[2] was running riot along those days ahead of her. Spring days, and summer days, and all sorts of days that would be her own. She breathed a quick prayer that life might be long. It was only yesterday she had thought with a shudder that life might be long.

20   She arose at length and opened the door to her sister's importunities. There was a feverish triumph in her eyes, and she carried herself unwittingly like a goddess of Victory. She clasped her sister's waist, and together they descended the stairs. Richards stood waiting for them at the bottom.

21   Some one was opening the front door with a latchkey. It was Brently Mallard who entered, a little travel-stained, composedly carrying his grip-sack and umbrella. He had been far from the scene of accident, and did not even know that there had been one. He stood amazed at Josephine's piercing cry; at Richards' quick motion to screen him from the view of his wife.

22   But Richards was too late.

23   When the doctors came, they said she had died of heart disease—of joy that kills.

PUBLISHED IN
1890

Reader's Response

1. Did you appreciate the ending of the story? Why or why not?

    _____

2. Did you like or dislike Louise Mallard? Why?

    _____

3. How is Louise Mallard's vision of marriage similar to or different from your own?

    _____

4. Which type of story do you prefer, a Gothic story, such as "The Tell-Tale Heart," a fairy tale, such as "The Nightingale and the Rose," or a more realistic narrative based on social concerns, such as "The Story of an Hour"?

    _____

5. What do you like about this type of story?

    _____

_____

2. *Fancy:* imagination.

∞ CLOSE READING

"The Story of an Hour" is filled with irony (the opposite of what is intended, expected or expressed). Sometimes the irony is explicit and you see it at first glance, but in other instances it is implicit. In the latter case, you have to think about the situation and what has happened to appreciate how ironic it is in a broader context. When you read the story a second time, see if you can discover all the examples of irony it contains. Before we look at irony in more depth, though, answer the following questions that are designed to help you begin your exploration of the text.

1. At the beginning of the story, why do Richards and Josephine try to break the bad news to Mrs. Mallard very gently?

2. According to the narrator, how would most women have reacted to such news?

3. How does Mrs. Mallard react at first?

4. Describe in your own words what Mrs. Mallard feels and realizes as she continues to gaze out the window.

5. What has Brently Mallard always felt toward his wife?

6. Why is Louise Mallard unhappy in her marriage?

7. What is the "crime" that Louise Mallard is talking about in the last sentence of paragraph 14 ("There would be no one to live for …")?

8. a) What does Louise Mallard recognize as "the strongest impulse of her being"?

   b) Explain what this means in your own words.

9. What happens at the end of the story?

10. What is the significance of the story's title?

✎ WRITER'S CRAFT

In preparation for the following questions, read the explanations on narration, style, setting, characterization and theme found in the Literary Elements section of this book.

Narration and Point of View

1. What point of view is used to tell the story?

   _____

2. What type of narrator is used to tell the story?

   _____

3. Does the narrator seem close to or distant from the reader in this story?

   _____

4. Think back to "The Tell-Tale Heart" by Edgar Allan Poe. Who narrates the story? Which type of narrator brings the reader closer to the characters in a story? (Circle the correct answer.)

   a) first person   b) third person

5. The story is told in the _____ tense, which also creates a certain distance.

   > The choice of narrator greatly affects the tone and atmosphere of a story. A first-person narrator can speak directly to the reader and involve him or her in a much more personal way. The reader gets a close-up view of what is happening. A third-person narrator usually keeps the reader at a greater distance from what is happening. The process is similar to watching something happen from afar.

Style and Literary Devices

▪ IRONY

1. Explain how the following features are ironic in the context of the story:

   a) the last sentence

   _____

   b) Richards' actions aimed at protecting Mrs. Mallard

   _____

   c) Mrs. Mallard's seeing her dead husband's body

   _____

   ➡

d) Mrs. Mallard's vision of her future

_____

e) Mrs. Mallard's exclamation: "Free! Body and soul free!"

_____

2. Find at least three more examples of irony in the story.

_____

_____

_____

▪ FORESHADOWING

> **Foreshadowing** is a literary device used by writers to hint at what will eventually happen in a story. The use of foreshadowing not only increases the enjoyment of the astute reader, who tries to predict what is going to happen, but also helps build tension and unity in the story.

1. Give two examples of foreshadowing in "The Story of an Hour."

_____

_____

▪ REPETITION

1. Like Poe and Wilde, Chopin uses repetition to create certain effects in "The Story of an Hour." She repeats two adjectives in particular that are full of significance to the story. What are they?

_____

2. Find another example of repetition of adjectives or concrete images in the story.

_____

▪ IMAGERY

1. "The Story of an Hour" is rich in imagery. For example, Mrs. Mallard's weeping is described as "the storm of grief" (paragraph 3). This powerful image appeals to at least two senses: sight and hearing. Give four examples from paragraphs 4 and 5 and identify the sense to which they appeal. (Each example must appeal to a different sense.)

_____

_____

_____

_____

> Chopin uses a great deal of contrastive imagery (images that are opposites) connected with details about the physical setting of the story. These opposite elements in the setting serve to bring out the theme.

2. Look at the following contrast:

Open   **versus**   **Closed**

open window (occurs twice)   shut in her room

Find three more images (spatial elements or gestures) associated with "open" and two more associated with "closed."

Open: _____

Closed: _____

3. Here is a list of words:

a) open   d) natural   g) dead

b) closed   e) repressed   h) alive

c) artificial   f) free

Which four words from the list best exemplify the kind of life that exists inside the Mallards' household?

_____   _____   _____   _____

Which four could be used to describe life outside the house?

_____   _____   _____   _____

4. Chopin uses the weather and the outdoors as an extended metaphor to parallel Mrs. Mallard's feelings. What do the following images represent? (Use your judgment when you are not quite sure.)

a) a storm of grief (paragraph 3)

_____

b) "... the tops of trees that were all aquiver with the new spring life. The delicious breath of rain..." (paragraph 5)

_____

c) "... patches of blue sky" showing through the clouds (paragraphs 6 and 8)

_____

d) "... something coming to her ... creeping out of the sky, reaching toward her through the sounds, the scents, the color that filled the air" (paragraph 9)

_____

e) "Spring days, and summer days ..." (paragraph 19)

_____

The contrasting images of "closed" and "open" and the extended metaphor of the weather add impact to the contrast between the "living death" and the repression that Mrs. Mallard experiences in her marriage and the sudden rush of aliveness and well-being she feels when she thinks she has been freed by her husband's death.

Theme and Characterization

1. Which of the characters in the story are round and which are flat? Justify your answer.

   _____

   _____

   _____

   _____

2. What is the main theme of the story?

   _____

VOCABULARY DEVELOPMENT

Identifying Stressed and Unstressed Syllables in Words

Pronounce each word in the following list and indicate the stresses by placing ' to the right of the stressed syllable. Then check your answers with a dictionary.

Example: triumph—tri'umph

afflicted	revealed	newspaper	breathed
disaster	intelligence	abandonment	keyhole
comfortable	assume	paralyzed	feverish
eaves	aquiver	suspension	self-assertion
develop	monstrous	fixed	recognized

Wrap-Up Activities

1. Write a paragraph about what freedom means to you.

2. Write about marriage from Mrs. Mallard's and/or Josephine's point of view.

3. Is marriage an outdated concept? Explain.

4. Discuss how Chopin uses irony effectively throughout the story.

5. Explain why Chopin can be considered a revolutionary author for her times.

6. Discuss Chopin's use of imagery in "The Story of an Hour."

7. How relevant is the main theme of "The Story of an Hour" to people today? Explain.

My eLab ACTIVITY

My eLab

Check your knowledge of some of the symbolism common to Western literature by doing the matching activity for this chapter.

An Occurrence at Owl Creek Bridge

CHAPTER CONTENTS

▸ **Reading:** "An Occurrence at Owl Creek Bridge" by Ambrose Bierce; author's biography; literary trends (Realism)

▸ **Literary Elements:** style and literary devices (imagery, foreshadowing and flashback); point of view (shift); setting

▸ **Vocabulary Development:** creating atmosphere with vivid present participles

▸ **My eLab Activity:** a crossword puzzle on the vivid vocabulary of this story

ABOUT THE AUTHOR

Ambrose Bierce was born in 1842 in the American Midwest. His family members had military careers and Bierce maintained the tradition by joining the Union army during the American Civil War. He fought in many battles, distinguishing himself, but he was eventually wounded and could no longer serve in the army. The brutality of the war horrified him, leaving him cynical and disillusioned about life. His cynicism and war experiences are seen throughout his stories. After leaving the army, he worked as a journalist for San Francisco newspaper tycoon William Randolph Hearst. He wrote many short stories that had a gothic flavour, and they made him immensely popular. He is best known for *The Devil's Dictionary* (1911), a humorous work. He died under mysterious circumstances. He went off to fight in the Mexican Civil War when he was in his seventies, and he was never heard from again. Although there are many theories about his death, most scholars believe that he died around 1914 while fighting.

LITERARY TRENDS OF THE TIMES

IN THE LATTER HALF OF THE NINETEENTH century, many American writers, such as Ambrose Bierce, Mark Twain, Henry James and Stephen Crane, embraced the tenets of Realism. Like many of their European counterparts, they put aside the Romantic view of the universe, with its emphasis on the supernatural, and turned their eyes to the world around them. These writers took a long hard look at the negative aspects and social ills of their times. Using vivid imagery, they tried to recreate the details of everyday life as realistically as possible in order to convey their point of view to their readers in a convincing way.

Many Americans were deeply affected by the devastation of the Civil War (1861-1865). Ambrose Bierce and his contemporaries no longer felt the need to look toward the supernatural or unusual to find horror; it surrounded them in their everyday lives. Many of Bierce's short stories are based on personal experience, exposing the almost indescribable horrors of war. The subject matter comes from real life, but Gothic elements in Bierce's vivid imagery give his stories a strong impact.

Warm-Up Discussion Topics

1. Have you read any stories about war? If so, which ones? Who wrote the stories?

_____

_____

2. Discuss your knowledge of the American Civil War and other wars that have taken place throughout history. What caused these wars? What were their outcomes?

_____

_____

✄ **INITIAL READING**

This story takes place during the American Civil War (1861-1865), which pitted the industrialized Northern states against eleven Southern states, whose agricultural, cash-crop economy was based on slave labour. The war started several months after the election of Abraham Lincoln, when eleven Southern states decided to secede from the Union (the United States of America) to form their own confederacy (the Confederate States of America) because they feared that the new president would abolish slavery. The Northern soldiers were known as Union soldiers (or Northern or Federal soldiers) and wore blue uniforms; the Southern soldiers (or Confederates) wore grey uniforms. After four years of bloodshed and conflict, the leader of the Confederate army, Robert E. Lee, was forced to surrender unconditionally to the Northern general, Ulysses S. Grant.

Read the story carefully and do not stop to look up unfamiliar words unless it is absolutely necessary. Try to visualize what is happening in the story as if you were watching a film.

An Occurrence at Owl Creek Bridge
Ambrose Bierce

1   A man stood upon a railroad bridge in northern Alabama, looking down into the swift water twenty feet below. The man's hands were behind his back, the wrists bound with a cord. A rope closely encircled his neck. It was attached to a stout cross-timber above his head and the slack fell to the level of his knees. Some loose boards laid upon the sleepers supporting the metals of the railway supplied a footing for him and his executioners—two private soldiers of the **Federal**[1] army, directed by a sergeant who in civil life may have been a deputy sheriff. At a short remove upon the same temporary platform was an officer in the uniform of his rank, armed. He was a captain. A sentinel at each end of the bridge stood with his rifle in the position known as "support," that is to say, vertical in front of the left shoulder, the hammer resting on the forearm thrown straight across the chest—a formal and unnatural position, enforcing an erect carriage of the body. It did not appear to be the duty of these two men to know what was occurring at the centre of the bridge; they merely blockaded the two ends of the foot planking that traversed it.

2   Beyond one of the sentinels nobody was in sight; the railroad ran straight away into a forest for a hundred yards, then, curving, was

---

1. *Federal:* Northen.

lost to view. Doubtless there was an outpost farther along. The other bank of the stream was open ground—a gentle acclivity topped with a stockade of vertical tree trunks, loopholed for rifles, with a single embrasure through which protruded the muzzle of a brass cannon commanding the bridge. Midway of the slope between the bridge and fort were the spectators—a single company of infantry in line, at "parade rest," the butts of the rifles on the ground, the barrels inclining slightly backward against the right shoulder, the hands crossed upon the stock. A lieutenant stood at the right of the line, the point of his sword upon the ground, his left hand resting upon his right. Excepting the group of four at the centre of the bridge, not a man moved. The company faced the bridge, staring stonily, motionless. The sentinels, facing the banks of the stream, might have been statues to adorn the bridge. The captain stood with folded arms, silent, observing the work of his subordinates, but making no sign. Death is a dignitary who, when he comes announced, is to be received with formal manifestations of respect, even by those most familiar with him. In the code of military etiquette, silence and fixity are forms of deference.

3 The man who was engaged in being hanged was apparently about thirty-five years of age. He was a civilian, if one might judge from his habit, which was that of a planter. His features were good—a straight nose, firm mouth, broad forehead, from which his long, dark hair was combed straight back, falling behind his ears to the collar of his well-fitting frock coat. He wore a moustache and pointed beard, but no whiskers; his eyes were large and dark grey, and had a kindly expression which one would hardly have expected in one whose neck was in the hemp. Evidently this was no vulgar assassin. The liberal military code makes provision for hanging many kinds of persons, and gentlemen are not excluded.

4 The preparations being complete, the two private soldiers stepped aside and each drew away the plank upon which he had been standing. The sergeant turned to the captain, saluted and placed himself immediately behind that officer, who in turn moved apart one pace. These movements left the condemned man and the sergeant standing on the two

ends of the same plank, which spanned three of the cross-ties of the bridge. The end upon which the civilian stood almost, but not quite, reached a fourth. This plank had been held in place by the weight of the captain; it was now held by that of the sergeant. At a signal from the former the latter would step aside, the plank would tilt and the condemned man go down between two ties. The arrangement commended itself to his judgment as simple and effective. His face had not been covered nor his eyes bandaged. He looked a moment at his "unsteadfast footing," then let his gaze wander to the swirling water of the stream racing madly beneath his feet. A piece of dancing driftwood caught his attention and his eyes followed it down the current. How slowly it appeared to move. What a sluggish stream!

5 He closed his eyes in order to fix his last thoughts upon his wife and children. The water, touched to gold by the early sun, the brooding mists under the banks at some distance down the stream, the fort, the soldiers, the piece of drift—all had distracted him. And now he became conscious of a new disturbance. Striking through the thought of his dear ones was a sound which he could neither ignore nor understand, a sharp, distinct, metallic percussion like the stroke of a blacksmith's hammer upon the anvil; it had the same ringing quality. He wondered what it was, and whether immeasurably distant or near by—it seemed both. Its recurrence was regular, but as slow as the tolling of a death knell. He awaited each stroke with impatience and—he knew not why—apprehension. The intervals of silence grew progressively longer; the delays became maddening. With their greater infrequency, the sounds increased in strength and sharpness. They hurt his ear like the thrust of a knife; he feared he would shriek. What he heard was the ticking of his watch.

6 He unclosed his eyes and saw again the water below him. "If I could free my hands," he thought, "I might throw off the noose and spring into the stream. By diving, I could evade the bullets and, swimming vigorously, reach the bank, take to the woods and get away home. My home, thank God, is as yet outside their lines; my wife and little ones are still beyond the invader's farthest advance.

5

7   As these thoughts, which have here to be set down in words, were flashed into the doomed man's brain, rather than evolved from it, the captain nodded to the sergeant. The sergeant stepped aside.

II

8   Peyton Farquhar was a well-to-do planter, of an old and highly respected Alabama family. Being a slave owner, and like other slave owners, a politician, he was naturally an original secessionist and ardently devoted to the Southern cause. Circumstances of an imperious nature, which it is unnecessary to relate here, had prevented him from taking service with the gallant army that had fought the disastrous campaigns ending with the fall of Corinth, and he chafed under the inglorious restraint, longing for the release of his energies, the larger life of the soldier, the opportunity for distinction. That opportunity, he felt, would come, as it comes to all in war time. Meanwhile he did what he could. No service was too humble for him to perform in aid of the South, no adventure too perilous for him to undertake if consistent with the character of a civilian who was at heart a soldier, and who in good faith and without too much qualification assented to at least a part of the frankly villainous dictum that all is fair in love and war.

9   One evening, while Farquhar and his wife were sitting on a rustic bench near the entrance to his grounds, a grey-clad soldier rode up to the gate and asked for a drink of water. Mrs. Farquhar was only too happy to serve him with her own white hands. While she was fetching the water, her husband approached the dusty horseman and inquired eagerly for news from the front.

10   "The Yanks are repairing the railroads," said the man, "and are getting ready for another advance. They have reached the Owl Creek bridge, put it in order and built a stockade on the north bank. The commandant has issued an order, which is posted everywhere, declaring that any civilian caught interfering with the railroad, its bridge, tunnels or trains will be summarily hanged. I saw the order."

11   "How far is it to the Owl Creek bridge?" Farquhar asked.

12   "About thirty miles."

13   "Is there no force on this side of the creek?"

14   "Only a picket post half a mile out, on the railroad, and a single sentinel at this end of the bridge."

15   "Suppose a man—a civilian and student of hanging—should elude the picket post and perhaps get the better of the sentinel," said Farquhar, smiling, "what could he acomplish?"

16   The soldier reflected. "I was there a month ago," he replied. "I observed that the flood of last winter had lodged a great quantity of driftwood against the wooden pier at this end of the bridge. It is now dry and would burn like tinder."

17   The lady had now brought the water, which the soldier drank. He thanked her ceremoniously, bowed to her husband and rode away. An hour later, after nightfall, he repassed the plantation, going northward in the direction from which he had come. He was a Federal scout.

III

18   As Peyton Farquhar fell straight downward through the bridge, he lost consciousness and was as one already dead. From this state he was awakened—ages later, it seemed to him— by the pain of a sharp pressure upon his throat, followed by a sense of suffocation. Keen, poignant agonies seemed to shoot from his neck downward through every fibre of his body and limbs. These pains appeared to flash along well-defined lines of ramification and to beat with an inconceivably rapid periodicity. They seemed like streams of pulsating fire heating him to an intolerable temperature. As to his head, he was conscious of nothing but a feeling of fullness—of congestion. These sensations were unaccompanied by thought. The intellectual part of his nature was already effaced; he had power only to feel, and feeling was torment. He was conscious of motion. Encompassed in a luminous cloud, of which he was now merely the fiery heart, without material substance, he swung through unthinkable arcs of oscillation, like a vast pendulum. Then all at once, with terrible suddenness, the light about him shot upward with the noise of a loud splash; a frightful roaring was in his ears, and all was

cold and dark. The power of thought was restored; he knew that the rope had broken and he had fallen into the stream. There was no additional strangulation; the noose about his neck was already suffocating him and kept the water from his lungs. To die of hanging at the bottom of a river!—the idea seemed to him ludicrous. He opened his eyes in the darkness and saw above him a gleam of light, but how distant, how inaccessible! He was still sinking, for the light became fainter and fainter until it was a mere glimmer. Then it began to grow and brighten, and he knew that he was rising toward the surface—knew it with reluctance, for he was now very comfortable. "To be hanged and drowned," he thought, "that is not so bad; but I do not wish to be shot. No; I will not be shot; that is not fair."

19   He was not conscious of an effort, but a sharp pain in his wrist apprised him that he was trying to free his hands. He gave the struggle his attention, as an idler might observe the feat of a juggler, without interest in the outcome. What splendid effort!—what magnificent, what superhuman strength! Ah, that was a fine endeavour! Bravo! The cord fell away; his arms parted and floated upward, the hands dimly seen on each side in the growing light. He watched them with a new interest as first one and then the other pounced upon the noose at his neck. They tore it away and thrust it fiercely aside, its undulations resembling those of a water snake. "Put it back, put it back!" He thought he shouted these words to his hands, for the undoing of the noose had been succeeded by the direst pang that he had yet experienced. His neck ached horribly; his brain was on fire; his heart, which had been fluttering faintly, gave a great leap, trying to force itself out at his mouth. His whole body was racked and wrenched with an insupportable anguish! But his disobedient hands gave no heed to the command. They beat the water vigorously with quick, downward strokes, forcing him to the surface. He felt his head emerge; his eyes were blinded by the sunlight; his chest expanded convulsively, and with a supreme and crowning agony, his lungs engulfed a great draught of air, which instantly he expelled in a shriek!

20   He was now in full possession of his physical senses. They were, indeed, preternaturally keen and alert. Something in the awful disturbance of his organic system had so exalted and refined them that they made record of things never before perceived. He felt the ripples upon his face and heard their separate sounds as they struck. He looked at the forest on the bank of the stream, saw the individual trees, the leaves and the veining of each leaf—saw the very insects upon them: the locusts, the brilliant-bodied flies, the grey spiders stretching their webs from twig to twig. He noted the prismatic colours in all the dewdrops upon a million blades of grass. The humming of the gnats that danced above the eddies of the stream, the beating of the dragon flies' wings, the strokes of the water-spiders' legs, like oars which had lifted their boat—all these made audible music. A fish slid along beneath his eyes and he heard the rush of its body parting the water.

21   He had come to the surface facing down the stream; in a moment the visible world seemed to wheel slowly round, himself the pivotal point, and he saw the bridge, the fort, the soldiers upon the bridge, the captain, the sergeant, the two privates, his executioners. They were in silhouette against the blue sky. They shouted and gesticulated, pointing at him. The captain had drawn his pistol, but did not fire; the others were unarmed. Their movements were grotesque and horrible, their forms gigantic.

22   Suddenly he heard a sharp report and something struck the water smartly within a few inches of his head, spattering his face with spray. He heard a second report, and saw one of the sentinels with his rifle at his shoulder, a light cloud of blue smoke rising from the muzzle. The man in the water saw the eye of the man on the bridge gazing into his own through the sights of the rifle. He observed that it was a grey eye and remembered having read that grey eyes were keenest, and that all famous marksmen had them. Nevertheless, this one had missed.

23   A counter-swirl had caught Farquhar and turned him half round; he was again looking into the forest on the bank opposite the fort. The sound of a clear, high voice in a monotonous singsong now rang out behind him and came across the water with a distinctness that

5

pierced and subdued all other sounds, even the beating of the ripples in his ears. Although no soldier, he had frequented camps enough to know the dread significance of that deliberate, drawling, aspirated chant; the lieutenant on shore was taking a part in the morning's work. How coldly and pitilessly—with what an even, calm intonation, presaging, and enforcing tranquillity in the men—with what accurately measured intervals fell those cruel words:

24 "Attention, company! . . . Shoulder arms! . . . Ready! . . . Aim! . . . Fire!"

25 Farquhar dived—dived as deeply as he could. The water roared in his ears like the voice of Niagara, yet he heard the dulled thunder of the volley and, rising again toward the surface, met shining bits of metal, singularly flattened, oscillating slowly downward. Some of them touched him on the face and hands, then fell away, continuing their descent. One lodged between his collar and neck; it was uncomfortably warm and he snatched it out.

26 As he rose to the surface, gasping for breath, he saw that he had been a long time under water; he was perceptibly farther down stream nearer to safety. The soldiers had almost finished reloading; the metal ramrods flashed all at once in the sunshine as they were drawn from the barrels, turned in the air, and thrust into their sockets. The two sentinels fired again, independently and ineffectually.

27 The hunted man saw all this over his shoulder; he was now swimming vigorously with the current. His brain was as energetic as his arms and legs; he thought with the rapidity of lightning.

28 "The officer," he reasoned, "will not make that martinet's error a second time. It is as easy to dodge a volley as a single shot. He has probably already given the command to fire at will. God help me, I cannot dodge them all!"

29 An appalling splash within two yards of him was followed by a loud, rushing sound, *diminuendo*, which seemed to travel back through the air to the fort and died in an explosion which stirred the very river to its deeps!

30 A rising sheet of water curved over him, fell down upon him, blinded him, strangled him! The cannon had taken a hand in the game. As he shook his head free from the commotion of the smitten water, he heard the deflected shot humming through the air ahead, and in an instant it was cracking and smashing the branches in the forest beyond.

31 "They will not do that again," he thought; "the next time they will use a charge of grape. I must keep my eye upon the gun; the smoke will apprise me—the report arrives too late; it lags behind the missile. That is a good gun."

32 Suddenly he felt himself whirled round and round—spinning like a top. The water, the banks, the forests, the now distant bridge, fort and men—all were commingled and blurred. Objects were represented by their colours only; circular horizontal streaks of colour—that was all he saw. He had been caught in a vortex and was being whirled on with a velocity of advance and gyration that made him giddy and sick. In a few moments he was flung upon the gravel at the foot of the left bank of the stream—the southern bank—and behind a projecting point which concealed him from his enemies. The sudden arrest of his motion, the abrasion of one of his hands on the gravel, restored him, and he wept with delight. He dug his fingers into the sand, threw it over himself in handfuls and audibly blessed it. It looked like diamonds, rubies, emeralds; he could think of nothing beautiful which it did not resemble. The trees upon the bank were giant garden plants; he noted a definite order in their arrangement, inhaled the fragrance of their blooms. A strange, roseate light shone through the spaces among their trunks and the wind made in their branches the music of Aeolian harps. He had no wish to perfect his escape—was content to remain in that enchanting spot until retaken.

33 A whiz and rattle of grapeshot among the branches high above his head roused him from his dream. The baffled cannoneer had fired him a random farewell. He sprang to his feet, rushed up the sloping bank, and plunged into the forest.

34 All that day he travelled, laying his course by the rounding sun. The forest seemed interminable; nowhere did he discover a break in it, not even a woodman's road. He had not known that he lived in so wild a

region. There was something uncanny in the revelation.

35 By nightfall, he was fatigued, footsore, famishing. The thought of his wife and children urged him on. At last he found a road which led him in what he knew to be the right direction. It was as wide and straight as a city street, yet it seemed untravelled. No fields bordered it, no dwelling anywhere. Not so much as the barking of a dog suggested human habitation. The black bodies of the trees formed a straight wall on both sides, terminating on the horizon in a point, like a diagram in a lesson in perspective. Overhead, as he looked up through this rift in the wood, shone great garden stars looking unfamiliar and grouped in strange constellations. He was sure they were arranged in some order which had a secret and malign significance. The wood on either side was full of singular noises, among which—once, twice, and again—he distinctly heard whispers in an unknown tongue.

36 His neck was in pain and lifting his hand to it, he found it horribly swollen. He knew that it had a circle of black where the rope had bruised it. His eyes felt congested; he could no longer close them. His tongue was swollen with thirst; he relieved its fever by thrusting it forward from between his teeth into the cold air. How softly the turf had carpeted the untravelled avenue— he could no longer feel the roadway beneath his feet!

37 Doubtless, despite his suffering, he had fallen asleep while walking, for now he sees another scene—perhaps he has merely recovered from a delirium. He stands at the gate of his own home. All is as he left it, and all bright and beautiful in the morning sunshine. He must have travelled the entire night. As he pushes open the gate and passes up the wide white walk, he sees a flutter of female garments; his wife, looking fresh and cool and sweet, steps down from the veranda to meet him. At the bottom of the steps, she stands waiting, with a smile of ineffable joy, an attitude of matchless grace and dignity. Ah, how beautiful she is! He springs forward with extended arms. As he is about to clasp her, he feels a stunning blow upon the back of the neck; a blinding white light blazes all about him with a sound like the shock of a cannon—then all is darkness and silence!

38 Peyton Farquhar was dead; his body, with a broken neck, swung gently from side to side beneath the timbers of the Owl Creek bridge.

PUBLISHED IN
1891

Reader's Response

1. What was your immediate reaction when you read the last lines of the story? Explain.

   _____

   _____

2. What other feelings or thoughts accompanied your reaction?

   _____

   _____

3. What did you like most about this story?

   _____

   _____

4. What did you like least?

   _____

   _____

✏ CLOSE READING

Now reread the story. Pay particular attention to how the story unwinds in a way that convinces the reader that Farquhar may actually be escaping.

1. Which army do the soldiers belong to, the North or the South?

   _____

2. What is the civil status of the man being hanged?

   _____

3. At the beginning of the story, why does it seem strange that this man is being hanged?

   _____

4. Who is Peyton Farquhar and which side is he on?

   _____

5. Why is Peyton Farquhar being hanged?

   _____

6. Why do Farquhar and his wife think the soldier who asks them for water is fighting for the South?

   _____

7. Which sentence in the following paragraph proves that the soldier has tricked Farquhar? (Circle the correct answer.)

   > The lady had now brought the water, which the soldier drank. He thanked her ceremoniously, bowed to her husband and rode away. An hour later, after nightfall, he repassed the plantation, going northward in the direction from which he had come. He was a Federal scout.

   a) first   b) second   c) third   d) fourth

8. In Part I, which paragraphs show that Farquhar's perceptions have altered and that his sense of time has slowed down immensely?

   a) paragraphs 1 and 2   b) paragraphs 3 and 4   c) paragraphs 5 and 6

9. In Part III, what does Farquhar want desperately to believe?

   _____

10. What happens to Farquhar at the end of the story?

✏ WRITER'S CRAFT

This story is remarkable for its vivid imagery and its intentionally misleading shifts in setting and point of view.

Style and Literary Devices

■ IMAGERY

1. In the paragraph below, find examples of imagery that relate to the following senses. (Write the first eleven words of each example.)

> He closed his eyes in order to fix his last thoughts upon his wife and children. The water, touched to gold by the early sun, the brooding mists under the banks at some distance down the stream, the fort, the soldiers, the piece of drift—all had distracted him. And now he became conscious of a new disturbance. Striking through the thought of his dear ones was a sound which he could neither ignore nor understand, a sharp, distinct, metallic percussion like the stroke of a blacksmith's hammer upon the anvil; it had the same ringing quality. He wondered what it was, and whether immeasurably distant or near by—it seemed both. Its recurrence was regular, but as slow as the tolling of a death knell. He awaited each stroke with impatience and—he knew not why—apprehension. The intervals of silence grew progressively longer; the delays became maddening. With their greater infrequency, the sounds increased in strength and sharpness. They hurt his ear like the thrust of a knife; he feared he would shriek. What he heard was the ticking of his watch.

Sight _____

Hearing _____

Touch _____

2. Which sense is predominant in the imagery of the preceding paragraph?

_____

3. The detailed description of the scene makes the story more _____ to the reader.

   a) abstract   b) realistic   c) distant

Narrative Structure: Foreshadowing and Flashback

1. Find a sentence in the preceding paragraph that is an example of foreshadowing.

_____

2. Find another example of foreshadowing in the first paragraph in Part III.

_____

3. What section of the story represents a flashback?

_____

4. What is the purpose of the flashback sequence in terms of what is happening at the beginning of the story?

_____

The flashback sequence interrupts the narrative structure and turns the reader's attention away from the immediacy of the events on the bridge. At the beginning of Part III, when the storyline returns to the bridge, the reader feels more empathy toward the man being hanged and hopes he will escape.

Point of View and Setting

1. What kind of narrator does the story have?

_____

2. At the beginning of the story, does the narrator seem close to or distant from the reader? Explain.

_____

_____

3. From whose point of view is the story being told in Part III? Does this make you feel close to or distant from the action?

_____

4. There are two types of setting in the story, an external setting and an internal one.

a) What is the external setting? _____

b) The internal setting is in Peyton Farquhar's _____.

5. The internal setting seems as realistic as the external setting because it is described in great _____.

The story shifts back and forth between the settings. As it shifts from the external to the internal setting, the narrator's point of view also shifts from distant to close, from objective to subjective. These shifts help fool the reader into believing that Farquhar has actually escaped.

6. Which tense is used to tell the story, excluding the second-to-last paragraph? _____.

7. In the second-to-last paragraph, the tense shifts to the _____.

The shift in tense gives an immediacy to the story and makes the reader feel even closer to the man trying to escape his fate. The author has effectively set the reader up for the shock of the final paragraph.

8. Based on the following paragraph, what is Farquhar's attitude toward the war?

Meanwhile he did what he could. No service was too humble for him to perform in aid of the South, no adventure too perilous for him to undertake if consistent with the character of a civilian who was at heart a soldier, and who in good faith and without too much qualification assented to at least a part of the frankly villainous dictum that all is fair in love and war.

_____

9. What is the attitude of the narrator toward Farquhar in the above paragraph?

    a) respectful   b) scornful

10. This attitude is conveyed through

    a) details.   b) exaggeration.   c) innuendos.   d) none of the answers

11. Find a word in the preceding paragraph that supports your answer. (*Hint*: it is repeated three

    times.)_____

12. Read the following paragraphs.

> He wore a moustache and pointed beard, but no whiskers; his eyes were large and dark grey, and had a kindly expression which one would hardly have expected in one whose neck was in the hemp. Evidently this was no vulgar assassin. The liberal military code makes provision for hanging many kinds of persons, and gentlemen are not excluded. (paragraph 5)
>
> One evening, while Farquhar and his wife were sitting on a rustic bench near the entrance to his grounds, a grey-clad soldier rode up to the gate and asked for a drink of water. Mrs. Farquhar was only too happy to serve him with her own white hands. While she was fetching the water, her husband approached the dusty horseman and inquired eagerly for news from the front. (paragraph 9)

    A. What is the narrator's attitude toward the military?

        a) neutral   b) respectful   c) scornful   d) none of the answers

    B. Which sentences in the above paragraphs support your answer? Write the first six words of the sentences.

    _____

    _____

13. Based on your reaction and understanding of the story, how do you think the author felt about war? Explain.

    _____

VOCABULARY DEVELOPMENT

Creating Atmosphere with Vivid Present Participles

Ambrose Bierce chose very powerful words to create mood and atmosphere in this story. The following present participles (-*ing* words) appear in the text. Bierce used them as adjectives or verbs to great effect.

Insert the following present participles correctly into the sentences and then circle the sense or senses they appeal to: **a)** sight; **b)** hearing; or **c)** touch.

Present Participles

beating	blinding	brooding	cracking	fluttering	gasping
gazing	hanging	humming	maddening	oscillating	pulsating
ringing	rising	roaring	sinking	smashing	spattering
spinning	striking	suffocating	tolling		

Sense

1. The _____ fog added a touch of melancholy and sadness to the   **a b c**
scene.

2. The mournful bell, _____ for the war dead, continued   **a b c**
_____ in my ears for hours.

3. The heat was _____, and he lay there _____ for breath.   **a b c**

4. With a thunderous _____, the tree fell, _____   **a b c**
everything beneath it.

5. After _____ one last time in the gentle breeze, the autumn leaves   **a b c**
looked like flocks of strange birds _____ to the ground.

6. The campers sat around the _____ fire _____ at the   **a b c**
dancing flames.

7. The young girl whipped the cream with the electric mixer, _____   **a b c**
everything close to the bowl.

8. The _____ drip-drip-drip of the leaky faucet and the monotonous   **a b c**
_____ of the refrigerator prevented him from sleeping.

9. Their legs moved in _____ rhythms to the _____ of the drum.   **a b c**

10. The lightning lit up the sky, _____ the church steeple in a last   **a b c**
_____ flash.

11. First _____ to the crest of the wave, then _____ out   **a b c**
of sight into its trough, the survivor, _____ from the floating
branch, was finally spotted by the rescue team.

Wrap-Up Activities

1. Describe a scene of your choice using evocative language. Try to appeal to at least three different senses.

2. Write about a terrifying situation using ten of the present participles from the Vocabulary Development activity above.

3. Describe how Bierce tricks the reader in this story.

4. Write about the impact of surprise endings.

5. Analyze the use of imagery in "An Occurrence at Owl Creek Bridge."

6. Analyze the theme of "An Occurrence at Owl Creek Bridge."

7. Write about the life of Ambrose Bierce and how his experiences are reflected in his works.

8. Write about the causes and the outcome of the American Civil War. How did this war affect the American people?

9. Write about the settings of the story. How are they developed? What devices are used by the author to create them?

10. Read other works by the same author. Give an oral presentation on the recurring themes, literary devices, etc., found in his works.

My eLab ACTIVITY

My eLab 📂

Check your understanding of the vivid vocabulary found in this story by doing the crossword puzzle for this chapter.

Hills Like White Elephants

CHAPTER CONTENTS

▸ **Reading:** "Hills Like White Elephants" by Ernest Hemingway; author's biography; literary trends (Realism, twentieth-century literature, experimentation)

▸ **Literary Elements:** characterization; dialogue; symbolism

▸ **Vocabulary Development:** foreign words

ABOUT THE AUTHOR

Ernest Hemingway (1899-1961) was one of the most influential American writers of the twentieth century. His writing style created a stir among literary circles for its ability to cut out extraneous language, leaving only the core. In other words, Hemingway used words sparingly, with minimal adjectives, creating an innovative writing style that many tried to copy. For this, as well as for the astounding works he produced, Hemingway was awarded the Nobel Prize for literature in 1954.

Hemingway was born the second child in a family of six children. His father was a doctor and the family belonged to the upper-middle class. He received a classical education in Chicago, where the family lived. He started writing while at high school, and some of his work was published in his school's literary magazine. Hemingway began his writing career as a journalist, first in Kansas and later in Toronto, where he wrote for the *Toronto Star*. It was as a journalist that Hemingway first started using the style that later made him famous. Journalism requires short sentences, short paragraphs and clarity.

At the start of World War I, Hemingway volunteered for service as an ambulance driver in Italy. He was soon wounded and spent time recuperating in a military hospital. There he fell in love with an American nurse, but the romance did not last. This experience, as well as other events, such as the Spanish Civil War and the Greek Civil War, which he covered as a journalist, allowed him to gain first-hand experience in the horrors of war. He was also able to collect material for his prize-winning stories and novels. His works, such as *Men Without Women* (1927), which contains "Hills Like White Elephants"; *A Farewell to Arms* (1929), a story about a love affair between an ambulance driver and a nurse; and *For Whom the Bell Tolls* (1940), which takes place during the Spanish Civil War, demonstrate that his wartime journalistic experiences influenced him greatly.

Hemingway had a very colourful personal life. He was a great sportsman and especially loved deep-sea fishing, which is the subject of his masterpiece, *The Old Man and the Sea* (1952). He was married four times. He committed suicide in 1961.

LITERARY TRENDS
OF THE TIMES

DURING THE FIRST HALF OF THE twentieth century, Realism remained a preoccupation of a great many writers, especially in North America. World War I, the Roaring Twenties, the Great Depression and World War II influenced many writers to continue exploring human experience through a realistic perspective. Some of the more famous writers include Willa Cather, John Steinbeck, William Faulkner, Shirley Jackson and Ernest Hemingway. Hemingway applied a bare, journalistic style to his short stories and novels to deal as realistically as possible with the subjects he chose to write about: war, hunting, death, masculinity and gender relations. However, like several of his contemporaries, Hemingway began to experiment with the form of the short story by eliminating the traditionally structured plot and focusing on more subtle psychological conflicts. The story you will read in this chapter exemplifies this new approach.

Warm-Up Discussion Topics

1. Have you ever been in a situation where the subject of discussion has made you feel uncomfortable? If so, how did you react to the situation?

2. Do you like to travel? Recount an interesting story that happened when you took a trip.

3. Would you rather travel than settle down in one place? Explain.

∞ INITIAL READING

The story "Hills Like White Elephants" is very short. Read it through the first time without worrying about unfamiliar vocabulary.

Hills Like White Elephants
Ernest Hemingway

1 THE hills across the valley of the Ebro were long and white. On this side there was no shade and no trees and the station was between two lines of rails in the sun. Close against the side of the station there was the warm shadow of the building and a curtain, made of strings of bamboo beads, hung across the open door into the bar, to keep out flies. The American and the girl with him sat at a table in the shade, outside the building. It was very hot and the express from Barcelona would come in forty minutes. It stopped at this junction for two minutes and went on to Madrid.

2   "What should we drink?" the girl asked. She had taken off her hat and put it on the table.

3   "It's pretty hot," the man said.

4   "Let's drink beer."

5   "Dos cervezas," the man said into the curtain.

6   "Big ones?" a woman asked from the doorway.

7   "Yes. Two big ones."

8   The woman brought two glasses of beer and two felt pads. She put the felt pads and the beer glasses on the table and looked at the man and the girl. The girl was looking off at the line of hills. They were white in the sun and the country was brown and dry.

9   "They look like white elephants," she said.

10   "I've never seen one," the man drank his beer.

11   "No, you wouldn't have."

12   "I might have," the man said. "Just because you say I wouldn't have doesn't prove anything."

13   The girl looked at the bead curtain. "They've painted something on it," she said. "What does it say?"

14   "Anis del Toro. It's a drink."

15   "Could we try it?"

16   The man called "Listen" through the curtain. The woman came out from the bar.

17   "Four reales."

18   "We want two Anis del Toro."

19   "With water?"

20   "Do you want it with water?"

21   "I don't know," the girl said. "Is it good with water?"

22   "It's all right."

23   "You want them with water?" asked the woman.

24   "Yes, with water."

25   "It tastes like licorice," the girl said and put the glass down.

26   "That's the way with everything."

27   "Yes," said the girl. "Everything tastes of licorice. Especially all the things you've waited so long for, like absinthe."

28   "Oh, cut it out."

29   "You started it," the girl said. "I was being amused. I was having a fine time."

30   "Well, let's try and have a fine time."

31   "All right. I was trying. I said the mountains looked like white elephants. Wasn't that bright?"

32   "That was bright."

33   "I wanted to try this new drink. That's all we do, isn't it—look at things and try new drinks?"

34   "I guess so."

35   The girl looked across at the hills.

36   "They're lovely hills," she said. "They don't really look like white elephants. I just meant the colouring of their skin through the trees."

37   "Should we have another drink?"

38   "All right."

39   The warm wind blew the bead curtain against the table.

40   "The beer's nice and cool," the man said.

41   "It's lovely," the girl said.

42   "It's really an awfully simple operation, Jig," the man said. "It's not really an operation at all."

43   The girl looked at the ground the table legs rested on.

44   "I know you wouldn't mind it, Jig. It's really not anything. It's just to let the air in."

45   The girl did not say anything.

46   "I'll go with you and I'll stay with you all the time. They just let the air in and then it's all perfectly natural."

47   "Then what will we do afterward?"

48   "We'll be fine afterward. Just like we were before."

49   "What makes you think so?"

50   "That's the only thing that bothers us. It's the only thing that's made us unhappy."

51   The girl looked at the bead curtain, put her hand out and took hold of two of the strings of beads.

52   "And you think then we'll be all right and be happy."

53   "I know we will. You don't have to be afraid. I've known lots of people that have done it."

54 "So have I," said the girl. "And afterward they were all so happy."

55 "Well," the man said, "if you don't want to, you don't have to. I wouldn't have you do it if you didn't want to. But I know it's perfectly simple."

56 "And you really want to?"

57 "I think it's the best thing to do. But I don't want you to do it if you don't really want to."

58 "And if I do it, you'll be happy and things will be like they were and you'll love me?"

59 "I love you now. You know I love you."

60 "I know. But if I do it, then it will be nice again if I say things are like white elephants, and you'll like it?"

61 "I'll love it. I love it now but I just can't think about it. You know how I get when I worry."

62 "If I do it, you won't ever worry?"

63 "I won't worry about that because it's perfectly simple."

64 "Then I'll do it. Because I don't care about me."

65 "What do you mean?"

66 "I don't care about me."

67 "Well, I care about you."

68 "Oh, yes. But I don't care about me. And I'll do it and then everything will be fine."

69 "I don't want you to do it if you feel that way."

70 The girl stood up and walked to the end of the station. Across, on the other side, were fields of grain and trees along the banks of the Ebro. Far away, beyond the river, were mountains. The shadow of a cloud moved across the field of grain and she saw the river through the trees.

71 "And we could have all this," she said. "And we could have everything and every day we make it more impossible."

72 "What did you say?"

73 "I said we could have everything."

74 "We can have everything."

75 "No, we can't."

76 "We can have the whole world."

77 "No, we can't."

78 "We can go everywhere."

79 "No, we can't. It isn't ours anymore."

80 "It's ours."

81 "No, it isn't. And once they take it away, you never get it back."

82 "But they haven't taken it away."

83 "We'll wait and see."

84 "Come on back in the shade," he said. "You mustn't feel that way."

85 "I don't feel any way," the girl said. "I just know things."

86 "I don't want you to do anything that you don't want to do—"

87 "Nor that isn't good for me," she said. "I know. Could we have another beer?"

88 "All right. But you've got to realize—"

89 "I realize," the girl said. "Can't we maybe stop talking?"

90 They sat down at the table and the girl looked across at the hills on the dry side of the valley and the man looked at her and at the table.

91 "You've got to realize," he said, "that I don't want you to do it if you don't want to. I'm perfectly willing to go through with it if it means anything to you."

92 "Doesn't it mean anything to you? We could get along."

93 "Of course it does. But I don't want anybody but you. I don't want anyone else. And I know it's perfectly simple."

94 "Yes, you know it's perfectly simple."

95 "It's all right for you to say that, but I do know it."

96 "Would you do something for me now?"

97 "I'd do anything for you."

98 "Would you please please please please please please please stop talking?"

99 He did not say anything but looked at the bags against the wall of the station. There were labels on them from all the hotels where they had spent nights.

100 "But I don't want you to," he said, "I don't care anything about it."

101 "I'll scream," the girl said.

102   The woman came out through the curtains with two glasses of beer and put them down on the damp felt pads. "The train comes in five minutes," she said.

103   "What did she say?" asked the girl.

104   "That the train is coming in five minutes."

105   The girl smiled brightly at the woman, to thank her.

106   "I'd better take the bags over to the other side of the station," the man said. She smiled at him.

107   "All right. Then come back and we'll finish the beer."

108   He picked up the two heavy bags and carried them around the station to the other tracks. He looked up the tracks but could not see the train. Coming back, he walked through the barroom, where people waiting for the train were drinking. He drank an Anis at the bar and looked at the people. They were all waiting reasonably for the train. He went out through the bead curtain. She was sitting at the table and smiled at him.

109   "Do you feel better?" he asked.

110   "I feel fine," she said. "There's nothing wrong with me. I feel fine."

PUBLISHED IN
1927

Reader's Response

1. What particular characteristics did you notice about the way this story is written?

_____

_____

2. What were your impressions about the two characters in the story?

_____

_____

∞ CLOSE READING

Reread the story, paying close attention to the subject of conversation between the two characters. You will need to read between the lines and make inferences to understand what they are talking about.

1. Who are the main characters in the story?

_____

_____

2. What is their relationship and attitude toward each other?

_____

_____

3. What is the general setting of the story?

_____

_____

4. What is the specific setting of the story?

_____

_____

5. Why is the couple at this place?

_____

_____

6. a) What is the major topic of their conflict?

_____

  b) Cite one or two sentences from the text that point to this topic—albeit indirectly.

_____

∽ WRITER'S CRAFT

Writers reveal characters in different ways. Refer to the Literary Elements section at the beginning of this book in order to understand the techniques writers use for character development.

> Hemingway uses two main methods to depict the characters and unfold the central conflict.
> - **Dialogue** refers to speech between two people.
> - **Symbolism** is the practice of using something concrete to represent something else (concrete or abstract).
>
> **Examples**
> Colours are often used to symbolize things. In Western cultures, for example, black often symbolizes death; white symbolizes innocence and purity; green symbolizes hope or rebirth.
>
> The seasons of the year can be used to symbolize different periods in a person's life.

Dialogue

1. This story is written in dialogue form. In your opinion, how effective is this form of writing?

_____

2. What kind of conversation does the couple have openly during the course of the story? (More than one answer.)

a) small talk   b) casual   c) light-hearted   d) profound

3. What do they talk about openly?

_____

4. Who starts the conversation—the man or the woman? _____

5. What is the tone of the conversation? (More than one answer.)

a) sarcastic   b) emotional   c) humorous   d) furtive   e) tense

6. What observation does the woman make about the scenery?

_____

7. What response does the man make to the woman's observation?

_____

8. a) What do their respective responses reveal about the characters?

Man: _____

Woman: _____

b) What do their responses reveal about their relationship?

_____

9. The woman makes the following statements in the story:

> "Everything tastes of licorice. Especially all the things you've waited so long for, like absinthe."

> "That's all we do, isn't it—look at things and try new drinks?"

These statements suggest that the characters' relationship is (more than one answer)

a) long term.   b) disappointing.   c) deep.   d) superficial.   e) happy.

10. Why is the major source of their conflict only implied?

_____

11. What does their conflict indicate about the relationship of the couple? In other words, what does the continuing dialogue reveal about how the couple communicates?

_____

12. How does this create tension in the story?

_____

13. a) What is the position of the man regarding the couple's major source of conflict?

_____

b) Why does he take this position?

_____

14. What is the woman's position?

_____

15. What three strategies or arguments does the man use in order to get his own way?

_____

■ ACTION VERSUS DIALOGUE

The dialogue technique is used to reveal what is said between the couple. However, what is unsaid is also very important for revealing characters. In the case of the couple, often their inability to communicate is shown by their desire to avoid being direct about their feelings. Their feelings are communicated by gestures or actions. For example, when the man starts to talk about the operation, the woman's actions show that she avoids the conversation. Moreover, often in real situations when people are uncomfortable about a subject, they show their reluctance to speak about the topic by fiddling or doing other body movements. Here is an example from the story:

> "The girl looked at the ground the table legs rested on."

> "The girl looked at the bead curtain, put her hand out and took hold of two of the strings of beads."

1. List two other actions by the man or the woman that indicate that each is trying to avoid discussing the subject directly.

   a) _____

   b) _____

2. Who do you think "wins" the argument? _____

   Support your answer with direct quotations from the text and explain your reasoning.

   _____

   _____

   _____

Symbolism

1. In the first paragraph of "Hills Like White Elephants," what form of writing does Hemingway use—descriptive, narrative, dialogue or monologue?

   _____

2. Write three different words used in the first paragraph that convey the dominant atmosphere of the setting.

   _____

3. The woman's observations about the scenery are symbolic when put into the perspective of the story's conflict. The expression "hills like white elephants" is also the title of the story. The hills are compared to elephants, meaning that they are big and round. Considering the woman's condition, what else can such a symbol represent?

   _____

4. The expression "white elephant" in English refers to something useless and unwanted. Again, considering the woman's condition, what could this expression symbolize in the story?

   _____

5. There are many other symbols in the story. Write an interpretation for the following symbols:

   a) the general scenery (It is hot, arid and without vegetation.)

   _____

   b) the train

   _____

   c) the fact that the couple is nameless (The woman is referred to once by her nickname, "Jig.")

   _____

   d) the use of the definite article instead of the indefinite article in reference to the couple (the American/the girl)

   _____

   e) where the couple is seated in comparison to the other people in the bar

   _____

6. List at least two other symbols and write your interpretation of them.

   a) _____

   b) _____

VOCABULARY DEVELOPMENT
Foreign Words

1. In "Hills Like White Elephants," Hemingway uses words that are not English.

   For each of the words below, find words or phrases in the story that explain their meanings. Write the English meanings beside the foreign words.

   a) "dos cervezas" _____

   b) "Anis del Toro" _____

   c) "four reales" (The meaning of this word is implied. Read the dialogue carefully to understand its meaning.)

   _____

2. Reread the part of the story where these words are found. Who is using these words?

   _____

3. What does it indicate about the other character and the couple's relationship?

   _____

4. What do the following words refer to?

   a) "Ebro" _____

   a) "Barcelona" and "Madrid" _____

5. How are Barcelona and Madrid a contrast to the place where the couple is waiting?

_____

_____

6. In your opinion, why did Hemingway use foreign language in this story?

_____

_____

Wrap-Up Activities

1. Explain the different symbols found in "Hills Like White Elephants." What do the symbols represent?

2. How effective is Hemingway's use of dialogue in creating a tense atmosphere between the characters?

3. The story "Hills Like White Elephants" is symbolic of the journey of life. Discuss.

4. Write a dialogue in which one person tries to convince another person to do something.

Beware of the Dog

A B O U T
THE **AUTHOR**

Roald Dahl (1916-1990) was born in Wales. He was a fighter pilot for the British Royal Air Force (RAF) during World War II. He was shot down in North Africa and later wrote a memoir about this experience. He wrote short stories, plays and children's novels. His first story appeared in *The Saturday Evening Post*. "Beware of the Dog" was originally published in 1944 in *Harper's*, an American magazine, and it was later published in *Over to You: Ten Stories of Flyers and Flying*. Dahl's stories were an instant success. Dahl wrote for both adults and children. His children's novels include *James and the Giant Peach* and *Charlie and the Chocolate Factory*. His stories for adults are full of suspense and include a twist at the end. He received many prizes for his written work, including the Edgar Allan Poe Award from the Mystery Writers of America, the Whitbread Award, the Federation of Children's Book Groups Award and many more.

LITERARY TRENDS
OF THE TIMES

WORLD WAR II LASTED FROM 1939-1945. It had tremendous consequences on the lives of people around the world. Literature of this period reflected the horrors of battle, the loss of family and country, and other wartime realities. Roald Dahl's short story, "Beware of the Dog," is semi-autobiographical and demonstrates the physical hardship and fear that soldiers had to endure while fighting in the war. Holocaust literature is also an important sub-genre of this period, demonstrating the horrors of war. For example, *The Diary of Anne Frank* shows the condition of Jews forced into hiding because of Nazi occupation.

Writers continued to be preoccupied with the tragedies of war after the end of World War II. Thus, they dwelled on themes of social injustice, death, violence, espionage, immigration, the Cold War, nuclear weapons and many other social ills. Like adult literature of this time, children's literature also reflected the realities of war and social issues such as love, death, violence and so on. For example, C. S. Lewis wrote *The Chronicles of Narnia*, in which the main characters have to leave their family home to avoid the bombing of London. In William Golding's famous novel *The Lord of the Flies*, the children are being evacuated from London when they are shipwrecked on an island.

Warm-Up Discussion Topics

1. Do you like reading stories or watching films about war? Write down some of your favourite stories or films and explain what you find interesting about them.

_____

_____

2. What do you know about World War II? In a few sentences, explain your knowledge of this war.

_____

_____

∞ INITIAL READING

The story "Beware of the Dog" is set during World War II. Read it through for the first time without worrying about unfamiliar vocabulary.

Beware of the Dog
Roald Dahl

1 DOWN below there was only a vast white undulating sea of cloud. Above there was the sun, and the sun was white like the clouds, because it is never yellow when one looks at it from high in the air.

2 He was still flying the Spitfire. His right hand was on the stick, and he was working the rudder bar with his left leg alone. It was quite easy. The machine was flying well, and he knew what he was doing.

3 Everything is fine, he thought. I'm doing all right. I'm doing nicely. I know my way home. I'll be there in half an hour. When I land, I shall taxi in and switch off my engine and I shall say, help me to get out, will you. I shall make my voice sound ordinary and natural and none of them will take any notice. Then I shall say, someone help me to get out. I can't do it alone because I've lost one of my legs. They'll all laugh and think that I'm joking, and I shall say, all right, come and have a look, you unbelieving bastards. Then Yorky will climb up onto the wing and look inside. He'll probably be sick because of all the blood and the mess. I shall laugh and say, for God's sake, help me out.

4 He glanced down again at his right leg. There was not much of it left. The cannon shell had taken him on the thigh, just above the knee, and now there was nothing but a great mess and a lot of blood. But there was no pain. When he looked down, he felt as though he were seeing something that did not belong to him. It had nothing to do with him. It was just a mess which happened to be there in the cockpit; something strange and unusual and rather interesting. It was like finding a dead cat on the sofa.

5 He really felt fine, and because he still felt fine, he felt excited and unafraid.

6 I won't even bother to call up on the radio for the blood wagon, he thought. It isn't necessary. And when I land, I'll sit there quite normally and say, some of you fellows come and help me out, will you, because I've lost one of my legs. That will be funny. I'll laugh a little while I'm saying it; I'll say it calmly and slowly, and they'll think I'm joking. When Yorky comes up onto the wing and gets sick, I'll say, Yorky, you old son of a bitch, have you fixed my car yet? Then when I get out, I'll make my report and

later I'll go up to London. I'll take that half bottle of whisky with me and I'll give it to Bluey. We'll sit in her room and drink it. I'll get the water out of the bathroom tap. I won't say much until it's time to go to bed; then I'll say, Bluey, I've got a surprise for you. I lost a leg today. But I don't mind so long as you don't. It doesn't even hurt. We'll go everywhere in cars. I always hated walking, except when I walked down the street of the coppersmiths in Baghdad, but I could go in a rickshaw. I could go home and chop wood, but the head always flies off the axe. Hot water, that's what it needs; put it in the bath and make the handle swell. I chopped lots of wood last time I went home, and I put the axe in the bath ...

7   Then he saw the sun shining on the engine cowling of his machine. He saw the rivets in the metal, and he remembered where he was. He realized that he was no longer feeling good; that he was sick and giddy. His head kept falling forward onto his chest because his neck seemed no longer to have any strength. But he knew that he was flying the Spitfire, and he could feel the handle of the stick between the fingers of his right hand.

8   I'm going to pass out, he thought. Any moment now I'm going to pass out.

9   He looked at his altimeter. Twenty-one thousand. To test himself he tried to read the hundreds as well as the thousands. Twenty-one thousand and what? As he looked, the dial became blurred, and he could not even see the needle. He knew then that he must bail out; that there was not a second to lose, otherwise he would become unconscious. Quickly, frantically, he tried to slide back the hood with his left hand, but he had not the strength. For a second he took his right hand off the stick, and with both hands he managed to push the hood back. The rush of cold air on his face seemed to help. He had a moment of great clearness, and his actions became orderly and precise. That is what happens with a good pilot. He took some quick deep breaths from his oxygen mask, and as he did so, he looked out over the side of the cockpit. Down below there was only a vast white sea of cloud, and he realized that he did not know where he was.

10   It'll be the Channel, he thought. I'm sure to fall in the drink.

11   He throttled back, pulled off his helmet, undid his straps and pushed the stick hard over to the left. The Spitfire dripped its port wing, and turned smoothly over onto its back. The pilot fell out.

12   As he fell, he opened his eyes, because he knew that he must not pass out before he had pulled the cord. On one side he saw the sun; on the other he saw the whiteness of the clouds, and as he fell, as he somersaulted in the air, the white clouds chased the sun and the sun chased the clouds. They chased each other in a small circle; they ran faster and faster, and there was the sun and the clouds and the clouds and the sun, and the clouds came nearer until suddenly there was no longer any sun, but only a great whiteness. The whole world was white, and there was nothing in it. It was so white that sometimes it looked black, and after a time it was either white or black, but mostly it was white. He watched it as it turned from white to black, and then back to white again, and the white stayed for a long time, but the black lasted only for a few seconds. He got into the habit of going to sleep during the white periods, and of waking up just in time to see the world when it was black. But the black was very quick. Sometimes it was only a flash, like someone switching off the light, and switching it on again at once, and so whenever it was white, he dozed off.

13   One day, when it was white, he put out a hand and he touched something. He took it between his fingers and crumpled it. For a time he lay there, idly letting the tips of his fingers play with the thing, which they had touched. Then slowly he opened his eyes, looked down at his hand, and saw that he was holding something, which was white. It was the edge of a sheet. He knew it was a sheet because he could see the texture of the material and the stitching on the hem. He screwed up his eyes, and opened them again quickly. This time he saw the room. He saw the bed in which he was lying; he saw the grey walls and the door and the green curtains over the window. There were some roses on the table by his bed.

14   Then he saw the basin on the table near the roses. It was a white enamel basin, and beside it there was a small medicine glass.

15   This is a hospital, he thought. I am in a hospital. But he could remember nothing. He lay back on his pillow, looking at the ceiling and wondering what had happened. He was gazing at the smooth greyness of the ceiling, which was so clean and grey, and then suddenly he saw a fly walking upon it. The sight of this fly, the suddenness of seeing this small black speck on a sea of grey, brushed the surface of his brain, and quickly, in that second, he remembered everything. He remembered the Spitfire and he remembered the altimeter showing twenty-one thousand feet. He remembered the pushing back of the hood with both hands, and he remembered the bailing out. He remembered his leg.

16   It seemed all right now. He looked down at the end of the bed, but he could not tell. He put one hand underneath the bedclothes and felt for his knees. He found one of them, but when he felt for the other, his hand touched something, which was soft and covered in bandages.

17   Just then the door opened and a nurse came in.

18   "Hello," she said. "So you've waked up at last."

19   She was not good-looking, but she was large and clean. She was between thirty and forty and she had fair hair. More than that he did not notice.

20   "Where am I?"

21   "You're a lucky fellow. You landed in a wood near the beach. You're in Brighton. They brought you in two days ago, and now you're all fixed up. You look fine."

22   "I've lost a leg," he said.

23   "That's nothing. We'll get you another one. Now you must go to sleep. The doctor will be coming to see you in about an hour." She picked up the basin and the medicine glass and went out.

24   But he did not sleep. He wanted to keep his eyes open because he was frightened that if he shut them again, everything would go away. He lay looking at the ceiling. The fly was still there. It was very energetic. It would run forward very fast for a few inches; then it would stop. Then it would run forward again, stop, run forward, stop, and every now and

then it would take off and buzz around viciously in small circles. It always landed back in the same place on the ceiling and started running and stopping all over again. He watched it for so long that after a while it was no longer a fly, but only a black speck upon a sea of grey, and he was still watching it when the nurse opened the door, and stood aside while the doctor came in. He was an Army doctor, a major, and he had some last war ribbons on his chest. He was bald and small, but he had a cheerful face and kind eyes.

25   "Well, well," he said. "So you've decided to wake up at last. How are you feeling?"

26   "I feel all right."

27   "That's the stuff. You'll be up and about in no time."

28   The doctor took his wrist to feel his pulse.

29   "By the way," he said, "some of the lads from your squadron were ringing up and asking about you. They wanted to come along and see you, but I said that they'd better wait a day or two. Told them you were all right, and that they could come and see you a little later on. Just lie quiet and take it easy for a bit. Got something to read?" He glanced at the table with the roses. "No. Well, nurse will look after you. She'll get you anything you want." With that he waved his hand and went out, followed by the large clean nurse.

30   When they had gone, he lay back and looked at the ceiling again. The fly was still there and as he lay watching it, he heard the noise of an airplane in the distance. He lay listening to the sound of its engines. It was a long way away. I wonder what it is, he thought. Let me see if I can place it. Suddenly he jerked his head sharply to one side. Anyone who has been bombed can tell the noise of a Junkers 88. They can tell most other German bombers for that matter, but especially a Junkers 88. The engines seem to sing a duet. There is a deep vibrating bass voice and with it there is a high-pitched tenor. It is the singing of the tenor which makes the sound of a JU-88 something which one cannot mistake.

31   He lay listening to the noise, and he felt quite certain about what it was. But where were

the sirens and where the guns? That German pilot certainly had a nerve coming near Brighton alone in daylight.

32 The aircraft was always far away, and soon the noise faded away into the distance. Later on there was another. This one, too, was far away, but there was the same deep undulating bass and the high singing tenor, and there was no mistaking it. He had heard that noise every day during the battle.

33 He was puzzled. There was a bell on the table by the bed. He reached out his hand and rang it. He heard the noise of footsteps down the corridor, and the nurse came in.

34 "Nurse, what were those airplanes?"

35 "I'm sure I don't know. I didn't hear them. Probably fighters or bombers. I expect they were returning from France. Why, what's the matter?"

36 "They were JU-88's. I'm sure they were JU-88's. I know the sound of the engines. There were two of them. What were they doing over here?"

37 The nurse came up to the side of his bed and began to straighten out the sheets and tuck them in under the mattress.

38 "Gracious me, what things you imagine. You mustn't worry about a thing like that. Would you like me to get you something to read?"

39 "No, thank you."

40 She patted his pillow and brushed back the hair from his forehead with her hand.

41 "They never come over in daylight any longer. You know that. They were probably Lancasters or Flying Fortresses."

42 "Nurse."

43 "Yes."

44 "Could I have a cigarette?"

45 "Why certainly you can."

46 She went out and came back almost at once with a packet of Players and some matches. She handed one to him and when he had put it in his mouth, she struck a match and lit it.

47 "If you want me again," she said, "just ring the bell," and she went out.

48 Once toward evening he heard the noise of another aircraft. It was far away, but even so he knew that it was a single-engined machine. But he could not place it. It was going fast; he could tell that. But it wasn't a Spit, and it wasn't a Hurricane. It did not sound like an American engine either. They make more noise. He did not know what it was, and it worried him greatly. Perhaps I am very ill, he thought. Perhaps I am imagining things. Perhaps I am a little delirious. I simply do not know what to think.

49 That evening the nurse came in with a basin of hot water and began to wash him.

50 "Well," she said, "I hope you don't still think that we're being bombed."

51 She had taken off his pyjama top and was soaping his right arm with a flannel. He did not answer.

52 She rinsed the flannel in the water, rubbed more soap on it and began to wash his chest.

53 "You're looking fine this evening," she said. "They operated on you as soon as you came in. They did a marvellous job. You'll be all right. I've got a brother in the RAF," she added. "Flying bombers."

54 He said, "I went to school in Brighton."

55 She looked up quickly. "Well, that's fine," she said. "I expect you'll know some people in the town."

56 "Yes," he said, "I know quite a few."

57 She had finished washing his chest and arms, and now she turned back the bedclothes so that his left leg was uncovered. She did it in such a way that his bandaged stump remained under the sheets. She undid the cord of his pyjama trousers and took them off. There was no trouble because they had cut off the right trouser leg, so that it could not interfere with the bandages. She began to wash his left leg and the rest of his body. This was the first time he had had a bed bath, and he was embarrassed. She laid a towel under his leg, and she was washing his foot with the flannel. She said, "This wretched soap won't lather at all. It's the water. It's as hard as nails."

58 He said, "None of the soap is very good now and, of course, with hard water it's hopeless."

As he said it, he remembered something. He remembered the baths which he used to take at school in Brighton, in the long stone-floored bathroom which had four baths in a room. He remembered how the water was so soft that you had to take a shower afterwards to get all the soap off your body, and he remembered how the foam used to float on the surface of the water, so that you could not see your legs underneath. He remembered that sometimes they were given calcium tablets because the school doctor used to say that soft water was bad for the teeth.

59 "In Brighton," he said, "the water isn't ..."

60 He did not finish the sentence. Something had occurred to him; something so fantastic and absurd that for a moment he felt like telling the nurse about it and having a good laugh.

61 She looked up. "The water isn't what?" she said.

62 "Nothing," he answered. "I was dreaming."

63 She rinsed the flannel in the basin, wiped the soap off his leg and dried him with a towel.

64 "It's nice to be washed," he said. "I feel better." He was feeling his face with his hands. "I need a shave."

65 "We'll do that tomorrow," she said. "Perhaps you can do it yourself then."

66 That night he could not sleep. He lay awake thinking of the Junkers 88's and of the hardness of the water. He could think of nothing else. They were JU-88's, he said to himself. I know they were. And yet it is not possible, because they would not be flying around so low over here in broad daylight. I know that it is true, and yet I know that it is impossible. Perhaps I am ill. Perhaps I am behaving like a fool and do not know what I am doing or saying. Perhaps I am delirious. For a long time he lay awake thinking these things, and once he sat up in bed and said aloud, "I will prove that I am not crazy. I will make a little speech about something complicated and intellectual. I will talk about what to do with Germany after the war." But before he had time to begin, he was asleep.

67 He woke just as the first light of day was showing through the slit in the curtains over the window. The room was still dark, but he could tell that it was already beginning to get light outside. He lay looking at the grey light, which was showing through the slit in the curtain, and as he lay there, he remembered the day before. He remembered the Junkers 88's and the hardness of the water; he remembered the large pleasant nurse and the kind doctor, and now the small grain of doubt took root in his mind and it began to grow.

68 He looked around the room. The nurse had taken the roses out the night before, and there was nothing except the table with a packet of cigarettes, a box of matches and an ashtray. Otherwise, it was bare. It was no longer warm or friendly. It was not even comfortable. It was cold and empty and very quiet.

69 Slowly the grain of doubt grew, and with it came fear, a light, dancing fear that warned but did not frighten; the kind of fear that one gets not because one is afraid, but because one feels that there is something wrong. Quickly the doubt and the fear grew so that he became restless and angry, and when he touched his forehead with his hand, he found that it was damp with sweat. He knew then that he must do something; that he must find some way of proving to himself that he was either right or wrong, and he looked up and saw again the window and the green curtains. From where he lay, that window was right in front of him, but it was fully ten yards away. Somehow he must reach it and look out. The idea became an obsession with him, and soon he could think of nothing except the window. But what about his leg? He put his hand underneath the bedclothes and felt the thick, bandaged stump, which was all that was left on the right-hand side. It seemed all right. It didn't hurt. But it would not be easy.

70 He sat up. Then he pushed the bedclothes aside and put his left leg on the floor. Slowly, carefully, he swung his body over until he had both hands on the floor as well; and then he was out of bed, kneeling on the carpet. He looked at the stump. It was very short and thick, covered with bandages. It was beginning to hurt and he could feel it throbbing. He wanted to collapse, lie down on the carpet and do nothing, but he knew that he must go on.

71 With two arms and one leg, he crawled over toward the window. He would reach forward

as far as he could with his arms; then he would give a little jump and slide his left leg along after them. Each time he did, it jarred his wound so that he gave a soft grunt of pain, but he continued to crawl across the floor on two hands and one knee. When he got to the window, he reached up, and one at a time he placed both hands on the sill. Slowly he raised himself up until he was standing on his left leg. Then quickly he pushed aside the curtains and looked out.

72   He saw a small house with a grey tiled roof standing alone beside a narrow lane, and immediately behind it there was a ploughed field. In front of the house there was an untidy garden, and there was a green hedge separating the garden from the lane. He was looking at the hedge when he saw the sign. It was just a piece of board nailed to the top of a short pole, and because the hedge had not been trimmed for a long time, the branches had grown out around the sign so that it seemed almost as though it had been placed in the middle of the hedge. There was something written on the board with white paint, and he pressed his head against the glass of the window, trying to read what it said. The first letter was a G, he could see that. The second was an A, and the third was an R. One after another he managed to see what the letters were. There were three words, and slowly he spelled the letters out aloud to himself as he managed to read them. G-A-R-E A-U C-H-I-E-N. Gare au chien. That is what it said.

73   He stood there balancing on one leg and holding tightly to the edges of the window sill with his hands, staring at the sign and at the whitewashed lettering of the words. For a moment he could think of nothing at all. He stood there looking at the sign, repeating the words over and over to himself, and then slowly he began to realize the full meaning of the thing. He looked up at the cottage and at the ploughed field. He looked at the small orchard on the left of the cottage and he looked at the green countryside beyond. "So this is France," he said. "I am in France."

74   Now the throbbing in his right thigh was very great. It felt as though someone was

pounding the end of his stump with a hammer, and suddenly the pain became so intense that it affected his head, and for a moment he thought he was going to fall. Quickly he knelt down again, crawled back to the bed and hoisted himself in. He pulled the bedclothes over himself and lay back on the pillow, exhausted. He could still think of nothing at all except the small sign by the hedge, and the ploughed field and the orchard. It was the words on the sign that he could not forget.

75   It was some time before the nurse came in. She came carrying a basin of hot water and she said, "Good morning, how are you today?"

76   He said, "Good morning, nurse."

77   The pain was still great under the bandages, but he did not wish to tell this woman anything. He looked at her as she busied herself with getting the washing things ready. He looked at her more carefully now. Her hair was very fair. She was tall and big-boned, and her face seemed pleasant. But there was something a little uneasy about her eyes. They were never still. They never looked at anything for more than a moment, and they moved too quickly from one place to another in the room. There was something about her movements also. They were too sharp and nervous to go well with the casual manner in which she spoke.

78   She set down the basin, took off his pyjama top and began to wash him.

79   "Did you sleep well?"

80   "Yes."

81   "Good," she said. She was washing his arms and his chest.

82   "I believe there's someone coming down to see you from the Air Ministry after breakfast," she went on. "They want a report or something. I expect you know all about it. How you got shot down and all that. I won't let him stay long, so don't worry."

83   He did not answer. She finished washing him, and gave him a toothbrush and some tooth powder. He brushed his teeth, rinsed his mouth and spat the water out into the basin.

84   Later she brought him his breakfast on a tray, but he did not want to eat. He was still

feeling weak and sick, and he wished only to lie still and think about what had happened. And there was a sentence running through his head. It was a sentence which Johnny, the Intelligence Officer of his squadron, always repeated to the pilots every day before they went out. He could see Johnny now, leaning against the wall of the dispersal hut with his pipe in his hand, saying, "And if they get you, don't forget, just your name, rank and number. Nothing else. For God's sake, say nothing else."

85 "There you are," she said as she put the tray on his lap. "I've got you an egg. Can you manage all right?"

86 "Yes."

87 She stood beside the bed. "Are you feeling all right?"

88 "Yes."

89 "Good. If you want another egg, I might be able to get you one."

90 "This is all right."

91 "Well, just ring the bell if you want any more." And she went out.

92 He had just finished eating when the nurse came in again.

93 She said, "Wing Commander Roberts is here. I've told him that he can only stay for a few minutes."

94 She beckoned with her hand and the Wing Commander came in.

95 "Sorry to bother you like this," he said.

96 He was an ordinary RAF officer, dressed in a uniform which was a little shabby, and he wore wings and a DFC. He was fairly tall and thin with plenty of black hair. His teeth, which were irregular and widely spaced, stuck out a little even when he closed his mouth. As he spoke, he took a printed form and a pencil from his pocket, and he pulled up a chair and sat down.

97 "How are you feeling?"

98 There was no answer.

99 "Tough luck about your leg. I know how you feel. I hear you put up a fine show before they got you."

100 The man in the bed was lying quite still, watching the man in the chair.

101 The man in the chair said, "Well, let's get this stuff over. I'm afraid you'll have to answer a few questions so that I can fill in this combat report. Let me see now, first of all, what was your squadron?"

102 The man in the bed did not move. He looked straight at the Wing Commander and he said, "My name is Peter Williamson. My rank is Squadron Leader and my number is nine seven two four five seven."

PUBLISHED IN
1943

Reader's Response

1. What is your opinion about the ending of the story?

_____

2. Do you think the pilot is correct in his conclusions?

_____

✂ CLOSE READING

Reread the story, paying close attention to the descriptions and clues that reveal the main character's situation.

1. Who is the main character in the story?

   _____

2. Where is he flying his plane?

   _____

3. What has happened to him before the beginning of the story?

   _____

4. What is his attitude toward what has happened to him?

   _____

5. What are his physical symptoms?

   _____

6. Where does he find himself when he regains consciousness after he jumps from the plane?

   _____

7. What initially causes him to be uneasy?

   _____

8. When he tells the nurse about his suspicions, how does she react?

   _____

9. What is the most important clue he sees to cause him distress about his situation?

   _____

10. How does he react when the RAF officer questions him?

    _____

11. Why does he react this way?

    _____

∞ WRITER'S CRAFT

Writers create tension in their works in different ways. Refer to the Literary Elements section at the beginning of this book to understand the techniques writers use to develop the dramatic structure of a story.

> The dramatic structure develops the tension in a work of fiction. In this story, Dahl uses a few techniques to build the tension of the story. He uses imagery, characterization, symbolism and narrative structure, all of which add to the suspenseful ambiance of the story.

▥ IMAGERY

1. The first two sentences immediately situate the main character. What words or phrases are the most effective in describing the setting at this time?

   _____

2. How do the first two sentences immediately create a feeling of tension in the story? Write down a few adjectives or nouns that come to mind when you recall this scene.

   _____

3. At the beginning of the story, the narrator describes the action in vivid detail. Find one example of description using the following senses in paragraphs 1 to 12 (before the pilot awakes in the hospital).

   Sight _____

   Sound _____

   Touch _____

4. Is the description of the pilot's accident realistic or abstract? _____

5. Based on your answer to the previous question, how does the description of the accident create tension in the story?

   _____

Symbolism

1. Reread paragraphs 1, 9 and 12. When the narrator is describing what is happening to the pilot, what words or phrases are repeated?

   _____

2. What colour is most often repeated in your answer to the previous question?

_____

3. What colour is predominantly used to describe what the pilot sees when he first wakes up in the hospital?

_____

4. What colour is the sign the pilot sees from his window?

_____

5. What connotations are usually associated with the colour that is the answer to the preceding two questions?

_____

6. In this story, how is the connotation of the colour reversed?

_____

7. Reread paragraphs 15 and 24.

a) What colour is the ceiling? It is _____. This colour is neither _____ nor _____, but something in between. It is also the colour of the walls.

b) What do you think this colour represents in the story?

_____

c) What does the pilot see on the ceiling and what do you think it symbolizes?

_____

Characterization

1. How does the narrator describe the pilot's perception of the nurse when he first sees her?

_____

2. What is the nurse's attitude toward the pilot?

_____

3. How does the nurse respond when the pilot asks her about the German planes?

_____

4. How does the nurse react when the pilot says he went to school in Brighton?

_____

5. How does the pilot's perception of the nurse change after he sees the sign written in French?

_____

Narrative Structure

1. Is the narrative structure of this story chronological or non-chronological?

_____

2. What is the purpose of the scenarios the pilot creates in his imagination about his friends waiting for him at the airport?

_____

3. Find one example of a flashback when the pilot is at the hospital. How does this flashback add to the tension of the story?

_____

_____

VOCABULARY DEVELOPMENT

Part I: Context Clues

Using context clues, choose the best possible answers for the following vocabulary questions.

1. In paragraph 3, *taxi in* means to
   a) catch a cab.   b) move slowly forward.   c) hire someone or something.

2. In paragraph 6, *rickshaw* refers to a
   a) vehicle.   b) park.   c) restaurant.

3. In paragraph 7, a synonym for *cowling* is
   a) a metal cover.   b) a baby cow.   c) an engine of an airplane.

4. In paragraph 32, *undulating* means
   a) silence.   b) noisy.   c) rising and falling.

5. In paragraph 57, *stump* means
   a) deception.   b) remaining part.   c) branch.

Part II: World War II Vocabulary

"Beware of the Dog" is set during World War II. As such, the story contains many references to specific vocabulary of this period. Research the following words. Then share your findings with your class.

1. Spitfire (paragraph 2)

2. Baghdad during World War II (paragraph 6)

3. Brighton, England (paragraph 21)

4. Junkers 88 (paragraph 30)

5. Lancasters (paragraph 41)

6. Flying Fortresses (paragraph 41)

7. Wing Commander (paragraph 93)

Wrap-Up Activities

1. Explain how Dahl's wartime experiences are reflected in the story "Beware of the Dog."

2. How is the title of the story meaningful?

3. Write about one aspect of World War II, such as the causes of the war, a particularly important battle, political leaders, and so on.

4. Compare the dramatic structure of this story with that of another story by Dahl.

5. Write about the theme of this story.

6. Research other wartime writers. What were some of the major works produced during this time?

7. How does Dahl successfully build tension in "Beware of the Dog"?

My eLab ACTIVITY

My eLab

Research the life and exploits of a pilot during World War II and then present your findings to your classmates. Check My eLab Documents for some Internet references you may wish to consult.

The Lottery

8

CHAPTER

ABOUT THE AUTHOR

Shirley Jackson (1919-1965) was born in San Francisco. Her mother, an ambitious socialite, was disappointed by her plain, intellectual daughter who showed little interest in social conventions. Later, Jackson attended university in New York State, where she met and married Stanley Edgar Hyman. Hyman became a literary critic, and the couple moved to a small town in Vermont when he was offered a teaching post at Bennington College. Jackson gave birth to four daughters, whom she both loved and neglected. Becoming more and more a non-conformist and suffering from bouts of depression, she was considered a very odd and controversial figure by many of the townspeople. She often felt persecuted, and some of the townspeople actually considered her to be a witch. Unfortunately, although Hyman was supportive of her radical ideas, Jackson, with her unkempt appearance and gaudy clothing, never seemed to fit in anywhere, even in her husband's liberal academic circles.

Jackson authored spectacularly different works. She wrote funny, banal stories for popular women's magazines of the fifties, while composing much darker and more disturbing pieces. "The Lottery," considered to be her masterpiece, stirred up more controversy than any other short story ever published in *The New Yorker* magazine.

As the years went by, Jackson became even more eccentric. She smoked, ate and drank too much, and popped pills from dawn to dusk. She died of heart failure during an afternoon nap at the relatively young age of 46.

LITERARY TRENDS
OF THE TIMES

AMERICAN PROSE LITERATURE SINCE World War II is difficult to categorize. Indeed, the narrative has become multi-dimensional due to international events, the development of technology that has resulted in world-altering inventions, such as television and personal computers, as well as societal changes, especially in male/female relationships.

World War II greatly influenced the literature in America in the 1940s and beyond. Writers such as Norman Mailer, author of *The Naked and the Dead* (1948), and Herman Wouk, author of *The Caine Mutiny* (1951), wrote about the brutality of war and the inhumanity that human beings can direct toward each other. War was described in a realistic style in order to capture the grim reality of its evil and violence. Writers effectively demonstrated that not only did violence negatively influence soldiers, it also influenced civilian society as well. Furthermore, writers explored the theme of individual freedom versus conformity in society. Beginning in the 1940s, they wrote about the conflict between individual growth and the need to belong.

Warm-Up Discussion Topics

1. Would you like to win a lottery? Why or why not?

_____

2. How important do you think it is for a society to keep its traditions?

_____

∞ **INITIAL READING**

Many people consider "The Lottery" to be a classic short story. Read on to discover why.

The Lottery
Shirley Jackson

1 THE morning of June 27th was clear and sunny, with the fresh warmth of a full-summer day; the flowers were blossoming profusely and the grass was richly green. The people of the village began to gather in the square, between the post office and the bank, around ten o'clock; in some towns there were so many people that the lottery took two days and had to be started on June 26th, but in this village, where there were only about three hundred people, the whole lottery took less than two hours, so it could begin at ten o'clock in the morning and still be through in time to allow the villagers to get home for noon dinner.

2 The children assembled first, of course. School was recently over for the summer, and the feeling of liberty sat uneasily on most of them; they tended to gather together quietly for a while before they broke into boisterous play, and their talk was still of the classroom and the teacher, of books and reprimands. Bobby Martin had already stuffed his pockets full of stones, and the other boys soon followed his example, selecting the smoothest and roundest stones; Bobby and Harry Jones and Dickie Delacroix—the villagers pronounced this name "Dellacroy"—eventually made a great pile of stones in one corner of the square and guarded it against the raids of the other boys. The girls stood aside, talking among themselves, looking over their shoulders at the boys, and the very small children rolled in the dust or clung to the hands of their older brothers or sisters.

3   Soon the men began to gather, surveying their own children, speaking of planting and rain, tractors and taxes. They stood together, away from the pile of stones in the corner, and their jokes were quiet and they smiled rather than laughed. The women, wearing faded house dresses and sweaters, came shortly after their menfolk. They greeted one another and exchanged bits of gossip as they went to join their husbands. Soon the women, standing by their husbands, began to call to their children, and the children came reluctantly, having to be called four or five times. Bobby Martin ducked under his mother's grasping hand and ran, laughing, back to the pile of stones. His father spoke up sharply, and Bobby came quickly and took his place between his father and his oldest brother.

4   The lottery was conducted—as were the square dances, the teenage club, the Halloween program—by Mr. Summers, who had time and energy to devote to civic activities. He was a round-faced, jovial man and he ran the coal business, and people were sorry for him, because he had no children and his wife was a scold. When he arrived in the square, carrying the black wooden box, there was a murmur of conversation among the villagers, and he waved and called, "Little late today, folks." The postmaster, Mr. Graves, followed him, carrying a three-legged stool, and the stool was put in the centre of the square and Mr. Summers set the black box down on it. The villagers kept their distance, leaving a space between themselves and the stool, and when Mr. Summers said, "Some of you fellows want to give me a hand?" there was a hesitation before two men, Mr. Martin and his oldest son, Baxter, came forward to hold the box steady on the stool while Mr. Summers stirred up the papers inside it.

5   The original paraphernalia for the lottery had been lost long ago, and the black box now resting on the stool had been put into use even before Old Man Warner, the oldest man in town, was born. Mr. Summers spoke frequently to the villagers about making a new box, but no one liked to upset even as much tradition as was represented by the black box. There was a story that the present box had been made with some pieces of the box that had preceded it, the one that had been constructed when the first people settled down to make a village here. Every year, after the lottery, Mr. Summers began talking again about a new box, but every year the subject was allowed to fade off without anything being done. The black box grew shabbier each year; by now it was no longer completely black but splintered badly along one side to show the original wood color, and in some places faded or stained.

6   Mr. Martin and his oldest son, Baxter, held the black box securely on the stool until Mr. Summers had stirred the papers thoroughly with his hand.

7   Because so much of the ritual had been forgotten or discarded, Mr. Summers had been successful in having slips of paper substituted for the chips of wood that had been used for generations. Chips of wood, Mr. Summers had argued, had been all very well when the village was tiny, but now that the population was more than three hundred and likely to keep on growing, it was necessary to use something that would fit more easily into the black box. The night before the lottery, Mr. Summers and Mr. Graves made up the slips of paper and put them in the box, and it was then taken to the safe of Mr. Summers' coal company and locked up until Mr. Summers was ready to take it to the square next morning. The rest of the year, the box was put away, sometimes one place, sometimes another; it had spent one year in Mr. Graves's barn and another year underfoot in the post office, and sometimes it was set on a shelf in the Martin grocery and left there.

8   There was a great deal of fussing to be done before Mr. Summers declared the lottery open. There were the lists to make up—of heads of families, heads of households in each family, members of each household in each family. There was the proper swearing-in of Mr. Summers by the postmaster, as the official of the lottery; at one time, some people remembered, there had been a recital of some sort, performed by the official of the lottery, a perfunctory, tuneless chant that had been rattled off duly each year; some people believed that the official of the lottery used to stand just so when he said or sang it; others

8

believed that he was supposed to walk among the people, but years and years ago this part of the ritual had been allowed to lapse. There had been, also, a ritual salute, which the official of the lottery had had to use in addressing each person who came up to draw from the box, but this also had changed with time, until now it was felt necessary only for the official to speak to each person approaching. Mr. Summers was very good at all this; in his clean white shirt and blue jeans, with one hand resting carelessly on the black box, he seemed very proper and important as he talked interminably to Mr. Graves and the Martins.

9   Just as Mr. Summers finally left off talking and turned to the assembled villagers, Mrs. Hutchinson came hurriedly along the path to the square, her sweater thrown over her shoulders, and slid into place in the back of the crowd. "Clean forgot what day it was," she said to Mrs. Delacroix, who stood next to her, and they both laughed softly. "Thought my old man was out back stacking wood." Mrs. Hutchinson went on, "and then I looked out the window and the kids were gone, and then I remembered it was the twenty-seventh and came a-running." She dried her hands on her apron, and Mrs. Delacroix said, "You're in time, though. They're still talking away up there."

10   Mrs. Hutchinson craned her neck to see through the crowd and found her husband and children standing near the front. She tapped Mrs. Delacroix on the arm as a farewell and began to make her way through the crowd. The people separated good-humouredly to let her through; two or three people said, in voices just loud enough to be heard across the crowd, "Here comes your Missus, Hutchinson," and "Bill, she made it after all." Mrs. Hutchinson reached her husband, and Mr. Summers, who had been waiting, said cheerfully, "Thought we were going to have to get on without you, Tessie." Mrs. Hutchinson said, grinning, "Wouldn't have me leave m'dishes in the sink, now, would you, Joe?" and soft laughter ran through the crowd as the people stirred back into position after Mrs. Hutchinson's arrival.

11   "Well, now," Mr. Summers said soberly, "guess we better get started, get this over with, so's we can go back to work. Anybody ain't here?"

12   "Dunbar," several people said. "Dunbar, Dunbar."

13   Mr. Summers consulted his list. "Clyde Dunbar," he said. "That's right. He's broke his leg, hasn't he? Who's drawing for him?"

14   "Me, I guess," a woman said, and Mr. Summers turned to look at her. "Wife draws for her husband," Mrs. Summers said. "Don't you have a grown boy to do it for you, Janey?" Although Mr. Summers and everyone else in the village knew the answer perfectly well, it was the business of the official of the lottery to ask such questions formally. Mr. Summers waited with an expression of polite interest while Mrs. Dunbar answered.

15   "Horace's not but sixteen yet," Mrs. Dunbar said regretfully. "Guess I gotta fill in for the old man this year."

16   "Right," Mr. Summers said. He made a note on the list he was holding. Then he asked, "Watson boy drawing this year?"

17   A tall boy in the crowd raised his hand. "Here," he said. "I'm drawing for m'mother and me." He blinked his eyes nervously and ducked his head as several voices in the crowd said things like "Good fellow, Jack," and "Glad to see your mother's got a man to do it."

18   "Well," Mr. Summers said, "guess that's everyone. Old Man Warner make it?"

19   "Here," a voice said, and Mr. Summers nodded.

20   A sudden hush fell on the crowd as Mr. Summers cleared his throat and looked at the list. "All ready?" he called. "Now, I'll read the names—heads of families first—and the men come up and take a paper out of the box. Keep the paper folded in your hand without looking at it until everyone has had a turn. Everything clear?"

21   The people had done it so many times that they only half listened to the directions; most of them were quiet, wetting their lips, not looking

around. Then Mr. Summers raised one hand high and said, "Adams." A man disengaged himself from the crowd and came forward. "Hi, Steve," Mr. Summers said, and Mr. Adams said, "Hi, Joe." They grinned at one another humourlessly and nervously. Then Mr. Adams reached into the black box and took out a folded paper. He held it firmly by one corner as he turned and went hastily back to his place in the crowd, where he stood a little apart from his family, not looking down at his hand.

22   "Allen," Mr. Summers said. "Anderson ... Bentham."

23   "Seems like there's no time at all between lotteries anymore," Mrs. Delacroix said to Mrs. Graves in the back row. "Seems like we got through with the last one only last week."

24   "Time sure goes fast," Mrs. Graves said.

25   "Clark ... Delacroix."

26   "There goes my old man," Mrs. Delacroix said. She held her breath while her husband went forward.

27   "Dunbar," Mr. Summers said, and Mrs. Dunbar went steadily to the box while one of the women said, "Go on, Janey," and another said, "There she goes."

28   "We're next," Mrs. Graves said. She watched while Mr. Graves came around from the side of the box, greeted Mr. Summers gravely and selected a slip of paper from the box. By now, all through the crowd there were men holding the small folded papers in their large hands, turning them over and over nervously. Mrs. Dunbar and her two sons stood together, Mrs. Dunbar holding the slip of paper.

29   "Harburt ... Hutchinson."

30   "Get up there, Bill," Mrs. Hutchinson said, and the people near her laughed.

31   "Jones."

32   "They do say," Mr. Adams said to Old Man Warner, who stood next to him, "that over in the north village they're talking of giving up the lottery."

33   Old Man Warner snorted. "Pack of crazy fools," he said. "Listening to the young folks, nothing's good enough for them. Next thing you know, they'll be wanting to go back to living in caves, nobody work anymore, live that way for a while. Used to be a saying about 'Lottery in June, corn be heavy soon.' First thing you know, we'd all be eating stewed chickweed and acorns. There's always been a lottery," he added petulantly. "Bad enough to see young Joe Summers up there joking with everybody."

34   "Some places have already quit lotteries," Mrs. Adams said.

35   "Nothing but trouble in that," Old Man Warner said stoutly. "Pack of young fools."

36   "Martin." And Bobby Martin watched his father go forward. "Overdyke ... Percy."

37   "I wish they'd hurry," Mrs. Dunbar said to her older son. "I wish they'd hurry."

38   "They're almost through," her son said.

39   "You get ready to run tell Dad," Mrs. Dunbar said.

40   Mr. Summers called his own name and then stepped forward precisely and selected a slip from the box. Then he called, "Warner."

41   "Seventy-seventh year I been in the lottery," Old Man Warner said as he went through the crowd. "Seventy-seventh time."

42   "Watson." The tall boy came awkwardly through the crowd. Someone said, "Don't be nervous, Jack," and Mr. Summers said, "Take your time, son."

43   "Zanini."

44   After that, there was a long pause, a breathless pause, until Mr. Summers holding his slip of paper in the air, said, "All right, fellows." For a minute, no one moved, and then all the slips of paper were opened. Suddenly, all the women began to speak at once, saying, "Who is it? Who's got it? Is it the Dunbars? Is it the Watsons?" Then the voices began to say, "It's Hutchinson. It's Bill. Bill Hutchinson's got it."

45   "Go tell your father," Mrs. Dunbar said to her older son.

46   People began to look around to see the Hutchinsons. Bill Hutchinson was standing

8

quiet, staring down at the paper in his hand. Suddenly, Tessie Hutchinson shouted to Mr. Summers, "You didn't give him time enough to take any paper he wanted. I saw you. It wasn't fair."

47 "Be a good sport, Tessie," Mrs. Delacroix called, and Mrs. Graves said, "All of us took the same chance."

48 "Shut up, Tessie," Bill Hutchinson said.

49 "Well, everyone," Mr. Summers said, "that was done pretty fast, and now we've got to be hurrying a little more to get done in time." He consulted his next list. "Bill," he said, "you draw for the Hutchinson family. You got any other households in the Hutchinsons?"

50 "There's Don and Eva," Mrs. Hutchinson yelled. "Make *them* take their chance!"

51 "Daughters draw with their husbands' families, Tessie," Mr. Summers said gently. "You know that as well as anyone else."

52 "It wasn't *fair*," Tessie said.

53 "I guess not, Joe," Bill Hutchinson said regretfully. "My daughter draws with her husband's family, that's only fair. And I've got no other family except the kids."

54 "Then, as far as drawing for families is concerned, it's you," Mr. Summers said in explanation, "and as far as drawing for households is concerned, that's you, too. Right?"

55 "Right," Bill Hutchinson said.

56 "How many kids, Bill?" Mr. Summers asked formally.

57 "Three," Bill Hutchinson said. "There's Bill, Jr., and Nancy, and little Dave. And Tessie and me."

58 "All right, then," Mr. Summers said. "Harry, you got their tickets back?"

59 Mr. Graves nodded and held up the slips of paper. "Put them in the box, then," Mr. Summers directed. "Take Bill's and put it in."

60 "I think we ought to start over," Mrs. Hutchinson said, as quietly as she could. "I tell you it wasn't *fair*. You didn't give him time enough to choose. *Every*body saw that."

61 Mr. Graves had selected the five slips and put them in the box, and he dropped all the papers but those onto the ground, where the breeze caught them and lifted them off.

62 "Listen, everybody," Mrs. Hutchinson was saying to the people around her.

63 "Ready, Bill?" Mr. Summers asked, and Bill Hutchinson, with one quick glance around at his wife and children, nodded.

64 "Remember," Mr. Summers said, "take the slips and keep them folded until each person has taken one. Harry, you help little Dave." Mr. Graves took the hand of the little boy, who came willingly with him up to the box. "Take a paper out of the box, Davy," Mr. Summers said. Davy put his hand into the box and laughed. "Take just *one* paper," Mr. Summers said. "Harry, you hold it for him." Mr. Graves took the child's hand and removed the folded paper from the tight fist and held it while little Dave stood next to him and looked up at him wonderingly.

65 "Nancy next," Mr. Summers said. Nancy was twelve, and her school friends breathed heavily as she went forward, switching her skirt and took a slip daintily from the box. "Bill, Jr.," Mr. Summers said, and Billy, his face red and his feet over-large, nearly knocked the box over as he got a paper out. "Tessie," Mr. Summers said. She hesitated for a minute, looking around defiantly, and then set her lips and went up to the box. She snatched a paper out and held it behind her.

66 "Bill," Mr. Summers said, and Bill Hutchinson reached into the box and felt around, bringing his hand out at last with the slip of paper in it.

67 The crowd was quiet. A girl whispered, "I hope it's not Nancy," and the sound of the whisper reached the edges of the crowd.

68 "It's not the way it used to be," Old Man Warner said clearly. "People ain't the way they used to be."

69 "All right," Mr. Summers said. "Open the papers. Harry, you open little Dave's."

70 Mr. Graves opened the slip of paper and there was a general sigh through the crowd as he held it up and everyone could see that it was

blank. Nancy and Bill, Jr., opened theirs at the same time, and both beamed and laughed, turning around to the crowd and holding their slips of paper above their heads.

71   "Tessie," Mr. Summers said. There was a pause, and then Mr. Summers looked at Bill Hutchinson, and Bill unfolded his paper and showed it. It was blank.

72   "It's Tessie," Mr. Summers said, and his voice was hushed. "Show us her paper, Bill."

73   Bill Hutchinson went over to his wife and forced the slip of paper out of her hand. It had a black spot on it, the black spot Mr. Summers had made the night before with the heavy pencil in the coal-company office. Bill Hutchinson held it up, and there was a stir in the crowd.

74   "All right, folks," Mr. Summers said. "Let's finish quickly."

75   Although the villagers had forgotten the ritual and lost the original black box, they still remembered to use stones. The pile of stones the boys had made earlier was ready; there were stones on the ground with the blowing scraps of paper that had come out of the box. Mrs. Delacroix selected a stone so large she had to pick it up with both hands and turned to Mrs. Dunbar. "Come on," she said. "Hurry up."

76   Mrs. Dunbar had small stones in both hands, and she said, gasping for breath, "I can't run at all. You'll have to go ahead and I'll catch up with you."

77   The children had stones already, and someone gave little Davy Hutchinson a few pebbles.

78   Tessie Hutchinson was in the centre of a cleared space by now, and she held her hands out desperately as the villagers moved in on her. "It isn't fair," she said. A stone hit her on the side of the head.

79   Old Man Warner was saying, "Come on, come on, everyone." Steve Adams was in the front of the crowd of villagers, with Mrs. Graves beside him.

80   "It isn't fair, it isn't right," Mrs. Hutchinson screamed, and then they were upon her.

PUBLISHED IN
1948

Reader's Response

1. What was your first reaction to this story?

_____

2. What questions came to mind when you read this story?

_____

∽ CLOSE READING

Answer the following questions.

1. What is the setting of the story?

_____

2. Why have the villagers gathered in the square?

_____

3. What are the villagers' attitudes toward the lottery at the beginning of the story?

a) Children's attitude _____

b) Men's attitude _____

c) Women's attitude _____

4. The lottery has a history. What are some of its rituals that are

a) still followed?

_____

_____

b) no longer followed?

_____

_____

5. What is Old Man Warner's opinion of the villages that have stopped holding a lottery?

_____

6. What is Tessie Hutchinson's initial attitude toward the lottery?

_____

7. How and why does Mrs. Hutchinson's attitude toward the lottery change?

_____

8. In your opinion, what is the theme of the story?

_____

_____

∞ WRITER'S CRAFT

Male/Female Stereotypes

1. Who is portrayed as more dominant in the story, males or females?

_____

2. List four examples from the story that support your answer to the previous question.

_____

_____

_____

_____

3. In your opinion, how do the male/female roles reinforce the meaning of the story?

_____

_____

Foreshadowing

Foreshadowing is a literary device that writers use to give clues to future events. Jackson uses many examples of foreshadowing in the story.

1. Give three examples of foreshadowing that you discover in the story.

_____

_____

_____

2. Explain how the examples given as answer to the previous question help to build tension in the plot.

_____

_____

_____

Irony

Irony refers to the device of writing or saying one thing but meaning the opposite.

1. How are the names of some of the characters, such as Mr. Summers, Old Man Warner and Mr. Graves ironic?

_____

_____

2. How is the setting of the story ironic?

_____

_____

3. How is the title of the story ironic?

_____

Narration and Point of View

1. a) What type of narrator tells this story and what is the narrator's point of view?

_____

b) How does the author's choice of narrator and point of view contrast with the horrific subject of the story?

_____

_____

VOCABULARY DEVELOPMENT
Synonyms

1. Find words in the appropriate paragraphs that have the following meanings.

a) plentiful, many (paragraph 1) _____

b) rough, noisy (paragraph 2) _____

c) criticize (paragraph 2) _____

d) a nag (paragraph 4) _____

e) articles, belongings (paragraph 5) _____

f) automatic, mechanical (paragraph 8) _____

g) irritably (paragraph 33) _____

h) openly challenging (paragraph 65) _____

2. List as many synonyms as possible for the words you listed in the preceding exercise. You may use a thesaurus or a dictionary to help you.

Wrap-Up Activities

1. Write about the ending of "The Lottery." Did you expect this ending?

2. Write about the importance of traditions in family and society.

3. Examine the use of traditional gender roles in "The Lottery."

4. What aspects of human nature are explored in "The Lottery"?

5. Discuss how Jackson uses characterization. Are the characters in the story fully developed, or are they stereotypes?

6. Compare the dramatic and narrative structures of the story.

7. What is the most important theme of the story? Support your opinion with strong arguments.

8. Discuss whether or not "The Lottery" is realistic.

My eLab ACTIVITY

My eLab

Research facts and myths about lotteries and present your findings to your classmates. Discuss whether lotteries are good for society.

The Veldt

CHAPTER
CONTENTS

▸ **Reading:** "The Veldt" by Ray Bradbury; author's biography; literary trends (post World War II; technological progress; science fiction)

▸ **Literary Elements:** plot (dramatic and narrative structure); setting; theme; symbolism; irony; psychological realism

▸ **Vocabulary Development:** vocabulary in context

▸ **My eLab Activity:** the impact of technology on modern societies

ABOUT
THE **AUTHOR**

Many consider **Ray Bradbury** to be one of the most influential writers in the development of science fiction as a genre. He was born in Waukegan, Illinois, U.S.A., in 1920, and by 1931, he was already writing short stories. Throughout the 1940s, he sold an increasing number of stories to various magazines, and with the publication of the *Martian Chronicles* in 1951 and *Fahrenheit 451* in 1953, he established himself as a prominent author of science fiction. He published more than five hundred works, including short stories, novels, plays, television and movie scripts, and poems. Over forty of his short stories were adapted for a television series (*The Ray Bradbury Television Theatre*), and several of his novels were turned into movies.

Many of Bradbury's best stories deal with the effects of technology on humankind. Although the action is set in the future, they actually deal with present life and concerns. Other themes include racism, nuclear war, censorship and the importance of human values and imagination.

Throughout his life Bradbury was awarded many prizes, including the O. Henry Memorial Award, the Benjamin Franklin Award, the World Fantasy Award for Lifetime Achievement and the Grand Master Award from the Science Fiction Writers of America. His animated film, *Icarus Montgolfier Wright*, about the history of aviation, was nominated for an Oscar. An Apollo astronaut named a crater on the moon Dandelion Crater in honour of Bradbury's novel, *Dandelion Wine*.

Apart from writing, Bradbury actively participated in the teams that designed, among others, the United States Pavilion at the 1964 New York World's Fair, the Spaceship Earth exhibition at the Epcot Centre in Florida and the Orbitron space ride at Euro-Disney in Paris.

LITERARY TRENDS
OF THE TIMES

THE SHORT STORY GENRE EVOLVED INTO more complex variations during the early twentieth century as more writers began to experiment with the form. Realism continued to be explored after World War II, the Cold War, the Civil Rights movement, the hippie revolution, the sexual revolution, the Vietnam War and the explosion of global communications. Writers of realistic stories tried to combine the best characteristics of both journalism and fiction: the detailed portrayal of real events married to the psychological truths and insights found in great fiction. This mixture often made stories more realistic in terms of setting and historical background, but it muddied the border between real and fictional people and events.

During the twentieth century, scientific and technological knowledge increased at exponential rates. Einstein's Theory of Relativity revolutionized the way many human beings viewed their place in the universe. Simple answers to the great questions of existence no longer appeared satisfactory in many cultures. Complexity and diversity increased as human populations exploded, and distances disappeared through the rise of global communications. In the 1950s, television, with its visual impact, displaced the written word as the best medium for

telling stories of physical adventure and suspenseful action. Many writers then turned inward to explore inner psychological turmoil and conflict, and worlds that did not exist. Influenced by scientific and technological developments, authors such as Ray Bradbury, Arthur C. Clarke, Frank Herbert, Robert Heinlein, Isaac Asimov and Ursula Le Guin, to name but a few, began to compose more fanciful types of stories set in imaginary times and places. Science fiction grew into its adolescence and adulthood. Although these science fiction writers let their imaginations roam, their stories often dealt with very real problems that humankind was either just beginning to face or seemed likely to face in the future.

Warm-Up Discussion Topics

1. Is the development of new technology synonymous with progress? Why or why not?

_____

2. How does new technology affect your life and habits?

_____

∞ INITIAL READING

Many people have dreamed of escaping the drudgery of household chores. They would then be free to spend their time on more important things. Read on to discover what such a life could be like.

The Veldt
Ray Bradbury

1 "GEORGE, I wish you'd look at the nursery."

2 "What's wrong with it?"

3 "I don't know."

4 "Well, then."

5 "I just want you to look at it, is all, or call a psychologist in to look at it."

6 "What would a psychologist want with a nursery?"

7 "You know very well what he'd want." His wife paused in the middle of the kitchen and watched the stove busy humming to itself, making supper for four.

8 "It's just that the nursery is different now than it was."

9 "All right, let's have a look."

10 They walked down the hall of their soundproofed, Happylife Home, which had cost them thirty thousand dollars installed,

9

this house which clothed and fed and rocked them to sleep and played and sang and was good to them. Their approach sensitized a switch somewhere and the nursery light flicked on when they came within ten feet of it. Similarly, behind them, in the halls, lights went on and off as they left them behind, with a soft automaticity.

11 "Well," said George Hadley.

12 They stood on the thatched floor of the nursery. It was forty feet across by forty feet long and thirty feet high—it had cost half again as much as the rest of the house. "But nothing's too good for our children," George had said.

13 The nursery was silent. It was empty as a jungle glade at hot high noon. The walls were blank and two dimensional. Now, as George and Lydia Hadley stood in the centre of the room, the walls began to purr and recede into crystalline distance, it seemed, and presently an African veldt appeared, in three dimensions; on all sides, in colours reproduced to the final pebble and bit of straw. The ceiling above them became a deep sky with a hot yellow sun.

14 George Hadley felt the perspiration start on his brow.

15 "Let's get out of the sun," he said. "This is a little too real. But I don't see anything wrong."

16 "Wait a moment, you'll see," said his wife.

17 Now the hidden odorophonics were beginning to blow a wind of odour at the two people in the middle of the baked veldtland. The hot straw smell of lion grass, the cool green smell of the hidden water hole, the great rusty smell of animals, the smell of dust like a red paprika in the hot air. And now the sounds: the thump of distant antelope feet on grassy sod, the papery rustling of vultures. A shadow passed through the sky. The shadow flickered on George Hadley's upturned, sweating face.

18 "Filthy creatures," he heard his wife say.

19 "The vultures."

20 "You see, there are the lions, far over, that way. Now they're on their way to the water hole. They've just been eating," said Lydia. "I don't know what."

21 "Some animal." George Hadley put his hand up to shield off the burning light from his squinted eyes. "A zebra or a baby giraffe, maybe."

22 "Are you sure?" His wife sounded peculiarly tense.

23 "No, it's a little late to be sure," he said, amused. "Nothing over there I can see but cleaned bone, and the vultures dropping for what's left."

24 "Did you hear that scream?" she asked.

25 "No."

26 "About a minute ago?"

27 "Sorry, no."

28 The lions were coming. And again George Hadley was filled with admiration for the mechanical genius who had conceived this room. A miracle of efficiency selling for an absurdly low price. Every home should have one. Oh, occasionally they frightened you with their clinical accuracy, they startled you, gave you a twinge, but most of the time what fun for everyone, not only your own son and daughter, but for yourself when you felt like a quick jaunt to a foreign land, a quick change of scenery. Well, here it was!

29 And here were the lions now, fifteen feet away, so real, so feverishly and startlingly real that you could feel the prickling fur on your hand, and your mouth was stuffed with the dusty upholstery smell of their heated pelts, and the yellow of them was in your eyes like the yellow of an exquisite French tapestry, the yellows of lions and summer grass, and the sound of the matted lion lungs exhaling on the silent noontide, and the smell of meat from the panting, dripping mouths.

30 The lions stood looking at George and Lydia Hadley with terrible green-yellow eyes.

31 "Watch out!" screamed Lydia.

32 The lions came running at them.

33 Lydia bolted and ran. Instinctively, George sprang after her. Outside, in the hall, with the door slammed, he was laughing and she was crying, and they both stood appalled at the other's reaction.

34 "George!"

35 "Lydia! Oh, my dear poor sweet Lydia!"

36   "They almost got us!"

37   "Walls, Lydia, remember; crystal walls, that's all they are. Oh, they look real, I must admit—Africa in your parlour—but it's all dimensional superreactionary, supersensitive colour film and mental tape film behind glass screens. It's all odorophonics and sonics, Lydia. Here's my handkerchief."

38   "I'm afraid." She came to him and put her body against him and cried steadily. "Did you see? Did you feel? It's too real."

39   "Now, Lydia ..."

40   "You've got to tell Wendy and Peter not to read any more on Africa."

41   "Of course—of course." He patted her.

42   "Promise?"

43   "Sure."

44   "And lock the nursery for a few days until I get my nerves settled."

45   "You know how difficult Peter is about that. When I punished him a month ago by locking the nursery for even a few hours—the tantrum he threw! And Wendy too. They live for the nursery."

46   "It's got to be locked, that's all there is to it."

47   "All right." Reluctantly he locked the huge door. "You've been working too hard. You need a rest."

48   "I don't know—I don't know," she said, blowing her nose, sitting down in a chair that immediately began to rock and comfort her. "Maybe I don't have enough to do. Maybe I have time to think too much. Why don't we shut the whole house off for a few days and take a vacation?"

49   "You mean you want to fry my eggs for me?"

50   "Yes." She nodded.

51   "And darn my socks?"

52   "Yes." A frantic, watery-eyed nodding.

53   "And sweep the house?"

54   "Yes, yes—oh, yes!"

55   "But I thought that's why we bought this house, so we wouldn't have to do anything?"

56   "That's just it. I feel like I don't belong here. The house is wife and mother now and nursemaid. Can I compete with an African veldt? Can I give a bath and scrub the children as efficiently or quickly as the automatic scrub bath can? I cannot. And it isn't just me. It's you. You've been awfully nervous lately."

57   "I suppose I have been smoking too much."

58   "You look as if you didn't know what to do with yourself in this house, either. You smoke a little more every morning and drink a little more every afternoon and need a little more sedative every night. You're beginning to feel unnecessary too."

59   "Am I?" He paused and tried to feel into himself to see what was really there.

60   "Oh, George!" She looked beyond him, at the nursery door. "Those lions can't get out of there, can they?"

61   He looked at the door and saw it tremble as if something had jumped against it from the other side.

62   "Of course not," he said.

63   At dinner they ate alone, for Wendy and Peter were at a special plastic carnival across town and had televised home to say they'd be late, to go ahead eating. So George Hadley, bemused, sat watching the dining-room table produce warm dishes of food from its mechanical interior.

64   "We forgot the ketchup," he said.

65   "Sorry," said a small voice within the table, and ketchup appeared.

66   As for the nursery, thought George Hadley, it won't hurt for the children to be locked out of it a while. Too much of anything isn't good for anyone. And it was clearly indicated that the children had been spending a little too much time on Africa. That sun. He could feel it on his neck, still, like a hot paw. And the lions. And the smell of blood. Remarkable how the nursery caught the telepathic emanations of the children's minds and created life to fill their every desire. The children thought lions, and there were lions. The children thought zebras, and there were zebras. Sun—sun. Giraffes— giraffes. Death and death.

67   That last. He chewed tastelessly on the meat that the table had cut for him. Death thoughts. They were awfully young, Wendy and Peter, for death thoughts. Or, no, you were

9

9

68   never too young, really. Long before you knew what death was you were wishing it on someone else. When you were two years old, you were shooting people with cap pistols.

68   But this—the long, hot African veldt—the awful death in the jaws of a lion. And repeated again and again.

69   "Where are you going?"

70   He didn't answer Lydia. Preoccupied, he let the lights glow softly on ahead of him, extinguished behind him as he padded to the nursery door. He listened against it. Far away, a lion roared.

71   He unlocked the door and opened it. Just before he stepped inside, he heard a faraway scream. And then another roar from the lions, which subsided quickly.

72   He stepped into Africa. How many times in the last year had he opened this door and found Wonderland, Alice, the Mock Turtle, or Aladdin and his Magical Lamp, or Jack Pumpkinhead of Oz, or Dr. Doolittle, or the cow jumping over a very real-appearing moon—all the delightful contraptions of a make-believe world. How often had he seen Pegasus flying in the sky ceiling, or seen fountains of red fireworks, or heard angel voices singing. But now, this yellow hot Africa, this bake oven with murder in the heat. Perhaps Lydia was right. Perhaps they needed a little vacation from the fantasy which was growing a bit too real for ten-year-old children. It was all right to exercise one's mind with gymnastic fantasies, but when the lively child mind settled on one pattern ...? It seemed that, at a distance, for the past month, he had heard lions roaring, and smelled their strong odour seeping as far away as his study door. But, being busy, he had paid it no attention.

73   George Hadley stood on the African grassland alone. The lions looked up from their feeding, watching him. The only flaw to the illusion was the open door through which he could see his wife, far down the dark hall, like a framed picture, eating her dinner abstractedly.

74   "Go away," he said to the lions.

75   They did not go.

76   He knew the principle of the room exactly. You sent out your thoughts. Whatever you thought would appear.

77   "Let's have Aladdin and his lamp," he snapped.

78   The veldtland remained; the lions remained.

79   "Come on, room! I demand Aladdin!" he said.

80   Nothing happened. The lions mumbled in their baked pelts.

81   "Aladdin!"

82   He went back to dinner. "The fool room's out of order," he said. "It won't respond."

83   "Or. "

84   "Or what?"

85   "Or it can't respond," said Lydia, "because the children have thought about Africa and lions and killing so many days that the room's in a rut."

86   "Could be."

87   "Or Peter's set it to remain that way."

88   "Set it?"

89   "He may have got into the machinery and fixed something."

90   "Peter doesn't know machinery."

91   "He's a wise one for ten. That I.Q. of his—"

92   "Nevertheless."

93   "Hello, Mom. Hello, Dad."

94   The Hadleys turned. Wendy and Peter were coming in the front door, cheeks like peppermint candy, eyes like bright blue agate marbles, a smell of ozone on their jumpers from their trip in the helicopter.

95   "You're just in time for supper," said both parents.

96   "We're full of strawberry ice cream and hot dogs," said the children, holding hands. "But we'll sit and watch."

97   "Yes, come tell us about the nursery," said George Hadley.

98   The brother and sister blinked at him and then at each other. "Nursery?"

99   "All about Africa and everything," said the father with false joviality.

100   "I don't understand," said Peter.

101   "Your mother and I were just travelling through Africa with rod and reel; Tom Swift and his Electric Lion," said George Hadley.

9

102 "There's no Africa in the nursery," said Peter simply.

103 "Oh, come now, Peter. We know better."

104 "I don't remember any Africa," said Peter to Wendy. "Do you?"

105 "No."

106 "Run see and come tell."

107 She obeyed.

108 "Wendy, come back here!" said George Hadley, but she was gone. The house lights followed her like a flock of fireflies. Too late, he realized he had forgotten to lock the nursery door after his last inspection.

109 "Wendy'll look and come tell us," said Peter.

110 "She doesn't have to tell me. I've seen it."

111. "I'm sure you're mistaken, Father."

112 "I'm not, Peter. Come along now."

113 But Wendy was back. "It's not Africa," she said breathlessly.

114 "We'll see about this," said George Hadley, and they all walked down the hall together and opened the nursery door.

115 There was a green, lovely forest, a lovely river, a purple mountain, high voices singing, and Rima, lovely and mysterious, lurking in the trees with colourful flights of butterflies, like animated bouquets, lingering on her long hair. The African veldtland was gone. The lions were gone. Only Rima was here now, singing a song so beautiful that it brought tears to your eyes.

116 George Hadley looked in at the changed scene. "Go to bed," he said to the children.

117 They opened their mouths.

118 "You heard me," he said.

119 They went off to the air closet, where a wind sucked them like brown leaves up the flue to their slumber rooms.

120 George Hadley walked through the singing glade and picked up something that lay in the corner near where the lions had been. He walked slowly back to his wife.

121 "What is that?" she asked.

122 He showed it to her. The smell of hot grass was on it and the smell of a lion. There were drops of saliva on it, it had been chewed and there were blood smears on both sides.

123 He closed the nursery door and locked it, tight.

124 In the middle of the night he was still awake and he knew his wife was awake. "Do you think Wendy changed it?" she said at last, in the dark room.

125 "Of course."

126 "Made it from a veldt into a forest and put Rima there instead of lions?"

127 "Yes."

128 "Why?"

129 "I don't know. But it's staying locked until I find out."

130 "How did your wallet get there?"

131 "I don't know anything," he said, "except that I'm beginning to be sorry we bought that room for the children. If children are neurotic at all, a room like that—"

132 "It's supposed to help them work off their neuroses in a healthful way."

133 "I'm starting to wonder." He stared at the ceiling.

134 "We've given the children everything they ever wanted. Is this our reward—secrecy, disobedience?"

135 "Who was it said, 'Children are carpets, they should be stepped on occasionally? We've never lifted a hand. They're insufferable—let's admit it. They come and go when they like; they treat us as if we were offspring. They're spoiled and we're spoiled."

136 "They've been acting funny ever since you forbade them to take the rocket to New York a few months ago."

137 "They're not old enough to do that alone, I explained."

138 "Nevertheless, I've noticed they've been decidedly cool toward us since."

139 "I think I'll have David McClean come tomorrow morning to have a look at Africa."

140 "But it's not Africa now; it's Green Mansions country and Rima."

141 "I have a feeling it'll be Africa again before then."

9

142   A moment later they heard the screams.

143   Two screams. Two people screaming from downstairs. And then a roar of lions.

144   "Wendy and Peter aren't in their rooms," said his wife.

145   He lay in his bed with his beating heart. "No," he said. "They've broken into the nursery."

146   "Those screams—they sound familiar."

147   "Do they?"

148   "Yes, awfully."

149   And although their beds tried very hard, the two adults couldn't be rocked to sleep for another hour. A smell of cats was in the night air.

150   "Father?" said Peter.

151   "Yes."

152   Peter looked at his shoes. He never looked at his father anymore, nor at his mother. "You aren't going to lock up the nursery for good, are you?"

153   "That all depends."

154   "On what?" snapped Peter.

155   "On you and your sister. If you intersperse this Africa with a little variety—oh, Sweden perhaps, or Denmark or China—"

156   "I thought we were free to play as we wished."

157   "You are, within reasonable bounds."

158   "What's wrong with Africa, Father?"

159   "Oh, so now you admit you have been conjuring up Africa, do you?"

160   "I wouldn't want the nursery locked up," said Peter coldly. "Ever."

161   "Matter of fact, we're thinking of turning the whole house off for about a month. Live sort of a carefree one-for-all existence."

162   "That sounds dreadful! Would I have to tie my own shoes instead of letting the shoe tier do it? And brush my own teeth and comb my hair and give myself a bath?"

163   "It would be fun for a change, don't you think?"

164   "No, it would be horrid. I didn't like it when you took out the picture painter last month."

165   "That's because I wanted you to learn to paint all by yourself, son."

166   "I don't want to do anything but look and listen and smell; what else is there to do?"

167   "All right, go play in Africa."

168   "Will you shut off the house sometime soon?"

169   "We're considering it."

170   "I don't think you'd better consider it anymore, Father."

171   "I won't have any threats from my son!"

172   "Very well." And Peter strolled off to the nursery.

173   "Am I on time?" said David McClean.

174   "Breakfast?" asked George Hadley.

175   "Thanks, had some. What's the trouble?"

176   "David, you're a psychologist."

177   "I should hope so."

178   "Well, then, have a look at our nursery. You saw it a year ago when you dropped by; did you notice anything peculiar about it then?"

179   "Can't say I did; the usual violences, a tendency toward a slight paranoia here or there, usual in children because they feel persecuted by parents constantly, but, oh, really nothing."

180   They walked down the hall. "I locked the nursery up," explained the father, "and the children broke back into it during the night. I let them stay so they could form the patterns for you to see."

181   There was a terrible screaming from the nursery.

182   "There it is," said George Hadley. "See what you make of it."

183   They walked in on the children without rapping.

184   The screams had faded. The lions were feeding.

9

185 "Run outside a moment, children," said George Hadley. "No, don't change the mental combination. Leave the walls as they are. Get!"

186 With the children gone, the two men stood studying the lions clustered at a distance, eating with great relish whatever it was they had caught.

187 "I wish I knew what it was," said George Hadley. "Sometimes I can almost see. Do you think if I brought high-powered binoculars here and—"

188 David McClean laughed dryly. "Hardly." He turned to study all four walls. "How long has this been going on?"

189 "A little over a month."

190 "It certainly doesn't feel good."

191 "I want facts, not feelings."

192 "My dear George, a psychologist never saw a fact in his life. He only hears about feelings; vague things. This doesn't feel good, I tell you. Trust my hunches and my instincts. I have a nose for something bad. This is very bad. My advice to you is to have the whole damn room torn down and your children brought to me every day during the next year for treatment."

193 "Is it that bad?"

194 "I'm afraid so. One of the original uses of these nurseries was so that we could study the patterns left on the walls by the child's mind, study at our leisure and help the child. In this case, however, the room has become a channel toward—destructive thoughts, instead of a release away from them."

195 "Didn't you sense this before?"

196 "I sensed only that you had spoiled your children more than most. And now you're letting them down in some way. What way?"

197 "I wouldn't let them go to New York."

198 "What else?"

199 "I've taken a few machines from the house and threatened them, a month ago, with closing up the nursery unless they did their homework. I did close it for a few days to show I meant business."

200 "Ah, ha!"

201 "Does that mean anything?"

202 "Everything. Where before they had a Santa Claus now they have a Scrooge. Children prefer Santas. You've let this room and this house replace you and your wife in your children's affections. This room is their mother and father, far more important in their lives than their real parents. And now you come along and want to shut it off. No wonder there's hatred here. You can feel it coming out of the sky. Feel that sun. George, you'll have to change your life. Like too many others, you've built it around creature comforts. Why, you'd starve tomorrow if something went wrong in your kitchen. You wouldn't know how to tap an egg. Nevertheless, turn everything off. Start anew. It'll take time. But we'll make good children out of bad in a year, wait and see."

203 "But won't the shock be too much for the children, shutting the room up abruptly, for good?"

204 "I don't want them going any deeper into this, that's all."

205 The lions were finished with their red feast.

206 The lions were standing on the edge of the clearing watching the two men.

207 "Now I'm feeling persecuted," said McClean. "Let's get out of here. I never have cared for these damned rooms. Make me nervous."

208 "The lions look real, don't they?" said George Hadley. "I don't suppose there's any way—"

209 "What?"

210 "—that they could become real?"

211 "Not that I know."

212 "Some flaw in the machinery, a tampering or something?"

213 "No."

214 They went to the door.

215 "I don't imagine the room will like being turned off," said the father.

216 "Nothing ever likes to die—even a room."

9

217   "I wonder if it hates me for wanting to switch it off."

218   "Paranoia is thick around here today," said David McClean. "You can follow it like a spoor. Hello." He bent and picked up a bloody scarf. "This yours?"

219   "No." George Hadley's face was rigid. "It belongs to Lydia."

220   They went to the fuse box together and threw the switch that killed the nursery.

221   The two children were in hysterics. They screamed and pranced and threw things. They yelled and sobbed and swore and jumped at the furniture.

222   "You can't do that to the nursery, you can't!"

223   "Now, children."

224   The children flung themselves onto a couch, weeping.

225   "George," said Lydia Hadley, "turn on the nursery, just for a few moments. You can't be so abrupt."

226   "No."

227   "You can't be so cruel."

228   "Lydia, it's off, and it stays off. And the whole damn house dies as of here and now. The more I see of the mess we've put ourselves in, the more it sickens me. We've been contemplating our mechanical, electronic navels for too long. My God, how we need a breath of honest air!"

229   And he marched about the house turning off the voice clocks, the stoves, the heaters, the shoe shiners, the shoe lacers, the body scrubbers and swabbers and massagers, and every other machine he could put his hand to.

230   The house was full of dead bodies, it seemed. It felt like a mechanical cemetery. So silent. None of the humming hidden energy of machines waiting to function at the tap of a button.

231   "Don't let them do it!" wailed Peter at the ceiling as if he was talking to the house, the nursery. "Don't let Father kill everything." He turned to his father. "Oh, I hate you!"

232   "Insults won't get you anywhere."

233   "I wish you were dead!"

234   "We were, for a long while. Now we're going to really start living. Instead of being handled and massaged, we're going to live."

235   Wendy was still crying and Peter joined her again. "Just a moment, just one moment, just another moment of nursery," they wailed.

236   "Oh, George," said the wife, "it can't hurt."

237   "All right—all right, if they'll only just shut up. One minute, mind you, and then off forever."

238   "Daddy, Daddy, Daddy!" sang the children, smiling with wet faces.

239   "And then we're going on a vacation. David McClean is coming back in half an hour to help us move out and get to the airport. I'm going to dress. You turn the nursery on for a minute, Lydia, just a minute, mind you."

240   And the three of them went babbling off while he let himself be vacuumed upstairs through the air flue and set about dressing himself. A minute later Lydia appeared.

241   "I'll be glad when we get away," she sighed.

242   "Did you leave them in the nursery?"

243   "I wanted to dress too. Oh, that horrid Africa. What can they see in it?"

244   "Well, in five minutes we'll be on our way to Iowa. Lord, how did we ever get in this house? What prompted us to buy a nightmare?"

245   "Pride, money, foolishness."

246   "I think we'd better get downstairs before those kids get engrossed with those damned beasts again."

247   Just then they heard the children calling, "Daddy, Mommy, come quick—quick!"

248   They went downstairs in the air flue and ran down the hall. The children were nowhere in sight. "Wendy? Peter!"

249   They ran into the nursery. The veldtland was empty save for the lions waiting, looking at them. "Peter, Wendy?"

250   The door slammed.

251 "Wendy, Peter!"

252 George Hadley and his wife whirled and ran back to the door.

253 "Open the door!" cried George Hadley, trying the knob. "Why, they've locked it from the outside! Peter!" He beat at the door.

254 "Open up!"

255 He heard Peter's voice outside, against the door.

256 "Don't let them switch off the nursery and the house," he was saying.

257 Mr. and Mrs. George Hadley beat at the door. "Now, don't be ridiculous, children. It's time to go. Mr. McClean'll be here in a minute and ... "

258 And then they heard the sounds.

259 The lions on three sides of them, in the yellow veldt grass, padding through the dry straw, rumbling and roaring in their throats.

260 The lions.

261 Mr. Hadley looked at his wife and they turned and looked back at the beasts edging slowly forward, crouching, tails stiff.

262 Mr. and Mrs. Hadley screamed.

263 And suddenly they realized why those other screams had sounded familiar.

264 "Well, here I am," said David McClean in the nursery doorway.

265 "Oh, hello." He stared at the two children seated in the centre of the open glade eating a little picnic lunch. Beyond them were the water hole and the yellow veldtland; above was the hot sun. He began to perspire. "Where are your father and mother?"

266 The children looked up and smiled. "Oh, they'll be here directly."

267 "Good, we must get going." At a distance Mr. McClean saw the lions fighting and clawing and then quieting down to feed in silence under the shady trees.

268 He squinted at the lions with his hand up to his eyes.

269 Now the lions were done feeding. They moved to the water hole to drink.

270 A shadow flickered over Mr. McClean's hot face. Many shadows flickered. The vultures were dropping down the blazing sky.

271 "A cup of tea?" asked Wendy in the silence.

PUBLISHED IN
1951

Reader's Response

1. How does this story compare to the other stories you have read so far (in terms of genre, theme, style, and so on)?

_____

_____

2. Did you find the end of this story effective? Why do you suppose this story ends the way it does?

_____

_____

∞ CLOSE READING

Answer the following questions.

1. What are some of the functions that the Hadleys' Happylife Home performs for the family? List at least five.

_____

_____

_____

2. Why is the home (in a general sense) a source of conflict between the parents and their children?

_____

_____

3. Why are George and Lydia worried about the nursery in particular?

_____

4. What is the original purpose of the nursery, according to the psychologist?

   a) It is an interactive room.

   b) It is a room where children can experience films in 3-D.

   c) It is a place that provides children with a release from destructive thoughts.

   d) It is a place for learning geography in a realistic way.

5. What have the children turned the nursery into?

_____

6. Why do the children prefer the Happylife Home to their parents?

_____

7. What does the psychologist advise in regard to the room and the children?

_____

8. Why do the parents fail to follow the psychologist's advice?

_____

_____

∞ WRITER'S CRAFT

Bradbury uses the elements of plot and setting in "The Veldt" to bring out the main themes of this story. Refer to the Literary Elements section to read about the characteristics of plot, setting and theme, all of which are important for the literary analysis of this piece.

9

Dramatic Structure

1. Summarize the plot of "The Veldt" in one sentence.

   _____

   _____

2. a) Is the dramatic structure traditional (incorporating exposition, complication, and so on) or experimental? _____

   b) Explain your answer.

   _____

   _____

3. At what point in the story does the climax occur?

   _____

4. Explain how Bradbury creates tension at the outset of the story.

   _____

5. a) Give three examples of foreshadowing in the story.

   _____

   _____

   b) Explain how the use of foreshadowing builds tension in the plot.

   _____

   _____

6. Give two examples of dialogue between the parents and children that are used to create tension in the story.

   _____

7. Here are two examples of descriptive narrative from the story.

> The lions were coming. And again George Hadley was filled with admiration for the mechanical genius who had conceived this room. A miracle of efficiency selling for an absurdly low price. Every home should have one.
>
> How many times in the last year had he opened this door and found Wonderland, Alice, the Mock Turtle, or Aladdin and his Magical Lamp, or Jack Pumpkinhead of Oz, or Dr. Doolittle, or the cow jumping over a very real-appearing moon—all the delightful contraptions of a make-believe world.

What kind of emotional charge or aura do they have? Circle the correct answer.
a) positive   b) negative   c) neutral

8. What purpose do the examples in the previous question serve to the context of the dramatic structure of the story?

_____

9. Reread the second example in question 7. Which literary device does it represent?

_____

> Bradbury is a master at building tension in the dramatic structure of a story through a variety of techniques, such as the creation of characters in conflict, the use of foreshadowing, flashback and dialogue, and the juxtaposition of sections containing contrasting emotional charges.

Narrative Structure

1. What form of writing does the majority of the story take?

_____

2. From what point of view is the story narrated?

_____

3. Two storylines are present in "The Veldt"; this technique is called "a story within a story."

a) What is the story within the main story? (*Hint*: it is repeated several times and takes place on the veldt.)

_____

b) What part of this story (the one taking place on the veldt) do George and Lydia see first, the beginning or the end? (*Hint*: reread ten lines beginning with paragraph 17.)

_____

In the story within the story, the action almost appears to move backward, while in the main story, it moves forward. The two stories intertwine at the climax and the main story winds down with the children calmly serving tea to the psychologist.

Setting

1. Describe the setting of the main story.

_____

_____

2. Describe the specific setting of the story within the main story.

_____

_____

3. How do the settings (places) contrast each other?

_____

_____

4. What literary device does the author constantly use to make the veldt and the lions seem real to the reader? (*Hint*: look at the paragraphs containing descriptions of the veldt.)

_____

5. What kind of overall atmosphere is generated by the setting of the story within the main story?

_____

Using a setting within a setting, Bradbury makes a vivid contrast between the vibrant and primal atmosphere of the African veldt and the artificial and technological atmosphere of the Happylife Home. This contrast leads directly to one of the major themes of the story.

Theme

1. What is the main theme of the story? (*Hint*: it concerns technology.)

_____

2. How does the plot (dramatic structure) bring out this theme?

_____

_____

3. How does the setting enhance the theme?

_____

_____

4. Cite at least one other theme in the story.

_____

Literary Devices

Ray Bradbury uses several other literary devices to give impact to the story. (Refer to the glossary as needed.)

■ SYMBOLISM

1. What do the following aspects of the story symbolize?
(N.B. Refer to the glossary at the end of the book or to previous chapters for definitions of symbolism and irony.)

a) the lions _____

b) the Happylife Home and the nursery _____

c) George and Lydia _____

d) the children _____

■ IRONY

1. What is ironic about the nursery?

_____

2. How are the following quotations from the story ironic?

a) "The house was full of dead bodies, it seemed. It felt like a mechanical cemetery."

_____

b) "I wish you were dead!"
"We were, for a long while. Now we're going to really start living. Instead of being handled and massaged, we're going to live.'"

_____

3. How is the last scene of the story ironic?

_____

Psychological Realism

Psychological realism means that the reasons underlying the characters' behaviour are plausible.

1. How do the interventions of David McClean bring psychological realism to the story?

_____

VOCABULARY DEVELOPMENT

Vocabulary in Context

Read the lines in which the following underlined words are found. Using contextual clues, define the words by choosing the best answer.

1. "with a soft <u>automaticity</u>" (last line in paragraph 10)
   a) regularly   b) working by itself   c) involuntarily

2. "on the <u>thatched</u> floor" (first line in paragraph 12)
   a) straw cover   b) hard cover   c) plastic cover

3. "a jungle <u>glade</u>" (second line in paragraph 13)
   a) grass   b) open space   c) tree

4. "recede into <u>crystalline</u> distance" (fifth line in paragraph 13)
   a) clear   b) cloudy   c) rainy

5. "an African <u>veldt</u>" (seventh line in paragraph 13)
   a) jungle   b) farm   c) grassland

6. "the hidden <u>odorophonics</u>" (first line in paragraph 17)
   a) scent machine   b) perfume   c) air conditioner

7. "a quick <u>jaunt</u>" (second last line in paragraph 28)
   a) shake   b) desire   c) trip

8. "Lydia <u>bolted</u>" (first line in paragraph 33)
   a) walked rapidly   b) stood still   c) moved suddenly

9. "a <u>frantic</u>, watery-eyed nodding" (paragraph 52)
   a) uncontrolled   b) smooth   c) slow

10. "George Hadley, <u>bemused</u>, sat watching" (fourth line in paragraph 63)
  a) waited patiently b) thought confusedly c) smiled constantly

11. "the telepathic <u>emanations</u>" (ninth line in paragraph 66)
  a) thoughts b) powers c) origins

12. "You can follow it like a <u>spoor</u>." (second line in paragraph 218)
  a) animal fur b) animal trap c) animal trail

Wrap-Up Activities

1. Which literary devices does Bradbury use to build suspense and tension in "The Veldt"?

2. Show how Bradbury uses imagery to create realistic and convincing scenes in "The Veldt."

3. Both Ernest Hemingway and Ray Bradbury use dialogue effectively in their short stories. Compare and contrast their use of this technique in the stories "Hills Like White Elephants" and "The Veldt."

4. Compare the theme of good versus evil in "The Lottery" and "The Veldt."

5. Compare the use of irony in "The Veldt" with the use of irony in another short story you have studied.

6. Compare Poe's and Bradbury's use of imagery to create atmosphere and mood in "The Tell-Tale Heart" and "The Veldt."

My eLab ACTIVITY

My eLab 🗁

Read the article entitled "The Nature of Technology" in My eLab Documents, taking note of the positive and negative impact of technology. Choose a new technology, such as the cellphone, and write an essay on how it affects human society in both positive and negative ways.

The Yellow Sweater

CHAPTER
CONTENTS

▸ **Reading:** "The Yellow Sweater" by Hugh Garner; author's biography; literary trends (post-World War II; social, urban subjects)

▸ **Literary Elements:** characterization; contrasting images; foreshadowing; symbols

▸ **Vocabulary Development:** denotations and connotations

▸ **My eLab Activity:** research hitchhiking in literature

ABOUT
THE **AUTHOR**

Hugh Garner (1913-1979) immigrated with his family to Toronto from England in 1919. Not long after, his father deserted the family, leaving his mother to raise four young children on her own. Garner grew up in the Cabbagetown district of the city, which provided the title for a novel he published in 1950. He later described his neighbourhood as "the largest Anglo-Saxon slum in North America."

He left school at the age of sixteen and worked at unskilled jobs during the Great Depression. Then he fought in both the Spanish Civil War and World War II. Afterwards, he became a full-time writer, producing novels such as *Storm Below* (1949), *The Silence on the Shore* (1962) and *The Intruders* (1976), as well as short stories. In 1963, he won a Governor General's Literary Award for *Hugh Garner's Best Stories*. He also wrote an autobiography and three police novels during the 1970s.

Garner's anti-establishment stance led him to write about the struggles of outsiders, such as alcoholics, displaced people and the dispossessed in Canadian society.

LITERARY TRENDS
OF THE TIMES

LITERATURE AFTER WORLD WAR II REFLECTED a variety of themes that touched upon social or urban subjects. Society and its values were rapidly changing, and young writers had to come to terms with the loss of innocence precipitated by the war with Nazi Germany, which had led to the attempted extermination of the handicapped, homosexuals and Gypsies, and had culminated in the death of six million Jews. Even before the war, writers such as James Joyce and Virginia Woolf experimented with new forms. William Golding explored the darker side of human nature in his Nobel Prize-winning work *The Lord of the Flies* (1954). Still others such as Graham Greene, Aldous Huxley and George Orwell made social or political statements in their works of fiction.

Like the literature of the United States and Britain, the literature of other English-speaking countries, such as Australia, Canada and India, developed rapidly after the war. Post-war English Canadian literature reflected urban experiences, concentrating on social issues, such as political separation, the immigrant experience, the breakdown of traditional values, racism, survival and isolation. Contemporary English Canadian writers, such as Hugh Garner, Margaret Atwood, Robertson Davies, Mordecai Richler, Leonard Cohen and Michael Ondaatje, gained international recognition and popularity.

Warm-Up Discussion Topics

1. Have you ever been in an uncomfortable or embarrassing situation with strangers?

_____

2. Have you ever hitchhiked? What are some positive and negative aspects of hitchhiking?

_____

✀ INITIAL READING

The subject of the next story was ahead of its time.

The Yellow Sweater
Hugh Garner

1   HE stepped on the gas when he reached the edge of town. The big car took hold of the pavement and began to eat up the miles on the straight, almost level, highway. With his elbow stuck through the open window he stared ahead at the shimmering greyness of the road. He felt heavy and pleasantly satiated after his good smalltown breakfast, and he shifted his bulk in the seat, at the same time brushing some cigar ash from the front of his salient vest. In another four hours he would be home—a day ahead of himself this trip, but with plenty to show the office for last week's work. He unconsciously patted the wallet resting in the inside pocket of his jacket as he thought of the orders he had taken.

2   Four thousand units to Slanders ... his secondbest line too ... four thousand at twelve percent ... four hundred and eighty dollars! He rolled the sum over in his mind as if tasting it, enjoying its tartness like a kid with a gumdrop.

3   He drove steadily for nearly an hour, ignorant of the smell of spring in the air, pushing the car ahead with his mind as well as with his foot against the pedal. The success of his trip and the feeling of power it gave him carried him along toward the triumph of his homecoming.

4   Outside a small village he was forced to slow down for a road repair crew. He punched twice on the horn as he passed them, basking in the stares of the yokels who looked up from their shovels, and smiling at the envy showing on their faces.

5   A rather down-at-heel young man carrying an army kitbag stepped out from the office of a filling station and gave him the thumb. He pretended not to see the gesture, and pressed down slightly on the gas so that the car began to purr along the free and open road.

6   It was easy to see that the warm weather was approaching, he thought. The roads were becoming cluttered up once more with hitchhikers. Why the government didn't clamp down on them was more than he could understand. Why should people pay taxes so that other lazy bums could fritter away their time roaming the country, getting free rides, going God knows where? They were dangerous too. It was only the week before that two of them had beaten up and robbed a man on this very same road. They stood a fat chance of him picking them up.

7   And yet they always thumbed him, or almost always. When they didn't, he felt cheated, as a person does when he makes up his mind not to answer another's greeting, only to have them pass by without noticing him.

8   He glanced at his face in the rear-view mirror. It was a typical middle-aged businessman's face, plump and well barbered, the shiny skin stretched taut across the cheeks. It was a face that was familiar to him not only from his possession of it, but because it was also the face of most of his friends. What was it the speaker at that service club luncheon had called it? "The physiognomy of success."

10

9   As he turned a bend in the road, he saw the girl about a quarter of a mile ahead. She was not on the pavement, but was walking slowly along the shoulder of the highway, bent over with the weight of the bag she was carrying. He slowed down, expecting her to turn and thumb him, but she plodded on as though impervious to his approach. He sized her up as he drew near. She was young by the look of her back ... stocking seams straight ... heels muddy but not rundown. As he passed, he stared at her face. She was a good-looking kid, probably eighteen or nineteen.

10 It was the first time in years that he had slowed down for a hiker. His reasons evaded him, and whether it was the feel of the morning, the fact of his going home or the girl's apparent independence, he could not tell. Perhaps it was a combination of all three, plus the boredom of a long drive. It might be fun to pick her up, to cross-examine her while she was trapped in the seat beside him.

11 Easing the big car to a stop about fifty yards in front of her, he looked back through the mirror. She kept glancing at the car, but her pace had not changed, and she came on as though she had expected him to stop. For a moment he was tempted to drive on again, angered by her indifference. She was not a regular hitchhiker or she would have waited at the edge of town instead of setting out to walk while carrying such a heavy bag. But there was something about her that compelled him to wait—something which aroused in him an almost forgotten sense of adventure, an eagerness not experienced for years.

12 She opened the right rear door, saying at the same time, "Thank you very much, sir," in a frightened little voice.

13 "Put your bag in the back. That's it, on the floor," he ordered, turning toward her with his hand along the back of the seat. "Come and sit up here."

14 She did as he commanded, sitting very stiff and straight against the door. She was small, almost fragile, with long dark hair that waved where it touched upon the collar of her light-coloured topcoat. Despite the warmth of the morning, the coat was buttoned, and she held it to her in a way that suggested modesty or fear.

15 "Are you going very far?" he asked, looking straight ahead through the windshield, trying not to let the question sound too friendly.

16 "To the city," she answered, with the politeness and eagerness of the recipient of a favour.

17 "For a job?"

18 "Well, not exactly—" she began. Then she said, "Yes, for a job."

19 As they passed the next group of farm buildings, she stared hard at them, her head turning with her eyes until they were too far back to be seen.

20 Something about her reminded him of his eldest daughter, but he shrugged off the comparison. It was silly of him to compare the two, one a hitchhiking farm skivvy and the other one soon to come home from finishing school. In his mind's eye he could see the photograph of his daughter Shirley that hung on the wall of the living room. It had been taken with a colour camera during the Easter vacation, and in it Shirley was wearing a bright yellow sweater.

21 "Do you live around here?" he asked, switching his thoughts back to the present.

22 "I was living about a mile down the road from where you picked me up."

23 "Sick of the farm?" he asked.

24 "No." She shook her head slowly, seriously.

25 "Have you anywhere to go in the city?"

26 "I'll get a job somewhere."

27 He turned then and got his first good look at her face. She was pretty, he saw, with the country girl's good complexion, her features small and even. "You're young to be leaving home like this," he said.

28 "That wasn't my home," she murmured. "I was living with my Aunt Bernice and her husband."

29 He noticed that she did not call the man her uncle.

30 "You sound as though you don't like the man your aunt is married to."

31 "I hate him!" she whispered vehemently.

32 To change the subject he said, "You've chosen a nice day to leave, anyhow."

10

33 "Yes."

34 He felt a slight tingling along his spine. It was the same feeling he had experienced once when sitting in the darkened interior of a movie house beside a strange yet, somehow, intimate young woman. The feeling that if he wished, he had only to let his hand fall along her leg …

35 "You're not very talkative," he said, more friendly now.

36 She turned quickly and faced him. "I'm sorry. I was thinking about—about a lot of things."

37 "It's too nice a morning to think of much," he said. "Tell me more about your reasons for leaving home."

38 "I wanted to get away, that's all."

39 He stared at her again, letting his eyes follow the contours of her body. "Don't tell me you're in trouble?" he asked.

40 She lowered her eyes to her hands. They were engaged in twisting the clasp on a cheap black handbag. "I'm not in trouble like that," she said slowly, although the tone of her voice belied her words.

41 He waited for her to continue. There was a sense of power in being able to question her like this without fear of having to answer any questions himself. He said, "There can't be much else wrong. Was it boy trouble?"

42 "Yes, that's it," she answered hastily.

43 "Where is the boy? Is he back there or in the city?"

44 "Back there," she answered.

45 He was aware of her nearness, of her young body beside him on the seat. "You're too pretty to worry about one boy," he said, trying to bridge the gap between them with unfamiliar flattery.

46 She did not answer him, but smiled nervously in homage to his remark.

47 They drove on through the morning, and by skilful questioning he got her to tell him more about her life. She had been born near the spot where he had picked her up, she said. She was an orphan, eighteen years old, who for the past three years had been living on her aunt's farm. On his part he told her a little about his job, but not too much. He spoke in generalities, yet let her see how important he was in his field.

48 They stopped for lunch at a drive-in restaurant outside a small town. While they were eating, he noticed that some of the other customers were staring at them. It angered him until he realized that they probably thought she was his mistress. This flattered him and he tried to imagine that it was true. During the meal, he became animated, and he laughed loudly at his *risqué* little jokes.

49 She ate sparingly, politely, not knowing what to do with her hands between courses. She smiled at the things he said, even at the remarks that were obviously beyond her.

50 After they had finished their lunch, he said to her jovially, "Here, we've been travelling together for two hours and we don't even know each other's names yet."

51 "Mine is Marie. Marie Edwards."

52 "You can call me Tom," he said expansively.

53 When he drew out his wallet to pay the check, he was careful to cover the initials G.G.M. with the palm of his hand.

54 As they headed down the highway once again, Marie seemed to have lost some of her timidity, and she talked and laughed with him as though he were an old friend. Once he stole a glance at her through the corner of his eye. She was staring ahead as if trying to unveil the future that was being overtaken by the onrushing car.

55 "A penny for your thoughts, Marie," he said.

56 "I was just thinking how nice it would be to keep going like this forever."

57 "Why?" he asked, her words revealing an unsuspected facet to her personality.

58 "I dunno," she answered, rubbing the palm of her hand along the upholstery of the seat in a gesture that was wholly feminine. "It seems so— safe here, somehow." She smiled as though apologizing for thinking such things. "It seems as if nothing bad could ever catch up to me again."

59 He gave her a quick glance before staring ahead at the road once more.

60 The afternoon was beautiful. The warm dampness of the fields bearing aloft the smell of uncovered earth and budding plants. The

sunwarmed pavement sang like muted violins beneath the spinning tires of the car. The clear air echoed the sound of life and growth and the urgency of spring.

61 As the miles clicked off, and they were brought closer to their inevitable parting, an idea took shape in his mind and grew with every passing minute. Why bother hurrying home, he asked himself. After all he hadn't notified his wife to expect him, and he wasn't due back until tomorrow.

62 He wondered how the girl would react if he should suggest postponing the rest of the trip overnight. He would make it worth her while. There was a tourist camp on the shore of a small lake about twenty miles north of the highway. No one would be the wiser, he told himself. They were both fancy free.

63 The idea excited him, yet he found himself too timid to suggest it. He tried to imagine how he must appear to the girl. The picture he conjured up was of a mature figure, inclined to stoutness, much older than she was in years but not in spirit. Many men his age had formed liaisons with young women. In fact, it was the accepted thing among some of the other salesmen he knew.

64 But there remained the voicing of the question. She appeared so guileless, so— innocent of his intentions. And yet it was hard to tell; she wasn't as innocent as she let on.

65 She interrupted his train of thought. "On an afternoon like this, I'd like to paddle my feet in a stream," she said.

66 "I'm afraid the water would be pretty cold."

67 "Yes, it would be cold, but it'd be nice too. When we were kids, we used to go paddling in the creek behind the schoolhouse. The water was strong with the spring freshet, and it would tug at our ankles and send a warm ticklish feeling up to our knees. The smooth pebbles on the bottom would make us twist our feet and we'd try to grab them with our toes ... I guess I must sound crazy," she finished.

68 No longer hesitant, he said, "I'm going to turn the car into one of these side roads, Marie. On a long trip I usually like to park for a while under some trees. It makes a little break in the journey."

69 She nodded her head happily. "That would be nice," she said.

70 He turned the car off the highway and they travelled north along the road that curved gently between wide stretches of steaming fields. The speed of the car was seemingly increased by the drumming of gravel against the inside of the fenders.

71 It was time to bring the conversation back to a more personal footing, so he asked. "What happened between you and your boyfriend, Marie?" He had to raise his voice above the noise of the hurtling stones.

72 "Nothing much," she answered, hesitating as if making up the answer. "We had a fight, that's all."

73 "Serious?"

74 "I guess so."

75 "What happened? Did he try to get a little gay maybe?"

76 She had dropped her head, and he could see the colour rising along her neck and into her hair behind her ears.

77 "Does that embarrass you?" he asked, taking his hand from the wheel and placing it along the collar of her coat.

78 She tensed herself at his touch and tried to draw away, but he grasped her shoulder and pulled her against him. He could feel the fragility of her beneath his hand and the trembling of her skin beneath the cloth of her coat. The odour of her hair and of some cheap scent filled his nostrils.

79 She cried, "Don't, please!" and broke away from the grip of his hand. She inched herself into the far corner of the seat again.

80 "You're a little touchy, aren't you?" he asked, trying to cover up his embarrassment at being repulsed so quickly.

81 "Why did you have to spoil it?"

82 His frustration kindled a feeling of anger against her. He knew her type all right. Pretending that butter wouldn't melt in her mouth, while all the time she was secretly laughing at him for being the sucker who picked her up, bought her a lunch and drove her into town. She couldn't fool him; he'd met her type before.

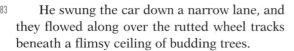

83 He swung the car down a narrow lane, and they flowed along over the rutted wheel tracks beneath a flimsy ceiling of budding trees.

84 "Where are we going?" she asked, her voice apprehensive now.

85 "Along here a piece," he answered, trying to keep his anger from showing.

86 "Where does this road lead?"

87 "I don't know. Maybe there's a stream you can paddle in."

88 There was a note of relief in her voice as she said, "Oh! I didn't mean for us—for you to find a stream."

89 "You don't seem to know *what* you mean, do you?"

90 She became silent then and seemed to shrink farther into the corner.

91 The trees got thicker, and soon they found themselves in the middle of a small wood. The branches of the hardwoods were mottled green, their buds flicking like fingers in the breeze. He brought the car to a stop against the side of the road.

92 The girl watched him, the corners of her mouth trembling with fear. She slid her hand up the door and grabbed the handle. He tried to make his voice matter-of-fact as he said, "Well, here we are."

93 Her eyes ate into his face like those of a mesmerized rabbit watching a snake.

94 He opened a glove compartment and pulled out a package of cigarettes. He offered the package to her, but she shook her head.

95 "Let's get going," she pleaded.

96 "What, already? Maybe we should make a day of it."

97 She did not speak, but the question stood in her eyes. He leaned back against the seat, puffing on his cigarette. "There's a tourist camp on a lake a few miles north of here. We could stay there and go on to the city tomorrow."

98 She stifled a gasp. "I can. I didn't think— I had no idea when we—"

99 He pressed his advantage. "Why can't you stay? Nobody'll know. I may be in a position to help you afterward. You'll need help, you know."

100 "No. No, I couldn't," she answered. Her eyes filled with tears.

101 He had not expected her to cry. Perhaps he had been wrong in his estimation of her. He felt suddenly bored with the whole business, and ashamed of the feelings she had ignited in him.

102 "Please take me back to the highway," she said, pulling a carefully folded handkerchief from her handbag.

103 "Sure. In a few minutes." He wanted time to think things out; to find some way of saving face.

104 "You're just like he was," she blurted out, her words distorted by her handkerchief. "You're all the same."

105 Her outburst frightened him. "Marie," he said, reaching over to her. He wanted to quiet her, to show her that his actions had been the result of an old man's foolish impulse.

106 As soon as his hand touched her shoulder, she gave a short cry and twisted the door handle. "No. No, please!" she cried.

107 "Marie, come here!" he shouted, trying to stop her. He grabbed her by the shoulder, but she tore herself from his grasp and fell through the door.

108 She jumped up from the road and staggered back through the grass into the belt of trees. Her stockings and the bottom of her coat were brown with mud.

109 "Don't follow me!" she yelled.

110 "I'm not going to follow you. Come back here and I'll drive you back to the city."

111 "No you don't! You're the same as he was!" she cried. "I know your tricks!"

112 He looked about him at the deserted stretch of trees, wondering if anybody could be listening. It would place him in a terrible position to be found with her like this. Pleading with her, he said, "Come on, Marie. I've got to go."

113 She began to laugh hysterically, her voice reverberating through the trees.

114 "Marie, come on," he coaxed. "I won't hurt you."

115 "No! Leave me alone. Please leave me alone!"

116 His pleas only seemed to make things worse. "I'm going," he said hurriedly, pulling the car door shut.

117 "Just leave me alone!" she cried. Then she began sobbing, "Bernice! Bernice!"

118 What dark fears had been released by his actions of the afternoon he did not know, but

they frightened and horrified him. He turned the car around in the narrow lane and let it idle for a moment as he waited, hoping she would change her mind. She pressed herself deeper into the trees, wailing at the top of her voice.

119   From behind him came a racking noise from down the road, and he looked back and saw a tractor coming around a bend. A man was driving it and there was another one riding behind. He put the car in gear and stepped on the gas.

120   Before the car reached the first turn beneath the trees, he looked back. The girl was standing in the middle of the road beside the tractor and she was pointing his way and talking to the men. He wondered if they had his license number, and what sort of a story she was telling them.

121   He had almost reached the highway again before he remembered her suitcase standing on the floor behind the front seat. His possession of it seemed to tie him to the girl; to make him partner to her terror. He pulled the car to a quick stop, leaned over the back of the seat and picked the suitcase up from the floor. Opening the door, he tossed it lightly to the side of the road with a

feeling of relief. The frail clasp on the cheap bag opened as it hit the ground and its contents spilled in the ditch. There was a framed photograph, some letters and papers held together with an elastic band, a comb and brush, and some clothing, including a girl's yellow sweater.

122   "I'm no thief," he said, pushing the car into motion again, trying to escape from the sight of the opened bag. He wasn't to blame for the things that had happened to her. It wasn't his fault that her stupid little life was spilled there in the ditch.

123   "I've done nothing wrong," he said, as if pleading his case with himself. But there was a feeling of obscene guilt beating his brain like a reiteration. Something of hers seemed to attach itself to his memory. Then suddenly he knew what it was—the sweater, the damned yellow sweater. His hands trembled around the wheel as he sent the car hurtling toward the safe anonymity of the city.

124   He tried to recapture his feelings of the morning, but when he looked at himself in the mirror, all he saw was the staring face of a fat, frightened old man.

PUBLISHED IN
1952

10

Reader's Response

1.   Write down adjectives that express your feelings after you read the story.

_____

2.   Discuss the following statement: we must pay a price for our actions.

_____

∽ CLOSE READING

1.   In paragraphs 1 to 8, how is the protagonist initially presented to the readers? Fill in the chart below to examine the characteristics of the protagonist.

Name	
Physical characteristics	
Attitudes	

2. What are the man's reasons for picking up the girl?

_____

3. In paragraph 10, which three words or phrases indicate how the man views himself in relation to the girl? (*Hint*: look at the last sentence of the paragraph.)

_____

4. How is the girl described?

_____

_____

5. What reasons does the girl give for leaving home?

_____

6. At the restaurant, the man is described as being "animated." Why does he behave in this manner?

_____

7. Why does the man lie to the girl about his name?

_____

8. What are the man's intentions regarding the girl when he initially suggests that they stop for a while by the side of the road?

_____

9. How does the girl react to the man when he touches her?

_____

10. How does the man rationalize the girl's reaction to his touch?

_____

11. Why does the girl's hysterical reaction in the woods frighten the man?

_____

12. What does the man do when he drives away?

_____

13. How does the man see himself at the end of the story?

_____

⌒ WRITER'S CRAFT

Hugh Garner's story "The Yellow Sweater" shows how words and their connotations (associated meanings) can depict characterization. Garner presents the domination and subjugation of his two main characters through the use of contrasting words and images.

Contrast

"The Yellow Sweater" is full of contrasting images. For example, the man is presented as "big and powerful," while the girl is portrayed as "fragile."

In paragraph 1, Garner depicts the power of the man through word associations such as "big car," "he felt heavy ... and pleasantly satiated," "shifted his bulk" and "patted the wallet."

In paragraph 3, other power-evoking word associations include "success of his trip," "the feeling of power" and "the triumph of his homecoming."

1. Find examples of the girl's fragility that contrast markedly with the man's power in the story.

_____

_____

_____

_____

2. The man's sense of power over the girl is also derived from his sexual interest in her. For example, the man's initial attitude is shown by the following statement.

"But there was something about her that compelled him to wait—something which aroused in him an almost forgotten sense of adventure, an eagerness not experienced for years."

Find at least four more examples of the man's powerful sexual attitude toward the girl.

_____

_____

_____

_____

_____

3. At what point in the story does the man begin to lose his power over the girl?

_____

4. Which words reveal that the man loses his power over the girl?

_____

5. How is the man's state at the end of the story different from his state at the beginning?

_____

Foreshadowing

1. a) Foreshadowing is also an important device in the story. Find an example of foreshadowing in paragraph 14.

_____

b) Explain how your choice represents an example of what is to come.

_____

2. Choose two other examples of foreshadowing in the story. Explain how each one predicts future events.

_____

_____

3. a) What is the girl's real reason for leaving her aunt's house?

_____

b) Find a sentence that supports your answer to the previous question.

_____

c) How does this sentence hint at what is to come in the story?

_____

Symbolism

Symbolism is a literary device in which one idea is represented by another idea. For example, a rose may be a symbol of love.

1. What is the significance of the yellow sweater in the story?

_____

2. Find at least two other symbols in the story and explain their significance.

_____

Theme

1. In your opinion, what is the theme of the story?

_____

VOCABULARY DEVELOPMENT

Connotations and Denotations

Garner uses vocabulary associated with the legal system to suggest the difference of power between the man and the girl.

1. Look at the following sentences and explain the denotative (dictionary) meaning and the connotative (associated, secondary) meaning of the words in italics.

It might be fun to pick her up, *to cross-examine* her while she was trapped in the seat beside him.

a) denotation _____

b) connotation _____

They drove on through the morning, and *by skilful questioning* he got her to tell him more about her life.

c) denotation _____

d) connotation _____

"I've done nothing wrong," he said, as *if pleading his case* with himself.

e) denotation _____

f) connotation _____

2. How does the vocabulary add to the theme of the story?

_____

Wrap-Up Activities

1. Write about gender stereotyping in "The Yellow Sweater."

2. Discuss Garner's use of legal vocabulary in "The Yellow Sweater."

3. Write about the power struggle that occurs in "The Yellow Sweater."

4. Explain the use of symbolism in "The Yellow Sweater."

5. Discuss characterization in "The Yellow Sweater."

6. How is the abuse of power presented in "The Yellow Sweater"? Does it represent various sorts of power abuse in society in general?

My eLab ACTIVITY

My eLab

Read the article about hitchhiking in literature and about the urban legend of the vanishing hitchhiker in My eLab. Answer the questions and write an essay or a story about a hitchhiker.

11

Newton

CHAPTER
CONTENTS

▸ **Reading:** "Newton" by Jeannette Winterson; author's biography; literary trends (twentieth-century scientific and technological discoveries: quantum mechanics, Einstein's theory of relativity, Freud's psychoanalytical theories; artistic experimentation: surrealism; major historical events)

▸ **Literary Elements:** style (breaking traditional rules); dialogue and interior monologue; Biblical references; imagery; symbols; black humour and irony; repetition

▸ **Vocabulary Development:** expanding your vocabulary; connotations

▸ **My eLab Activity:** read about eccentricity.

ABOUT THE AUTHOR

Jeannette Winterson was born in Manchester, England, in 1959. She was adopted into a deeply religious family, and religion was a dominant factor in her childhood. As a result of this strong religious environment, Jeanette Winterson started her writing career as a preacher and sermon writer. Her novel *Oranges Are Not the Only Fruit* (1985) is somewhat autobiographical, dealing with her time as a preacher. Soon after joining, she left the Church due to a difference in values about personal lifestyle choices such as homosexuality. She then obtained a Bachelor of Arts degree from Oxford University in 1981 and started to write fiction, essays and articles. She has received many awards for her work, including the Whitbread Prize in 1985 for the best fiction by a first-time writer for her novel *Oranges Are Not the Only Fruit* and the John Llewelyn Rhys Memorial Prize for her novel *The Passion* (1987). She also won the American Academy of Arts and Letters' E. M. Forster Award for her book *Sexing the Cherry*. She has gained immense popularity in recent years.

TWENTIETH-CENTURY TRENDS

BEFORE THE TWENTIETH CENTURY, THE scientific and Judeo-Christian thinking patterns prevalent in Western societies led people to think about reality in terms of absolutes. Classical physics explained the behaviour of many natural phenomena in terms of laws affecting the two basic elements of the universe: energy (emitted in continuous waves) and matter (concerning solid particles, the smallest of which was the atom). Many scientists were confident that they were close to unlocking the basic secrets of a mechanical universe. Judeo-Christian religions, which had dominated Western societies for centuries, required individuals to conform to precepts presented as laws handed down to humankind by the Creator. In the Judeo-Christian view of creation, a caring

God had placed human beings at the centre of a meaningful universe that they were destined to control. Such concepts as "right and wrong," "good and evil" and "true and false" were considered absolutes; time was seen as a constant that would never be subject to change.

The twentieth century witnessed spectacular discoveries and inventions in science and technology that would shake the foundations of western beliefs, not only about the universe but also about human society. In 1900, the German physicist Max Planck hypothesized that energy was emitted in discrete bundles or quanta; in other words, it behaved like matter (particles) in some instances and like waves in others. Quantum mechanics, which focused on the study of minute particles, developed over the next thirty years and carved itself an important place next to classical physics (which continued to explain large-scale events). In 1905, Albert Einstein formulated his famous theory of relativity, which proposed that the space/time continuum became curved as something approached the speed of light. No longer was time an absolute. As scientists made new discoveries, they realized that the universe was far more complex than they had thought. Gradually many people began to see everything around them in relative rather than absolute terms.

Sigmund Freud's psychoanalytical theories, which divided the human psyche into the ego (the conscious) and the id (the subconscious), began to take hold early in the century. Previously, most influential thinkers had held to a strictly rationalist view of human behaviour, maintaining that a person's reason was very much in control of everything that the person did. Freud's theories placed the rationalist view in doubt. He theorized that people were not aware of many subconscious elements that influenced their behaviour. Later thinkers postulated that imagination and artistic creativity both originated in the subconscious.

As the scientific world was undergoing a revolution, the universe of the arts was also changing. A vast amount of experimentation took place, transcending traditional artistic boundaries. Art changed radically and rapidly with the advent of many new movements. Visual art became less representational and more abstract. **Surrealism** was an artistic movement that flourished in Europe between the world wars, affecting both painting and literature. Its adherents included painters Salvador Dali and René Magritte, and authors Gertrude Stein and James Joyce. Surrealists fought against the rationalist view of reality and tried to express the role of the subconscious. They juxtaposed fantastic and realistic imagery in incongruous combinations that gave a dreamlike quality to much of their work. This combination of the conscious and unconscious realms was termed *surreality* by poet André Breton.

Politically, the twentieth century witnessed major events and clashes in ideologies that sent shockwaves through most strata of human endeavour: World War I (1914-1918)—supposedly the war to end all wars—the Roaring Twenties, the Great Depression (1929-1939), the rise of Nazism, World War II (1939-1945) and the hydrogen bomb, the Cold War of the 1950s, the wars in southeast Asia in the latter half of the century, the rise of Communism early in the century and its fall in the 1990s, the break-up of the great empires and various genocide attempts throughout the century. These and other historical events served to shake the foundations of Western society's faith in rationalism, which disturbed the understanding and vision of occidental values and the role of Western society in the universe.

This brief overview of the twentieth century is very incomplete because it concentrates only on the aspects important to the study of the short story in this section. We encourage you to consult other sources of information to further your study of this fascinating century.

Warm-Up Discussion Topics

1. People use art and science and technology to explore and interpret reality. What basic differences can you see between the approaches these fields use in their exploration and interpretations?

2. Are you more attracted to art or to science and technology? Give reasons for your answer.

11

∞ INITIAL READING

The story you are about to read exemplifies some of the experimentation that has emerged in the short story genre in recent years. Its surrealistic approach and its challenge to tradition make the story an excellent example of the spirit that reigned during much of the twentieth century. Moreover, it points out some very common but rather serious human failings in a humorous, light-hearted manner.

Before continuing, make sure that you read the section above entitled "Twentieth-Century Trends," as it will add to your understanding and enjoyment of the piece. Here is further information to prepare you for this challenging story.

Reading Hints

As you read the story, try to visualize the scenes and enjoy them as surreal art. Many of them are full of **black humour** (see definition in the Writer's Craft section following the story) and irony. Also, note the writer's use of **repetition**, recurring **imagery** and **contrast** between opposites, as these will hint at the story's theme(s). Try to appreciate the **connotations** associated with words, phrases, images, and so on, because they will also lead you to the underlying meanings in the work.

Newton
Jeannette Winterson

1   THIS is the story of Tom.
This is the story of Tom and his neighbours.
This is the story of Tom and his neighbours and his neighbour's garden.
This is the story of Tom.

2   "All of my neighbours are Classical Physicists," said Tom. "Their laws of motion are determined. They rise at 7 a.m. and leave for work at 8 a.m. The women take coffee at 10 a.m. If you see a body on the street between 1 and 2 p.m., lunchtime, it can only be the doctor, it can only be the undertaker, it can only be the stranger."

"I am the stranger," said Tom.

3   "What is the First Law of Thermodynamics?" said Tom.

"You can't transfer heat from a colder to a hotter. I've never known any warmth from my neighbours so I would reckon this is true. Here in Newton we don't talk much. That is, my neighbours talk all the time, they swap gossip, but I never have any, although sometimes I am some."

. . .

4   "What is the Second Law of Thermodynamics?" said Tom.

"Everything tends toward the condition of entropy. That is, the energy is still there

somewhere but for all useful purposes it is lost. Take a look at my neighbours here in Newton and you'll see what it means."

My neighbour has a garden full of plastic flowers. "It's easy," she says, "and so nice." When her husband died, she had him laminated, and he stands outside now, hands on his hips, carefully watching the sky.

"What's the matter, Tom?" she says, her head bobbing along the fence like a duck in a shooting parlour.

"Why don't you get married? In my day nobody had any trouble finding someone. We just did it and made the best of it. There were no screwballs then."

5   "What none?"

6   She bobbed faster and faster, gathering a bosom-load of underwear from the washing line. I knew she wanted me to stare at it, she wants to prove that I am a screwball. After all, if it's me, it's not her, it's not the others. You can't have more than one per block.

She wheeled round, ready to bob back up the other way, knickers popping from every pore.

"Tom, we were glad to be normal. In those days it was something good, something to be proud of."

. . .

7   Tom the screwball. Here I am with my paperback foreign editions and my corduroy trousers ("You got something against Levi's?" he asked me,

before he was laminated). All the men round here wear Levi's, denims or chinos. The only stylistic difference is whether they pack their stomach inside or outside the waistband.

They suspect me of being a homosexual. I wouldn't care. I wouldn't care what I was if only I were something.

8   "What do you want to be when you grow up?" said my mother, a long time ago, many times a long time ago.

"A fireman, an astronaut, a spy, a train driver, a hard hat, an inventor, a deep sea diver, a doctor and a nurse."

"What do you want to be when you grow up?" I ask myself in the mirror most days.

"Myself. I want to be myself."

And who is that, Tom?

9   Into the clockwork universe the quantum child. Why doesn't every mother believe her child can change the world? The child can. This is the joke. Here we are still looking for a saviour and hundreds are being born every second. Look at it, this tiny capsule of new life, indifferent to your prejudices, your miseries, unmindful of the world already made. Make it again? They could if we let them, but we make sure they grow up just like us, fearful like us. Don't let them know the potential that they are. Don't let them hear the grass singing. Let them live and die in Newton, ticktock, the last breath.

10 There was a knock at my door, I hid my Camus in the fridge and peered through the frosted glass. Of course I can't see anything. They never remind you of that when you fit frosted glass.

"Tom? Tom?" RAP RAP.

It's my neighbour. I shuffled to the door, feet bare, shirt loose. There she is, her hair coiled on her head like a wreath on a war memorial. She was dressed solely in pink.

"I'm not interrupting, am I, Tom?" she said, her eyes shoving past me into the kitchen.

"I was reading."

"That's what I thought. I said to myself, poor Tom will be reading. He won't be busy. I'll ask him to help me out. You know how difficult it is for a woman to manage alone. Since my husband was laminated, I haven't had it easy, Tom."

11 She smelled of woman; warm, perfumed, slightly threatening. I had to be careful not to act like a screwball. I offered her coffee. She seemed pleased, although she kept glancing at my bare feet and loose shirt. Never mind, she needs me to help her with something in the house. That's normal, that's nice, I want to be normal and nice.

12 "My mother's here. Will you help me get her into the house?"

"Now? Shall we go now?"

"She's had a long journey. She can rest in the truck a while. Shall we have that coffee you offered me first?"

13 I don't love my neighbour but still my hand trembled over the sugar spoon. They've made me feel odd and outside for so long, that now even the simplest things feel strange.

How does a normal person make coffee? What is it about me that worries them so much? I'm clean. I have a job.

14 "Tom, tell me, is it the modern thing to keep books in the refrigerator?"

In cheap crime novels, you often read the line, "He spun round." It makes me laugh to imagine a human being so animated, but when she asked me that question, I spun. One second I was facing the sink, the next second I was facing her, and she was facing me, holding my copy of Camus.

"I was just fetching out the milk, Tom. Who is Albert K. Mew?" She pronounced it like an enraged cat.

"He's a Frenchman. A French writer. I don't know how he came to be in the fridge."

She repeated my words slowly as though I had just offered her a universal truth.

"You don't know how he came to be in the fridge?" I shrugged and smiled and tried to disarm her.

"It's a big fridge. Don't you ever find things in the fridge you had forgotten about?"

"No, Tom. Never. I store cheese at the top, and then beer and bacon underneath, and underneath those I keep my weekend chicken, and at the bottom I have salad things and eggs. Those are the rules. It was the same when my husband was alive and it is the same now."

I was beginning to regard her with a new respect. The Grim Reaper came to call. He took her husband from the bed but left the weekend chicken on the shelf.

O Death, where is thy sting?

15 My neighbour, still holding my Camus, leaned forward confidentially, her arms resting on the table. She looked intimate, soft, I could see the beginning of her breasts.

"Tom, have you ever wondered whether you need help?" She said HELP with four capital letters, like a doorstep evangelist.

"If you mean the fridge, anyone can make a mistake." She leaned forward a little further. More breast.

"Tom, I'm going to be tough with you. You know what your problem is? You read too many geniuses. I don't know if Mr. K. Mew is a genius, but the other day you were seen in the main square reading Picasso's notebooks. Children were coming out of school and you were reading Picasso. Miss Fin at the library tells me that all you ever borrow are works of genius. She has no record of you ever ordering a sea story. Now that's unhealthy. Why is it unhealthy? You yourself are not a genius, if you were we would have found out by now. You are ordinary like the rest of us and ordinary people should lead ordinary lives. Like the rest of us, here in Tranquil Gardens."

She leaned back, her bosom with her.

"Shall we go and help your mother?" I said.

Outside, my neighbour walked toward a closed van parked in front of her house. I'd seen her mother a couple of years previously, but I couldn't see her now.

"She's in the back, Tom. Go round the back."

My neighbour flung open the back doors of the hired van and certainly there was her mother, sitting upright in the wheelchair that had been her home and her car. She was smiling a fearful plasticy smile, her teeth as perfect as a cheetah's.

"Haven't they done a wonderful job, Tom? She's even better than Doug, and he was pretty advanced at the time. I wish she could see herself. She never guessed I'd laminate her. She'd be so proud."

"Are those her own teeth?"

"They are now, Tom."

"Where will you put her?"

"In the garden with the flowers. She loved flowers."

· · ·

16 Slowly, slowly, we heaved down mother. We wheeled her over the swept pavement to the whitewashed house. It was afternoon coffee time and a lot of neighbours had been invited to pay their respects. They were so respectful that we were outside talking plastic until the men came home. My neighbour gets an incentive voucher for every successful lamination she introduces. She reckons that if Newton will only

do it her way, she'll have 75 percent of her own lamination costs paid by the time she dies.

"I've seen you hanging around the cemetery, Tom. It's not hygienic."

What does she think I am? A ghoul? I've told her before that my mother is buried there, but she just shakes her head and tells me that young couples need the land.

"Until we learn to stop dying, Tom, we have to live with the consequences. There's no room for the dead unless you treat them as ornamental."

I have tried to tell her that if we stop dying, all the cemeteries in the world can never release enough land for the bulging, ageing population. She doesn't listen, she just looks dreamy and thinks about the married couples.

Newton is jammed with married couples. We need one-way streets to let the singles through. I hate going shopping in Newton. I hate clubbing my way through the crocodile files, two by two in Main Street, as though the ark has landed. Complacent shoulder blades, battered baby buggies. DIY stores crammed with HIM and shopping malls heaving with HER. Don't they know that too much role playing is bad for the health? Imagine being a wife and saying "Honey, have you got time to fix the toilet?" Imagine being a husband and figuring out how to clean the toilet when she's left you.

Why are they married? It's normal, it's nice. They do it the way they do everything else in Newton. Ticktock says the clock.

17 "Tom, thank you, Tom," she cooed at me when her mother was safely settled beside the duck pond. The ducks are bath-time yellow with chirpy red beaks and their pond has real water with a bit of chlorine in it just in case. I had never been in my neighbour's garden before. It was quiet. No rustling in the undergrowth. No undergrowth to rustle in. No birds yammering. She tells me that peace is what the countryside is all about.

"If you were a genius, Tom, you could work here. The silence. The air. I have a unit you know, filters the air as it enters the garden."

It was autumn and there were a few plastic leaves scattered about on the AstroTurf. At the bottom of the garden, my neighbour has a shed, made of imitation wood, where she keeps her stocks for the changing of the seasons. She has told me many times that a garden must have variety and in her ventilated Aladdin's cave are the reassuring copies of nature. Tulips, red and white, hang meekly upside down by their stems. Daffodils in bright bunches are jumbled with loose camellia blooms, waiting to be slotted into the everlasting tree. She even has a row of squirrels clutching identical nuts.

"Those are going out soon, along with the autumn creeper." She has Virginia Creeper cascading down the house. It's still green. This is the burnt and blazing version.

"Mine's turning already," I said.

"Too early," she said. "You can't depend on nature. I don't like leaves falling. They don't fall where they should. If you don't regulate nature, why, she'll just go ahead and do what she likes. We have to regulate her. If we don't, it's volcanoes and forest fires and floods and death and bodies scattered everywhere, just like leaves."

Like leaves. Just like leaves. Don't you like them just a little where they fall? Don't you turn them over to see what is written on the other side? I like that. I like the simple text that can be read or not, that lies beneath your feet and mine, read or not. That falls, rain and wind, though nobody scoops it up to take it home. Life fell at your feet and you kicked her away and she bled on your shoes and when you came home, your mother said, "Look at you, covered in leaves."

You were covered in leaves. You peeled them off one by one, exposing the raw skin beneath. All those leavings. And when what had to fall was fallen, you picked it up and read what was written on the other side. It made no sense to you. You screwed it up in your pocket where it burned like a live coal. Tell me why they left you, one by one, the ones you loved? Didn't they like you? Didn't they, like you, need a heart that was a book with no last page? Turn the leaves.

"The leaves are turning," said Tom.

18 She asked me back to supper as a thank you, and I thought I should go because that's what normal people do; eat with their neighbours, even though it is boring and the food is horrible. I searched for a tie and wore it.

19 "Tom, come in, what a lovely surprise!"

She must mean what a lovely surprise for me. It can hardly be a surprise for her, she's been cooking all afternoon.

Once inside the dining room, I know she means me. I know that because the entire population of Newton is already seated at the dinner table, a table that begins crammed up against the display cabinet of Capodimonte and

extends ... and extends ... through a jagged hole blown in the side of the house, out and on toward the bus station.

"I think you know everyone, Tom," says my neighbour. "Sit here, by me, in Doug's place. You're about his height." Do I know everyone? It's hard to say, since beyond the hole, all is lost.

"Tom, take a plate. We're having chicken cooked in bacon strips and stuffed with hardboiled eggs. There's a salad I made and plenty of cheese and beer in the fridge if you want it."

She drifted away from me, her dress clinging to her like a drowned man. Nobody looked up from their plates. They were eating chicken, denims and chinos all, eating the three or four hundred fowl laid on the table, half a dozen eggs per ass. I was still trying to work out the roasting details, the oven size, when BAM, one of the chickens exploded, pelting my neighbour with eggs like hand grenades. One of her arms flew off but luckily for her, not the one she needed for her fork. Nobody noticed. I wanted to speak, I wanted to act, I began to speak, to act, just as my neighbour herself returned carrying a covered silver dish.

"It's for you Tom," she says, as the table falls silent. Already on my feet I was able to lift the huge lid with some dignity. Underneath was a chicken.

"It's your chicken, Tom."

She's telling the truth. Poking out of the ass of the chicken, I can see my copy of *L'Étranger* by Albert Camus. It hasn't been shredded, so I can take it out. When I open it, I see that there are no words left on any of the pages. The pages are blank.

"We wanted to help you, Tom." Her eyes are full of tears. "Not just me. All of us. A helping hand for Tom."

Slowly the table starts to clap, faster and louder. The table shakes, the dishes roll from side to side like the drunken tableware in a sea story. This is a sea story. The captain and the crew have gone mad and I am the only passenger. Reeling, I ran from the dining room into the kitchen and slammed the door behind me. Here was peace. Hygienic enamelled peace.

Tom slid to the floor and cried.

20 Time passed. In Newton it always does and everyone knows how long it takes for time to pass and so nobody gets confused. Tom didn't know how much time had passed. He woke from an aching sleep and put his fist through the frosted glass kitchen door. He went home and took his

big coat and filled the pockets with books and the books seemed like live coals to him. He walked away from Newton, but he did look back once, and what he saw was a table stretching out past the bend in the road and on through the streets and houses joining them together in an orgy of matching cutlery. World without end.

"But now," says Tom, "the hills are ripe and the water leaps at my throat when I shave."

Ticktock says the clock in Newton.

PUBLISHED IN
1992

Reader's Response

1. What was your first reaction to this story? List some adjectives to indicate what you were thinking and feeling.

2. In your opinion, are the attitudes of the Newton townspeople similar to or different from the attitudes of other people in contemporary society? Explain your answer.

✦ CLOSE READING

1. When does the story take place?
   a) in the future (realistic)
   b) in contemporary times
   c) in the past
   d) in a time beyond time (surrealistic)

   Justify your answer.

2. Why does Tom compare his neighbours to Classical Physicists? Give one direct example from the text to support your answer.

3. Describe how Tom applies the first and second law of thermodynamics in order to portray his neighbours' characteristics.

4. In your opinion, is Tom's opinion of his neighbours positive or negative? Justify your answer.

5. During the neighbour's conversation with Tom, it is obvious that her attitude toward life is different from Tom's. Find three things that Tom's neighbour says or does that reveal this difference.

6. Find three different characteristics about Tom that reveal how he is different from his neighbours.

_____

_____

_____

11

7. What does Tom say he wants to be when he grows up? How is this significant to the story?

_____

8. Find the statement made by the neighbour that reveals the purpose behind her dinner invitation to Tom.

_____

9. What does the neighbour do to Tom's book?

_____

10. Why does she do this to Tom's book?

_____

11. What is Tom's reaction to the events during dinner? What does he think and feel when he sees his book?

_____

_____

12. Why does Tom leave Newton?

_____

✌ WRITER'S CRAFT

Biblical References

1. Paragraph 14 ends with the question: "O Death, where is thy sting?" This celebrated phrase comes from the Bible (I Corinthians 15-55) and is used ironically here to imply that Tom's neighbour seems more upset about someone breaking "refrigerator rules" than about her husband's death. There are several other biblical references in the paragraphs indicated below. Quote the references.

a) paragraph 9: _____

b) paragraph 13: _____

c) paragraph 15: _____

d) paragraph 20: _____

2. Tom's neighbour's garden evokes a famous biblical garden (although very ironically). Which one?

_____

3. The scene of the banquet evokes another famous biblical image. Which one?

_____

Winterson, like many artists in the Occident, uses biblical imagery to add depth and meaning to her work. For example, a reference to the Garden of Eden often indicates humankind's relentless search for paradise—for an ideal existence.

Literary Devices

■ IMAGERY

1. Which sense is solicited most often by Winterson's imagery? _____

2. Describe one image that you find to be particularly surreal.

_____

"Newton" abounds in surrealistic imagery designed to breathe life into the story and to stimulate the reader's imagination and thoughts. The use of surreal elements is characteristic of many of Winterson's short stories.

■ SYMBOLISM

Many writers use **symbols** as clues to help the reader understand the overall meaning of the story. Many symbols in this story help to reveal its principal theme.

1. Write your interpretation of the following symbols.

a) Tom's neighbour's garden (Find something different from the ironic religious evocation.)

_____

_____

b) *L'Étranger* by Albert Camus

_____

c) The name of the town—Newton

_____

2. Find two other symbols indicating that Tom does not fit into Newton.

a) _____

b) _____

■ BLACK HUMOUR AND IRONY

Black humour can be defined as the deliberate placing of the morbid or absurd side by side with the comical. This is often meant to shock or disturb the intended audience.

1. Explain how the idea of laminating the dead can be seen as black humour and surreal at the same time.

   _____

2. Describe another example of Winterson's use of black humour in the story.

   _____

3. How is the situation where Tom's neighbour finds his book in the fridge ironic?

   _____

   _____

> Winterson uses many instances of black humour and irony in this story to underline meaning and to add to the overall surreal quality of the piece. This also adds an element of surprise and unpredictability to the story—the reader is unable to guess what aspect of traditional Western society she will attack next. This gives impact to the style.

■ REPETITION

> In this story, Winterson uses repetition of words, sentences, images and ideas as a means of enhancing the story's theme.

1. The first four sentences of the story are repetitions. What do you notice about their structure or patterns?

   _____

2. Which clause is repeated most often in the first four sentences? What is its significance?

   _____

3. The second paragraph also includes repetition. What is being repeated in this paragraph?

   _____

4. What does the use of repetition suggest about the townspeople's lifestyle?

   _____

5. Tom's neighbour is characterized through repetition. List two things in reference to the neighbour that indicate repetition.

   _____

   _____

6. When talking to Tom in his house, the neighbour tells him what she keeps in her refrigerator. What is significant about the way she organizes her fridge?

   _____

7. The story is full of repeated references to the classical physicist Newton and his scientific laws. Read the last sentence of the story. What can you imply from it regarding the author's vision of traditional society as exemplified by the town?

   _____

11

■ DIALOGUE AND INTERIOR MONOLOGUE

> **Interior monologue** consists of a directly written presentation of the ongoing thoughts and emotions a character experiences in his or her head (although they are often disjointed and disorganized).

The theme is also revealed through **dialogue** and **interior monologue**, or Tom's thoughts.

1. Reread the conversation between Tom and his neighbour, beginning with the line, "'Tom? Tom?' RAP RAP." and ending with the sentence, "She leaned back, her bosom with her."

   In your opinion, what is the central conflict between Tom and his neighbour as indicated during their conversation?

   _____

   _____

2. Write one sentence from this dialogue to support your answer to the preceding question.

   _____

3. Tom reveals his own philosophy on life when he reflects on the neighbour's words and actions.

   Write an interpretation of the following quotations (which represent Tom's thoughts).

   > Into the clockwork universe the quantum child. Why doesn't every mother believe her child can change the world? The child can. This is the joke. Here we are still looking for a saviour and hundreds are being born every second. Look at it, this tiny capsule of new life, indifferent to your prejudices, your miseries, unmindful of the world already made. Make it again? They could if we let them, but we make sure they grow up just like us, fearful like us. Don't let them know the potential that they are. Don't let them hear the grass singing. Let them live and die in Newton, ticktock, the last breath.

   _____

   _____

   > Like leaves. Just like leaves. Don't you like them just a little where they fall? Don't you turn them over to see what is written on the other side? I like that. I like the simple text that can be read or not, that lies beneath your feet and mine, read or not. That falls, rain and wind, though nobody scoops it up to take it home. Life fell at your feet and you kicked her away and she bled on your shoes and when you came home, your mother said, "Look at you, covered in leaves."

   _____

   _____

4. Find another example of interior monologue that reveals Tom's opinions and attitudes.

   _____

   _____

5. What is the central conflict of the story?

   _____

6. In your opinion, what is the principal theme of the story?

_____

7. What images does the author use to describe the following items in the last paragraph of the story?

    a) books: _____

    b) hills: _____

    c) shaving: _____

8. What characteristic do the images share?

_____

9. If we consider that Tom symbolizes "the artist," what statement could Winterson be making about the effects of science and technology on human civilization?

_____

Breaking Traditional Rules

1. In this story, Winterson deliberately breaks many traditional rules for stylistic reasons. (This is called artistic license and is not something a student should imitate in the context of academic writing!) Explain how Winterson goes beyond traditional rules and usage with regard to the following aspects of the story.

    a) paragraph format and content:

    _____

    b) narration and point of view:

    _____

    c) flow of the main storyline:

    _____

    d) sequence of tenses:

    _____

    e) characterization:

    _____

2. Read the following excerpt.

> Tell me why they left you, one by one, the ones you loved? Didn't they like you? Didn't they, like you, need a heart that was a book with no last page? Turn the leaves.
>
> "The leaves are turning," said Tom.

    a) Note three words that the author repeats with a different meaning each time.

    _____

    b) Note one punctuation mark that changes the meaning of the sentence when it is added (twice).

    _____

Winterson plays with the multiple meanings of words and grammatical structure to evoke meaningful associations and connections in the mind of the reader.

VOCABULARY DEVELOPMENT
Expanding Your Vocabulary

Match the word or phrase in the left-hand column with its definition/explanation in the right-hand column.

1. knickers _____

2. Grim Reaper _____

3. Newton _____

4. entropy _____

5. screwball _____

6. Camus _____

7. DIY _____

a) the hypothetical tendency for all matter and energy in the universe to evolve to a state of inert uniformity

b) a weird, eccentric person *(slang)*

c) underwear; panties *(British English)*

d) a home improvement centre

e) a seventeenth-century English mathematician and scientist who invented differential calculus and formulated "laws" concerning the physical universe

f) the personification of Death—dressed in a black hooded cloak and holding a scythe

g) a twentieth-century French writer and philosopher whose book *L'Étranger* deals with the absurdity of the human condition

Connotations

Connotations are meanings or ideas suggested by or associated with words.

Match the following words, phrases and sentences from the story with their connotations.

1. Newton _____

2. plastic _____

3. nature _____

4. normal _____

5. Tick-tock says the clock. _____

6. leaves _____

7. a saviour _____

8. frosted glass _____

9. chicken _____

10. quantum child _____

a) Judeo-Christian religions; help from outside

b) mechanical; predictable; unthinking repetition

c) hiding reality from view; obscuring clear vision

d) accepted norm; everyday; what ordinary people eat

e) uncontrolled; unregulated; unpredictable; alive

f) conformist; society's expectations

g) dead; artificial; an imitation

h) scientific laws; rules

i) trees (nature); pages (a book, a story, art); quits (changes)

j) new direction; possibility for change; potential

Jeanette Winterson uses connotations extensively in this story to add layers of meaning to the work.

Wrap-Up Activities

1. Compare the different values of the two main characters in "Newton."

2. How is nature portrayed in "Newton" and what is its significance?

3. Discuss the appropriateness of the story's title.

4. Explain Winterson's vision of the artist as exemplified by Tom.

5. What is Winterson's view of science and technology as shown by the short story?

6. How are Louise Mallard in "The Story of an Hour" and Tom in "Newton" similar and/or different?

7. The story "Newton" is a warning about what contemporary society is and has the potential to become. Discuss.

8. The story "Newton" is about conflict. Highlight the different conflicts found in the story and explain their significance in terms of the theme.

9. Analyze Jeannette Winterson's writing style.

10. How does Winterson's use of imagery enhance the theme of "Newton"?

11. Compare the settings of "The Veldt" and "Newton" and discuss their importance in terms of tone and atmosphere, as well as theme.

My eLab ACTIVITY

My eLab

Read about eccentricity and discuss whether Tom, the protagonist in "Newton," is eccentric. Are all artists eccentric? Write an essay on the place of eccentrics in Western society.

Tapka

CHAPTER
CONTENTS

▸ **Reading:** "Tapka" by David Bezmozgis; author's biography; literary trends (the immigrant experience)

▸ **Literary Elements:** narration and point of view (first person); style and literary devices (repetition, imagery, metaphor and simile, contrast of opposites and irony)

▸ **Vocabulary Development:** vivid present participles and gerunds ("-ing" words)

ABOUT THE **AUTHOR**

David Bezmozgis is both an author and a documentary filmmaker. Born in Latvia in 1973, he immigrated to Toronto, Canada, seven years later with his Russian-speaking parents. He studied literature at McGill University in Montreal, followed by fine arts at the University of Southern California's School of Cinema-Television. His first book, *Natasha and Other Stories*, published in 2004, quickly became a bestseller and won the Canadian Jewish Book Award and the Danuta Gleed Award in 2005. It was also short-listed for other prestigious awards and has been translated into twelve languages. David Bezmozgis is currently living and writing in Toronto.

In *Natasha and Other Stories*, Bezmozgis presents episodes from the life of Mark, a young Russian-speaking immigrant, as he grows up in Toronto during the 1970s. "Tapka" is the first short story in the collection and introduces the reader to six-year-old Mark.

LITERARY TRENDS
OF THE TIMES

THE POPULATIONS OF CANADA AND THE United States are mainly composed of immigrants. Originally, migrants to North America came from Britain, France and Spain, but they were soon joined by people from many different countries. During the last few decades, Canada, as well as the United States, to a lesser degree, has continued to accept large numbers of newcomers. Canadians pride themselves on their openness to multiculturalism, but immigrants must still learn English (or French in Quebec) in order to function in Canadian society. The city of Toronto receives about 45 percent of the new immigrants to Canada, so it is little wonder that many new voices in Canadian literature hail from this city.

Settling in a new country is not an easy process. Newcomers must first discover the basic tenets and workings of their new society. They must learn the official language, find a place to live, acquire a job, send their children to school and develop a social network. Before an individual is integrated into the social fabric, he or she must adapt—a process that can be long and difficult. Mordecai Richler, Joy Kogawa, Neil Bissoondath, Rohinton Mistry, Ken Wiwa, Anna Porter, Shyam Selvadurai, Alberto Manguel, Dany Laferrière, Michelle Berry and Ying Chen are just some of the many Canadian writers who have written about the immigration experience. With *Natasha and Other Stories*, David Bezmozgis joins them as a new and articulate voice.

Warm-Up Discussion Topics

1. Describe a childhood experience that taught you an important lesson.

2. What are the challenges facing a newcomer to your community?

3. Would you like to start a new life in another country where you don't speak the language? Why or why not?

∞ INITIAL READING

First read the story, told from a child's point of view, for personal enjoyment.

Tapka
David Bezmozgis

1 GOLDFINCH was **flapping** clotheslines, a tenement delirious with striving. 6030 Bathurst: insomniac **scheming** Odessa. Cedarcroft: **reeking** borscht in the hallways. My parents, Baltic aristocrats, took an apartment at 715 Finch fronting a ravine and across from an elementary school—one respectable block away from the Russian swarm. We lived on the fifth floor, my cousin, aunt and uncle directly below us on the fourth. Except for the Nahumovskys, a couple in their fifties, there were no other Russians in the building. For this privilege, my parents paid twenty extra dollars a month in rent.

2 In March of 1980, near the end of the school year but only three weeks after our arrival in Toronto, I was enrolled in Charles H. Best Elementary. Each morning, with our house key hanging from a brown shoelace around my neck, I kissed my parents goodbye and, along with my cousin, Jana, tramped across the ravine—I to the first grade, she to the second. At three o'clock, **bearing** the germs of a new vocabulary, we tramped back home. Together, we then waited until six for our parents to return from George Brown City College, where they were taking their obligatory classes in English.

3 In the evenings, we assembled and compiled our linguistic bounty.

4 Hello, havaryew?

5 Red, yellow, green, blue.

6 May I please go to the washroom?

7 Seventeen, eighteen, nineteen, twenny.

8 Joining us most nights were the Nahumovskys. They attended the same English classes and travelled with my parents on the same bus. Rita Nahumovsky was a beautician, her face spackled with makeup, and Misha Nahumovsky was a tool and die maker. They came from Minsk and didn't know a soul in Canada. With **abounding** enthusiasm, they incorporated themselves into our family. My parents were glad to have them. Our life was tough, we had it hard—but the Nahumovskys had it harder. They were alone, they were older, they were stupefied by the demands of language. Being essentially helpless themselves, my parents found it gratifying to help the more helpless Nahumovskys.

9 After dinner, as we gathered on cheap stools around our table, my mother repeated the day's lessons for the benefit of the Nahumovskys and, to a slightly lesser degree, for the benefit of my father. My mother had always been a dedicated student and she extended this dedication to George Brown City College. My father and the Nahumovskys

12

came to rely on her detailed notes and her understanding of the curriculum. For as long as they could, they listened attentively and groped toward comprehension. When this became too frustrating, my father put on the kettle, Rita painted my mother's nails and Misha told Soviet jokes.

10   In a first-grade classroom, a teacher calls on her students and inquires after their nationality. "Sasha," she says. Sasha: "Russian." "Very good," says the teacher. "Arnan," she says. Arnan says, "Armenian." "Very good," says the teacher. "Lubka," she says. Lubka says, "Ukrainian." "Very good," says the teacher. And then she asks Dima. Dima says, "Jewish." "What a shame," says the teacher, "so young and already a Jew."

· · ·

11   The Nahumovskys had no children, only a white Lhasa-Apso named Tapka. The dog had lived with them for years before they emigrated and then travelled with them from Minsk to Vienna, from Vienna to Rome and from Rome to Toronto. During our first month in the building, Tapka was in quarantine and I saw her only in photographs. Rita had dedicated an entire album to the dog, and to dampen the pangs of separation, she consulted the album daily. There were shots of Tapka in the Nahumovskys' old Minsk apartment, seated on the cushions of faux Louis XIV furniture; there was Tapka on the steps of a famous Viennese palace; Tapka at the Vatican; in front of the Coliseum; at the Sistine Chapel; and under the Leaning Tower of Pisa. My mother—despite having grown up with goats and chickens in her yard—didn't like animals and found it impossible to feign interest in Rita's dog. Shown a picture of Tapka, my mother wrinkled her nose and said "foo." My father also couldn't be bothered. With no English, no money, no job and only a murky conception of what the future held, he wasn't equipped to admire Tapka on the Italian Riviera. Only I cared. Through the photographs I became attached to Tapka and projected upon her the ideal traits of the dog I did not have. Like Rita, I counted the days until Tapka's liberation.

12   The day Tapka was to be released from quarantine, Rita prepared an elaborate dinner. My family was invited to celebrate the dog's arrival. While Rita cooked, Misha was banished from their apartment. For distraction, he seated himself at our table with a deck of cards. As my mother reviewed sentence construction, Misha played hand after hand of Durak with me.

13   "The woman loves this dog more than me. A taxi to the customs facility is going to cost us ten, maybe fifteen dollars. But what can I do? The dog is truly a sweet little dog."

14   When it came time to collect the dog, my mother went with Misha and Rita to act as their interpreter. With my nose to the window, I watched the taxi take them away. Every few minutes, I reapplied my nose to the window. Three hours later the taxi pulled into our parking lot and Rita emerged from the back seat **cradling** animated fur. She set the fur down on the pavement, where it assumed the shape of a dog. The length of its coat concealed its legs, and as it hovered around Rita's ankles, it appeared to have either a thousand tiny legs or none at all. My head ringing "Tapka, Tapka, Tapka," I raced into the hallway to meet the elevator.

15   That evening Misha toasted the dog:

16   "This last month, for the first time in years, I have enjoyed my wife's undivided attention. But I believe no man, not even one as perfect as me, can survive so much attention from his wife. So I say, with all my heart, thank God our Tapka is back home with us. Another day and I fear I may have requested a divorce."

17   Before he drank, Misha dipped his pinkie finger into his vodka glass and offered it to the dog. Obediently, Tapka gave Misha's finger a thorough licking. Duly impressed, my uncle declared her a good Russian dog. He also gave her a lick of his vodka. I gave her a piece of my chicken. Jana rolled her a pellet of bread. Misha taught us how to dangle food just out of Tapka's reach and thereby induce her to perform a charming little dance. Rita also produced "Clonchik," a red and yellow rag clown. She tossed Clonchik under the table, onto the couch, down the hallway, and into the kitchen; over and over Rita called, "Tapka, get Clonchik," and without fail, Tapka got Clonchik. Everyone delighted in Tapka's antics except for my mother, who sat stiffly in her chair, her feet slightly off the ground, as though preparing herself for a mild electric shock.

18 After the dinner, when we returned home, my mother announced that she would no longer set foot in the Nahumovskys' apartment. She liked Rita, she liked Misha, but she couldn't sympathize with their attachment to the dog. She understood that the attachment was a consequence of their lack of sophistication and also·their childlessness. They were simple people. Rita had never attended university. She could derive contentment from talking to a dog, brushing its coat, putting ribbons in its hair and repeatedly throwing a rag clown across the apartment. And Misha, although very lively and a genius with his hands, was also not an intellectual. They were good people, but a dog ruled their lives.

19 Rita and Misha were sensitive to my mother's attitude toward Tapka. As a result, and to the detriment of her progress with English, Rita stopped visiting our apartment. Nightly, Misha would arrive alone while Rita attended to the dog. Tapka never set foot in our home. This meant that in order to see her, I spent more and more time at the Nahumovskys'. Each evening, after I had finished my homework, I went to play with Tapka. My heart soared every time Rita opened the door and Tapka raced to greet me. The dog knew no hierarchy of affection. Her excitement was infectious. In Tapka's presence, I resonated with doglike glee.

20 Because of my devotion to the dog and their lack of an alternative, Misha and Rita added their house key to the shoelace hanging around my neck. Every day, during our lunch break and again after school, Jana and I were charged with·caring for Tapka. Our task was simple: put Tapka on her leash, walk her to the ravine, release her to chase Clonchik and then bring her home.

21 Every day, sitting in my classroom, understanding little, effectively friendless, I counted down the minutes to lunchtime. When the bell rang, I met Jana on the playground and we sprinted across the grass toward our building. In the hall, our approaching footsteps elicited **panting** and scratching. When I inserted the key into the lock, I felt emanations of love through the door. And once the door was open, Tapka hurled herself at us, her entire body consumed with an ecstasy of **wagging**. Jana and I took turns embracing her, petting her, covertly **vying** for her favour. Free of Rita's scrutiny, we also satisfied certain anatomical curiosities. We examined Tapka's ears, her paws, her teeth, the roots of her fur and her doggy genitals. We poked and prodded her, we threw her up in the air, rolled her over and over, and swung her by her front legs. I felt such **overwhelming** love for Tapka that sometimes when hugging her, I had to restrain myself from **squeezing** too hard and **crushing** her little bones.

22 It was April when we began to care for Tapka. Snow melted in the ravine; sometimes it rained. April became May. Grass absorbed the thaw, turned green; dandelions and wildflowers sprouted yellow and blue; birds and insects flew, crawled and made their characteristic noises. Faithfully and reliably, Jana and I attended to Tapka. We walked her across the parking lot and down into the ravine. We threw Clonchik and said "Tapka, get Clonchik." Tapka always got Clonchik. Everyone was proud of us. My mother and my aunt wiped tears from their eyes while talking about how responsible we were. Rita and Misha rewarded us with praise and chocolates. Jana was seven and I was six; much had been asked of us, but we had risen to the challenge.

23 Inspired by everyone's confidence, we grew confident. Whereas at first we made sure to walk thirty paces into the ravine before releasing Tapka, we gradually reduced that requirement to ten paces, then five paces, until finally we released her at the grassy border between the parking lot and ravine. We did this not out of laziness or recklessness but because we wanted proof of Tapka's love. That she came when we called was evidence of her love, that she didn't piss in the elevator was evidence of her love, that she offered up her belly for scratching was evidence of her love, all of this was evidence, but it wasn't proof. Proof could come only in one form. We had intuited an elemental truth: love needs no leash.

24 That first spring, even though most of what was said around me remained a mystery, a thin rivulet of meaning trickled into my cerebral catch basin and collected into a little pool of knowledge. By the end of May, I could sing the ABC song. Television taught me to say "What's up, Doc?" and "super-duper." The playground introduced me to "shithead," "mental case" and "gaylord," and I sought every opportunity to apply my new knowledge.

25 One afternoon, after spending nearly an hour in the ravine throwing Clonchik in a thousand different directions, Jana and I lolled in the sunlit pollen. I called her "shithead," "mental case" and "gaylord," and she responded by calling me "gaylord," "shithead" and "mental case."

26 "Shithead."

27 "Gaylord."

28 "Mental case."

29 "Tapka, get Clonchik."

30 "Shithead."

31 "Gaylord."

32 "Come, Tapka-lapka.'

33 "Mental case."

34 We went on like this, over and over, until Jana threw the clown and said, "Shithead, get Clonchik." Initially, I couldn't tell if she had said this on purpose or if it had merely been a blip in her rhythm. But when I looked at Jana, her smile was triumphant.

35 "Mental case, get Clonchik."

36 For the first time, as I watched Tapka **bounding** happily after Clonchik, the profanity sounded profane.

37 "Don't say that to the dog."

38 "Why not?"

39 "It's not right."

40 "But she doesn't understand."

41 "You shouldn't say it."

42 "Don't be a baby. Come, shithead, come, my dear one."

43 Her tail wagging with accomplishment, Tapka dropped Clonchik at my feet.

44 "You see, she likes it."

45 I held Clonchik as Tapka pawed frantically at my shins.

46 "Call her shithead. Throw the clown."

47 "I'm not calling her shithead."

48 "What are you afraid of, shithead?"

49 I aimed the clown at Jana's head and missed.

50 "Shithead, get Clonchik."

51 As the clown left my hand, Tapka, a white shining blur, oblivious to insult, was already cutting through the grass. I wanted to believe that I had intended the "shithead" exclusively for Jana, but I knew it wasn't true.

52 "I told you, gaylord, she doesn't care."

53 I couldn't help thinking, "Poor Tapka," and looked around for some sign of recrimination. The day, however, persisted in unimpeachable brilliance: sparrows winged overhead; bumblebees levitated above flowers; beside a lilac shrub, Tapka clamped down on Clonchik. I was amazed at the absence of consequences.

54 Jana said, "I'm going home."

55 As she started for home, I saw that she was still holding Tapka's leash. It swung insouciantly from her hand. I called after her just as, once again, Tapka deposited Clonchik at my feet.

56 "I need the leash."

57 "Why?"

58 "Don't be stupid. I need the leash."

59 "No you don't. She comes when we call her. Even shithead. She won't run away."

60 Jana turned her back on me and proceeded toward our building. I called her again, but she refused to turn around. Her receding back was a blatant provocation. Guided more by anger than by logic, I decided that if Tapka was closer to Jana, then the onus of responsibility would become hers. I picked up the doll and threw it as far as I could into the parking lot.

61 "Tapka, get Clonchik."

62 Clonchik tumbled through the air. I had put everything in my six-year-old arm behind the throw, which still meant that the doll wasn't going very far. Its trajectory promised a drop no more than twenty feet from the edge of the ravine. Running, her head arched to the sky, Tapka tracked the flying clown. As the doll reached its apex, it crossed paths with a sparrow. The bird veered off toward Finch Avenue and the clown plummeted to the asphalt. When the doll hit the ground, Tapka raced past it after the bird.

63 A thousand times we had thrown Clonchik and a thousand times Tapka had retrieved him. But who knows what passes for a thought in the mind of a dog? One moment a Clonchik is a

12

Clonchik, and the next moment a sparrow is a Clonchik.

64   I shouted at Jana to catch Tapka and then watched as the dog, her attention fixed on the sparrow, skirted past Jana and into traffic. From the slope of the ravine, I couldn't see what had happened. I saw only that Jana had broken into a sprint and I heard the **caterwauling** of tires followed by a shrill fractured yip.

65   By the time I reached the street, a line of cars was already stretched a block beyond Goldfinch. At the front of the line were a brown station wagon and a pale blue sedan blistered with rust. As I neared, I noted the chrome letters on the back of the sedan: D-U-S-T-E-R. In front of the sedan, Jana kneeled in a tight semicircle with a pimply young man and an older woman wearing very large sunglasses. Tapka lay panting on her side at the centre of their circle. She stared at me, at Jana. Except for a hind leg **twitching** at the sky at an impossible angle, she looked much as she did when she rested on the rug at the Nahumovskys' apartment after a romp in the ravine.

66   Seeing her this way, barely mangled, I started to convince myself that things weren't as bad as I had feared, and I edged forward to pet her. The woman in the sunglasses said something in a restrictive tone that I neither understood nor heeded. I placed my hand on Tapka's head and she responded by turning her face and allowing a trickle of blood to escape onto the asphalt. This was the first time I had ever seen dog blood and I was struck by the depth of its colour. I hadn't expected it to be red, although I also hadn't expected it to be not-red. Set against the grey asphalt and her white coat, Tapka's blood was the red I envisioned when I closed my eyes and thought: red.

67   I sat with Tapka until several dozen car horns demanded that we clear the way. The woman with the large sunglasses ran to her station wagon, returned with a blanket and scooped Tapka off the street. The pimply young man stammered a few sentences of which I understood nothing except the word "sorry." Then we were in the back seat of the station wagon with Tapka in Jana's lap. The woman kept talking until she realized that we couldn't understand her at all. As we started to drive, Jana remembered something.

I motioned for the woman to stop the car and scrambled out. Above the atonal chorus of car horns I heard:

68   "Mark, get Clonchik."

69   I ran and got Clonchik.

70   For two hours Jana and I sat in the reception area of a small veterinary clinic in an unfamiliar part of town. In another room, with a menagerie of various afflicted creatures, Tapka lay in traction, connected to a blinking machine by a series of tubes. Jana and I had been allowed to see her once but were rushed out when we both burst into tears. Tapka's doctor, a woman in a white coat and furry slippers resembling bear paws, tried to calm us down. Again, we could neither explain ourselves nor understand what she was saying. We managed only to establish that Tapka was not our dog. The doctor gave us colouring books, stickers and access to the phone. Every fifteen minutes we called home. Between phone calls we absently flipped pages and sniffled for Tapka and for ourselves. We had no idea what would happen to Tapka, all we knew was that she wasn't dead. As for ourselves, we already felt punished and knew only that more punishment was to come.

71   "Why did you throw Clonchik?"

72   "Why didn't you give me the leash?"

73   "You could have held on to her collar."

74   "You shouldn't have called her shithead."

75   At six-thirty, my mother picked up the phone. I could hear the agitation in her voice. The ten minutes she had spent at home not knowing where I was had taken their toll. For ten minutes, she had been the mother of a dead child. I explained to her about the dog and felt a twinge of resentment when she said, "So it's just the dog?" Behind her, I heard other voices. It sounded as though everyone was speaking at once, pursuing personal agendas, translating the phone conversation from Russian to Russian until one anguished voice separated itself, "My God, what happened?" Rita.

76   After getting the address from the veterinarian, my mother hung up and ordered another expensive taxi. Within a half-hour, my parents, my aunt and Misha and Rita pulled up at the clinic. Jana and I waited for them on the sidewalk. As

soon as the taxi doors opened, we began to sob. Partly out of relief but mainly in the hope of **eliciting** sympathy. As I ran to my mother, I caught sight of Rita's face. Her face made me regret that I also hadn't been hit by a car.

77 As we clung to our mothers, Rita descended upon us.

78 "Children, what oh what have you done?"

79 She pinched compulsively at the loose skin of her neck, raising a cluster of pink marks.

80 While Misha methodically counted individual bills for the taxi driver, we swore on our lives that Tapka had simply gotten away from us. That we had minded her as always, but, inexplicably, she had seen a bird and bolted from the ravine and into the road. We had done everything in our power to catch her, but she had surprised us, eluded us, been too fast.

81 Rita considered our story.

82 "You are liars. Liars!"

83 She uttered the words with such hatred that we again burst into sobs.

84 My father spoke in our defence.

85 Rita Borisovna, how can you say this? They are children.

86 They are liars. I know my Tapka. Tapka never chased birds. Tapka never ran from the ravine.

87 "Maybe today she did?"

88 "Liars."

89 Having delivered her verdict, she had nothing more to say. She waited anxiously for Misha to finish paying the driver.

90 "Misha, enough already. Count it a hundred times, it will still be the same."

91 Inside the clinic there was no longer anyone at the reception desk. During our time there, Jana and I had watched a procession of dyspeptic cats and lethargic parakeets disappear into the back rooms for examination and diagnosis. One after another they had come and gone until, by the time of our parents' arrival, the waiting area was entirely empty and the clinic officially closed. The only people remaining were a night nurse and the doctor in the bear paw slippers who had stayed expressly for our sake.

92 Looking desperately around the room, Rita screamed: "Doctor! Doctor!" But when the doctor appeared, she was incapable of making

herself understood. Haltingly, with my mother's help, it was communicated to the doctor that Rita wanted to see her dog.

93 Pointing vigorously at herself, Rita asserted: "Tapka. Mine dog."

94 The doctor led Rita and Misha into the veterinary version of an intensive care ward. Tapka lay on her little bed, Clonchik resting directly beside her. At the sight of Rita and Misha, Tapka weakly wagged her tail. Little more than an hour had elapsed since I had seen her last, but somehow over the course of that time, Tapka had shrunk considerably. She had always been a small dog, but now she looked desiccated. Rita started to cry, grotesquely **smearing** her mascara. With trembling hands, and with sublime tenderness, she stroked Tapka's head.

95 "My God, my God, what has happened to you, my Tapkachka?"

96 Through my mother, and with the aid of pen and paper, the doctor provided the answer. Tapka required two operations. One for her leg. Another to stop internal bleeding. An organ had been damaged. For now, a machine was helping her, but without the machine, she would die. On the paper the doctor drew a picture of a scalpel, of a dog, of a leg, of an organ. She made an arrow pointing at the organ and drew a teardrop and coloured it in to represent "blood." She also wrote down a number preceded by a dollar sign. The number was 1,500.

97 At the sight of the number, Rita let out a low animal moan and steadied herself against Tapka's little bed. My parents exchanged a glance. I looked at the floor. Misha said, "My dear God." The Nahumovskys and my parents each took in less than five hundred dollars a month. We had arrived in Canada with almost nothing, a few hundred dollars, but that had all but disappeared on furniture. There were no savings. Fifteen hundred dollars. The doctor could just as well have written a million.

98 In the middle of the intensive care ward, Rita slid down to the floor. Her head thrown back, she appealed to the fluorescent lights: "Nu, Tapkachka, what is going to become of us?"

99 I looked up from my feet and saw horror and bewilderment on the doctor's face. She tried to put a hand on Rita's shoulder, but Rita violently shrugged it off.

100   My father attempted to intercede.

101   "Nu, Rita Borisovna, I understand that it is painful, but it is not the end of the world."

102   "And what do you know about it?"

103   "I know that it must be hard, but soon you will see … Even tomorrow we could go and help you find a new one."

104   My father looked to my mother for approval, to ensure that he had not promised too much.

105   "A new one? What do you mean a new one? I don't want a new one. Why don't you get yourself a new son? A new little liar? How about that? New. Everything we have now is new."

106   On the linoleum floor, Rita keened, rocking back and forth. She hiccupped, as though hyperventilating. Pausing for a moment, she looked up at my mother and told her to translate to the doctor. To tell her that she would not let Tapka die.

107   "I will sit here on this floor forever. And if the police come to drag me out, I will bite them."

108   "Ritachka, this is crazy."

109   "Why is it crazy? My Tapka's life is worth more than fifteen hundred dollars. Because we don't have the money, she should die here? It's not her fault."

110   Seeking rationality, my mother turned to Misha. Misha, who had said nothing all this time except "My dear God."

111   "Misha, do you want me to tell the doctor what Rita said?"

112   Misha shrugged philosophically.

113   "Tell her or don't tell her, you see my wife has made up her mind. The doctor will figure it out soon enough."

114   "And you think this is reasonable?"

115   "Sure. Why not? I'll sit on the floor too. The police can take us both to jail. Besides Tapka, what else do we have?"

116   Misha sat on the floor beside his wife.

117   I watched as my mother struggled to explain to the doctor what was happening. With a mixture of words and gesticulations, she got the point across. The doctor, after considering her options, sat down on the floor beside Rita and Misha. Once again she tried to put her hand on Rita's shoulder. This time, Rita, who was still rocking back and forth, allowed it. Misha rocked in time to his wife's rhythm. So did the doctor. The three of them sat in a line, swaying together like campers at a campfire. Nobody said anything. We looked at each other. I watched Rita, Misha and the doctor swaying and swaying. I became mesmerized by the swaying. I wanted to know what would happen to Tapka; the swaying answered me.

118   The swaying said: "Listen, shithead, Tapka will live. The doctor will perform the operation. Either money will be found or money will not be necessary."

119   I said to the swaying: "This is very good. I love Tapka. I meant her no harm. I want to be forgiven."

120   The swaying replied: "There is reality and then there is truth. The reality is that Tapka will live. But let's be honest, the truth is you killed Tapka. Look at Rita; look at Misha. You see, who are you kidding? You killed Tapka and you will never be forgiven."

PUBLISHED IN
2004

Reader's Response

1. Could you relate to some of the situations described in the story? Which ones?

2. Which character in the story aroused the most sympathy in you? Why?

3. How did you feel at the end of the story?

✂ CLOSE READING

Reread the story, paying close attention to the author's depiction of the various characters and the details he uses to create the mood, atmosphere and meaning.

1. Who is telling the story? How old is the narrator at the time of the events in the story?

_____

2. What is the setting of the story?

_____

3. Who are the Nahumovskys and why do they rely heavily on the narrator's mother?

_____

4. Who is Tapka and where is she at the beginning of the story?

_____

_____

5. Why does the relationship between the narrator's mother and the Nahumovskys deteriorate?

_____

6. Where does the narrator go every evening?

_____

7. What responsibility do the Nahumovskys give the narrator and his cousin in April?

_____

8. Why do the narrator and Jana gradually change their way of looking after Tapka?

_____

_____

9. What happens when the children call Tapka insulting names and what do they conclude?

_____

_____

10. How does the accident happen and who is responsible for it?

_____

_____

11. How does the veterinarian communicate with the Nahumovskys?

_____

12. Why is Tapka so important to the Nahumovskys?

_____

13. How do the Nahumovskys convince the veterinarian to save Tapka even though they are unable to pay the fee?

_____

14. Does the story have a happy ending? Explain.

_____

15. a) How have the narrator's language skills evolved since the period depicted in the story?

_____

   b) What conclusion can you draw from this concerning the narrator's integration into his new society?

_____

✎ WRITER'S CRAFT

Dramatic Structure, Narration and Point of View

1. What point of view is used to tell the story?

_____

2. Does the narrator tell the story through the eyes of a child, an adult or both? Find one or more examples in the story to support your answer.

_____

_____

3. Is the dramatic structure of the story traditional or innovative? Indicate the points of rising tension, climax and falling tension.

_____

_____

> The last paragraph of the story is an **epiphany** for the narrator—that is, it represents a sudden, meaningful insight about life. This makes the reader stop and re-evaluate the nature of the story, which seemed fairly simple and straightforward up to this point.

Setting

1. How does the author use the specific setting in the first paragraph to help define the difference in social status between the narrator's family and the other Russian immigrants?

_____

2. In what month and year does the story begin? _____

3. What kind of life do the narrator, his family and the Nahumovskys have at this point in time? What are the main reasons for this?

_____

_____

4. What kind of atmosphere is created by the author's descriptions of the month of April?

_____

_____

5. How is this atmosphere reflected in the way the adults view the children?

_____

_____

6. What spring-related metaphor depicts the narrator's progress in learning English two months after the story begins?

_____

_____

7. What kind of language do the narrator and his cousin start using with each other at this time?

_____

_____

8. How do Tapka and the natural world react when the children start calling the dog swear words?

_____

_____

9. How does the narrator feel about referring to the dog with profanities?

_____

_____

> Bezmozgis uses specific details concerning the setting to create atmosphere, to reveal the theme and characters and to add realism to the story.

Characterization

> The central character of the story is a small, white Lhasa-Apso named Tapka. Notice how the author brings out the other characters in the story through their relationships with Tapka.

1. a) Why does the narrator's mother dislike Tapka and what does she conclude about Rita and Misha?

_____

_____

b) What does this reveal about her personality, values and perception of social status?

_____

2. How does the narrator relate to Tapka at first?

_____

3. a) What does the use of profanity reveal about the change in the way the narrator and Jana relate to Tapka?

_____

b) Why do you think the narrator feels ill at ease about referring to Tapka with profanity but Jana does not?

_____

4. a) When they find out how much the operation will cost, how does the narrator's father try to comfort the Nahumovskys?

_____

b) What does this reveal about him?

_____

5. a) How important is Tapka to Rita and Misha? Support your answer with proof from the story.

_____

_____

b) How do they communicate this importance to the veterinarian?

_____

c) What is the veterinarian's response and what does this reveal about her?

_____

Style and Literary Devices

David Bezmozgis uses very succinct images and precise, incisive language to tell the story. His writing has been compared to that of Chekhov.

■ IMAGERY

1. Which sense is dominant in the story? _____

Give two particularly effective examples from the text.

_____

_____

2. a) What other senses does the author appeal to?

_____

b) Give an example from the story for each sense.

_____

_____

3. Find an example of a

a) metaphor (other than the one you gave in question 6 of the "Setting" section above).

_____

b) simile.

_____

■ DICTION

1. How would you describe the sentences in this story? (Circle the correct answer.) They are

a) mostly long.   c) fragments and run-ons.

b) mostly short.   d) of varying lengths and types.

2. The language is mainly

a) formal.   b) informal.   c) concrete.   d) abstract.

3. How does Bezmozgis make the reader more aware of the fact that the characters in the story are Russian? Provide three examples from the text.

_____

_____

_____

4. Which of the following adjectives characterize the descriptions of spring in the story?

a) long and drawn out   b) short and dense   c) flowery

5. Give an example of humour from the text.

_____

6. How would you characterize the author's use of humour? It is

a) subtle.   b) understated.   c) sarcastic.   d) slapstick.

■ SYMBOLISM

1. What do you think the colour red symbolizes for the narrator when he first sees Tapka's blood in paragraph 66?

_____

2. What do you think the leash symbolizes? (Reread paragraph 23.)

_____

3. Reread the last paragraph. What do you think Tapka symbolizes?

_____

Theme

This story contains themes about immigration, responsibility and language and communication—among others. Succinctly state **two** of the themes. (Remember, a theme has to show an opinion or a value judgement.)

_____

_____

12

VOCABULARY DEVELOPMENT

Vivid Present Participles and Gerunds ("-ing" Words)

Read the words in their context in the story. Write the infinitive of the verb in the second column and then match the word with its meaning in the third column.

Participle/Gerund	Infinitive	Meaning	
1. **flapping** clotheslines	_____	_____	a) to carry
2. **scheming** Odessa (Ukraine)	_____	_____	b) to move jerkily and involuntarily
3. **reeking** borscht (beet soup)	_____	_____	c) short, loud breathing
4. **bearing** the germs	_____	_____	d) to be filled; to be great in amount
5. **abounding** enthusiasm	_____	_____	e) to compete
6. **cradling** animated fur	_____	_____	f) to hug; to break into little pieces
7. our … footsteps elicited **panting**	_____	_____	g) to wave; to flutter in the wind
8. an ecstasy of **wagging**	_____	_____	h) to exert pressure on
9. **vying** for her favour	_____	_____	i) to leap; to jump
10. **overwhelming** love	_____	_____	j) to give off a strong odour
11. **squeezing** too hard	_____	_____	k) to submerge
12. **crushing** her little bones	_____	_____	l) to bring; to draw out
13. **bounding** happily	_____	_____	m) to hold protectively (like a baby)
14. **caterwauling** of tires	_____	_____	n) to spread messily
15. a hind leg **twitching**	_____	_____	o) to plot; to plan
16. **eliciting** sympathy	_____	_____	p) to screech; screeching
17. **smearing** her mascara	_____	_____	q) to move from side to side

Wrap-Up Activities

1. How does the author bring out the characters in "Tapka"?

2. What does the author seem to be saying about the relationship between language and communication in "Tapka"?

3. Discuss how the author portrays certain aspects of the immigrant experience in "Tapka."

4. Compare "Tapka" to another story from *Natasha and Other Stories*.

5. Discuss the psychological realism of the characters in "Tapka."

6. Explain the meaning of the story's ending.

   > The swaying replied: "There is reality and then there is truth. The reality is that Tapka will live. But let's be honest, the truth is you killed Tapka. Look at Rita; look at Misha. You see, who are you kidding? You killed Tapka and you will never be forgiven."

7. Discuss the author's use of imagery and symbolism in "Tapka."

"The New Food," "Bread" and "Old Habits Die Hard"

CHAPTER
CONTENTS

▸ **Reading:** "The New Food" by Stephen Leacock; "Bread" by Margaret Atwood; "Old Habits Die Hard" by Makeda Silvera; authors' biographies; literary trends (genre of the short short story and experimentation)

▸ **Literary Elements:** style; diction (figurative devices: similes, metaphors, personification and imagery, and rhythm, alliteration, assonance, consonance and repetition); level of language; syntax; tone; attitude; different literary styles

▸ **Vocabulary Development:** denotation and connotation

▸ **My eLab Activity:** creative writing activity: flash fiction

In this chapter, we will approach the short story from a new angle. To complete our study of the various elements comprising a traditional analysis of a short story, we will compare and contrast the styles used to write three rather unique—and very short—short stories: "The New Food" by Stephen Leacock, "Bread" by Margaret Atwood and "Old Habits Die Hard" by Makeda Silvera.

ABOUT
THE **AUTHORS**

Stephen Leacock (1869-1944) was considered to be the best-known humorist in the English-speaking world between 1915 and 1925. Leacock was a professor of economics and political science at McGill University in Montreal, but he is best remembered for his witty, satirical short stories. He also wrote literary essays and articles on social issues. Many of his works were first published in magazines and later collected in more than sixty books. *Sunshine Sketches of a Little Town* (1912) and *Arcadian Adventures with the Idle Rich* (1914) are two of Leacock's most famous works.

Margaret Atwood, born in Ottawa in 1939, is one of Canada's most widely read authors. A prolific author of more than twenty-five literary works, she has received international recognition for her novels and poetry, and has also written several children's books and television scripts. Atwood, who has received many prestigious awards and honours, including the Governor General's Award, the Centennial Medal from Harvard University, The Sunday Times Award for Literary Excellence in the U.K., as well as numerous honorary doctorates, was also elected Chevalier de l'Ordre des Arts et des Lettres en France. Her recent works include *Morning in the Burned House* (1995), *Alias Grace* (1996), *Oryx and Crake* (2003), *The Tent* (2006) and *The Penelopiad: The Myth of Penelope and Odysseus* (2006).

13

Makeda Silvera was born in Jamaica in 1955 and came to Canada in 1967. She is very active in developing and promoting the writing of women of colour. She has edited and written material for many anthologies, while publishing collections of her own short stories: *Remembering G* (1991), *Her Head a Village* (1994) and *The Heat Does not Bend* (2002). Much of Silvera's work deals with the issue of being the "other" in society—that is, being a Black, a lesbian, an immigrant, a member of the working class or an elderly person.

THE GENRE OF THE
SHORT SHORT STORY

SHORT STORIES VARY WIDELY IN LENGTH. Many twentieth-century writers experimented with extremely short versions of the genre. In spite of their brevity, these stories can be quite complex and often have a powerful impact on the reader. Although experimental, many follow in the realist and romantic traditions. Some deal succinctly with important social issues, while others take a more fanciful or poetic approach to the universal human experience.

Warm-Up Discussion Topics

1. Could new techniques that allow for faster food preparation—or "instant food"—benefit people today?

2. What does the word *bread* bring to mind? Brainstorm five associations.

3. What do you think your life will be like when you are an elderly person? What kind of conditions do you hope for?

∞ INITIAL READING

Each of the following stories deals with a social issue that is relevant today. However, each story has a unique style and approach. Read and savour each one before continuing on to the next. Pay particular attention to the writing style that the author has chosen to use.

The New Food
Stephen Leacock

1 I see from the current columns of the daily press that "Professor Plumb, of the University of Chicago, has just invented a highly concentrated form of food. All the essential nutritive elements are put together in the form of pellets, each of which contains from one to two hundred times as much nourishment as an ounce of an ordinary article of diet. These pellets, diluted with water, will form all that is necessary to support life. The professor looks forward confidently to revolutionizing the present food system."

2 Now this kind of thing may be all very well in its way, but it is going to have its drawbacks as well. In the bright future anticipated by Professor Plumb, we can easily imagine such incidents as the following:

3 The smiling family were gathered round the hospitable board. The table was plenteously laid with a soup plate in front of each beaming child, a bucket of hot water before the Christmas dinner of the happy home, warmly covered by a thimble and resting on a poker chip. The expectant whispers of the little ones were hushed as the father, rising from his chair, lifted the thimble and disclosed a small pill of concentrated nourishment on the chip before him. Christmas turkey, cranberry sauce, plum pudding, mince pie, it was all there, all jammed into that little pill and only waiting to expand. Then the father with deep reverence, and a devout eye alternating between the pill and heaven, lifted his voice in a benediction.

4 At this moment there was an agonized cry from the mother.

5 "Oh, Henry, quick! Baby has snatched the pill!" It was too true. Dear little Gustavus Adolphus, the golden-haired baby boy, had grabbed the whole Christmas dinner off the poker chip and bolted it. Three hundred and fifty pounds of concentrated nourishment passed down the oesophagus of the unthinking child.

6 "Clap him on the back!" cried the distracted mother. "Give him water!"

7 The idea was fatal. The water striking the pill caused it to expand. There was a dull rumbling sound and then, with an awful bang, Gustavus Adolphus exploded into fragments!

8 And when they gathered the little corpse together, the baby lips were parted in a lingering smile that could only be worn by a child who had eaten thirteen Christmas dinners.

PUBLISHED IN
1910

Bread
Margaret Atwood

1 IMAGINE a piece of bread. You don't have to imagine it—it's right here in the kitchen, on the bread board, in its plastic bag, lying beside the bread knife. The bread knife is an old one you picked up at an auction; it has the word BREAD carved into the wooden handle. You open the bag, pull back the wrapper, cut yourself a slice. You put butter on it, then peanut butter, then honey and you fold it over. Some of the honey runs out onto your fingers and you lick it off. It takes you about a minute to eat the bread. This bread happens to be brown, but there is also white bread, in the refrigerator, and a heel of rye you got last week, round as a full stomach then, now going mouldy. Occasionally you make bread. You think of it as something relaxing to do with your hands.

13

2   Imagine a famine. Now imagine a piece of bread. Both of these things are real, but you happen to be in the same room with only one of them. Put yourself into a different room—that's what the mind is for. You are now lying on a thin mattress in a hot room. The walls are made of dried earth and your sister, who is younger than you are, is in the room with you. She is starving, her belly is bloated, flies land on her eyes; you brush them off with your hand. You have a cloth too, filthy but damp, and you press it to her lips and forehead. The piece of bread is the bread you've been saving, for days it seems. You are as hungry as she is, but not yet as weak. How long does this take? When will someone come with more bread? You think of going out to see if you might find something that could be eaten, but outside the streets are infested with scavengers and the stink of corpses is everywhere.

3   Should you share the bread or give the whole piece to your sister? Should you eat the piece of bread yourself? After all, you have a better chance of living; you're stronger. How long does it take to decide?

4   Imagine a prison. There is something you know that you have not yet told. Those in control of the prison know that you know. So do those not in control. If you tell, thirty or forty or a hundred of your friends, your comrades, will be caught and will die. If you refuse to tell, tonight will be like last night. They always choose the night. You don't think about the night, however, but about the piece of bread they offered you. How long does it take? The piece of bread was brown and fresh and reminded you of sunlight falling across a wooden floor. It reminded you of a bowl, a yellow bowl that was once in your home. It held apples and pears; it stood on a

table you can also remember. It's not the hunger or the pain that is killing you but the absence of the yellow bowl. If you could only hold the bowl in your hands, right here, you could withstand anything, you tell yourself. The bread they offered you is subversive; it's treacherous; it does not mean life.

5   There were once two sisters. One was rich and had no children; the other had five children and was a widow, so poor that she no longer had any food left. She went to her sister and asked her for a mouthful of bread. "My children are dying," she said. The rich sister said, "I do not have enough for myself," and drove her away from the door. Then the husband of the rich sister came home and wanted to cut himself a piece of bread; but when he made the first cut, out flowed red blood.

6   Everyone knew what that meant.

7   This is a traditional German fairytale.

8   The loaf of bread I have conjured for you floats about a foot above your kitchen table. The table is normal; there are no trap doors in it. A blue towel floats beneath the bread, and there are no strings attaching the cloth to the bread or the bread to the ceiling or the table to the cloth—you've proved it by passing your hand above and below. You didn't touch the bread, though. What stopped you?

9   You don't want to know whether the bread is real or whether it's just a hallucination I've somehow duped you into seeing. There's no doubt that you can see the bread; you can even smell it; it smells like yeast, and it looks solid enough, solid as your own arm. But can you trust it? Can you eat it? You don't want to know, imagine that.

PUBLISHED IN
1983

Old Habits Die Hard
Makeda Silvera

1 OLD man, skin scaly tree bark to touch. Rust eyes, water hazy. The iron is gone. Legs, arms, ready kindling. Bedbug. Bedridden. Bedlam. Bedpan. Bedraggled. Bedfast.

2 Faeces don't give ear to him anymore. Old man in diapers. Old man in white gown. Mashed potatoes with milk is all he can eat. Old man needs steady hand to feed him. Out of habit, old woman folds clean, neatly ironed pyjamas. Clean towel. Wash rag. Enamel carrier filled with mashed potatoes.

3 Disordered eyes. Looks past visors. Old man recollects just one, old woman. The others bear no memory. Disappointed, you can see it on their faces, the tight turn of the lips, the begging in the eyes. Talk to us. Touch us. Remember us. He only sits, no teeth to his grin. Old man looks and looks. Memory escapes. No longer father, husband, grandfather, uncle, brother, friend.

4 Old man pulls toward old woman. Grab him, he'll shit, piss on the floors, run around like a madman, a bedlamite. The visitors approve of the restraint. We love him, they say. Old man

wants to run, old man wants to go home. The visitors go. Room too depressing: some stringy flowers in a mug, a plastic balloon the only grace, a heavy curtain shuts out the light.

5 Old woman stays behind. She feeds him potatoes, eggs, milk through a straw. She talks to him. He cannot answer. She tells him things, answers for him. His hands are cool. She pulls the blankets closer to his body. His face sweet like dark plums. Time to leave. Keepers in white come to lead her out. She kisses old man. Water in his eyes. He stares. He stares. Night is a black sheet. Old man pass away, old man dead, old man gone. She had felt it. Hands cool, getting cold, heat leaving the face, purple turning black, eyes turning.

6 The mourners come, eat, sing, cry, drink, help to bury him. They go home. Old woman must bury him a second time: clothes to give away—Salvation Army, Goodwill; mattress to turn over; bank account to settle; pension to straighten out. One pot to cook, one mouth to feed. Out of habit, old woman does the wash, folds her nightgown. She always irons it. Washes towels, washes rags, folds them. Those go into the suitcase. Changes her bed sheets. Best pillowcase; lovely lace, that. Lies down. Pulls up the black sheet of night.

PUBLISHED IN
1994

Reader's Response

1. Which of the three stories did you prefer? Why?

_____

∞ CLOSE READING

Answer the following questions.

1. Identify the type of narration (first person or third person) used in each story. (*Hint*: two of the stories contain two types of narration because they contain a story or stories within the main story.)

 a) "The New Food" _____

 b) "Bread" _____

 c) "Old Habits Die Hard" _____

2. In which story does the narrator ask the reader directly to imagine a scene or scenes? (Circle the correct answer.)

a) "The New Food"   b) "Bread"   c) "Old Habits Die Hard"

3. In which story is the narrator most distant from the reader?

a) "The New Food"   b) "Bread"   c) "Old Habits Die Hard"

4. Explain your answer to the previous question.

_____

_____

_____

5. Which story needs to assume that the reader comes from a privileged society, such as middle-class North America, in order to have the greatest impact?

a) "The New Food"   b) "Bread"   c) "Old Habits Die Hard"

Give details from the story to support your answer.

_____

_____

6. Match each story with one of its main themes.

"The New Food"   _____   a) Nothing is left in the lives of old couples as they await death.

"Bread"   _____   b) People can use their imaginations to imagine anything.

"Old Habits Die Hard"   _____   c) Technological advances are not necessarily positive.

d) Love and compassion are very much alive in banal daily actions.

e) New technology can transform people's lives in many ways.

f) Materially well-off people take important aspects of their lives for granted.

✂ WRITER'S CRAFT

Style

> The essence of literature lies in its distinct use of language for a specific purpose. Each writer has his or her unique **style** of writing, and it is this style that creates the overall effect of the literary work on the reader. Writers often craft short short stories by paying particular attention to style. In such stories, the style of the piece is intrinsically connected with the theme.

The three short stories in this chapter have very different styles. Each serves to illustrate the importance of style in enhancing meaning and aesthetic enjoyment. Refer to the Literary Elements section at the beginning of this book (as well as the glossary) in order to understand the various aspects of style that appear as headings in the following section.

When analyzing the style of a work, two aspects are usually taken into consideration: **diction** (choice of words) and **syntax** (arrangement of words, phrases and sentences). The questions that follow in this section are designed to help you learn to analyze these aspects in order to write convincingly about them. Moreover, you may wish to apply some of the concepts to your own writing.

■ DICTION

Figurative Devices: Metaphors and Imagery

All three stories use metaphors and powerful imagery to create meaning. (Refer to the glossary for a definition of these terms.) "Bread" and "Old Habits Die Hard" literally abound in figurative devices. Note how these aspects of style draw readers into the story, engaging both their imagination and interest, thus adding impact to the theme(s) of the work.

1. How might "the new food" in Leacock's story be seen as a metaphor for technological development?

   _____

2. a) Briefly explain why imagery is an important aspect of "Bread."

   _____

   b) Give an example of an image that you found particularly effective in "Bread."

   _____

3. How might "Bread" be seen as a metaphor for fiction or a work of art?

   _____

   _____

4. a) Quote the metaphor concerning night that appears twice (with slightly different syntax) in "Old Habits Die Hard."

   _____

   b) What do you think this metaphor signals the first time it appears?

   _____

   c) What do you think it signifies at the end of the story?

   _____

Rhythm: Alliteration, Assonance, Consonance and Repetition
(Refer to the glossary for definitions of these terms.)

1. Which two stories utilize very rhythmical language?

   a) "The New Food"   b) "Bread"   c) "Old Habits Die Hard"

2. Find six words from the first paragraph of one of the stories to exemplify how the author exploits several different rhythmical devices both cleverly and effectively.

   _____

   Note how the author cleverly depicts the essence of the person's situation in just two lines at the beginning of "Old Habits Die Hard." Although these lines are presented as prose, they have a very poetic feel to them. This poetic quality characterizes the entire piece.

Level of Language

1. Which story is written in a journalistic, factual style at the beginning?

   a) "The New Food"   b) "Bread"   c) "Old Habits Die Hard"

2. Which story is written as a conversation (using fairly informal language)?

   a) "The New Food"   b) "Bread"   c) "Old Habits Die Hard"

3. Which story contrasts formal and informal levels of language and includes dialogue to add both humour and realism to the piece?

   a) "The New Food"   b) "Bread"   c) "Old Habits Die Hard"

■ SYNTAX

1. Which story utilizes short sentences and sentence fragments (sentences that lack a main verb and are not a complete idea)?

   a) "The New Food"   b) "Bread"   c) "Old Habits Die Hard"

2. Explain how this "bared-to-the-bone" writing style fits the characters and situations depicted in the story.

   _____

   _____

3. What can the reader conclude about the essence of the relationship between the two main characters in this story?

   _____

   _____

4. Read the last paragraph of "Bread."

   a) What do you notice about the length and grammatical construction of the second sentence?

   _____

   b) Comment on the length and kinds of sentences that follow the second sentence of the last paragraph.

   _____

   > In "Old Habits Die Hard," Makeda Silvera consistently uses short sentences and sentence fragments. This syntax fits the characterizations and essence of the story. In "Bread," Margaret Atwood uses long sentences, which transport the reader to another world, with short, troubling questions that bring the reader back to earth, creating an uneasiness that permeates the story.

■ TONE

1. The three stories all describe awful situations. Which story does this using exaggeration and irony?

   a) "The New Food"   b) "Bread"   c) "Old Habits Die Hard"

2. What effect does this have on the tone of this story?

   _____

3. Which story has a general tone of

a) sadness?　　　　　_____

b) uneasiness?　　　 _____

■ ATTITUDE

1. In which story does the attitude of the author seem clear and unambiguous?

a) "The New Food"　　　 b) "Bread"　　　 c) "Old Habits Die Hard"

2. How does this attitude fit the overall style of the piece?

_____

_____

VOCABULARY DEVELOPMENT

Denotation and Connotation

Denotation refers to the dictionary meaning of a term. **Connotation** refers to additional implications of the word's meaning.

1. What is the predominant connotation of "scientific progress" in most people's minds?

a) positive　　　　　 b) negative　　　 c) neutral

2. Which story was written to refute this connotation?

a) "The New Food"　　　 b) "Bread"　　　 c) "Old Habits Die Hard"

3. Which word in the second paragraph of the story chosen above indicates that the narrator is planning to refute this?

_____

4. What are the connotations concerning "bread" that are developed in the middle and at the end of the story? Complete each phrase with qualitative adjectives. Then state whether the connotations are positive, negative or both. (See the following example for the beginning of the story.)

Examples
Beginning: middle-class kitchen—connotations concerning bread: common, everyday and abundant; taken for granted; something we can waste; baking bread is a pleasurable activity and not a necessity—positive or neutral connotations.

Middle: famine situation—connotations concerning bread: _____

_____

prison situation—connotations concerning bread: _____

_____

German fairy-tale—connotations concerning bread: _____

_____

End: middle-class kitchen—connotations concerning bread: _____

_____

5. What possible effects do these contrasting connotations have on the reader by the end of the story?

_____

Wrap-Up Activities

1. Discuss how the style of Leacock's story differs from that of one of the other stories.

2. Explore various interpretations and layers of meaning in the story "Bread."

3. Discuss whether each of the stories is part of the realist tradition.

4. Discuss another story by one of the authors.

5. Choose one of the short stories in this chapter and write about how the author's style infuses meaning and aesthetic enjoyment in the story.

6. Analyze how the narrator in "Bread" manipulates the reader.

7. Discuss the importance and role of the reader's imagination in "Bread."

8. Explain the techniques Silvera used to breathe life into the two main characters in "Old Habits Die Hard."

9. Compare the writing styles of two authors as illustrated by the short stories in this chapter.

10. Write about the relevance of the theme in one of the short stories in this chapter.

11. Write a short story in which you experiment with some of the aspects of style.

My eLab ACTIVITY

My eLab 📂

Read about flash fiction and consult some sample stories. Follow the proposed steps and write a flash fiction story not longer than 350 words.

POETRY

After a brief summary of poetic elements and the evolution of poetry from a historical perspective, Chapter 14 introduces traditional approaches for analyzing several older works. Chapter 15, by contrast, uses other types of approaches to explore more contemporary poems.

Poetic Elements

Poetry, like prose, may be enjoyed in many different ways. Some people like to listen to poems as triggers for their own ideas or writing. Others enjoy the imaginative way in which the poet plays with words and creates images that evoke meaning. Still others enjoy reading poems for the musicality of the language, the creativity of the words the poet has chosen and the special rhythms they work together to create.

Through the activities in this section, you will be invited to communicate your thoughts and feelings about particular poems, both orally and in writing. Sometimes you will be asked to give a personal response to a poem; other times you will be expected to use a more traditional analytic approach, looking at various poetic aspects, such as the poet's use of imagery, meter, rhyme, and so on.

At this point, you may ask: what is the purpose of analyzing a poem? In fact, there are several purposes. Poets are highly skilled craftspeople. Their imaginative insights and honed language skills enable them to communicate important ideas and feelings in striking ways. Not only do poems illustrate the most beautiful elements in a language, they also make you look at things in new ways. Examining how a poet uses words heightens your own awareness of language and its possibilities, helping you become a better communicator and writer.

Every well-written poem is like a skilfully cut diamond: every facet is polished and perfect, standing at just the right angle. There is nothing extra in the poem; there is nothing out of place. A well-crafted poem, like a diamond, is unique and complete unto itself. One of your tasks in the analysis of poetry is to reveal the artistry and craft underlying a particular poem. To do this effectively, you need to look at how the poet uses language and imagery; then you need to describe this particular usage with appropriate language. Certain terms describing poetry date back to the ancient Greeks. These terms—and others that have been added over the years—are presented and explained in the following paragraphs.

Traditionally, you use certain language to talk about poetry. A **poem** consists of a series of **lines** (sometimes referred to as "verses") presented on a page. It usually has both a visual and a rhythmical pattern, but not always. A poem is often divided into sections called **stanzas**, which are composed of two or more lines; these sections often have similar patterns. A **refrain** is a line, phrase or stanza that is repeated at intervals throughout the poem (or song). The word **verse** refers to the fact that a line of poetry usually has a particular rhythm and/or rhyme. (Please note that it is preferable to use the word *stanza* to refer to the different divisions of a poem, but in a song, this is not the case.) Finally, a poem often contains a **persona**, which is the voice or speaker in the poem. Sometimes the persona represents the voice of the poet, but often the persona is a fictional character.

Poetry can be analyzed from a number of different approaches: traditional, formalist, reader response, historical, sociological, structuralist, feminist, minority, Marxist and so on. In all of these approaches, the analyst pays attention to two basic elements: 1) **theme** (the significant or overall meaning of the poem); and 2) **mood** (the overall feeling or atmosphere of the poem that the poet creates and the reader experiences).

To illustrate the elements of poetry, let us look at Alfred Edward Housman's "When I Was One and Twenty." Our illustration comprises three parts: Reading and Responding to the Poem, Examining Poetic Devices and Writing about Poetry. Each part includes at least one exercise.

∽ READING AND RESPONDING TO THE POEM

Form small groups. Then have somebody in your group read the poem aloud, or listen to the recorded version in My eLab Documents. Finally, discuss your reactions as a group using the questions provided as a guideline.

When I Was One-and-Twenty
A. E. Housman

When I was one-and-twenty
I heard a wise man say,
"Give crowns and pounds and guineas
But not your heart away;
Give pearls away and rubies
But keep your **fancy**[1] free."
But I was one-and-twenty,
No use to talk to me.

When I was one-and-twenty
I heard him say again,
"The heart out of the bosom
Was never given in vain;
'Tis paid with sighs a plenty
And sold for endless **rue**[2]."
And I am two-and-twenty,
And oh, 'tis true, 'tis true.

1. Fancy: love or romantic attachment
2. Rue: sorrow or remorse

PUBLISHED IN
1896

EXERCISE 1 Responding to the Poem

1. What did you like about the poem? Give specific examples from the poem.

_____

_____

2. What did you not understand about the poem? Be specific.

_____

_____

3. Paraphrase the poem in your own words and then discuss possible interpretations of the poem with your group. Ask your instructor for help if there are still some aspects of the poem (for example, vocabulary or theme) that you did not understand.

4. How does the title set the stage for the rest of the poem?

_____

_____

5. Who is the persona (or speaker) in the poem and to whom is he or she speaking?

_____

6. What kind of advice does the speaker in the poem give the reader?

_____

7. Does the speaker take his own advice? How do you know?

_____

_____

8. Is the speaker's advice relevant to today's youth? Why or why not?

_____

_____

∞ POETIC DEVICES

Diction

Poets use poetic devices to help create the theme and mood of a poem. Many of these devices are common to all literary genres, so you are likely to be familiar with some of them already, especially since many appear in a chart in the Literary Elements section (p. 13). However, there are several additions that should prove particularly useful when studying poetry: suggestion, apostrophe, hyperbole, antithesis, oxymoron, paradox, pun and symbolism. Read their definitions carefully and study the examples in the following table. Then complete the exercise that follows.

Diction

Diction: refers to a writer's choice and use of words and to the words' characteristics.

Denotation: the literal or dictionary meaning, for example, a house

Connotation: the associated or suggested meaning, for example, a home (connotations: warmth, comfort, safety, and so on.)

Concreteness: perceptible by the senses, for example, a book

Abstractness: not perceptible by the senses, only by the mind, for example, generosity

Figures of speech: expressions used in a non-literal way to create special effect or extend meaning
    Simile: a comparison of two objects using "like" or "as," for example, "The boat is as big as a whale."
    Metaphor: an implied comparison of two objects without using "like" or "as," for example, "My love is the universe."
    Personification: giving human qualities to inanimate or abstract objects, for example, "The boat smiled into the wind."
    Apostrophe: used to address a dead person as if he or she is still alive, for example, "Shakespeare, you are a rare poet."
    Hyperbole: an exaggeration, for example, "I'm as hungry as a horse."
    Oxymoron: a juxtaposition of two ideas that reflect opposite concepts, for example, "cruel kindness"

Level of language
    Formal, for example, "child"
    Informal, for example, "kid"
    Slang, for example, "squirt"

Imagery: descriptive word pictures usually related to one or more of the five senses, for example, "The sand on the beach shone silver and gold."

Suggestion: the choice of evocative words to enhance meaning, for example, referring to "my home" rather than "my house"

Symbolism: the use of an object or action to signify more than its literal meaning, for example, a rose symbolizing love

Antithesis: linking two contrasting ideas in order to make each one more vivid, for example, "Believing nothing or believing all" (Dryden)

Paradox: a statement that seems to be contradictory and true at the same time, for example, "The silence of midnight, to speak truly, though apparently a paradox, rung in my ears." (Mary Shelley)

Irony: the opposite to what is intended or expressed, for example, "Experience is something you don't get until just after you need it."

Pun: a play on words creating two different meanings simultaneously, for example, "The bottom is part of the 'hole.'"

Let us examine the diction the poet uses in the poem "When I Was One-and-Twenty."

1. What kind of diction does Housman use in this poem? Is the language
   a) concrete?   b) abstract?

2. Is the level of language...
   a) formal?   b) informal?
   (*Hint*: think about when the poem was written and look at the length of the words.)

3. a) What metaphors does the poet use regarding the heart?

   _____

   b) What kind of value do these metaphors bestow upon the heart?

   _____

4. Cite two words in the second stanza that are connected to the imagery used in the third and fifth lines of the first stanza.

   _____   _____

   > Note how this choice of vocabulary extends the metaphor and how the poet places the metaphor so that it contrasts with the idea of "giving away one's heart." This enhances the theme of the poem and adds to its unity.

5. What is ironic about the last two lines of the second stanza?

   _____

   > Housman uses irony to point to a basic "truth" about life.

Structure, Form, Sound and Rhythm

Next, let us look at other aspects of poetry that bring out the meaning and theme of a poem by creating atmosphere and mood. The **structure** of a poem can be based on its form, its theme or an interaction of the two. **Formal structure** represents the way a poem is divided into stanzas and how the stanzas fit together. **Thematic structure** represents the way the main idea or theme of the poem is presented. Similar to a plot in a short story, the theme is revealed through the way in which conflicts, ambiguities, tensions or uncertainties are presented and then resolved in the poem.

The **form** of a poem refers to its overall design (the look of a poem on a page) and its patterns (rhythm, meter and stanzas). A poem can have a highly structured (or closed) form or an open form with few or no apparent patterns. The form of a poem influences our perception of it and the mood it creates. For example, the form may be a traditional, four-line verse with lines of equal length, or it may have a very special shape on the page. Some common traditional forms are the **couplet**, consisting of two consecutive lines of verse that rhyme and are approximately the same length, and the **quatrain** (or four-line stanza). Others include **tercets** (three-line stanzas), **sestets** (six-line stanzas) and **octaves** (eight-line stanzas). Some poets, such as e. e. cummings, also experiment with capitalization, or the lack thereof, and punctuation; these visual factors influence how a reader experiences a poem.

Common types of closed poems include the **sonnet** and the **sestina**. A **ballad** is a narrative, folk poem. A **lyric** is a short poem similar to a song and expresses an idea or emotion. An **ode** is a more serious, longer poem on a noble subject. An **epic** is a long poem about someone heroic or noble who performs an important action.

Of course, among the most important aspects of poetry are the actual **sound** of the words and the underlying **rhythms** the poet creates. The sound and rhythm of the words, phrases and sentences often resemble musical patterns. Both serve to create a general mood in the poem. The following table presents some basic sound devices.

Basic Sound Devices

Euphony: words that sound harmonious

Cacophony: words that sound harsh and jarring when used together

Soft and hard consonant sounds: the letters *l, m, n* and *v* produce soft sounds, for example, "love," "longing" and "mother," while the letters *b, p, t, k* and *g* produce harsh sounds, for example, "bucket" and "puck." Soft and harsh sounds influence the **tone** of the poem (the attitude of the poet toward the subject), and the sounds chosen can reflect different degrees of pleasantness and unpleasantness.

Alliteration: a repetition of the initial letter or sound in a series of words, which can emphasize the musicality of the series and the last word in the series, for example, "big, brown bicycle"

Onomatopoeia: a sound that imitates the meaning of words, for example, "buzzing bee"

Rhyme: a repetition of identical sounds in words usually appearing at the end of a poetic line, for example, "The rain in Spain stays mainly on the plain." The **rhyme scheme** refers to the pattern of rhyme in a poem, for example, "a-b-b-a/c-d-d-c" (each letter refers to a different rhyme and the letters are grouped according to the stanzas).

Assonance: a repetition of identical vowel sounds, for example, "The wind is whispering its intentions."

Consonance: a repetition of identical consonant sounds, for example, "Sally saw serpents slithering slowly."

Repetition: a repetition of certain words, phrases or elements in order to create a special effect, for example, "True!—nervous—very, very dreadfully nervous I had been and am!"

When we speak about the **rhythms** of English, we are referring to the patterns of **stressed** and **unstressed** syllables in the spoken language. That is to say, stressed syllables are emphasized or accented in natural speech, whereas unstressed syllables are not emphasized or are unaccented. For example, the word *table* has two syllables. The first syllable, *ta*, is stressed, which is shown by the following accent mark: . The second syllable, *ble*, is unstressed, which is shown by the following accent mark: ˘ (or a dot). The accent marks are placed above the appropriate syllables, for example, "ta ble." (Please note that dictionaries indicate stressed syllables in slightly different ways.)

Poetry uses **meter** (the way in which rhythm is measured) as an intrinsic form of its structure. English poetry measures rhythm or meter in terms of groups of stressed and unstressed syllables called **poetic feet**. Meter is established by the dominant poetic foot, or the stress that is repeated the most often in a line of poetry. English poetry has five main types of stressed and unstressed syllables, or poetic feet.

Poetic Feet			
Name	**Syllables**	**Stress Pattern**	**Example**
Iambic foot *	2	unstressed, stressed	pre pare, re pair
Trochaic foot	2	stressed, unstressed	im age, po et
Anapestic foot	3	unstressed, unstressed, stressed	in hu mane, un der stand
Dactylic foot	3	stressed, unstressed, unstressed	e le phant, cro co dile
Spondaic foot	2	stressed, stressed	short cut, hide out

* The **iambic foot** is considered by many to be the basic rhythmic element in English, so it is not unusual for many English poems to use this form. An iambic foot (or **iamb**) consists of an unstressed syllable followed by a stressed syllable, for example, "pre pare " and "de feat ." It is found in most of Shakespeare's works. The number of feet per line is counted and this determines the metrical pattern. The feet are separated by a slash. The number of poetic feet found in a verse line is referred to by a Greek name. There are a number of different lengths (see table below).

Meter		
Name	**Definition**	**Example**
Monometer	Consists of one foot, but this form is rare in poetry.	The sun
Dimeter	Consists of two feet and is also very rare in poetry.	The sun / shone down.
Trimeter	Consists of three feet.	The sun / shone warm / and bright.
Tetrameter	Consists of four feet.	The sun / shone warm / and bright / on the town.
Pentameter*	Consists of five feet.	The sun / shone warm / and bright / on the town / today.
Hexameter	Consists of six feet (also referred to as an "Alexandrian").	The gold / en sun / shone warm / and bright / on the town / today.
Heptameter	Consists of seven feet and is very rare in poetry.	Again / the gold / en sun / shone warm / and bright / on the town / today.

* The most common type of foot in English poetry is the iambic pentameter, that is, there are five repetitions of unstressed, stressed syllables per line. A good example is Shakespeare's sonnet "XLVI," whose first line is: "Mine eye and heart are at a mortal war ..."

Prior to the twentieth century, most poems in English were written in metrical verse (in other words, the poetry employed rhyme and meter). However, early in the twentieth century, many poets, like other artists, started to experiment with both the form and content of their works. Many began to write in **free verse**, which did not conform to any fixed pattern (in either rhyme or rhythm), but which often had its own subtle rhythms.

Note: Shakespeare wrote his plays in **blank verse**, which has meter but no rhyme scheme. Here is an example from *Richard III* (Act 5, Scene 5) in iambic pentameter.

> My conscience hath a thousand several tongues,
> And every tongue brings in a several tale,
> And every tale condemns me for a villain.

EXERCISE 3 Examining Structure, Form, Sound and Rhythm

Once again, let us return to "When I Was One-and-Twenty" to see how the author uses structure, form, sound and rhythm to his advantage in the poem.

1. Describe the form, rhythm (meter) and rhyme scheme of the poem.

    _____

    > Note how the poet uses the words "one-and-twenty" and "two-and-twenty" (as opposed to "twenty-one" and "twenty-two") in order to keep the rhythm more regular and poetic.

2. What do you notice about the rhyme scheme of the stanzas in terms of similarities and differences?

    _____

3. Quote a line in which the normal grammatical structure (for example, the word order) was changed in order to fit the poetic meter chosen by the poet.

    _____

4. Give one example of each of the following sound devices.
    a) assonance: _____
    b) consonance: _____
    c) alliteration: _____
    d) repetition: _____

5. How would you qualify the tone of the poem?
    a) deadly serious   b) light-hearted   c) happy   d) depressed   e) comical

6. Justify your answer to the previous question.

    _____

7. Is this poem **a)** an ode, **b)** a lyric or **c)** an epic?

8. Is this poem written using **a)** a traditional or **b)** an experimental approach and structure?

    > The use of rhyme, regular rhythms and various poetic devices gives this poem a song-like quality that is in keeping with the poem's theme and mood.

✏ WRITING ABOUT POETRY

Now read this sample essay about "When I Was One-and-Twenty." It is an example of an interpretation and analysis of a poem that you can use as a model. You will see that it combines and organizes into a whole much of the information from the questions you answered in previous sections.

1   "When I Was One-and-Twenty" is a lyric poem written by A. E. Housman in which the speaker or persona, a young man in his early twenties, received some good advice from a "wise" person. This wise, and we can assume "older," person told him to never give his heart away. The speaker did not listen to the advice, even when he was told a second time, and now that he is a year older, he is feeling very sad and remorseful. We can surmise either that the person he fell in love with did not feel the same about him or that the love affair didn't last very long. Either way, it was not worth the trouble and pain.

2   This poem touches on two universal themes: the first is the fact that often young people do not listen to the advice of an older, more experienced person. Parents always seem to be complaining of this, one of the basic ironies of life! The second theme is that a person's love is very valuable. To show the value of love, the poet uses the metaphors of, first, currency—"crowns and pounds and guineas"—and, then, jewels—"pearls and rubies"—both of which are considered very valuable. The wise man states that it would be better to give these very valuable items away than to give away one's heart. The metaphors are extended into the second stanza when the wise man uses the words "paid" and "sold" regarding what happens when one gives one's heart away; that is, one pays for it dearly. The extended metaphors serve to add unity to the poem as well as to enhance the meaning. Moreover, the contrast is made between concrete and material things—money and jewels—and something abstract—love—that one can't buy. The wise man seems to be saying, in effect, that it is worse to suffer because of lost or misplaced love than to be poor.

3   The poem has a traditional form consisting of two stanzas of eight lines of iambic trimeter and a rhyme scheme: *abcbcdad aefeagag*. The lines are short and lyrical and fit well with the theme which is treated in a light-hearted way. The language in the poem is concrete and informal but much of it seems old-fashioned for present-day readers. For the most part, it is based upon simple, conversational language that Housman has transformed into a poem using several poetic techniques. First of all, in the title he reverses "twenty-one" to read "one-and-twenty," which has a better rhythmic fit and sounds musical. "When I Was One-and-Twenty" is repeated at the beginning of each stanza and then transformed into "And I am two-and-twenty," in the second last line. In the last line, the words "'tis true" are repeated and this also is like a song. The use of alliteration— "'tis true,"—consonance—"two-and-twenty"—and assonance—"crowns and pounds"—also adds to the musicality of the language. In conclusion, the regular rhythms and use of rhyme and various poetic devices give this poem a song-like quality that is in keeping with the poem's themes and light-hearted mood. Although this poem was written over a century ago, people today can still relate to the themes and appreciate the ironies expressed in this poem.

EXERCISE 4 Verifying Knowledge of Terms

Indicate in which paragraph of the sample interpretation above you find mention of the following poetic devices.

a) metaphor   1 - 2 - 3

b) sound effects   1 - 2 - 3

c) identification of the persona   1 - 2 - 3

d)	the title of the poem	1 - 2 - 3
e)	figures of speech	1 - 2 - 3
f)	meter and rhyme	1 - 2 - 3
g)	literal meaning	1 - 2 - 3
h)	diction	1 - 2 - 3
i)	type of poem	1 - 2 - 3
j)	theme	1 - 2 - 3
k)	irony	1 - 2 - 3

USEFUL QUESTIONS FOR WRITING ABOUT POETRY

- What is the meaning and relevance of the poem's title?
- Who is the speaker or persona in the poem and what tone of voice does he or she use?
- Who is the speaker addressing as an audience?
- What is the literal meaning of the poem?
- What aspects of the setting, topic and/or voice seem most important to the poem's meaning?
- What imagery is used in the poem? Is there any symbolism? How are imagery and symbolism used effectively?
- What is the main theme of the poem? Is it implicit or explicit?
- What type of diction is found in the poem? Are the connotations of words important? Are the words abstract or concrete? Is the language formal, informal or slang?
- What is the structure of the poem? How are its parts organized—formally or thematically? How does the structure and organization help develop the theme of the poem?
- Is the poem's form open or closed? Does it have a regular meter and rhyme scheme? How do these aspects contribute to the theme of the poem? Is the poem a particular type?
- What types of sound effects are used in the poem—repetition, alliteration, onomatopoeia, and so on? Is the general effect one of euphony or cacophony?
- Are there any special uses of capital letters, punctuation or spelling in the poem? If so, are they relevant to the poem's meaning or form?
- What are the most outstanding features of the poem? What is their significance?

My eLab

The poems in the next two chapters have been recorded in My eLab Documents so that you can listen to them before doing the activities. Hearing them recited should both increase your enjoyment and help you in your interpretations.

History *of* **Poetry**

The following section contains a short historical approach to the development of poetry. That is, poetry is examined within the context of history. This is not to suggest that poetry or any other form of literature develops within neat categories of socio-historical trends. Rather, the creation and development of literature is fluid. It can be argued that such categories of literary trends are artificial. However, the socio-historical approach to studying literature can give a general overview of the spirit of the times and help the reader understand the meaning and purpose of the writings of a particular era.

The words *poet* and *poetry* have a long history that dates back to ancient Greece: the Greek word *poietes* means *maker* or *composer* and comes from the verb *poiein—to create*. We can therefore say that poetry is word art designed to express elevated thought and emotion in a striking and imaginative way through the musicality of language. Poetry has often been intimately connected with song, as the two have many things in common, such as rhythmic patterns and the use of evocative words and images.

Literature is first introduced to children in the form of poetry. For example, songs, lullabies and nursery rhymes are all a part of the earliest childhood recollections. Literature at its earliest period in the history of civilization took the form of poetry. These stories were told in the earliest times of recorded history and had a certain musicality to them. It is believed that the structure of a repetitive rhythmic pattern made it easier for storytellers to remember their tales. The oral tradition was strong before the invention of the printing press and before literacy was common.

Early English poetry originated in the oral tradition. Pre-fifteenth-century poetry was usually sung by court minstrels and focused on the great deeds of chieftains or lords. These records of heroic deeds were called "epics" or "romances." Some records and texts of English poetry before the fifteenth century have been lost because records and texts of literary works, which were written on parchment and kept in monasteries, have disappeared. However, some poetry survived, and we can actually categorize it by form. Much of it was religious poetry that adapted warrior imagery from a Germanic tradition. It did not rhyme until it adopted the French and Latin traditions, after the Norman Conquest of 1066. It did, however, have strong rhythm and alliteration. It is important to remember that the language of this period was not English as we know it. With the Norman Conquest of England in 1066, Old French became the language of the English court and the aristocracy. Thus, the poetry of that time would not be very understandable to the modern reader.

It might be argued that English poetry started with Geoffrey Chaucer (1340/43-1400). Before his era, much of the existing poetry in Britain was written with short lines that had a break in the middle, alliteration and no end rhyme. Medieval poetry was concerned with the themes of courtly love and religion. Chaucer wrote in English, which by his time had become the official language of the land. He changed the usual metric form to the iambic pentameter—a pattern that has remained popular in English poetry. Chaucer's most famous work is called *The Canterbury Tales*. This masterpiece consists of a series of poetic tales or stories about common themes. The stories usually consist of rhyming couplets, containing elements of comedy, irony, satire and love.

"Poetry lifts the veil from the hidden beauty of the world and makes familiar objects be as if they were not familiar ..."

The fifteenth century was not a particularly innovative period for English poetry. However, a corpus of anonymous poetry shows that two trends in poetry can be identified: popular poems, such as ballads and folk songs (which had their origins in traditional songs and tales), and historical or religious poems usually written by clerics.

By contrast, the sixteenth century marked the Golden Age of English poetry. It was the century of Queen Elizabeth I and William Shakespeare (1564-1616). Life in England during this period was greatly influenced by the Italian Renaissance ("rebirth"). The Renaissance brought new heights to the arts, and artists were elevated to the status of creator or master. The beauty of man, nature and God was exalted. Elizabeth I was a grand patron of the arts, and under her support, Renaissance poetry and drama became formidable vehicles of literature. Themes of idealized love, passion, beauty and virtue were popular topics in literature. The artist also looked back to the ancients in Greek and Roman mythology for inspiration. Shakespeare, though known mainly for his plays, excelled in writing sonnets.

During the seventeenth century (the late Renaissance), two streams of poetry predominated. In one, poets attempted to write with smooth, rhythmic versification, using colourful imagery, pastoral scenes and portrayals of nature. Poetry of this school also showed a more reflective mood. The second popular stream of poetry was called "metaphysical." It reflected a desire on the part of poets to experiment with verse, form and rhyme. Imagery was bizarre and powerful. Poets of this school wrote satires, epigrams, elegies and sonnets. Their topics emphasized the new scientific discoveries and knowledge of their era. John Donne (1572-1631) was the most famous of these metaphysical poets. His sonnets reflected passionate love—not the idealized love that the earlier Renaissance poets wrote about—but the realistic love and passion between a man and a woman.

> *"When power narrows the areas of man's concerns, poetry reminds him of the richness and diversity of his existence. When power corrupts, poetry cleanses."*
>
> —John F. Kennedy, speech, October 26th, 1963

It was also at this time that poetry, which had always been concerned with humanity's relationship with God, started to reflect more rational themes about human beings and their world. Poets wrote about human beings' ability to reason about themselves and their place in the universe. For example, even though Donne wrote many poems and sermons with religious themes, he also questioned humanity's role in terms of its existence. Another great poet of this age was John Milton (1607-1674), whose epic poem *Paradise Lost* was about the fall of Adam and Eve from grace and their removal from the Garden of Eden. In his work, Milton tried to rationalize God's divine plan for mankind.

By the eighteenth century, the rationalistic approach was very much the norm. This period is often called the Age of Reason or the Age of Enlightenment because faith in science and great scientific discoveries dominated. It was believed that humanity's greatest virtue was its ability to reason, and literary themes reflected this new sensibility. Indeed, good sense and non-extravagance were the ideals. The eighteenth century was also the age of great prose writing. The novel and the essay became very popular forms of literature during this period, and consequently, poetry had competition in conveying the newly popular themes of rationality. Indeed, Jonathan Swift (1667-1745), best known for his novel *Gulliver's Travels*, wrote poetry as well. Another famous poet of this period was Alexander Pope (1688-1744). The underlying theme of his works incorporated philosophical musing, self-analysis and satire. Thomas Gray (1716-1771) elevated the elegy, a poem characterized by five or six feet to a line, depicting melancholic pondering.

The first part of the nineteenth century was called the Age of Romanticism (1798-1832). Romanticism turned away from cold, dispassionate ways of looking at humanity and society, instead stressing passionate attempts at creativity. This philosophy was seen within the genre of poetry. The Romantic period began with William Wordsworth's publication of *Lyrical Ballads* in 1798. Poets such as William Blake (1757-1827), William Wordsworth (1770-1850) and Samuel Taylor Coleridge (1772-1834) were also influenced by political events. The French Revolution stirred in these men a belief that great new political and social changes were about to take place. The spirit of the times focused on human rights and social justice.

Poetry in the Romantic Age reflected the belief that inspiration came from inside the individual and not from the outside world. Wordsworth and others thought that poems should reflect inner feelings and emotions. Accordingly, the lyrical poem, utilizing the first-person narrator, became the most popular type of poem. Romantic poets also believed in inspiration and relied on spontaneous composition. They gained their inspiration from nature and the surrounding landscape. The Romantics were fascinated with the supernatural. For example, Coleridge wrote poems such as *Christabel* and *Kubla Khan*, which dealt with magic and the supernatural. Lord Byron (1788-1824) was very much interested in the occult; some of his poems included satanic figures as their major characters.

The Romantic poets lauded the artistic quest to experience all of the human senses profoundly. Lord Byron, Percy Bysshe Shelley (1792-1822) and John Keats (1795-1821) were considered to be Romantic poets of the second generation. Shelley's poetry expressed a rejection of political tyranny and extolled the inherent goodness in human nature. Keats's poetry speculated on the nature of beauty and the emotions of human experience.

The Victorian Age (1832-1901), the years of Queen Victoria's reign in Great Britain, was characterized by great social and economic change. First, industrialization altered the structure of English society from one based on an agrarian model to one based on commerce. There was a tremendous shift in the population as people migrated from farms to cities, where most worked in factories or mines under very poor conditions. Second, science and technology experienced a great growth spurt, especially in the areas of engineering, architecture and experimental science. Charles Darwin's treatises *The Origins of the Species* (1859) and *The Descent of Man* (1871) caused great controversy and led to an ongoing debate regarding the existence of God.

The Victorian poets reacted to these social changes. Alfred, Lord Tennyson (1809-1892) wrote poems concerned with social and economic ills, the nature of humanity and our relationship with God. Other Victorian poets, such as Robert Browning (1812-1889) and Matthew Arnold (1822-1888), reflected on the nature of human relationships and the social ills confronting English society due to changing economic structures. Elizabeth Barrett Browning (1806-1861), Robert Browning's wife, also wrote on the nature of love, as well as on the condition of women and children.

In the United States, the publication of *Leaves of Grass* in 1855 by Walt Whitman (1819-1892) marked the emergence of American poetry. In this lengthy poem, written in unconventional meter and rhyme, the poet celebrated his self and his country. During the latter half of the nineteenth century, a shy recluse named Emily Dickinson (1830-1886) wrote over one thousand poems. The first were printed in 1890, and only then did people begin to discover the richness and intensity of her poetry. Her very individual techniques influenced many poets who followed her.

The twentieth century was characterized by great changes in social attitudes, norms and political and economic relations. The century saw two world wars and rapid advances in science and technology. Poetry reflected the nature of these changes. Poets at the beginning of the century, influenced by the French Symbolist writers Charles Baudelaire (1821-1867), Stéphane Mallarmé (1842-1898) and Arthur Rimbaud (1854-1891), experimented with images, focusing on the object in its concrete form and avoiding anything that confused the image. Language came under great scrutiny as poetry turned away from past verse forms in order to reflect a more conversational style. Ezra Pound (1885-1972) and T. S. Eliot (1888-1965), two American expatriates, exemplified the new modernist spirit. Eliot experimented with concrete images and symbols, adding ironic elements when discussing social disillusion. His poem "The Waste Land" is considered by many to be the most famous poem of the twentieth century. William Butler Yeats (1865-1939) wrote poetry based on Irish mythology and metaphysical speculations.

Meanwhile, Wallace Stevens (1879-1955) exalted the imagination, and Marianne Moore (1887-1972) experimented with the use of quotations in her brilliant verse. The poetry of many American poets of this period shows a preoccupation with realism and common speech. William Carlos Williams (1883-1963) wrote about the everyday in a very spare but vivid style, and e. e. cummings (1894-1962) presented his lyrical poems using very unusual typographical forms. In the British Isles, poetry by Siegfried Sassoon (1886-1967) and Robert Graves (1895-1985) reflected concerns about the violence of war, life and death, and questioned traditional religious beliefs. Dylan Thomas (1914-1953), C. D. Lewis (1904-1972) and W. H. Auden (1907-1973) wrote about death, tyranny and man's inhumanity.

After World War II, a new generation of British poets, including Philip Larkin (1922-1985) and Ted Hughes (1930-1999), continued to write themes reflecting the social concerns of modern times. Unlike many of the modernists, Robert Frost (1874-1963), perhaps the best-loved of the twentieth-century American poets, wrote poems using traditional devices, although he too wrote in the vernacular. Poets such as Sylvia Plath (1933-1963) wrote about love and death.

In the 1950s and 1960s, many American poets, such as Allen Ginsberg (1926-1997) and his collaborator, poet/publisher Lawrence Ferlinghetti (1919-), both of whom exemplified the Beat Generation and its concerns, experimented with poetic form and meaning. Ginsberg's live reading of his poem "Howl" electrified audiences. Since the late 1960s, post-modernist John Ashbery has been the dominant figure in American poetry. Many contemporary poets seem to fall into two distinct schools: the traditionalists, who build on past conventions, and the innovators (often part of the "L=A=N=G=U=A=G=E" writing movement), who reject past conventions and occupy themselves with the act of writing itself and with creating work that consists of nonsense and unmeaning.

English poetry also flourished outside of Britain and the United States during the twentieth-century. Poets from Canada, Australia, New Zealand, the Caribbean and India continue to experiment with the creation of poetry that is both unique and deeply universal. In Canada, Earle Birney (1904-1995), Margaret Avison (1918-), Irving Layton (1912-2006) and Leonard Cohen (1934-) wrote poetry that was concerned with social themes.

Pre-1900 Poetry

CHAPTER
CONTENTS

▸ **Listening:** "To Every Thing There Is a Season"; "A Divine Image"; "I Wandered Lonely as a Cloud"; "She Walks in Beauty"; "Sonnet XLIII"; "The Eagle: A Fragment"; "Time"; "Tell All the Truth but Tell It Slant"; "A Man Said to the Universe"

▸ **Reading:** authors' biographies; background information on the sonnet

▸ **Writing:** using traditional approaches for interpreting poetry

▸ **Speaking:** group discussions about poems

An Ancient Poem

To Every Thing There Is a Season

A B O U T
THE **AUTHOR**

For centuries, people attributed the writings of Ecclesiastes to a Hebrew king named **Solomon**, who lived during the tenth century BC and was famous for his wisdom. However, modern scholars now think that the text in Ecclesiastes might have been taken from the notebook of a Hebrew wisdom teacher who probably lived in the third century BC, some six hundred years later.

BACKGROUND
INFORMATION

ECCLESIASTES IS A BOOK OF WISDOM and poetry found in the Old Testament of the Hebrew and Christian Bible. "Ekklesiastes," a Greek word that means "a speaker before an assembly," is a translation of the original Hebrew word "Qohelet," which means "a wise man who preaches to the young." Thus, the author of the work was probably a teacher (or preacher) of some sort. This person questioned the meaning and purpose of human existence, noting that what happened to people during life did not depend on their actions but on time and chance. From this, he concluded that the ways

of God were unknowable and that we should enjoy the small day-to-day pleasures in life.

The Bible has influenced occidental literature throughout the centuries and continues to do so today. The content and style of Ecclesiastes is reflected in the work of such modern authors as Walt Whitman, Jean-Paul Sartre, Albert Camus and Ernest Hemingway. Moreover, the stanzas you will study in this chapter form the basis of a popular song by The Byrds, an influential band in the 1960s. The song is called "Turn! Turn! Turn!"

Warm-Up Discussion Topics

Discuss your answers to the following questions with another student.

1. In your opinion, do human beings control what happens to them in life? Explain your answer.

_____

_____

_____

2. What kind of influence do you think a holy book, such as the Bible, has on people's lives?

_____

_____

_____

∽ INITIAL READING

Read the poem for the first time for personal enjoyment.

To Every Thing There Is a Season
The Bible, King James Version: Ecclesiastes 3:1-8

To every thing there is a season,

And a time to every purpose under heaven:

A time to be born, and a time to die;

A time to plant, and a time to pluck up that which is planted;

A time to kill, and a time to heal;

A time to break down, and a time to build up;

A time to weep, and a time to laugh;

A time to mourn, and a time to dance;

A time to cast away stones, and a time to gather stones together;

A time to embrace, and a time to refrain from embracing;

A time to get, and a time to lose;

A time to keep, and a time to cast away;

A time to rend, and a time to sew;

A time to keep silence, and a time to speak;

A time to love, and a time to hate;

A time of war, and a time of peace.

© PEARSON LONGMAN — REPRODUCTION PROHIBITED

SECTION 2 • POETRY   193

Reader's Response

Exchange your answers to the following questions with another student.

1. What was your initial reaction to this poem?

_____

2. What are the main characteristics that struck you about this poem?

_____

✆ CLOSE READING

1. Which of the following best summarizes the poem? (Circle the most appropriate answer.) The poem

   a) is a sonnet.

   b) comprises a series of unrelated images.

   c) mainly consists of a play on the different meanings of the word "time."

   d) presents a series of contrasting actions.

2. What has the poet observed? The poet has observed that

   a) it is hard to understand what time actually is.

   b) there seems to be a basic rhythm to people's life experiences.

   c) you cannot predict the future or know the past.

   d) things happen randomly in a person's life.

3. One could say that the first line in the poem presents the author's _____

_____ .(*Hint*: it is essential in an essay but usually not seen in a poem.)

4. Which of the following best paraphrases the title and the first line of the poem?

   a) You can associate a season with each stage in a person's life.

   b) The seasons are always changing.

   c) There is an appropriate time for everything in human experience.

   d) The seasons pass, and so does time.

5. From the context, what does the word "rend" in the last line mean?

   a) mend   b) tear   c) fix   d) reap

6. What is the main idea of this poem? It

   a) expresses experiences that are unique to one person.

   b) only concerns people who are very religious.

   c) emphasizes the universality of human experience.

   d) connects one generation to the next.

✺ WRITER'S CRAFT

1. How does the poet develop a sense of meter and rhythm in the work? He uses

   a) similes.   b) irony.   c) personification.   d) repetition.

2. What is the predominant poetic foot (meter) used in the poem?

   It is an ———————————————— foot.

3. What kinds of words are predominantly used in the poem?

   a) one-syllable words

   b) two-syllable words

   c) three-syllable words

   d) both long and short words, more or less equally

   The poet's choice of words and rhythm gives the poem a lyrical, song-like quality.

4. What is the dominant figure of speech that recurs in this poem? (*Hint*: the linking of two contrasting ideas in order to make each one more vivid)

   a) oxymoron   b) hyperbole   c) apostrophe   d) antithesis

5. What could the poet's use of opposites express? It could express the

   a) duality of human experience.

   b) ironies of life.

   c) naturalness of changing seasons.

6. In your own words, express the principal theme of the poem.

   _____

   _____

Wrap-Up Activities

1. Write a personal interpretation of "To Every Thing There Is a Season."

2. Compare "To Every Thing There Is a Season" to "Song of Songs," also found in Ecclesiastes and attributed to King Solomon.

3. Compare "To Every Thing There Is a Season" to the song "Turn! Turn! Turn!" by The Byrds.

4. Compare the authors' use of the word "time" in "To Every Thing There Is a Season" and "Time" by Frederick George Scott.

5. The statement "Vanity of vanities! All is vanity!" is found both at the beginning and end of Ecclesiastes. Write a personal interpretation of the statement.

6. Write about another poem from the Bible or a different religious book.

7. Write about how the content or style of Ecclesiastes has influenced a modern author.

8. Write about King Solomon and his times.

The Romantics and the Victorians

A Divine Image

14

ABOUT THE **AUTHOR**

William Blake was born in 1757 and died in 1827. By the age of 14, he had started to draw and to write poetry. He married at 24 years of age. His wife was illiterate, but he taught her to read and write. He was also said to have been an emotional and jealous husband. He published his first book of poetry, *Poetical Sketches,* at the age of 26. These poems were lyrical, and in his works, he experimented with rhyme and rhythm. His most famous poems were published under the title *Songs of Innocence* (1789); later they were combined with others to comprise *Songs of Innocence and Experience* (1794). These poems demonstrate Blake's philosophy that the state of human existence is divided into two contrary groups roughly equivalent to the qualities of reason and passion. Blake was also an engraver and illustrated his writings. By the twentieth century, he was recognized as one of the great poets of English literature.

Warm-Up Discussion Topics

1. What are some common human emotions?

   _____

2. In your opinion, are human beings basically good or selfish in nature? Discuss.

   _____

∾ **INITIAL READING**

Read the poem for the first time for personal enjoyment.

A Divine Image
William Blake

Cruelty has a Human Heart,
And Jealousy a Human Face,
Terror, the Human Form Divine,
And Secrecy, the Human Dress.

The Human Dress is forged Iron,
The Human Form, a fiery Forge,
The Human Face, a Furnace seal'd,
The Human Heart, its hungry Gorge.

PUBLISHED IN
1790-1791

Reader's Response

1. What were your initial impressions of the poem?

    _____

2. Which word or phrase in the poem do you think is the most important? Explain your answer.

    _____

14

∞ CLOSE READING AND WRITER'S CRAFT

1. What type of form does the poem have?

    a) couplet   b) tercet   c) free verse   d) quatrain

2. What type of poetic foot and meter does the poem have?

    _____

3. Is there any use of repetition in the poem? If yes, give examples.

    _____

    _____

4. Explain the rhyming pattern of the poem.

    _____

5. Find examples of the following from the poem.

	Alliteration	Assonance	Consonance
Stanza 1			
Stanza 2			

6. What kind of image does the poet paint of humankind? (more than one answer)

    a) positive   b) negative   c) neutral   d) gentle   e) fierce   f) loving

    Cite two examples from the text to justify your answer.

    _____

    _____

    Note how the poet's use of repeated short sentences, alliteration, assonance and consonance highlights his view of humankind.

7. What is the mood of the poem?

    a) happy   b) sad   c) indifferent   d) harsh

8. What is the significance of the poem's title? Do you think it is meant to be ironic?

Wrap-Up Activities

1. Discuss William Blake's vision of humanity using "Divine Image" and one of his other poems.

2. Research William Blake's life and times. Do this poem and some of his other works represent the spirit of the times?

3. Research William Blake's *Songs of Innocence and Experience* and discuss the themes and contrasting points of view in the collections of poems.

4. What is special about William Blake as an artist? What major themes did he try to represent in his engravings and poetry? Are they relevant to people today?

I Wandered Lonely as a Cloud

ABOUT
THE **AUTHOR**

Romantic poet **William Wordsworth** was born in 1770 in West Cumberland, England. He went to Cambridge University and graduated in 1791. While attending university, Wordsworth vacationed in France and became a supporter of the French Revolution. At the same time, he fell in love with a Frenchwoman, Annette Vallon, with whom he had a daughter out of wedlock. They planned to marry, but the French Revolution prevented them from joining each other, and the marriage never took place. Wordsworth began writing poetry while at Cambridge and published a volume of poetry in 1798 entitled *Lyrical Ballads, with a Few Other Poems*. This volume was a joint effort with another great poet, Samuel Taylor Coleridge, and is considered to be the first major statement of Romanticism in England. In 1805, Wordsworth married Mary Hutchinson, a friend from childhood. In 1807, his *Poems in Two Volumes* was published, bringing him great recognition. His themes reflected his reminiscences of his past—of moments in time that were meaningful to him. He became the Poet Laureate in 1843 and died in 1850.

Warm-Up Discussion Topics

1. Where do you go to relax? Describe your favourite place to relax.

2. Describe an object of beauty. It might be a painting, a scene from nature, a building, a piece of music and so on. Give details so that the others in your group can visualize it clearly.

∞ INITIAL READING

Read the poem for the first time for personal enjoyment.

I Wandered Lonely as a Cloud
William Wordsworth

14

I wandered lonely as a cloud
That floats on high o'er vales and hills,
When all at once I saw a crowd,
A host, of golden daffodils;
Beside the lake, beneath the trees,
Fluttering and dancing in the breeze.

Continuous as the stars that shine
And twinkle on the milky way,
They stretched in never-ending line
Along the margin of a bay:
Ten thousand saw I at a glance,
Tossing their heads in sprightly dance.

The waves beside them danced; but they
Out-did the sparkling waves in glee:
A poet could not but be gay,
In such a jocund company:
I gazed—and gazed—but little thought
What wealth the show to me had brought:

For oft, when on my couch I lie
In vacant or in pensive mood,
They flash upon that inward eye
Which is the bliss of solitude;
And then my heart with pleasure fills,
And dances with the daffodils.

PUBLISHED IN
1804

Reader's Response

1. How did this poem make you feel? Explain.

_____

2. What sorts of images or ideas came to mind as you read this poem?

_____

∞ CLOSE READING AND WRITER'S CRAFT

1. What does the poet see on his walk?

_____

2. What feelings do the flowers evoke in the poet? Write two or three lines to support your answer.

_____

_____

_____

3. How do the feelings or emotions of the persona (narrator) change throughout the poem? Give examples.

_____

4. What are the structure (note the number of lines and verses) and rhyming scheme of the poem?

_____

5. What is the rhythm or meter of the poem? (If you cannot remember examples of rhythm and meter, refer to the Poetic Elements section.)

_____

6. Use the chart below to indicate the sound patterns of this poem.

	Alliteration	Assonance	Consonance
Stanza 1			
Stanza 2			
Stanza 3			

7. Would you describe the overall sound effects in the poem as **a)** euphony or **b)** cacophony?

8. Find examples of the following figures of speech.

a) similes

_____

b) metaphors

_____

c) personification

_____

d) imagery

_____

e) symbols

_____

9. What is the theme of the poem?

_____

_____

10. What is the mood of the poem?

_____

11. How do the form and structure influence the mood of the poem?

_____

_____

12. What kind of poem is this?

    a) an ode   b) a lyric   c) an epic   d) a sonnet   e) blank verse

14

Wrap-Up Activities

1. How do the imagery, sound effects and diction in "I Wandered Lonely as a Cloud" contribute to the development of the poem's mood and theme?

2. In the preface to *Lyrical Ballads* (1798), Wordsworth wrote: "It is the honourable characteristic of Poetry that its materials are to be found in every subject which can interest the human mind." Discuss Wordsworth's assessment and give examples from one or more of his poems.

3. Compare the following two poems by Wordsworth: "I Wandered Lonely as a Cloud" and "The Sun Has Long Been Set." (http://www.bartleby.com/145/ww205.html; http://www.everypoet.com/archive/poetry/William_Wordsworth/william_wordsworth_205.htm) Discuss how Wordsworth uses nature and natural phenomena to express his feelings.

4. Why is William Wordsworth referred to as the first Romantic poet? How does his poetry exhibit the characteristics of the Romantic period?

She Walks in Beauty

ABOUT
THE **AUTHOR**

George Gordon, Lord Byron was born in 1788 in London, England, but lived with his mother in Aberdeen, Scotland, for most of his childhood. He is considered by many to be one of the greatest of the Romantic poets. His whole life was a series of love affairs. Being a restive individual, he spent much of his adult life travelling around Europe, spending a great deal of time in Italy and Greece. His love affairs and travels gave him material for much of his poetry. His first volume of poetry, *Fugitive Pieces*, was published in 1806, and in 1807, he published *Poems on Various Occasions*. His poetry brought him instant recognition, but his long poems, *Childe Harolde* and *Don Juan*, brought him great fame. He wrote many poems in his short lifetime. He died of fever in Greece in 1824.

Warm-Up Discussion Topics

1. In your opinion, have romance and chivalry been replaced by cynicism and individualism in contemporary times? Give reasons for your answer.

2. Do you like to read romantic literature, such as historical romances or poetry? If so, what are your favourite works? What aspects about these works do you like most?

∞ INITIAL READING

Read the poem for the first time for personal enjoyment.

She Walks in Beauty
Lord Byron

She walks in beauty, like the night
Of cloudless climes and starry skies;
And all that's best of dark and bright
Meet in her aspect and her eyes:
Thus mellow'd to that tender light
Which heaven to gaudy day denies.

One shade the more, one ray the less,
Had half impair'd the nameless grace
Which waves in every raven tress,
Or softly lightens o'er her face;
Where thoughts serenely sweet express
How pure, how dear their dwelling-place.

And on that cheek, and o'er that brow,
So soft, so calm, yet eloquent,
The smiles that win, the tints that glow,
But tell of days in goodness spent,
A mind at peace with all below,
A heart whose love is innocent!

PUBLISHED IN
1815

Reader's Response

1. What did you visualize as you read this poem?

_____

2. Did you have positive or negative feelings while reading the poem? Explain.

_____

14

❧ CLOSE READING AND WRITER'S CRAFT

1. What or who is the subject of this poem?

_____

2. In the first stanza, to what is the narrator comparing the subject of the poem?

   a) darkness   b) a beautiful night   c) stars   d) the moon

3. In the first stanza, find one example of a simile.

_____

4. In the first stanza, what part of the woman's face reflects the beauty that the narrator talks about?

_____

5. Which colour forms the dominant image in the poem?

   a) black   b) white   c) silver   d) gold

6. In the first stanza, what adjective is used to describe "day"?

_____

7. Does this word have a positive or negative connotation?

_____

8. In stanza 2, which parts of the body does the narrator refer to?

_____

9. In stanza 2, what adjective does the narrator use to describe the nature of the woman in the poem?

_____

10. In stanza 3, which parts of the body does the narrator refer to?

_____

_____

11. In stanza 3, what is "innocent"?

_____

12. Give examples of the following.

    a) alliteration   _____

    b) consonance   _____

    c) assonance   _____

    d) repetition   _____

13. Does this poem have mainly **a)** soft- or **b)** hard-sounding words and phrases? Give a convincing example of a phrase (with at least four words) to support your answer.

_____

14. What is the meter and rhyme scheme of the poem?

_____

15. Does it have an **a)** open or **b)** closed form?

Wrap-Up Activities

1. How do the imagery and contrasting images in "She Walks in Beauty" affect the poem's theme?

2. How do the form and sound of "She Walks in Beauty" develop the mood of the poem?

3. Lord Byron was writing about his cousin in this poem. Describe how he feels about her physical and psychological qualities. Support your ideas with concrete examples from the poem.

4. Compare "She Walks in Beauty" to another poem by Lord Byron.

5. Research details about Lord Byron's life and times and discuss why he was a hero in Greece when he died but not in England (he was refused burial in Westminster Abbey). Explain how his life reflects the Romantic period.

Sonnet XLIII

ABOUT
THE **AUTHOR**

Elizabeth Barrett Browning was born is 1806 in England and achieved great fame in her lifetime as a poet. During her childhood, she received the same education as her brothers, something unusual for a woman at the time. She studied Latin, Greek, history, philosophy and literature. She published her first book of poetry when she was only thirteen years old. As she grew older, she was troubled by ill health. Also, her over-protective father had forbidden any of his children to marry, so she lived for many years as an invalid, not receiving any suitors. By the age of 39, she was a very famous poet. It was at that time that she met, fell in love with and married Robert Browning, another famous Romantic poet. Her love poems about their relationship are world famous. Of these is a collection of poems, *Sonnets from the Portuguese,* which includes "Sonnet XLIII." Elizabeth Barrett Browning died in 1861.

Warm-Up Discussion Topics

1. Can true love exist without passion? Explain your thoughts.

_____

2. What is love? What makes people fall in love?

_____

14

∞ INITIAL READING

Read the poem for the first time for personal enjoyment.

Sonnet XLIII
Elizabeth Barrett Browning

How do I love thee? Let me count the ways.
I love thee to the depth and breadth and height
My soul can reach, when feeling out of sight
For the ends of Being and ideal Grace.
I love thee to the level of every day's
Most quiet need, by sun and candle-light.
I love thee freely, as men strive for Right;
I love thee purely, as they turn from Praise.
I love thee with the passion put to use
In my old griefs, and with my childhood's faith.
I love thee with a love I seemed to lose
With my lost saints—I love thee with the breath,
Smiles, tears, of all my life!—and, if God choose,
I shall but love thee better after death.

PUBLISHED IN
1850

Reader's Response

1. This poem and the previous one, "She Walks in Beauty," are about love. Which one do you find to be more passionate? State your reasons.

_____

2. Have you ever felt the same way as Elizabeth Barrett Browning about someone? If so, how long did this feeling last? If not, would you like to experience this feeling? Why or why not?

_____

THE SONNET

THE SONNET HAS BEEN ONE OF THE most popular forms of poetry since the Renaissance. It reached its height of popularity in sixteenth-century England when William Shakespeare perfected the form. The sonnet follows strict rules of versification. Each sonnet consists of fourteen lines in iambic pentameter (five poetic feet of unstressed, stressed syllables).

Two types of sonnet forms exist. The Petrarchan, named after the Italian poet, has eight lines (called an octave) with the rhyming pattern *abbaabba*, and six lines (called a sestet) with the rhyming pattern *xyzxyz*. The English or Elizabethan sonnet consists of three quatrains with the rhyming pattern *abab cdcd efef*, and a couplet at the end with the rhyming pattern *gg*.

∞ CLOSE READING AND WRITER'S CRAFT

1. This poem is

   a) an Elizabethan sonnet.

   b) a Petrarchan sonnet.

2. The poetic foot of this poem is _____.

3. The verse length of this poem is _____.

4. What type of narrator does the poem have?

   _____

5. Who are the characters in this poem?

   _____

6. Which two sentences show the specific purpose of this poem?

   _____

7. Which literary device occurs most frequently in this poem?

   a) similes   b) metaphors   c) repetition   d) imagery

8. The line, "I love thee freely, as men strive for Right," is an example of

   a) personification.   b) symbolism.   c) a metaphor.   d) a simile.

9. One possible theme of this poem is that

   a) the narrator's love is so strong that she hopes it will continue after death.

   b) the narrator feels that there are too many ways in which she can love.

   c) the narrator believes that love is too strong an emotion.

   d) the narrator cannot express her emotions clearly enough.

14

10. List four sound devices used in this poem and give examples of each.

_____

_____

_____

_____

11. What is the predominant type of sound in the words and phrases in this poem? How does this sound affect the poem's mood?

_____

12. How does the mood of the poem reinforce the theme?

_____

Wrap-Up Activities

1. How does Browning use repetition and other sound effects to develop the mood and theme of "Sonnet XLIII"?

2. Compare "She Walks in Beauty" to "Sonnet XLIII."

3. Write about the Romantic period in English literature.

4. Who were the major Romantic poets and what influenced their life and works?

5. Compare the themes of two poems from the Romantic era.

The Eagle: A Fragment

ABOUT THE AUTHOR

Alfred, Lord Tennyson (1809-1892) is considered to be one of the greatest and most popular of the Victorian poets. Tennyson was the fourth in a family of twelve children. He had a difficult childhood; his father was a temperamental alcoholic. Tennyson went to Cambridge but had to leave before he graduated because of his family's financial difficulties. He started to write poetry at a young age and became renowned for his elegiac poem _In Memoriam,_ which he wrote in 1850. He became Poet Laureate in 1850, and in that same year, he married his sweetheart of twenty years, Emily Sellwood. In 1884, he was given a peerage that granted him the title of Lord. Tennyson's poetry was concerned with the themes of progress, an important phenomenon of the industrial revolution. He is also known for his philosophical renderings on life and death.

Warm-Up Discussion Topics

1. Which animals and birds do you find the most breathtaking? Why?

   _____

2. Humans have always dreamed of being able to fly like birds. Comment on this statement.

   _____

∞ INITIAL READING

Read the poem for the first time for personal enjoyment.

The Eagle: a Fragment
Alfred, Lord Tennyson

*H*e clasps the crag with crooked hands;
Close to the sun in lonely lands,
Ringed with the azure world, he stands.

The wrinkled sea beneath him crawls;
He watches from his mountain walls,
And like a thunderbolt he falls.

PUBLISHED IN
ca. 1850

Reader's Response

1. List some adjectives and adverbs that came to mind as you read this poem.

   _____

2. Are there any other wild animals that Tennyson's poem could be used to describe? Explain your answer.

   _____

∞ CLOSE READING AND WRITER'S CRAFT

1. What is the setting of the poem?

   _____

2. In the poem, what moment in the eagle's daily activities does the poet describe?

_____

3. In your opinion, why does the poet call the poem "a fragment"?

_____

4. What is the relationship between the title of the poem and the poem itself?

_____

14

5. What type of form (couplet, tercet, quatrain, and so on) and rhyming scheme does the poem have?

_____

6. What is the rhythm or meter (number of poetic feet) of the poem?

_____

7. What are the sound patterns in this poem? Fill in the table below.

	Alliteration	Assonance	Consonance
Stanza 1			
Stanza 2			

8. Find three examples of words with hard sounds in each stanza of this poem.

Stanza 1: _____

Stanza 2: _____

9. What kind of mood or atmosphere do these hard sounds help create in the poem?

_____

10. Find examples of the following poetic devices.

a) similes _____

b) metaphors _____

c) personification _____

d) imagery _____

e) symbols _____

11. Which three of the following adjectives best describe the eagle as depicted in the poem?

a) strong   d) magnificent

b) tyrannical   e) powerful

c) independent   f) man-like

Wrap-Up Activities

1. How do the formal elements (length, structure, poetic and sound devices, and so on) and subject of the poem complement and reinforce each other?

2. Read "The Kraken" by Tennyson (http://www.online-literature.com/tennyson/930/; http://www.web-books.com/Classics/Poetry/Anthology/Tennyson/Kraken.htm) and compare it to "The Eagle: A Fragment."

3. Choose another poem by Tennyson and do a literary analysis of it.

4. Research the literary trends of the time and describe how Tennyson's poetry reflected them.

5. Write a poem about an animal you feel strongly about. Use imagery and sound patterns to express your feelings toward the animal.

Early North American Poets

Time

ABOUT
THE **AUTHOR**

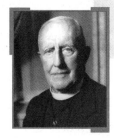

Frederick George Scott was born in 1861 in Montreal, Quebec. He spent more than forty years as the rector of a church in Quebec City. During World War I, he served as chaplain for Canadian soldiers. He wrote poetry, including sonnets, on themes of war and nature. He was also concerned with God and religion in the life of mankind. He died in 1944.

Warm-Up Discussion Topics

1. In your opinion, why is contemporary society so preoccupied with looking young?

_____

2. Western society is often accused of being disrespectful toward the elderly. Do you agree or disagree with this statement? Explain your opinion.

_____

∞ INITIAL READING

Read the poem for the first time for personal enjoyment.

Time
Frederick George Scott

I saw Time in his workshop carving faces;

Scattered around his tools lay, blunting griefs,

Sharp cares that cut out deeply in reliefs

Of light and shade; sorrows that smooth the traces

Of what were smiles. Nor yet without fresh graces

His handiwork, for oft times rough were ground

And polished, oft the pinched made smooth and round;

The calm look, too, the impetuous fire replaces.

Long time I stood and watched; with hideous grin

He took each heedless face between his knees,

And graved and scarred and bleached with boiling tears.

I wondering turned to go, when, lo! my skin

Feels crumpled, and in glass my own face sees

Itself all changed, scarred, careworn, white with years.

PUBLISHED IN
1886

Reader's Response

1. What were your initial feelings when you read this poem?

_____

2. Have you ever thought about getting older? How do you think aging will affect your body and lifestyle as you reach middle and old age?

_____

∞ CLOSE READING AND WRITER'S CRAFT

1. The poem is divided into _____ parts. The _____ part has _____ lines and the _____ part has _____ lines.

2. "Time" is

a) a Petrarchan sonnet.

b) an Elizabethan sonnet.

14

3. Analyze the rhyming scheme of the poem. It is _____.

4. The poem has

   a) a third-person narrator.

   b) a second-person narrator.

   c) multiple narrators.

   d) a first-person narrator.

5. Who or what is the subject of the first half of the poem?

   _____

6. Who or what is the subject of the second half of the poem?

   _____

7. What is the narrator doing in the second half of the poem?

   _____

8. What does the narrator discover about himself at the end of the poem?

   _____

9. What figurative device is used to enhance the subject's characteristics in the first half of the poem?

   a) a metaphor

   b) a simile

   c) personification

   d) imagery

10. Where is Time found in the poem?

   _____

11. According to line 1, what is Time doing there?

   _____

12. List one thing that his tools are used for (line 2).

   _____

13. In the second half of the poem, find an example that reveals Time's attitude toward his work?

   _____

14. What is the attitude of the narrator as he sees Time in action?

   _____

15. For each of the following quotations, state whether it represents assonance, consonance or alliteration.

a) "... **s**orrows that **s**mooth the traces of what were **s**miles ..." _____

b) "... S**c**attered **ar**ound ... Shar**p** **c**ares ..." _____

c) "... his tools **l**ay, b**l**unting ..." _____

16. In your opinion, what is the principal theme of the poem?

_____

Wrap-Up Activities

1. Write an analysis on whether or not "Time" meets the traditional criteria of a sonnet.

2. How does the effective use of imagery and other poetic devices bring out the theme in the poem?

3. Choose Frederick George Scott or another Canadian poet from this period (Bliss Carman, Isabella Valancy Crawford, Archibald Lampman, Charles G. D. Roberts, and so on) and write an essay about his / her works and major influences.

4. Write an essay about the historical development of Canadian poetry.

5. Write a sonnet and read it aloud to your classmates.

Tell All the Truth but Tell It Slant

ABOUT THE AUTHOR

Emily Dickinson is considered to be one of the greatest poets of American literature. She was born in Amherst, Massachusetts, in 1830. She came from a well-off, educated family of high social standing in the community. Her grandfather, Samuel Dickinson, had founded Amherst College. Emily had one brother and one sister. The children were well educated. Emily started to write poetry at an early age. She was a prolific writer who produced around 1,800 poems. However, only ten of the poems were published in her lifetime; the rest were found by her sister after the poet's death. Emily Dickinson died in 1886 after living most of her life as a recluse.

Warm-Up Discussion Topics

1. Is it important for people to be truthful? Why or why not?

_____

2. In your opinion, are there any circumstances in which one should not tell the truth? Or should one tell the truth at all times? Give reasons for your answer.

_____

∞ INITIAL READING

Read the poem for the first time for personal enjoyment.

Tell All the Truth but Tell It Slant
Emily Dickinson

Tell all the Truth but tell it slant—
Success in Circuit lies
Too bright for our infirm Delight
The Truth's superb surprise
As Lightning to the Children eased
With explanation kind
The Truth must dazzle gradually
Or every man be blind—

PUBLICATION YEAR
unknown

Reader's Response

1. What was your first reaction to this poem?

_____

2. Did you feel the poem contained contradictions when you first read it? Explain.

_____

∞ CLOSE READING AND WRITER'S CRAFT

1. From the context of the poem, a synonym for the word "slant" is

   a) slope.   b) lie.   c) disparagingly.   d) indirectly.

2. The poem states that adults should tell the truth the way they

   a) lie.   b) exaggerate.   c) give children explanations.   d) none of the answers

3. Find three words in the poem that are examples of sight imagery.

_____

4. What is the rhyming scheme of this poem?

_____

5. Describe the rhythm in terms of feet and meters.

Odd lines: _____

Even lines: _____

6. Give four examples of lines that contain alliteration and/or consonance.

_____

7. Give two examples of lines that contain assonance or internal rhymes.

_____

8. Find an example of onomatopoeia in the poem.

_____

9. What does the poet mean by the last two lines of the poem?

_____

10. In this poem, the poet seems to be saying that people are

    a) naïve.   b) clever.   c) unintelligent.   d) educated.

Wrap-Up Activities

1. How do imagery and sound patterns help develop the theme in "Tell All the Truth but Tell It Slant"?

2. Using examples from "Tell All the Truth but Tell It Slant," show how the poem exemplifies Dickinson's attention to the rhythm and musicality of language.

3. Do you agree or disagree with Dickinson's advice in this poem? Is she addressing children or adults? Give reasons for your answers.

4. Write about the life and works of Dickinson using a biographical approach. How did her life experience affect her work?

5. Write an analysis of this poem.

6. Read other poems by Emily Dickinson and analyze their themes.

A Man Said to the Universe

ABOUT THE AUTHOR

Stephen Crane was born in 1871 into a middle-class family in Newark, New Jersey. He lived a very flamboyant, short life filled with many adventures. He wrote both prose and poetry, using his real-life experiences from his career as a war journalist for much of his works. For example, by 1898, Crane was a reporter covering the Spanish-American War. He also covered the Greco-Turkish war. He published many works. His first story, *The King's Favour*, was published in the *New York Tribune* in 1891. In 1893, he self-published *Maggie: A Girl of the Streets*. In the same year, he wrote *The Red Badge of Courage*, his most famous novel. He died in 1900 from tuberculosis.

14

Warm-Up Discussion Topics

1. How important do you think human beings are in the universe? Are they more important than other creatures and plants?

_____

2. How important is it for humans to have a reason for their existence? Explain.

_____

❧ INITIAL READING

Read the poem for the first time for personal enjoyment.

A Man Said to the Universe
Stephen Crane

A man said to the universe:

"Sir, I exist!"

"However," replied the universe,

"The fact has not created in me

A sense of obligation."

PUBLISHED IN
1894

Reader's Response

1. What reactions did you have to this poem?

_____

2. What do you like about this poem? What do you dislike?

_____

✂ CLOSE READING AND WRITER'S CRAFT

14

1. What figure of speech does Crane use in the poem?

    a) a metaphor   b) onomatopoeia   c) a simile   d) personification

2. a) What other word could Crane have used instead of "universe"?

_____

    b) What kind of connotation would this word have given the poem?

_____

    c) What kind of connotation does the word "universe" have?

_____

3. a) Does this poem have a rhyme scheme and regular rhythmical pattern (feet and meter)?

    yes ☐ no ☐

    b) If so, what are they? _____

    c) If not, do you think that "A Man Said to the Universe" constitutes poetry? Justify your answer.

_____

4. What are the poet's thoughts concerning human beings in this poem?

    a) People have a sense of obligation to each other.

    b) People seem to think they are important.

    c) People do not care about each other.

    d) People love each other.

5. In this poem, the world is seen as

    a) safe.   b) kind.   c) cold.   d) dark.

6. What is humankind's role in the scheme of things according to this poem?

    a) insignificant   b) important   c) central   d) none of the answers

7. Explain the reasoning behind your answer to the previous question.

_____

8. In what ways is this poem different from the other poems in this chapter?

_____

Wrap-Up Activities

1. Write about the role of humankind in the universe as presented in "A Man Said to the Universe." Is it the same as or different from contemporary ideas in society? Consider the following questions. In this poem, do human beings represent the centre of the universe? What is humankind's place in the universe? During Crane's lifetime, what did society think about the role of human beings in the universe?

2. Compare the two poets, Crane and Dickinson, and their attitude toward humankind as seen in their respective poems.

3. Research the life of Stephen Crane. How did his experiences affect his work?

4. Read another poem by Stephen Crane and compare it to "A Man Said to the Universe."

5. Explain how Stephen Crane's works set the stage for the twentieth century.

6. Research another American poet of this period, such as Ralph Waldo Emerson, Henry Wadsworth Longfellow, Walt Whitman, Edgar Allan Poe, Herman Melville, and so on, and make an oral presentation about his or her life and works.

Post-1900 Poetry

15

CHAPTER

APPROACHES TO ANALYZING POETRY

THERE ARE DIFFERENT APPROACHES TO analyzing a poem. Sometimes an analysis combines more than one approach.

The **reader response** approach maintains that the reader plays a central role in the literary work. The feelings of the reader must be considered and valued when analyzing a work of literature. In this approach, the reader brings his or her own personal knowledge to the poem in order to develop a connection with the work and to understand his or her response to it. When analyzing a poem from a reader response perspective, the reader might want to compare his or her own life experience to that of the poet by telling an anecdote or by discussing the emotional responses the poem evokes in the reader.

The **formalist** approach to literary criticism developed in the early twentieth century. Formalists examine the form or language of a poem and study how it is used to gain insight into the meaning of the work. Formalists look at the varied structural and linguistic parts to understand how the writer created the poem. This type of criticism only considers the work itself and does not give too much importance to external factors, such as historical context and the author's personal experiences. Thus, formalism concentrates on rhetoric and linguistics when analyzing a poem. Formalist

critics created a concrete methodology akin to that of the sciences in order to study poetry.

The **historical** approach studies literature within the context of history. For example, the historical events and the spirit of the times are considered to be very important as background to the literary work. Critics who use this approach feel that a writer cannot be divorced from the events that he or she is writing about; therefore, a writer is necessarily influenced by the ideas of the times. For example, the poem "In Flanders Fields" was written by a Canadian doctor serving in Flanders, Belgium, during World War I. The subject matter of the poem grew out of the poet's experience that resulted from the events of the war.

The **sociological** approach is similar to the historical approach in that it too considers the spirit of the times, but from a sociological perspective. This approach includes the following types of factors: economic status, education, male/female roles, adult versus child, group versus individual, urban versus rural, as well as social groups, such as family, tribe, community, country and nationality.

The **structuralist** approach attempts to discover the unifying elements of a work by looking at its different parts. In this sense, it is similar to the formalist approach; however, it diverges by trying to understand the deeper structures of the work itself. For example, an important concern of a structuralist is to look at the relationship between *langue* (language as a whole) and *parole* (utterance—a particular use of individual components of language). Structuralist scholars look at language as a sign system. They also consider literary works in two ways: in terms of language and in terms of the system used to create the text, such as its genre. The structuralist approach employs the Jungian idea of archetypes in order to understand the deeper, universal elements of literature. Karl Jung was a German philosopher who theorized that all literary works may be categorized into universal archetypes or primary models.

The **feminist** approach presumes that literature favours a masculine perspective and disfavours the importance of feminine characters and their concerns. This approach attempts to reverse the male-dominant perspective in order to study literature through the perspective of women and their role in the creation of literature.

The **minority** approach utilizes the perspective of issues facing minorities to analyze literature. These issues include marginalization, racism, discrimination, intolerance of other cultural perspectives and value systems, majority versus minority and so on.

The **Marxist** approach to literary analysis seeks to analyze literature in terms of class struggle. Characters are evaluated through an analysis of their class status. A Marxist approach observes whether characters are rich or poor, and examines the effects of political and economic trends on the work itself.

Modern Poetry

The poems in this and the next section were written between the end of World War I and the present. As you respond to and analyze these works, you will integrate what you have learned from the activities in the previous sections of this book. You will be asked to incorporate the reader response approach and elements from other types of approaches in your essays. Be sure to consult the Poetic Elements section on page 178 as well as units 1 to 5 in Part 2 of this book; they offer explanations and examples to help you with many aspects of your analysis.

Your instructor may choose to have you listen to, read and discuss the poems in small groups before you actually write about them.

In Flanders Fields

ABOUT
THE **AUTHOR**

John McCrae was born in 1872 in Guelph, Ontario, Canada. He was a respected university teacher and doctor in Montreal before joining the army and fighting in World War I on the western front in 1914. He was eventually sent to France with a medical unit. It was during his service that he experienced firsthand the horrible nature of war. In 1918, he died of pneumonia in France while still on active duty. His book of poetry, *In Flanders Fields and Other Poems*, was published in 1919.

In Flanders Fields
John McCrae

In Flanders fields the poppies blow
Between the crosses, row on row,
That mark our place; and in the sky
The larks, still bravely singing, fly
Scarce heard amid the guns below.

We are the Dead. Short days ago
We lived, felt dawn, saw sunset glow,
Loved, and were loved, and now we lie
          In Flanders fields.

Take up our quarrel with the foe:
To you from failing hands we throw
The torch; be yours to hold it high.
If ye break faith with us who die
We shall not sleep, though poppies grow
          In Flanders fields.

PUBLISHED IN
1919

15

✐ WRITING ACTIVITY

Write an interpretation of "In Flanders Fields." Incorporate the reader response approach and elements from the formalist, structuralist and historical approaches.

Here are some suggestions to help you begin your analysis of the poem. They are intended to help you articulate your written response.

1 READER'S RESPONSE

- What were your reactions when you first read the poem?

- Write a list of adjectives that describe your feelings about the poem.

Underline three words or phrases from the poem that are the most meaningful to you. Explain why you chose them.

- Read the poem again.
  - Make a list of meaningful words that relate to the subject of the poem.
  - From your list, state some general conclusions about how the words illustrate the meaning or theme of the poem. Then compare your answers with those of other students.

- McCrae wrote "In Flanders Fields" in response to a personal experience that moved him deeply. In your experience, has an image or an event ever moved you profoundly? If so, what were your reactions to the image or event?

2 FORMAL AND STRUCTURAL ELEMENTS

- What is the rhyming scheme of the poem?
- How would you define the rhythm or meter of the poem?
- Use the chart below to indicate the sound patterns of the poem.

	Alliteration	Assonance	Consonance
Stanza 1			
Stanza 2			
Stanza 3			

- Find examples of the following poetic devices in the poem.
  - similes
  - metaphors
  - personification
  - imagery
  - symbols
- What is the dominant mood of the poem?
- How does the form of the poem reinforce the mood?

3 HISTORICAL AND SOCIOLOGICAL DIMENSIONS

- What is the setting of the poem?
- Who is the speaker (or persona) in the poem?
- What, if any, assumptions can you make about the speaker?
- What is the subject of the poem?
- What is the theme of the poem?
- What, if any, assumptions can you make about the social groups involved in the poem, such as community, country and nationality?

Wrap-Up Activities

1. Write about the major themes in "In Flanders Fields." What message is the poet trying to communicate to the reader?

2. Research the historical context in "In Flanders Fields." Write about the poem in terms of its historical context.

3. Research the origins of Remembrance Day.

4. Consider the use of imagery and symbolism (especially Christian symbolism) in "In Flanders Fields."

5. Research and write about the life of John McCrae.

6. What is political poetry? Is "In Flanders Fields" a political poem?

7. Compare "In Flanders Fields" to another poem about war by poets such as Wilfred Owen (1893–1918), Rupert Brooke (1887–1915), Isaac Rosenberg (1890–1918), Siegfried Sassoon (1886–1967) and Edward Thomas (1878–1917).

Stopping by Woods on a Snowy Evening

ABOUT
THE **AUTHOR**

Robert Frost (1874–1963) is undoubtedly considered to be one of America's best poets. Frost often set his poems in the New England countryside, and he followed in the steps of the Romantic poets. His very individualistic and lyrical poetry is simple on the surface but can be read at many different levels. "Stopping by Woods on a Snowy Evening" was published in his collection of poetry called *New Hampshire*. Frost received the Pulitzer Prize four times.

Stopping by Woods on a Snowy Evening
Robert Frost

Whose woods these are I think I know.
His house is in the village though;
He will not see me stopping here
To watch his woods fill up with snow.

My little horse must think it queer
To stop without a farmhouse near
Between the woods and frozen lake
The darkest evening of the year.

He gives his harness bells a shake
To ask if there is some mistake.
The only other sound's the sweep
Of easy wind and downy flake.

The woods are lovely, dark, and deep,
But I have promises to keep,
And miles to go before I sleep,
And miles to go before I sleep.

PUBLISHED IN
1923

∞ WRITING ACTIVITY

Write an essay that interprets the poem. Incorporate the reader response and structuralist approaches.

Here are some suggestions to help you begin your analysis of "Stopping by Woods on a Snowy Evening."

1 READER'S RESPONSE

- What were your initial reactions to the poem?
- Where do you go to relax? Describe a peaceful place.
- What are the general feelings of the speaker in the poem? Provide examples from the poem to support your answer.
- Do you view the message of the poem as positive or negative? Explain your answer.

2 FORMAL AND STRUCTURAL ELEMENTS

- What is the meter and rhyme scheme of the poem?
- Find examples of alliteration, assonance and consonance in the poem.
- Think about the diction of the poem.
- Does the poem contain more concrete or abstract words?
- How do the formal elements create the mood of the poem?

3 POETIC DEVICES

Consider the imagery.

- Think about the images connected with the poem's setting. How do they affect the mood and theme of the poem?
- What senses are used to describe the images?

Consider the symbolism.

- How are the woods portrayed?
  - What kind of mood or feeling is connected with them?
  - What do you think they symbolize?
  - Who do you think they belong to?
- How is the horse portrayed and what does he do?
  - What does he add to the poem?
  - What is his role in the poem?
  - What does he represent?

4 SETTING

- What is the season?
  - Which words are associated with the season?
  - What feelings are engendered by the seasonal imagery?
  - What do you think the season symbolizes?
- What time of day is it?
  - What do you think the time of day symbolizes?

5 THE PROMISES

- What do you think the promises are?
- What do you think they signify?

6 THE SPEAKER

- Who is the speaker in the poem?
- What feelings is he experiencing?

Wrap-Up Activities

1. Write about how the mood is created in each stanza of "Stopping by Woods on a Snowy Evening," and how it helps to develop the theme of the poem.

2. Write about connotations that certain words and phrases evoke in "Stopping by Woods on a Snowy Evening." How do these connotations help develop the theme of the poem?

3. Write about how Frost creates tension in "Stopping by Woods on a Snowy Evening."

4. How does Frost's use of traditional rhythm and rhyming patterns contrast with the profoundly philosophical themes in his poetry, such as death, isolation, joy, freedom and so on?

5. Compare "Stopping by Woods on a Snowy Evening" to another poem by Frost.

6. Write about the importance of nature in the poetry of Robert Frost.

Contemporary Poetry

David

ABOUT
THE **AUTHOR**

Earle Birney was born in 1904 in Calgary, Alberta, and he spent much of his youth on an isolated farm in British Columbia. He studied at universities in Canada, California and London. He was also very politically active and served in the Canadian Army during World War II. He won the Governor General's Award for Literature, first for *David and Other Poems* in 1942, and then for *Now Is Time* in 1945. Although he is best known for his poetry, he also wrote novels, plays and radio dramas. He established Canada's first university-level, creative writing program, and he was made an Officer of the Order of Canada in 1970. He died at the age of 91.

David
Earle Birney

David and I that summer cut trails on the Survey,
All week in the valley for wages, in air that was steeped
In the wail of mosquitoes, but over the sunalive weekends
We climbed, to get from the ruck of the camp, the surly

Poker, the wrangling, the snoring under the fetid
Tents, and because we had joy in our lengthening coltish.
Muscles, and mountains for David were made to see over,
Stairs from the valleys and steps to the sun's retreats.

II

Our first was Mount Gleam. We hiked in the long afternoon
To a curling lake and lost the lure of the faceted
Cone in the swell of its sprawling shoulders. Past
The inlet we grilled our bacon, the strips festooned

On a poplar prong, in the hurrying slant of the sunset.
Then the two of us rolled in the blanket while round us the cold
Pines thrust at the stars. The dawn was a floating
Of mists till we reached to the slopes above timber, and won

To snow like fire in the sunlight. The peak was upthrust
Like a fist in a frozen ocean of rock that swirled
Into valleys the moon could be rolled in. Remotely unfurling
Eastward the alien prairie glittered. Down through the dusty

Skree on the west we descended, and David showed me
How to use the give of shale for giant incredible
Strides. I remember, before the larches' edge,
That I jumped a long green surf of juniper flowing

Away from the wind, and landed in gentian and saxifrage
Spilled on the moss. Then the darkening firs
And the sudden whirring of water that knifed down a fern-hidden
Cliff and splashed unseen into mist in the shadows.

III

One Sunday on Rampart's arête a rainsquall caught us,
And passed, and we clung by our blueing fingers and bootnails
An endless hour in the sun, not daring to move
Till the ice had steamed from the slate. And David taught me

How time on a knife-edge can pass with the guessing of fragments
Remembered from poets, the naming of strata beside one,
And matching of stories from schooldays ... We crawled astride
The peak to feast on the marching ranges flagged

By the fading shreds of the shattered stormcloud. Lingering
There it was David who spied to the south, remote,
And unmapped, a sunlit spire on Sawback, an overhang
Crooked like a talon. David named it the Finger.

That day we chanced on the skull and the splayed white ribs
Of a mountain goat underneath a cliff-face, caught
On a rock. Around were the silken feathers of hawks.
And that was the first I knew that a goat could slip.

IV

And then Inglismaldie. Now I remember only
The long ascent of the lonely valley, the live
Pine spirally scarred by lightning, the slicing pipe
Of invisible pika, and great prints, by the lowest

Snow, of a grizzly. There it was too that David
Taught me to read the scroll of coral in limestone
And the beetle-seal in the shale of ghostly trilobites,
Letters delivered to man from the Cambrian waves.

V

On Sundance we tried from the col and the going was hard.
The air howled from our feet to the smudged rocks
And the papery lake below. At an outthrust we balked
Till David clung with his left to a dint in the scarp,

Lobbed the ice-axe over the rocky lip,
Slipped from his holds and hung by the quivering pick,
Twisted his long legs up into space and kicked
To the crest. Then grinning, he reached with his freckled wrist

And drew me up after. We set a new time for that climb.
That day returning we found a robin gyrating
In grass, wingbroken. I caught it to tame but David
Took and killed it, and said, "Could you teach it to fly?"

VI

In August, the second attempt, we ascended The Fortress.
By the forks of the Spray we caught five trout and fried them,
Over a balsam fire. The woods were alive
With the vaulting of muledeer and drenched with clouds all the morning

Till we burst at noon to the flashing and floating round
Of the peaks. Coming down we picked in our hats the bright
And sun-hot raspberries, eating them under a mighty
Spruce, while a marten moving like quicksilver scouted us.

VII

But always we talked of the Finger on Sawback, unknown
And hooked, till the first afternoon in September we slogged
Through the musky woods, past a swamp that quivered with frogsong,
And camped by a bottlegreen lake. But under the cold

Breath of the glacier sleep would not come, the moonlight
Etching the Finger. We rose and trod past the feathery
Larch, while the stars went out, and the quiet heather
Flushed, and the skyline pulsed with the surging bloom

Of incredible dawn in the Rockies. David spotted
Bighorns across the moraine and sent them leaping

With yodels the ramparts redoubled and rolled to the peaks,
And the peaks to the sun. The ice in the morning thaw

Was a gurgling world of crystal and cold blue chasms,
And seracs that shone like frozen salt-green waves.
At the base of the Finger we tried once and failed. Then David
Edged to the west and discovered the chimney; the last

Hundred feet we fought the rock and shouldered and kneed
Our way for an hour and made it. Unroping we formed
A cairn on the rotting tip. Then I turned to look north
At the glistening wedge of giant Assiniboine, heedless

Of handhold. And one foot gave. I swayed and shouted.
David turned sharp and reached out his arm and steadied me,
Turning again with a grin and his lips ready
To jest. But the strain crumbled his foothold. Without

A gasp he was gone. I froze to the sound of grating
Edge-nails and fingers, the slither of stones, the lone
Second of silence, the nightmare thud. Then only
The wind and the muted beat of unknowing cascades.

VIII

Somehow I worked down the fifty impossible feet
To the ledge, calling and getting no answer but echoes
Released in the cirque, and trying not to reflect
What an answer would mean. He lay still, with his lean

Young face upturned and strangely unmarred, but his legs
Splayed beneath him, beside the final drop,
Six hundred feet sheer to the ice. My throat stopped
When I reached him, for he was alive. He opened his grey

Straight eyes and brokenly murmured "over … over."
And I, feeling beneath him a cruel fang
Of the ledge thrust in his back, but not understanding,
Mumbled stupidly, "Best not to move," and spoke

Of his pain. But he said, "I can't move … If only I felt
Some pain." Then my shame stung the tears to my eyes
As I crouched, and I cursed myself, but he cried,
Louder, "No, Bobbie! Don't ever blame yourself.

I didn't test my foothold." He shut the lids
Of his eyes to the stare of the sky, while I moistened his lips
From our water flask and tearing my shirt into strips
I swabbed the shredded hands. But the blood slid

From his side and stained the stone and the thirsting lichens,
And yet I dared not lift him up from the gore
Of the rock. Then he whispered, "Bob, I want to go over!"
This time I knew what he meant and I grasped for a lie

And said, "I'll be back here by midnight with ropes
And men from the camp and we'll cradle you out." But I knew
That the day and the night must pass and the old dews
Of another morning before such men unknowing

The ways of mountains could win to the chimney's top.
And then, how long? And he knew ... and the hell of hours
After that, if he lived till we came, roping him out.
But I curled beside him and whispered, "The bleeding will stop.

You can last." He said only, "Perhaps ... For what? A wheelchair,
Bob?" His eyes brightening with fever upbraided me.
I could not look at him more and said, "Then I'll stay
With you." But he did not speak, for the clouding fever.

I lay dazed and stared at the long valley,
The glistening hair of a creek on the rug stretched
By the firs, while the sun leaned round and flooded the ledge,
The moss, and David still as a broken doll.

I hunched to my knees to leave, but he called and his voice
Now was sharpened with fear. "For Christ's sake push me over!
If I could move ... Or die ..." The sweat ran from his forehead,
But only his eyes moved. A hawk was buoying

Blackly its wings over the wrinkled ice.
The purr of a waterfall rose and sank with the wind.
Above us climbed the last joint of the Finger
Beckoning bleakly the wide indifferent sky.

Even then in the sun it grew cold lying there ... And I knew
He had tested his holds. It was I who had not ... I looked
At the blood on the ledge, and the far valley. I looked
At last in his eyes. He breathed, "I'd do it for you, Bob."

IX

I will not remember how nor why I could twist
Up the wind-devilled peak, and down through the chimney's empty
Horror, and over the traverse alone. I remember
Only the pounding fear I would stumble on It

When I came to the grave-cold maw of the bergschrund ... reeling
Over the sun-cankered snow-bridge, shying the caves
In the névé ... fear, and the need to make sure It was there
On the ice, the running and falling and running, leaping

Of gaping green-throated crevasses, alone and pursued
By the Finger's lengthening shadow. At last through the fanged
And blinding seracs I slid to the milky wrangling
Falls at the glacier's snout, through the rocks piled huge

On the humped moraine, and into the spectral larches,
Alone. By the glooming lake I sank and chilled
My mouth but I could not rest and stumbled still
To the valley, losing my way in the ragged marsh.

I was glad of the mire that covered the stains, on my ripped
Boots, of his blood, but panic was on me, the reek
Of the bog, the purple glimmer of toadstools obscene
In the twilight. I staggered clear to a fire waste, tripped

And fell with a shriek on my shoulder. It somehow eased
My heart to know I was hurt, but I did not faint

And I could not stop while over me hung the range
Of the Sawback. In blackness I searched for the trail by the creek

And found it ... My feet squelched a slug and horror
Rose again in my nostrils. I hurled myself
Down the path. In the woods behind some animal yelped.
Then I saw the glimmer of tents and babbled my story.

I said that he fell straight to the ice where they found him,
And none but the sun and incurious clouds have lingered
Around the marks of that day on the ledge of the Finger,
That day, the last of my youth, on the last of our mountains.

PUBLISHED IN
1940

∞ WRITING ACTIVITY

Write an essay that interprets the poem. Incorporate the reader response approach and the structuralist, formalist, sociological and biographical approaches.

Here are some suggestions to help you with your analysis.

1 READER'S RESPONSE

- What were your thoughts about the narrator's decision? Do you agree or disagree with his actions?

- Write down some descriptive imagery that you found most engaging in the poem. Explain your choices.

- Do you have a best friend? What characteristics about him or her do you consider the most important?

- In your opinion, what is the message of the poem?

2 CHARACTERIZATION

- Who are the major characters in the poem?

- What is their relationship to each other?

- What do they share in common?

- How are they similar or dissimilar?

- How is each character introduced in the poem?

- At what point in the poem is the narrator named? How is this significant?

3 STRUCTURE AND FORM

- What kind of poem is "David"—a lyrical poem, an elegy, an ode, and so on?

- Analyze the verse, rhythm and rhyming scheme of the poem.

- What is the dramatic structure of the poem?

- Does the poem contain a dominant sound? Give examples of alliteration, assonance and consonance from the poem.

4 SETTING

- Describe the setting of "David."

- Give examples of the passage of time in the poem.

- How does the time element influence characterization in the poem?

5 POETIC DEVICES

- Find examples of metaphors and similes in the poem.

- Which senses create the poem's dominant mood? Find examples of imagery related to these senses.

- Find examples of foreshadowing in the poem.

- Find examples of personification in the poem.

- Find examples of descriptive words created by the poet. Are these effective in creating the atmosphere of the poem?

- Find examples of repetition in the poem.

- Find examples of contrasting words and images in the poem.

6 SOCIOLOGICAL AND HISTORICAL FACTORS

- How has society traditionally viewed euthanasia?

- What were the morals and values of society at the time the poem was written?

- Research euthanasia from a legal point of view.

- Think about the poet's lifestyle, religious philosophy, interests and personal problems.

- Think about different influences on the poet.

- What was taking place in the world around him?

Wrap-Up Activities

1. What is the theme of the poem? Is there more than one theme in the poem? If so, do the themes relate to each other?

2. Contrast the poem's form and structure with its content.

3. Research the life of the poet. How do elements in the poet's life shape the poem?

4. How is euthanasia viewed in our society? Discuss the ethical controversy of this subject.

5. Does the narrator have a choice regarding the character of David?

6. Discuss whether "David" falls into the traditional definition of an elegy—a poem that grieves for someone who is dead.

7. Analyze the literary period during which the poem was written.

The Swimmer

ABOUT THE **AUTHOR**

Irving Layton (Israel Pincu Lazarovitch) was born in 1912 in Romania. In 1913, his family immigrated to Montreal, where he completed a university degree in agriculture while honing skills as a debater and poet. He worked at a variety of odd jobs before completing an M.A. in Political Science and becoming a teacher. Apparently, Layton wrote "The Swimmer" in an exalted frenzy in a Montreal restaurant in 1944, deciding then and there that he would devote his life to poetry. Layton was a prolific poet, whose work breathed vigour and life into the Canadian poetry scene. He was nominated for a Nobel Prize in the early 1980s. He married several times and died in January, 2006.

The Swimmer
Irving Layton

The afternoon foreclosing, see
The swimmer plunges from his raft,
Opening the spray corollas by his act of war
The snake heads strike
Quickly and are silent.

Emerging see how for a moment
A brown weed with marvellous bulbs,
He lies imminent upon the water
While light and sound come with a sharp passion
From the gonad sea around the Poles
And break in bright cockle-shells about his ears.

He dives, floats, goes under like a thief
Where his blood sings to the tiger shadows
In the scentless greenery that leads him home,
A male salmon down fretted stairways
Through underwater slums …

Stunned by the memory of lost gills
He frames gestures of self-absorption
Upon the skull-like beach;
Observes with instigated eyes
The sun that empties itself upon the water,
And the last wave romping in
To throw its boyhood on the marble sand.

PUBLISHED IN
1944

✑ WRITING ACTIVITY

Write an essay that interprets the poem. Use elements from the reader response, sociological, formalist and structuralist approaches. Here are some suggestions to help you.

1 READER'S RESPONSE

- What was your reaction to the poem? How did it make you feel?

- Which images in the poem did you find most powerful?

- Did anything strike you about the swimmer himself?

- How would you categorize the type and level of activity portrayed by the poem? Is it mainly physical or mental?

2 SETTING

- Describe the setting of the poem. Is the time as important as the place?

- What aspects of the setting are important to the mood of the poem?

- What kind of relationship does the swimmer have with the water?

- What are the biological origins of the swimmer according to the poem?

3 IMAGERY

- Look for examples of maleness, power and sexuality in the poem.

- Are there female elements in the poem? Are they seductive?

- What metaphors and similes are present in the poem and how do they enhance the
  – description of the swimmer?
  – mood of the poem?

- Which senses are addressed by the imagery? Which senses are dominant?

- Look for opposing images in the poem and the types of tension they create.

- What do you think "instigated eyes" are like?

- What characteristic(s) do the "skull-like beach" and the "marble sand" have in common?
  – How is this imagery different from the other imagery in the poem?

- How does the imagery affect the mood of the poem?

4 DICTION

- Is the language in the poem concrete or abstract?

- Look for unusual words and phrases in the poem. What is their impact on the flow of language and effectiveness of imagery?

- What do you associate with the word "Poles"? Why do you think this word is capitalized in the poem?

5 STRUCTURE AND FORM

- Analyze the verse and rhyme scheme of the poem.

- What is the purpose of each stanza? What does each one portray?

- How is each stanza different from the others in the poem? Is there a progression of some sort?

15

6 THEME

- What relationships, in terms of power, can you see in the poem?

- Consider this statement: "He frames gestures of self-absorption." What do you think this means? Whom do you think this statement refers to in a broader sense?

- What do you think the swimmer and the water symbolize in the poem?

Wrap-Up Activities

1. How does the poet use nature to express the theme in "The Swimmer"?

2. Explore the metaphor of the swimmer in the poem.

3. What was the impact of Irving Layton's poetry on the evolution of Canadian poetry?

4. Discuss how Irving Layton uses imagery and diction to bring out the underlying theme of "The Swimmer."

5. Compare the portrayal of nature in "The Swimmer" to that in "David" by Earle Birney.

6. Describe Irving Layton's vision of mankind as presented in "The Swimmer."

7. Compare two of Irving Layton's poems and relate them to his life.

A Kite Is a Victim

ABOUT THE **AUTHOR**

Leonard Cohen (1934-) is one of Canada's most popular and influential poets and songwriters. He wrote his first book of poems in 1956 as an undergraduate at McGill University. "A Kite Is a Victim" appears in his second book of poems, *The Spice Box of Earth*, published in 1961. By 1966, Cohen had published a third book of poems, as well as two novels, and had begun to write and record songs. His works contain themes about religion, sex, poetry, death, beauty and power, and he has been called a "black romantic." Cohen's awards include two Governor General's Awards and five Juno Awards. In 2003, he was named Companion of the Order of Canada. He continues to write poetry and songs.

A Kite Is a Victim
Leonard Cohen

A kite is a victim you are sure of.
You love it because it pulls
gentle enough to call you master,
strong enough to call you fool;
because it lives
like a desperate trained falcon
in the high sweet air,
and you can always haul it down
to tame it in your drawer.

A kite is a fish you have already caught
in a pool where no fish come,
so you play him carefully and long,
and hope he won't give up,
or the wind die down.

A kite is the last poem you've written,
so you give it to the wind,
but you don't let it go
until someone finds you
something else to do.

A kite is a contract of glory
that must be made with the sun,
so you make friends with the field
the river and the wind,
then you pray the whole cold night before,
under the travelling cordless moon,
to make you worthy and lyric and pure.

PUBLISHED IN
1961

✍ WRITING ACTIVITY

Write an interpretation of the poem. Incorporate the reader response approach and the formalist and structuralist approaches.

Use the following suggestions of activities and questions for inspiration.

1 READER'S RESPONSE

- Make a list of five things you notice about the poem.

- What does a kite make you think of?
  – Which eight words (not present in the poem) do you associate with a kite?
  – What kind of feelings do you experience when you think of a kite?

- What does the poem's title make you think of?

2 POINT OF VIEW AND AUDIENCE

- Who is the persona or speaker in the poem?

- Who does "you" refer to?

- Who is the intended audience?

- What is the speaker's purpose?

3 IMAGERY

- What four metaphors does the author use to talk about a kite?
  – How is a kite similar to and different from each of the four metaphorical images?
  – How are the metaphors similar to and different from each other?

- What contrasting images and words are presented in the first stanza of the poem?

- How is the "travelling cordless moon" different from the kite?
  – What do you think this image symbolizes?

- Which words or phrases are repeated in the poem?
  – What role or purpose does this repetition play?

4 DICTION

- What level of language does the poem use?

- How concrete/abstract is the language?

- Are denotation and connotation important to the poem?

5 STRUCTURE

- Does the poem have an open or closed structure?

- Consider the order of the stanzas. Why do you think the author chose to place them in this order?

- What is the effect of using different metaphors in the poem?
  – Do you think the poem's meaning would be generated as effectively with only one metaphor?

- How does the poet create tension in each stanza of the poem? How does he create tension in the poem as a whole?
  – How does this tension help develop the theme of the poem?

- Make a list of the opposites that appear in the poem.
  – What effects do these opposing words and images create in the mind of the reader?

6 THEME

- Can the reader discover the poem's meaning directly?
  – If not, how is the meaning discovered indirectly?
  – What role do opposing associations play in conveying the poem's meaning?

- What two opposing forces are at play when you fly a kite?
  – Are these forces present in the imagery of the poem?

- What do you think the kite and its opposing associations represent in the poem?

Wrap-Up Activities

1. Describe how imagery and other poetic devices bring out the theme of "A Kite Is a Victim."

2. How does Leonard Cohen use different perspectives and opposites to develop the theme of "A Kite Is a Victim"?

3. Some people believe that Leonard Cohen is the voice of his generation. Write an essay about this idea.

4. Compare one of Leonard Cohen's poems to one of his songs.

5. Compare "A Kite Is a Victim" to "Eagle: A Fragment" by Alfred Lord Tennyson or "The Swimmer" by Irving Layton.

6. Write about the influences found in Leonard Cohen's poems and songs.

7. Write a poem about a common object using different perspectives to generate meaning.

Calamity

ABOUT
THE **AUTHOR**

F. R. Scott was born in Quebec City in 1899. He studied as a Rhodes Scholar at Oxford University. Scott was a social philosopher and became a law professor and then Dean of Law at McGill University in Montreal. He often wrote about social concerns. His poem "Calamity" is part of *The Collected Poems of F. R. Scott* (1981), which won a Governor General's Award. Scott died in Montreal in 1985.

Calamity

F. R. Scott

A laundry truck

Rolled down the hill

And crashed into my maple tree.

It was a truly North American calamity.

Three cans of beer fell out

(Which in itself was revealing)

And a jumble of skirts and shirts

Spilled onto the ploughed grass.

Dogs barked, and the children

Sprouted like dandelions on my lawn.

Normally we do not speak to one another on this avenue,

But the excitement made us suddenly neighbours.

People exchanged remarks

Who had never been introduced

And for a while we were quite human.

Then the policeman came—

Sedately, for this was Westmount—

And carefully took down all names and numbers.

The towing truck soon followed,

Order was restored.

The starch came raining down.

PUBLISHED IN
1981

✍ WRITING ACTIVITY

Write an interpretation of the poem that incorporates elements from the reader response, sociological and formalist approaches. Here are some suggestions to help you with your interpretation.

1 READER'S RESPONSE

- Describe a similar "calamity" that you have witnessed.

- In your opinion, what is the significance of the poem's title?

- Do you agree with the poet that North American culture promotes isolation?

2 NARRATION

- Who is the poem's narrator?

- What do you know about him or her?

- Why is he or she telling you about this incident?

- Who are the other characters in the poem?

3 POETIC ELEMENTS

- Analyze the ironic elements of the poem: the title, the situation and the last line.

- Consider the humorous and journalistic elements of the poem.

- Look at the types of images found in the poem, their significance and the way they are placed and used in the poem.

- Consider the poem's plot or storyline.

- How does the poet use narrative techniques (a story with a dramatic structure) to make his point?

4 DICTION AND FORM

- Is the language in the poem concrete or abstract?

- What kind of sound devices are used in the poem and what is their effect?
  – Does the poem contain examples of consonance, assonance, onomatopoeia, rhyme, euphony or cacophony?

- Does the poem have a regular rhythmic structure (meter)?

- Is "Calamity" an open or closed poem?

- Do the diction and form fit the mood and theme of the poem?

5 SOCIOLOGICAL DIMENSION

- What can you deduce about Westmount (an autonomous municipality within the urban community of Montreal, Quebec)?

- How does the narrator feel about Westmount?
  – How do you know? Choose a line from the poem to support your answer.

- What can you deduce about the narrator's view of North American society?

Wrap-Up Activities

1. Write about the principal and secondary themes of "Calamity."

2. What kind of sociological judgments does "Calamity" make?

3. Analyze the form and structure of the poem. How do they help develop the theme of the poem?

4. Compare North American culture to that of another country. How are we similar to or different from other cultures in terms of our attitudes regarding family and community?

5. How does the poet use irony and other poetic devices to bring out the theme in "Calamity"?

6. Write a poem of your own that has a sociological perspective brought out by humour or irony.

Treblinka Gas Chamber

ABOUT
THE **AUTHOR**

Phyllis Webb was born in 1927 in Victoria, British Columbia. She has taught at the University of British Columbia and worked at CBC Radio. She has also written many works of poetry. Webb is known for her poems about philosophical and social concerns. The poem "Treblinka Gas Chamber" was originally published in 1982 in a book called *The Vision Tree: Selected Poems,* which won her a Governor General's Award. For many years, she has also been exploring visual arts as a painter and collage artist.

Treblinka Gas Chamber
Phyllis Webb

Klostermayer ordered another count of the children.
Then their stars were snipped off and thrown into
the centre of the courtyard. It looked like a field of
buttercups. —JOSEPH HYAMS. *A Field of Buttercups*

fallingstars
    a field of
        buttercups
      yellow stars
          of David
            falling
the prisoners
      the children
        falling
     in heaps
       on one another
          they go down
Thanatos
    showers
      his dirty breath
         they must breathe
           him in
   they see stars
       behind their
          eyes
David's
  a field of
     buttercups
    a metaphor
      where all that's
        left lies down

PUBLISHED IN
1982

∞ WRITING ACTIVITY

Write an interpretation of the poem that incorporates elements from the reader response, historical and formalist approaches.

You will have noticed that the poem is arranged into a special shape. It is an example of a poem issuing from an international, post-World War II movement known as **concrete** poetry. Concrete poets rejected conventional forms and structure, experimenting with unusual grammatical structures, spelling, sounds and the actual shape of the written poem on a page.

Here are some aspects to examine in your approach to the poem.

1 READER'S RESPONSE

- What do you associate with the title of the poem?

- Which words or phrases in the poem are the most meaningful to you? List them.
  – Explain why they are meaningful to you.

2 HISTORICAL CONTEXT

- Research the historical context of Treblinka.

- Who does the name "David" refer to?

- Who is Thanatos?

3 FORM AND STRUCTURE

- What is the significance of the special shape of the poem?

- Why do you think the author chose this particular form for the poem?

- Is the form of the poem effective? Why or why not?
  – What feeling of movement is created by the shape of the poem?

- What do you notice about grammatical conventions, such as sentence structure, in the poem?

- What do you think this signifies?

- What effect does the recurrent use of the image of a field of buttercups have on the unity of the poem?

- What do you perceive to be the poet's reaction to the events she describes? What makes you think this?

4 POETIC ELEMENTS

- Find examples of metaphors in the poem.

- What are the different uses and meanings of the word "stars"?

- Are there other symbols in the poem?

- Find links between the images.

- Find examples of repetition in the poem.

- What is the predominant colour in the poem?

Wrap-Up Activities

1. Write about the connection between the form and theme of "Treblinka Gas Chamber."

2. Write about the historical context of the poem.

3. How does the poet use imagery and symbolism to bring out the theme of "Treblinka Gas Chamber"?

4. Discuss the contrast in diction between the title and the rest of the poem.

5. Write about the poet's influences and link them to the poem.

6. Compare "Treblinka Gas Chamber" to another artistic portrayal (in a novel, short story, film, and so on) on the same subject. Explain the impact that the portrayal has on the reader, viewer, etc.

Meditations on the Declension of Beauty by the Girl with the Flying Cheekbones

ABOUT
THE **AUTHOR**

Marlene Nourbese Philip was born in 1947 and grew up in Tobago, but now lives in Canada. She is a lawyer, writer and poet. Her poetry and other writings have been published widely in Canada, the United States and Great Britain. The award-winning Toronto author was named a Guggenheim Fellow in poetry in 1990.

Meditations on the Declension of Beauty by the Girl with the Flying Cheekbones

M. Nourbese Philip

If not   If not   If
Not
If not in yours
                In whose
In whose language
Am I
If not in yours
                In whose

In whose language
Am I   I am
                    If not in yours
In whose
            Am I
(if not in yours)
            I am yours
In whose language
                Am I not
Am I not   I am yours
If not in yours
If not in yours
                    In whose
In whose language
            Am I ...

Girl with the flying cheekbones:

She is

I am

Woman with the behind that drives men mad

And if not in yours

Where is the woman with a nose broad

As her strength

If not in yours

In whose language

Is the man with the full-moon lips

Carrying the midnight of colour

Split by the stars—a smile

If not in yours

                    In whose

In whose language

                Am I
                Am I not
                Am I   I am yours
                Am I not   I am yours
                Am I   I am
If not in yours
                    In whose
In whose language
            Am I
If not in yours
            Beautiful

PUBLISHED IN
1989

✎ WRITING ACTIVITY

Write an interpretation of the poem that incorporates elements from the reader response, formalist, feminist and minority approaches.

Here are some suggestions to help you analyze the poem.

1 READER'S RESPONSE

- What was the first characteristic of the poem that captured your attention?
- Define beauty. What are some characteristics of beauty?
- In your opinion, what is the poet trying to say in the poem?

2 FORM AND STRUCTURE

- Examine the use of repetition and other rhythmic elements in the poem.
- Why does the narrator of the poem repetitively ask questions?
- Examine the form of the poem.
- Consider wordplay and its importance to the meaning and form of the poem, for example, "Am I—Am I not" and "Am I—I am."
- How effective are the form and wordplay in the poem?

3 IMAGERY

- What are some examples of imagery in the poem?
- What senses do the images employ?
- What are the effects of the imagery on the meaning of the poem?

4 SOCIOLOGICAL AND FEMINIST ELEMENTS

- What does the author mean by the word "language"?
- What is the underlying question that the poem asks?
- What is the significance of the final word of the poem?
- How are men and women portrayed in our society?
- Discuss racial and gender stereotypes.

Wrap-Up Activities

1. Write about the implicit message of "Meditations on the Declension of Beauty by the Girl with the Flying Cheekbones" and link it to contemporary issues surrounding gender and race.

2. Write about the form, structure and language of the poem. How are these elements used to develop the poem's theme?

3. Discuss the appropriateness of the title "Meditations on the Declension of Beauty by the Girl with the Flying Cheekbones" in regard to the rest of the poem.

4. Research the historical forces that brought Caribbean immigrants to North America.

5. Compare this poem to another poem written by the same author.

6. Is the poem only meaningful to women, or is it meaningful to both sexes?

Levels

ABOUT
THE **AUTHOR**

Robyn Sarah was born in New York City in 1949. She graduated from the Quebec Conservatory of Music and also completed an M.A. in English at McGill University in Montreal after majoring in philosophy. Robyn Sarah's first published poems appeared in the 1970s, and since then, she has written seven collections of poems, numerous essays and two books of short stories. Her work, known for its musicality and thoughtful insights, has been widely published in Canada and the United States, and she has won awards for her poetry and fiction. Robyn Sarah taught English for twenty years at the college level and presently lives in Montreal, where she continues to write.

15

Levels
Robyn Sarah

In this city the hospitals
are on the hill, the sick look down
from their high place, upon the tortuous
peregrinations of the well,
or they look up, they gaze on the serene
procession of clouds. And theirs
is the realm between.

I think of you up there,
remote behind your allocated pane,
your porthole on the man-swarm
and eternity. No way to know
which way you're facing now,
what side you'll exit on, this time,
how much you think on it, or care.

A life is a life. What
will we make of that?
What is the real world?
Privately, no one believes
he's living in it.

We are about to begin the descent,
the voice says. We say: *I've paid my dues.*

Sunset. It is the hour when hospital windows
Beam gold into the eyes
Of runners on the upper avenues.

PUBLISHED IN
2003

∾ **WRITING ACTIVITY**

Write an interpretation of the poem that incorporates elements from the reader response, formalist and sociological approaches.

Here are some aspects for you to consider.

1 READER'S RESPONSE

- What did the poem make you think about in general?
  – How did it make you feel?

- Which image in the poem did you find most striking? Why?

2 POINT OF VIEW AND SOCIOLOGICAL CONSIDERATIONS

- Who does "they" refer to in the first stanza of the poem?

- Who is the poem about ("you" in the second stanza)?
  – Why is this person in the place he or she is?

- From what point of view is the poem narrated?
  – What can we assume about the speaker's state of health?

- Who does "he" refer to in the third stanza?

- Who is "We" in the third and fourth stanzas?

- Who do you think "the voice" refers to?

- What groups of people are contrasted in the poem?
  – How do these contrasts bring out the poem's theme?

- What are the characteristics of the different "levels" in the poem?

3 SETTING

- What is the setting of the poem?

- Where is the protagonist of the poem situated?
  – What can the protagonist see above and below himself or herself?

- Where is the speaker situated in the poem?
  – What is the distance between the speaker and the protagonist?

- How does the title add to the meaning of the poem?

4 IMAGERY

- What contrasting images can you find in the poem?

- What explicit and implicit metaphors are used in the poem?

- What natural imagery does the poem incorporate and what is its significance?

5 DICTION

- What do the following refer to?
  – "tortuous peregrinations"
  – "realm in between"
  – "your allocated pane, your porthole"
  – "I've paid my dues."

15

6 STRUCTURE AND THEME

- What is the role of each stanza in the overall context of the poem?
  - First summarize the meaning of each stanza, and then look at each stanza's importance to the poem.
- What creates tension in the poem?
  - What possible meanings does this tension bring out in the poem?
- What is the speaker wondering literally?
  - What broader "wondering" do you think this represents?
  - Who do you think the protagonist of the poem represents?
- What is the main theme of the poem?

Wrap-Up Activities

1. How do spatial imagery and perspective bring out the theme of "Levels"?

2. Explain the significance of "Levels" as a title for the poem.

3. How does the poet use the image of a hospital and sickness to explore the human condition in "Levels"?

4. In "Levels," the speaker asks two philosophical questions: "A life is a life. What will we make of that?" and "What is the real world?" Explain the role of these questions in the poem.

5. Compare "Levels" to one of the other poems that you have read in this chapter.

6. Write an essay about how poetry can be used to explore important aspects of human experience.

✐ GENERAL ESSAY TOPICS

1. Compare an example of a song from a specific period to a poem from the same era.

2. Compare a short story and a poem from the Romantic Age. Illustrate how they reflect the preoccupations of artists of the period.

3. Compare a contemporary poem in which the poet experiments with form to a short story in which the author experiments with short story conventions.

4. There are many types of poems in English poetry, including the narrative, lyric, ode, elegy, epic and ballad. Research the form and development of one of these types of poems.

5. Choose a poem that you particularly like. Read it aloud and present an oral analysis to the class.

6. Write about Canada's involvement in World War I or World War II. Relate it to either "In Flanders Fields" or "Treblinka Gas Chamber."

7. Write a poem.

8. Alone or with a partner, choose a period of time and a selection of poets from the era. Study their backgrounds and read their key works. Prepare a short report. Your instructor may want you to present the report to a group, to the class or in written form.

9. Do a more in-depth analysis about a literary period. Research the social, political and economic trends of the time, taking note of their influence on poetry. Prepare a short report. Your instructor may want you to present the report to a group, to the class or in written form.

10. Choose a favourite poem and create a work of art inspired by it. Present your work to the class.

11. Research in detail the life and works of a particular poet.

You may choose one of the poets mentioned in this book or someone else. Here is a list of some poets to consider.

Maya Angelou	Phyllis Gottlieb	Al Purdy
Matthew Arnold	Langston Hughes	Christina Rossetti
Margaret Atwood	Ted Hughes	Edna St. Vincent Millay
W. H. Auden	John Keats	William Shakespeare
Gwendolyn Brooks	Joy Kogawa	Percy Bysshe Shelly
Robert Browning	Rudyard Kipling	Gertrude Stein
Geoffrey Chaucer	Dorothy Livesay	Wallace Stevens
George Elliot Clarke	Longfellow	Marc Strand
Lucille Clifton	John Milton	Dylan Thomas
Billy Collins	Marianne Moore	Miriam Waddington
e. e. cummings	Ogden Nash	Henry Wadsworth
Walter de la Mare	P. K. Page	Alice Walker
Louis Dudek	Sylvia Plath	Anne Wilkinson
George Eliot	Alexander Pope	William Carlos Williams
T. S. Eliot	Ezra Pound	W. B. Yeats

DRAMA

After a brief summary of the evolution of drama from a historical perspective, Chapter 16 discusses the elements of a Shakespearean tragedy. Chapter 17, by contrast, introduces elements of a one-act play.

History *of* Drama

Drama as a literary form has existed since the earliest civilizations. The ancient Greeks and Romans developed highly sophisticated forms of the genre, writing both tragedies and comedies. Tragedy originated in Greece in the fifth century BC, and stressed the relationships between human beings and the gods, who offered insightful advice on problems. For example, the great playwright Sophocles (496-406 or 405 BC) showed humans to be both weak and heroic, as is evident in *Oedipus Rex*. In this tragedy, Oedipus becomes king by unknowingly slaying his father and marrying his mother, thereby fulfilling his destiny. The tragedies of Euripides (480-406 BC) questioned the traditional role of the gods. In plays such as *Electra* (date unknown), *Helen* (412 BC) and *Iphigenia in Aulis*, (405 BC), he showed great psychological insights through his characters.

Ancient Roman theatre was influenced by Greek tragedy and comedy. The early Roman dramas were performed at festivals and sporting contests. The first theatre in Rome was built of stone in the first century BC. Comedies, farces and slapsticks were favourites of the public. Among some ancient Roman comedy playwrights were Levius Andronicus and Plautus, both of whom wrote fanciful plays about love or slightly mocking plays about the upper classes, incorporating slapstick humour and music into their works. During the reign of Julius Caesar, audiences also enjoyed mimes. Eventually, large spectacular shows replaced traditional tragedies and comedies in the Roman Empire.

English drama can trace its origins to Rome. The popularity of drama lessened after the fall of the Roman Empire until about the tenth century. Two factors contributed to this decline. First, the early Christians and the infant church condemned play-acting because drama was associated with paganism. Second, non-Christian invaders throughout Europe kept indigenous forms of theatre from flourishing because of the uncertainty of life due to war, and as a result, there are few remaining records of early Christian plays. The absence of documents makes it impossible for us to know which types of plays were performed at that time. It is likely that travelling players went from town to town, entertaining the public. These travelling players told their tales through song, acting and mime.

By the twelfth century, these players had become so popular that the Church, which had condemned drama as a pagan tradition, started using it for its own purposes. Drama became a means for promoting religious activities. Liturgical plays (dramatizations of prayers and stories from the Bible) were performed at Easter and Christmas, depicting the traditional stories of Creation and the birth of Christ, eventually leading to the advent of non-devotional drama.

In addition, the miracle play (portraying stories of miracles) developed from liturgical drama and led to the creation of the morality play. In morality plays, different characters represented abstract qualities. The most famous morality play of the time was *Everyman* (*c*.1500), the story of a character by the same name who meets Death and finds that his friends forsake him. Only his friend Good Deeds follows Everyman on his journey to meet Death. Morality plays showed the progression from liturgical plays, which were wholly religious in theme, to those that were less religious and more didactic and political.

By the sixteenth century, drama had adopted realism, becoming concerned with humanistic themes. Humanism emphasized the ability of people to reason and evaluated the place of man in the universe. The movement was founded in northern Italy in the fourteenth century, when Petrarch (1304-1374), a theologian and man of letters, created an educational system emphasizing the study of classical literature, such as poetry, grammar, rhetoric and history. The sixteenth century also popularized two forms of theatre that had their roots in classical traditions: the tragedy and the comedy. Tragedy was influenced by the Roman writer Seneca (*c.* 4 BC-65 AD), who lived around the time of Nero, while comedy was influenced by the Roman writers Plautus (*c.* 254-184 BC) and Terence (*c.* 190-159 BC). The first comedy to develop along the lines of classical theory was *Ralph Roister Doister* (1553-1554) by Nicholas Udall. Another form that gained popularity at that time was the chronicle play (a popular presentation of history), for example, *The True Chronicle of King Leir* (1594, 1604), a predecessor to Shakespeare's *King Lear*.

Elizabethan England marked a flowering of drama. It was the time of the great playwrights Thomas Kyd (1558-1594), Christopher Marlowe (1564-1593) and, of course, William Shakespeare (1564-1616). Shakespeare, considered to be one of the greatest writers of all time, wrote histories, tragedies and comedies. The heroes of his historical dramas were ancient kings and queens. The heroes of his tragedies were men who, despite possessing great character, suffered weaknesses that either lead to their corruption or landed them in impossible situations. His comedies always revolved around love. *Henry VI* (1588) was Shakespeare's earliest work, and since then, his plays have withstood the test of time. They have been produced on stage throughout the centuries, and in modern times, most, if not all, of his plays have been adapted for film and television, for example, *Romeo and Juliet*, *King Lear*, *The Tempest* and *Hamlet*, just to name a few.

Although theatre flourished during the time of Elizabeth I, by the time Charles I (1600-1649) came to the throne in 1625, it was in serious peril. This was due largely to a new religious climate in England called Puritanism, a form of worship within the confines of Protestantism. Puritans believed that excess ceremony or ritual in religion was wrong. Holding that the extravagance of the Catholic Church, which relied heavily on ceremony, was evil, they abolished all rites and rituals from their system of worship. Puritans were a powerful and influential group in England at the time. Charles I had many conflicts with Parliament, which was largely composed of Puritans. When he tried to impose the liturgy of the Anglican Church or the Church of England, which consisted of ritual and ceremony, over England and Scotland, there was a large public outcry. Other problems, for example, money, caused him to struggle against Parliament and eventually led to civil war. Charles I was beheaded in 1649, and the monarchy was not restored until 1660. In the interregnum, the English government was run by Parliament. Life was very strict for the general population during those years. The Puritans banned all forms of entertainment and amusement, especially theatre, because they considered it to be not only disrespectful toward political figures but also the work of the devil.

After the restoration of the monarchy in 1660, a period known in English history as the "Restoration," theatre redeveloped. However, during that time, plays were often obscene and raucous in reaction to the period of the ban. Drama eventually developed into two branches during that era: heroic plays and comedies of manners. The latter took a cynical and witty look at court life and was greatly influenced by the French playwright

Molière. In England, John Dryden (1631-1700) was perhaps the most important playwright of the day, writing many comedies, heroic plays and tragedies.

The period from the eighteenth century to the mid-nineteenth century represented a quieter era for drama. Certain forms, such as the drama of sensibility, as well as parodies and burlesques, were prominent. Important playwrights of that period included Henry Fielding, author of *Tom Thumb* (1730), John Gay, author of *The Beggar's Opera* (1728), a ballad-opera play, and Oliver Goldsmith, author of *She Stoops to Conquer* (1773), a comedy.

The second half of the nineteenth century experienced another revival of the theatre. Two of its greatest representatives were Oscar Wilde (1854-1900) and George Bernard Shaw (1856-1950). Wilde wrote comedies such as *An Ideal Husband* (1895) and *The Importance of Being Earnest* (1895). Shaw criticized social evils, for example, prostitution, in *Mrs. Warren's Profession* (1894). In *Arms and the Man* (1894), he was concerned with the romantic image of the soldier. His comedies *Man and Superman* (1901) and *Pygmalion* (1912) were social commentaries.

The twentieth century was rich in new and traditional forms of theatre. Dramatists expressed social concerns and dealt with poverty, realism of war and its effects on the general population, political issues, the fast pace of changing technology and the individual's ability to adapt. Some famous playwrights included Somerset Maugham, author of *Caesar's Wife* (1919), and Noel Coward, author of *Private Lives* (1930). Both playwrights were known for their sophisticated comedies of manners. Other important dramatists included T. S. Eliot, who wrote *Murder in the Cathedral* (1935), W. H. Auden, who experimented with verse form in *The Ascent of F6* (1936), and Harold Pinter, who was awarded the Nobel Prize for literature in 2005.

English-language drama of the twentieth century also developed rich traditions outside Great Britain. Here is a list of some playwrights from different countries.

Australia: Elizabeth Backhouse, Kylie Tennant, Ray Lawler, Patrick White, Hal Porter, Douglas Stewart, David Williamson

Canada: James Reaney, George Ryga, Drew Hayden Taylor, Robert Lepage, Carol Bolt, Sharon Pollock, Betty Lambert, Joanna Glass, Judith Thompson

Caribbean: Derek Walcott, Kwame Dawes

India: Mahesh Dattani, Ninaz Khodaiji, Gurcharan Das, Girish Karnad

New Zealand: Barbara Anderson, Jeff Addison, Dave Armstrong

South Africa: Athol Fugard, Brett Bailey, Lesego Rampolokeng, Xoli Norman, Rajesh Gopie

United States: Arthur Miller, Tennessee Williams, David Mamet, Lanford Wilson, Sam Shepard, Ntozake Shange, Marsha Norman, Beth Henley, Tina Howe, Wendy Wasserstein, August Wilson

Dramatic Terms

Match the dramatic terms in column A with the definitions in column B.

A	B
____ **1** mime	**a)** a long speech delivered by one character
____ **2** play	**b)** a light piece of satire
____ **3** sketch	**c)** a performance of a story on stage
____ **4** improvisation	**d)** dramatic expression without speech
____ **5** monologue	**e)** a spontaneous expression of speech and actions to present a story
____ **6** pantomime	**f)** a farcical drama performed with mimicry
____ **7** skit	**g)** a short play—often musical
____ **8** oration	**h)** a drama with exaggerated emotions and a happy ending
____ **9** melodrama	**i)** a formal speech
____ **10** soliloquy	**j)** talking without regard to listeners

Discussion

1. Which do you prefer to see, a film or a play? Explain your answer.

_____

_____

2. Have you ever seen any plays performed? If so, make a list of the plays and write down what you liked and/or disliked about them.

_____

_____

_____

_____

_____

_____

_____

_____

_____

CHAPTER 16

The Tragedy of Othello, the Moor of Venice

CHAPTER CONTENTS

▸ **Reading:** *The Tragedy of Othello, the Moor of Venice* by William Shakespeare; author's biography; literary trends (realism and experimentalism)

▸ **Literary Elements:** mood; setting; characterization; dialogue; symbolism; irony; dramatic structure

▸ **Vocabulary Development:** Shakespearean English and modern English

▸ **My eLab Activity:** famous quotations from Shakespeare

ABOUT THE AUTHOR

William Shakespeare was born on April 23rd, 1564, in Stratford-upon-Avon, England. He married Anne Hathaway on November 28th, 1582, when he was 18. The couple had a daughter and then twins, a boy and a girl. Shakespeare went to London in around 1592 and tried to establish a career as an actor and playwright. By 1594, he had gained a reputation as both while writing for Lord Chamberlain's Men, a theatre company, in which he also held a managing partnership. The company's reputation was soon established, and its productions became favourites of royalty. Shakespeare died in April of 1616.

Although the precise dates of Shakespeare's works are uncertain, he wrote *The Tragedy of Othello, the Moor of Venice* around 1604. The play shows how tragedy results when reason submits to human emotions. The story of Othello was apparently taken from an Italian model. The fact that *The Tragedy of Othello, the Moor of Venice* and the rest of Shakespeare's works continue to be read in their original and translated versions shows that he is indeed one of the greatest writers the world has ever known.

STRUCTURE OF A PLAY

IN TRADITIONAL DRAMA, SUCH AS THE tragedies of William Shakespeare, a play is divided into five **acts**, each of which consists of several **scenes**. Acts are thirty minutes to an hour in length, whereas scenes can last for as little as a few seconds, or as long as thirty minutes. An entire performance takes two to three hours. In contemporary times, most plays have fewer acts and do not last as long as Shakespeare's (although Robert Lepage's seven-act *The Seven Streams of the Ota River* actually runs for seven hours and has two intermissions, including an hour-long meal break!).

Scenes are made up of **dialogue** (the actors' **lines** and/or **speeches**) and **stage directions** (instructions to the actors and director about the set, gestures, entrances and exits, and so on).

The plot (dramatic structure) of a traditional play is similar in structure to that of a traditional short story, **rising in action** to a climax in Act III; then **falling** until the final **dénouement** at the end of Act V. In the first parts of the play, the life and fortunes of the leading character (a **tragic hero** or **protagonist**) would go well, a conflict would then arise leading to a climax and tragedy, usually with much loss of life, including the protagonist's, at the end of the play.

Although the presence of a narrator is not entirely precluded, usually there is no narrator because the action is meant to unfold in the presence of the audience. We analyze drama using many of the same literary terms found in the analysis of short stories and novels: theme, plot, setting, characterization, irony, foreshadowing, imagery, dialogue and so on. Review these terms in the Literary Elements section at the beginning of the book.

16

LITERARY TRENDS
OF THE TIMES

THEATRE IN ENGLAND DURING THE Renaissance had a somewhat unsavoury reputation. The London city authorities refused to give permission for theatres to perform plays. Therefore, theatres were found outside the city limits. Most often, plays were performed in the afternoon because there was no electric lighting in those days. As it was considered somewhat improper for noble women to attend the theatre, they often wore masks to hide their identity while watching plays. Moreover, the roles of women and girls in plays were played by boys.

Elizabeth I, however, was a great patron of the arts. She gave her support to the playwrights and their productions. The London citizenry also enjoyed theatre performances. Elizabeth's reign included some of the most prolific writers of all time. Christopher Marlowe (1564-1593), a contemporary of William Shakespeare, enjoyed popular success. Ben Johnson (1572-1637) was another eminent dramatist of the times.

Most plays of the era can be categorized into three types: histories, tragedies and comedies. For example, Shakespeare wrote many historical plays about kings, such as *Henry V* and *Richard III*. Marlowe also wrote a historical play called *Edward II*. Tragedies were also popular subjects of plays. Marlowe's *Dr. Faustus*, Thomas Kyd's *The Spanish Tragedy* and Shakespeare's *Hamlet*, *King Lear* and *The Tragedy of Othello, the Moor of Venice* are just a few tragedies that enjoyed immense popularity. Comedies, such as Thomas Middleton's *A Chaste Maid in Cheapside* and Shakespeare's *As You Like It* and *Twelfth Night*, were also very successful.

Warm-Up Discussion Topics

1. Do you prefer watching theatrical performances or reading novels? Explain your answer.

2. Have you ever seen or read a work by William Shakespeare? If yes, what was your opinion of it?

∞ INITIAL READING

Go to My eLab Documents to print out and read the complete play, *The Tragedy of Othello, the Moor of Venice* by William Shakespeare (1604-1605).

Reader's Response

1. Did you find *The Tragedy of Othello, the Moor of Venice* difficult to understand when you first read it? If yes, what were the major sources of your difficulties?

_____

2. Which character in the play did you find most interesting and why?

_____

∞ CLOSE READING

Act 1

1. What is the setting of the act?

_____

2. What is the relationship between the following characters in the play?

Othello _____

Desdemona _____

Emilia _____

Iago _____

Michael Cassio _____

Roderigo _____

3. Iago hates Othello because

   a) he wants Othello's job.

   b) Desdemona was mean to Iago's wife.

   c) Iago is in love with Desdemona.

   d) Othello passed Iago over for promotion.

4. Which of the following statements is **false**?

   a) Roderigo is in love with Desdemona.

   b) Roderigo is a Venetian navy captain.

   c) Roderigo hates Othello.

   d) Roderigo plots with Iago against Othello.

5. Brabantio thinks that Othello has married Desdemona through

   a) love.

   b) bribery.

   c) witchcraft.

   d) all of the answers

6. How does Othello claim he has won Desdemona's love and affection?

   _____

7. Which of the following statements is **false**?

   a) Desdemona loves Othello.

   b) Brabantio eventually accepts his daughter's marriage.

   c) Othello married Desdemona because she was rich.

   d) The Venetian senators accept Othello's reasons for the marriage.

Act 2

8. What is the setting of the act?

   _____

9. Why is Othello sent there?

   _____

10. Iago plots against Cassio by

   a) getting him drunk.

   b) pretending to be his friend.

   c) instigating Cassio to fight against Montano.

   d) all of the answers

11. What is Othello's reaction when he hears that Cassio has been fighting with Montano?

   _____

12. What decision does Othello make regarding Cassio as a result of the fight?

   _____

13. Which of the following statements is **true**?

   a) Desdemona and Cassio are lovers.

   b) Cassio asks Desdemona to help him become friends with Othello.

   c) Othello hates Cassio.

   d) Desdemona hates Cassio.

14. Iago tells Cassio to seek Desdemona's help with Othello because Iago

    a) wants Cassio to succeed professionally.

    b) thinks that Othello was unjust to dismiss Cassio.

    c) believes that Cassio will give him a job promotion.

    d) wants Othello to become jealous of Cassio.

Act 3

15. In Scene 3, how does Iago plant the first seeds of doubt in Othello's mind regarding Desdemona?

_____

16. Iago wants Roderigo to stay in Cyprus because

    a) they are good friends.

    b) he wants Roderigo to seduce Desdemona.

    c) he wants Othello to hire Roderigo.

    d) Roderigo is his brother-in-law.

17. Why does Desdemona ask Othello to invite Cassio for dinner?

_____

18. Does Othello invite Cassio for dinner? _____

19. In Scene 3, Iago says to Othello, "She did deceive her father, marrying you; and when she seem'd to shake and fear your looks, she lov'd them most." In this statement, what is Iago implying about Desdemona?

_____

20. Why does Iago want Emilia to steal Desdemona's handkerchief?

_____

21. What lie does Iago tell Othello about the handkerchief?

_____

22. Does Othello believe Iago's lies? _____

23. The handkerchief is important to Othello because it

    a) was his mother's.

    b) is said to have magical qualities.

    c) was the first present he gave his wife.

    d) all of the answers

Act 4

24. Who does Cassio give the handkerchief to? _____

25. Which of the following statements is **false**?

    a) Othello believes that Desdemona gave the handkerchief to Cassio.

    b) Othello believes Iago is lying.

    c) Othello wants to kill Cassio.

    d) Othello wants to kill Desdemona.

26. What does Othello do to cause Lodovico to question his status as "the noble Moor"?

_____

27. How does Emilia react when Othello questions her about Desdemona?

    a) She defends her.

    b) She condemns her.

    c) She lies to Othello.

    d) all of the answers

28. What does Iago want Roderigo to do to Cassio?

_____

Act 5

29. What happens to Desdemona?

_____

30. Who tells the truth about Iago? _____

31. What happens to Emilia?

_____

32. What happens to Iago?

_____

33. What does Othello do when he finds out about Iago's treachery?

_____

✍ WRITER'S CRAFT

Symbolism

■ COLOUR

1. Read the following excerpt from the play.

> Your heart is burst, you have lost half your soul;
> Even now, now, very now, an old black ram
> Is tupping your white ewe. Arise, arise;
> Awake the snorting citizens with the bell,
> Or else the devil will make a grandsire of you:
> Arise I say.

(Act 1, Scene 1, 87-92)

a) What two colours are found in this excerpt?

_____

_____

b) Which characters in the play do the colours refer to?

_____

_____

2. Read the following excerpt from the play.

> Whether a maid so tender, fair, and happy,
> So opposite to marriage that she shunn'd
> The wealthy curled darlings of our nation,
> Would ever have, to incur a general mock,
> Run from her guardage to the sooty bosom
> Of such a thing as thou [...]

(Act 1, Scene 2, 66-71)

a) What colour connotation does the word "fair" have? _____

b) Who is the "fair ewe"? _____

c) What colour connotation does the phrase "sooty bosom" have? _____

d) Who does "sooty bosom" refer to? _____

e) Does "fair ewe" have a positive or negative connotation? _____

f) Does "sooty bosom" have a positive or negative connotation? _____

The colours black and white are used throughout the play to develop characterization.

3. Find a quotation that characterizes Othello differently from what people have said about him earlier. (*Hint*: look at Act 1, Scene 3.)

_____

4. How is Iago viewed by Othello and the other characters in the play?

a) positively

b) negatively

5. What is Iago's true colour (his characteristics)?

a) black

b) white

■ WATER

6. Read the following excerpt from the play.

> But that I love the gentle Desdemona,
> I would not my unhoused free condition
> Put into circumscription and confine
> For the sea's worth.[…]
>
> (Act 1, Scene 2, 25-27)

a) Othello is saying that his _____ for Desdemona is as deep

as the _____ .

7. Read the following excerpt from the play.

> […] O my soul's joy!
> If after every tempest come such calms,
> May the winds blow till they have waken'd death!
> And let the labouring bark climb hill of seas
> Olympus-high […]
>
> (Act 2, Scene 1, 187-191)

What do these words reveal about Othello's feelings for Desdemona?

a) He loves her passionately.

b) He does not want her to be there.

c) He does not love her.

8. Consider the settings of the play—Venice and Cyprus. Water is important to both places because Venice is built on water and Cyprus is an island.

a) How is Desdemona reunited with Othello on Cyprus?

_____

b) Water symbolizes the _____ that Desdemona and Othello feel for each other.

■ **HANDKERCHIEF**

9. To whom did the handkerchief originally belong? _____

10. What special powers is the handkerchief supposed to possess?

_____

11. What would happen if the owner of the handkerchief lost it?

_____

12. Why did Othello give the handkerchief to Desdemona?

_____

13. What does the handkerchief symbolize in the play?

_____

> *The Tragedy of Othello, the Moor of Venice* incorporates many different literary devices, such as irony, foreshadowing, metaphors and symbolism. We have only been able to consider a few of the more obvious symbols in the play.

VOCABULARY DEVELOPMENT

Shakespearean English and Modern English

Read the following soliloquy by Iago in Act 2, Scene 3 and answer the questions that follow.

> If I can fasten but one cup upon him,
> With that which he hath drunk to-night already,
> He'll be as full of quarrel and offence
> As my young mistress' dog. Now, my sick fool Roderigo,
> Whom love hath turn'd almost the wrong side out,
> To Desdemona hath to-night caroused
> Potations pottle-deep; and he's to watch:
> Three lads of Cyprus, noble swelling spirits,
> That hold their honours in a wary distance,
> The very elements of this warlike isle,
> Have I to-night fluster'd with flowing cups,
> And they watch too. Now, 'mongst this flock of drunkards,
> Am I to put our Cassio in some action
> That may offend the isle.—But here they come:
> If consequence do but approve my dream,
> My boat sails freely, both with wind and stream.

1. Who is the subject of Iago's speech?_____

2. What does Iago want to do?

_____

16

3. How is this speech significant in regard to Othello's tragic outcome?

_____

4. Rewrite the soliloquy using modern English.

Wrap-Up Activities

1. Write about the relationship between Othello and Iago.

2. Write about the literary techniques Shakespeare used to create his villain. Is Iago a round or flat character?

3. Act out a scene from *The Tragedy of Othello, the Moor of Venice*.

4. Write about the role of women in Shakespeare's work and times.

5. Research the political and social situations in Venice and Cyprus at the time the play was written. Is there any historical accuracy of the events mentioned in the play, such as the Turkish invasion of Cyprus?

6. Compare and contrast the characters of Desdemona and Emilia.

7. Write about the literary devices found in *The Tragedy of Othello, the Moor of Venice*, such as irony, imagery, symbolism, and so on.

My eLab ACTIVITY

My eLab

You are likely familiar with many famous quotations from Shakespeare. In My eLab, match each quotation with its meaning.

17

Trifles

CHAPTER
CONTENTS

▸ **Reading:** *Trifles* by Susan Glaspell; author's biography; structure of a play; literary trends (realism and experimentalism)

▸ **Literary Elements:** mood; setting; characterization; dialogue; symbolism; irony; dramatic structure

▸ **Vocabulary Development:** slang versus standard English

▸ **My eLab Activity:** from stage to radio

ABOUT
THE **AUTHOR**

Susan Glaspell was born in 1876 in Davenport, Iowa. Although she wrote short stories and articles, she was especially well known as a playwright. She and her writer husband, George Cram Cook, founded the Provincetown Players, a theatre group that experimented with new trends. The group moved to New York in 1916 and staged two one-act plays by Eugene O'Neill, as well as Glaspell's own play, *Trifles*. The productions were a great success. Glaspell's works reflected her own interests in women's issues, and *Trifles* provides a good example of her exploration of women's lives. Her other works include *The Outside* (1918), *Woman's Honour* (1918), *Bernice* (1919), *The Inheritors* (1921) and *The Verge* (1921). Glaspell won the Pulitzer Prize for drama in 1931 for *Alison's House* (1930).

LITERARY TRENDS
OF THE TIMES: THE ONE-ACT PLAY

THE SHORT PLAY BECAME POPULAR IN the eighteenth century as an after-show diversion. Its real purpose was often to entice latecomers to theatrical productions at reduced admissions prices. By the nineteenth century, the one-act play changed its objective and became the curtain-raiser (the pre-show entertainment) for the main attraction. Commercial theatre owners used short plays as a means to entertain the audience as they waited for latecomers to be seated.

Moreover, by the late nineteenth century, repertory theatres—non-commercial theatres that encouraged experimental forms in drama and poetry—grew rapidly. By the early twentieth century, many playwrights, such as

Eugene O'Neill, Susan Glaspell and George S. Kaufmann, used sketches and one-act plays as an opportunity to stage their works in repertory and small community theatres. Furthermore, vaudeville theatre, consisting of variety entertainment that included music-hall sketches, songs, dances and famous actors, developed as another popular form of entertainment in the early twentieth century. Great actors used one-act plays as vehicles to showcase their talents in variety theatre.

The one-act play is not just a shortened version of a full-length play—it can stand up in its own right as a complete literary work. The one-act play is defined as a short uninterrupted dramatic presentation. Like the short story, it has a unified structure that captures and develops a single episode in a character's life. In addition, like the short story, which must have a unity of effect, the one-act play conforms to a logical sequence in time, place and plot. The problem is quickly exposed and resolved. The one-act play continues to be a successful and popular dramatic form.

Warm-Up Discussion Topics

1. What sorts of things do you associate with the word *trifles*?

2. If someone learns that another person did something terrible, should the first person inform the authorities, regardless of the circumstances?

⟡ INITIAL READING

Reading a play involves different reading strategies from those associated with reading a novel or short story. Plays are meant to be performed by the players and seen by the audience. Every aspect of the performance—the set, costumes and physical appearance and gestures of the characters—are as important to the meaning of the play as the actual words spoken. Therefore, when reading a play, it is important that the reader pay careful attention to the characters' physical descriptions, gestures and actions, the sound effects, the props, such as furniture and other items seen on the set, and any other explanations or directions provided by the playwright.

As you read the following play for the first time, try to imagine the setting, as well as the characters and their voices. You may wish to read the play aloud with other students to add realism and enjoyment to the experience.

Trifles
Susan Glaspell

SCENE: *The kitchen in the now abandoned farmhouse of John Wright, a gloomy kitchen, and left without having been put in order—unwashed pans under the sink, a loaf of bread outside the breadbox, a dish towel on the table—other signs of uncompleted work. At the rear, the outer door opens, and the Sheriff comes in, followed by the county Attorney and Hale. The Sheriff and Hale are men in middle life, the county* Attorney *is a young man; all are much bundled up and go at once to the stove. They are followed by the two women—the Sheriff's Wife first; she is a slight, wiry woman with a thin, nervous face. Mrs. Hale is larger and would ordinarily be called more comfortable looking, but she is disturbed now and looks fearfully about as she enters. The women have come in slowly and stand close together near the door.*

COUNTY ATTORNEY (*rubbing his hands*). This feels good. Come up to the fire, ladies.

MRS. PETERS (*after taking a step forward*). I'm not—cold.

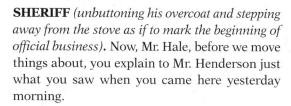

SHERIFF *(unbuttoning his overcoat and stepping away from the stove as if to mark the beginning of official business)*. Now, Mr. Hale, before we move things about, you explain to Mr. Henderson just what you saw when you came here yesterday morning.

COUNTY ATTORNEY. By the way, has anything been moved? Are things just as you left them yesterday?

SHERIFF *(looking about)*. It's just the same. When it dropped below zero last night, I thought I'd better send Frank out this morning to make a fire for us—no use getting pneumonia with a big case on; but I told him not to touch anything except the stove—and you know Frank.

COUNTY ATTORNEY. Somebody should have been left here yesterday.

SHERIFF. Oh—yesterday. When I had to send Frank to Morris Centre for that man who went crazy—I want you to know I had my hands full yesterday. I knew you could get back from Omaha by today, and as long as I went over everything here myself—

COUNTY ATTORNEY. Well, Mr. Hale, tell just what happened when you came here yesterday morning.

HALE. Harry and I had started to town with a load of potatoes. We came along the road from my place; and as I got here, I said, "I'm going to see if I can't get John Wright to go in with me on a party telephone." *[People in the country had to share a telephone line; in other words, they could listen to the conversations of all the people sharing the line.]* I spoke to Wright about it once before, and he put me off, saying folks talked too much anyway, and all he asked was peace and quiet—I guess you know about how much he talked himself; but I thought maybe if I went to the house and talked about it before his wife, though I said to Harry that I didn't know as what his wife wanted made much difference to John—

COUNTY ATTORNEY. Let's talk about that later, Mr. Hale. I do want to talk about that, but tell now just what happened when you got to the house.

HALE. I didn't hear or see anything; I knocked at the door, and still it was all quiet inside. I knew they must be up, it was past eight o'clock. So I knocked again, and I thought I heard somebody say, "Come in." I wasn't sure, I'm not sure yet, but I opened the door—this door *(indicating the door by which the two women are still standing)*, and there in that rocker—*(pointing to it)* sat Mrs. Wright. *(They all look at the rocker.)*

COUNTY ATTORNEY. What—was she doing?

HALE. She was rockin' back and forth. She had her apron in her hand and was kind of—pleating it.

COUNTY ATTORNEY. And how did she—look?

HALE. Well, she looked queer.

COUNTY ATTORNEY. How do you mean—queer?

HALE. Well, as if she didn't know what she was going to do next. And kind of done up.

COUNTY ATTORNEY. How did she seem to feel about your coming?

HALE. Why, I don't think she minded—one way or other. She didn't pay much attention. I said, "How do, Mrs. Wright, it's cold, ain't it?" And she said, "Is it?"—and went on kind of pleating at her apron. Well, I was surprised; she didn't ask me to come up to the stove, or to set down, but just sat there, not even looking at me, so I said, "I want to see John." And then she—laughed. I guess you would call it a laugh. I thought of Harry and the team outside, so I said a little sharp: "Can't I see John?" "No," she says, kind o' dull like. "Ain't he home?" says I. "Yes," says she, "he's home." "Then why can't I see him?" I asked her, out of patience. "'Cause he's dead," says she. "*Dead*?" says I. She just nodded her head, not getting a bit excited, but rockin' back and forth. "Why—where is he?" says I, not knowing what to say. She just pointed upstairs—like that *(himself pointing to the room above)*. I got up, with the idea of going up there. I talked from there to here—then I says, "Why, what did he die of?" "He died of a rope around his neck," says she, and just went on pleatin' at her apron. Well, I went out and called Harry. I thought I might—need help. We went upstairs, and there he was lying—

COUNTY ATTORNEY. I think I'd rather have you go into that upstairs, where you can point it all out. Just go on now with the rest of the story.

HALE. Well, my first thought was to get that rope off. I looked *(stops, his face twitches)* …

but Harry, he went up to him, and he said, "No, he's dead all right, and we'd better not touch anything." So we went back downstairs. She was still sitting that same way. "Has anybody been notified?" I asked." "No," says she, unconcerned. "Who did this, Mrs. Wright?" said Harry. He said it business-like—and she stopped pleatin' her apron. "I don't know," she says. "You don't know?" says Harry. "No," says she. "Weren't you sleepin' in the bed with him?" says Harry. "Yes," says she, "but I was on the inside." "Somebody slipped a rope round his neck and strangled him, and you didn't wake up?" says Harry. "I didn't wake up," she said after him. We must 'a looked as if we didn't see how that could be, for after a minute she said, "I sleep sound." Harry was going to ask her more questions, but I said maybe we ought to let her tell her story first to the coroner, or the sheriff, so Harry went fast as he could to Rivers' place, where there's a telephone.

COUNTY ATTORNEY. And what did Mrs. Wright do when she knew that you had gone for the coroner?

HALE. She moved from that chair to this one over here (*pointing to a small chair in the corner*) ... and just sat there with her hands held together and looking down. I got a feeling that I ought to make some conversation, so I said I had come in to see if John wanted to put in a telephone, and at that she started to laugh, and then she stopped and looked at me—scared.

(*The County Attorney, who has had his notebook out, makes a note.*) I dunno, maybe it wasn't scared. I wouldn't like to say it was. Soon Harry got back, and then Dr. Lloyd came, and you, Mr. Peters, and so I guess that's all I know that you don't.

COUNTY ATTORNEY (*looking around*). I guess we'll go upstairs first—and then out to the barn and around there. (*to the Sheriff*) You're convinced that there was nothing important here—nothing that would point to any motive?

SHERIFF. Nothing here but kitchen things.

(*The County Attorney, after again looking around the kitchen, opens the door of a cupboard closet. He gets up on a chair and looks on a shelf, pulls his hand away, sticky.*)

COUNTY ATTORNEY. Here's a nice mess. (*The women draw nearer.*)

MRS. PETERS (*to the other woman*). Oh, her fruit; it did freeze. (*to the Lawyer*) She worried about that when it turned so cold. She said the fire'd go out and her jars would break.

SHERIFF. Well, can you beat the women! Held for murder and worryin' about her preserves.

COUNTY ATTORNEY. I guess before we're through, she may have something more serious than preserves to worry about.

HALE. Well, women are used to worrying over trifles. (*The two women move a little closer together.*)

COUNTY ATTORNEY (*with the gallantry of a young politician*). And yet, for all their worries, what would we do without the ladies? (*The women do not unbend. He goes to the sink, takes a dipperful of water from the pail and, pouring it into a basin, washes his hands. Starts to wipe them on the roller towel, turns it for a cleaner place*) Dirty towels! (*kicks his foot against the pans under the sink*) Not much of a housekeeper, would you say, ladies?

MRS. HALE (*stiffly*). There's a great deal of work to be done on a farm.

COUNTY ATTORNEY. To be sure. And yet (*with a little bow to her*) ... I know there are some Dickson county farmhouses which do not have such roller towels. (*He gives it a pull to expose its full length again.*)

MRS. HALE. Those towels get dirty awful quick. Men's hands aren't always as clean as they might be.

COUNTY ATTORNEY. Ah, loyal to your sex, I see. But you and Mrs. Wright were neighbours. I suppose you were friends, too.

MRS. HALE (*shaking her head*). I've not seen much of her of late years. I've not been in this house—it's more than a year.

COUNTY ATTORNEY. And why was that? You didn't like her?

MRS. HALE. I liked her all well enough. Farmers' wives have their hands full, Mr. Henderson. And then—

COUNTY ATTORNEY. Yes—?

MRS. HALE (*looking about*). It never seemed a very cheerful place.

COUNTY ATTORNEY. No—it's not cheerful. I shouldn't say she had the homemaking instinct.

MRS. HALE. Well, I don't know as Wright had, either.

COUNTY ATTORNEY. You mean that they didn't get on very well?

MRS. HALE. No, I don't mean anything. But I don't think a place'd be any cheerfuller for John Wright's being in it.

COUNTY ATTORNEY. I'd like to talk more of that a little later. I want to get the lay of things upstairs now. *(He goes to the left, where three steps lead to a stair door.)*

SHERIFF. I suppose anything Mrs. Peters does'll be all right. She was to take in some clothes for her, you know, and a few little things. We left in such a hurry yesterday.

COUNTY ATTORNEY. Yes, but I would like to see what you take, Mrs. Peters, and keep an eye out for anything that might be of use to us.

MRS. PETERS. Yes, Mr. Henderson. *(The women listen to the men's steps on the stairs, then look about the kitchen.)*

MRS. HALE. I'd hate to have men coming into my kitchen, snooping around and criticizing. *(She arranges the pans under sink which the Lawyer had shoved out of place.)*

MRS. PETERS. Of course, it's no more than their duty.

MRS. HALE. Duty's all right, but I guess that deputy sheriff that came out to make the fire might have got a little of this on. *(gives the roller towel a pull)* Wish I'd thought of that sooner. Seems mean to talk about her for not having things slicked up when she had to come away in such a hurry.

MRS. PETERS *(who has gone to a small table in the left rear corner of the room, and lifted one end of a towel that covers a pan)*. She had bread set. *(stands still)*

MRS. HALE *(eyes fixed on a loaf of bread beside the breadbox, which is on a low shelf at the other side of the room. Moves slowly toward it)*. She was going to put this in there. *(picks up the loaf, then abruptly drops it. In a manner of returning to familiar things)* It's a shame about her fruit. I wonder if it's all gone. *(gets up on the chair and looks)* I think there's some here that's all right, Mrs. Peters. Yes—here. *(holding it toward the window)* This is cherries, too. *(looking again)* I declare I believe that's the only one. *(gets down, bottle in her hand. Goes to the sink and wipes it off on the outside)* She'll feel awful bad after all her hard work in the hot weather. I remember the afternoon I put up my cherries last summer. *(She puts the bottle on the big kitchen table, centre of the room, front table. With a sigh, is about to sit down in the rocking chair. Before she is seated, she realizes what chair it is; with a slow look at it, she steps back. The chair, which she has touched, rocks back and forth.)*

MRS. PETERS. Well, I must get those things from the front room closet. *(She goes to the door at the right, but after looking into the other room, steps back.)* You coming with me, Mrs. Hale? You could help me carry them. *(They go into the other room; reappear, Mrs. Peters carrying a dress and skirt, Mrs. Hale following with a pair of shoes.)*

MRS. PETERS. My, it's cold in there. *(She puts the clothes on the big table, and hurries to the stove.)*

MRS. HALE *(examining the skirt)*. Wright was close. I think maybe that's why she kept so much to herself. She didn't even belong to the Ladies' Aid. I suppose she felt she couldn't do her part, and then you don't enjoy things when you feel shabby. She used to wear pretty clothes and be lively, when she was Minnie Foster, one of the town girls singing in the choir. But that— oh, that was thirty years ago. This all you was to take in?

MRS. PETERS. She said she wanted an apron. Funny thing to want, for there isn't much to get you dirty in jail, goodness knows. But I suppose just to make her feel more natural. She said they was in the top drawer in this cupboard. Yes, here. And then her little shawl that always hung behind the door. *(opens stair door and looks)* Yes, here it is. *(quickly shuts door leading upstairs)*

MRS. HALE *(abruptly moving toward her)*. Mrs. Peters?

MRS. PETERS. Yes, Mrs. Hale?

MRS. HALE. Do you think she did it?

MRS. PETERS *(in a frightened voice)*. Oh, I don't know.

MRS. HALE. Well, I don't think she did. Asking for an apron and her little shawl. Worrying about her fruit.

MRS. PETERS (starts to speak, glances up, where footsteps are heard in the room above. In a low voice). Mr. Peters says it looks bad for her. Mr. Henderson is awful sarcastic in speech, and he'll make fun of her sayin' she didn't wake up.

MRS. HALE. Well, I guess John Wright didn't wake when they was slipping that rope under his neck.

MRS. PETERS. No, it's strange. It must have been done awful crafty and still. They say it was such a—funny way to kill a man, rigging it all up like that.

MRS. HALE. That's just what Mr. Hale said. There was a gun in the house. He says that's what he can't understand.

MRS. PETERS. Mr. Henderson said coming out that what was needed for the case was a motive; something to show anger or—sudden feeling.

MRS. HALE (who is standing by the table). Well, I don't see any signs of anger around here. (She puts her hand on the dish towel, which lies on the table, stands looking down at the table, one half of which is clean, the other half messy.) It's wiped here. (makes a move as if to finish work, then turns and looks at the loaf of bread outside the breadbox. Drops towel. In that voice of coming back to familiar things) Wonder how they are finding things upstairs? I hope she had it a little more red-up up there. You know, it seems kind of sneaking. Locking her up in town and then coming out here and trying to get her own house to turn against her!

MRS. PETERS. But, Mrs. Hale, the law is the law.

MRS. HALE. I s'pose 'tis. (unbuttoning her coat) Better loosen up your things, Mrs. Peters. You won't feel them when you go out. (Mrs. Peters takes off her fur tippet, goes to hang it on hook at back of room, stands looking at the under part of the small corner table.)

MRS. PETERS. She was piecing a quilt. (She brings the large sewing basket, and they look at the bright pieces.)

MRS. HALE. It's log cabin pattern. Pretty, isn't it? I wonder if she was goin' to quilt or just knot it? (Footsteps have been heard coming down the stairs. The Sheriff enters, followed by Hale and the County Attorney.)

SHERIFF. They wonder if she was going to quilt it or just knot it. (The men laugh; the women look abashed.)

COUNTY ATTORNEY (rubbing his hands over the stove). Frank's fire didn't do much up there, did it? Well, let's go out to the barn and get that cleared up. (The men go outside.)

MRS. HALE (resentfully). I don't know as there's anything so strange, our takin' up our time with little things while we're waiting for them to get the evidence. (She sits down at the big table, smoothing out a block with decision.) I don't see as it's anything to laugh about.

MRS. PETERS (apologetically). Of course, they've got awful important things on their minds. (pulls up a chair and joins Mrs. Hale at the table)

MRS. HALE (examining another block). Mrs. Peters, look at this one. Here, this is the one she was working on, and look at the sewing! All the rest of it has been so nice and even. And look at this! It's all over the place! Why, it looks as if she didn't know what she was about! (After she has said this, they look at each other, then start to glance back at the door. After an instant, Mrs. Hale has pulled at a knot and ripped the sewing.)

MRS. PETERS. Oh, what are you doing, Mrs. Hale?

MRS. HALE (mildly). Just pulling out a stitch or two that's not sewed very good. (threading a needle) Bad sewing always made me fidgety.

MRS. PETERS (nervously). I don't think we ought to touch things.

MRS. HALE. I'll just finish up this end. (suddenly stopping and leaning forward) Mrs. Peters?

MRS. PETERS. Yes, Mrs. Hale?

MRS. HALE. What do you suppose she was so nervous about?

MRS. PETERS. Oh—I don't know. I don't know as she was nervous. I sometimes sew awful queer when I'm just tired. (Mrs. Hale starts to say something, looks at Mrs. Peters, then goes on sewing.) Well, I must get these things wrapped up. They may be through sooner than we think.

(putting apron and other things together) I wonder where I can find a piece of paper, and string.

MRS. HALE. In that cupboard, maybe.

MRS. PETER *(looking in cupboard)*. Why, here's a birdcage. *(holds it up)* Did she have a bird, Mrs. Hale?

MRS. HALE. Why, I don't know whether she did or not—I've not been here for so long. There was a man around last year selling canaries cheap, but I don't know as she took one; maybe she did. She used to sing real pretty herself.

MRS. PETERS *(glancing around)*. Seems funny to think of a bird here. But she must have had one, or why should she have a cage? I wonder what happened to it?

MRS. HALE. I s'pose maybe the cat got it.

MRS. PETERS. No, she didn't have a cat. She's got that feeling some people have about cats—being afraid of them. My cat got in her room, and she was real upset and asked me to take it out.

MRS. HALE. My sister Bessie was like that. Queer, ain't it?

MRS. PETERS *(examining the cage)*. Why, look at this door. It's broke. One hinge is pulled apart.

MRS. HALE *(looking, too)*. Looks as if someone must have been rough with it.

MRS. PETERS. Why, yes. *(She brings the cage forward and puts it on the table.)*

MRS. HALE. I wish if they're going to find any evidence they'd be about it. I don't like this place.

MRS. PETERS. But I'm awful glad you came with me, Mrs. Hale. It would be lonesome of me sitting here alone.

MRS. HALE. It would, wouldn't it? *(dropping her sewing)* But I tell you what I do wish, Mrs. Peters. I wish I had come over sometimes she was here. I—*(looking around the room)*—wish I had.

MRS. PETERS. But of course you were awful busy, Mrs. Hale—your house and your children.

MRS. HALE. I could've come. I stayed away because it weren't cheerful—and that's why I ought to have come. I—I've never liked this place. Maybe because it's down in a hollow, and you don't see the road. I dunno what it is, but it's a lonesome place and always was. I wish I had come over to see Minnie Foster sometimes. I can see now—*(shakes her head)*.

MRS. PETERS. Well, you mustn't reproach yourself, Mrs. Hale. Somehow we just don't see how it is with other folks until—something comes up.

MRS. HALE. Not having children makes less work—but it makes a quiet house, and Wright out to work all day, and no company when he did come in. Did you know John Wright, Mrs. Peters?

MRS. PETERS. Not to know him; I've seen him in town. They say he was a good man.

MRS. HALE. Yes—good; he didn't drink, and kept his word as well as most, I guess, and paid his debts. But he was a hard man, Mrs. Peters. Just to pass the time of day with him. *(shivers)* Like a raw wind that gets to the bone. *(pauses, her eye falling on the cage)* I should think she would 'a wanted a bird. But what do you suppose went with it?

MRS. PETERS. I don't know, unless it got sick and died. *(She reaches over and swings the broken door, swings it again; both women watch it.)*

MRS. HALE. She—come to think of it, she was kind of like a bird herself—real sweet and pretty, but kind of timid and—fluttery. How—she—did—change. *(silence; then as if struck by a happy thought and relieved to get back to everyday things)* Tell you what, Mrs. Peters, why don't you take the quilt in with you? It might take up her mind.

MRS. PETERS. Why, I think that's a real nice idea, Mrs. Hale. There couldn't possibly be any objection to it, could there? Now, just what would I take? I wonder if her patches are in here—and her things. *(They look in the sewing basket.)*

MRS. HALE. Here's some red. I expect this has got sewing things in it. *(brings out a fancy box)* What a pretty box. Looks like something somebody would give you. Maybe her scissors are in here. *(opens box; suddenly puts her hand to her nose)* Why—*(Mrs. Peters bends nearer, then turns her face away.)* There's something wrapped up in this piece of silk.

MRS. PETERS. Why, this isn't her scissors.

MRS. HALE (*lifting the silk*). Oh, Mrs. Peters—it's—(*Mrs. Peters bends closer.*)

MRS. PETERS. It's the bird.

MRS. HALE (*jumping up*). But, Mrs. Peters—look at it. Its neck! Look at its neck! It's all—to the other side.

MRS. PETERS. Somebody—wrung—its neck.

(*Their eyes meet. A look of growing comprehension, of horror. Steps are heard outside. Mrs. Hale slips the box under the quilt pieces and sinks into her chair. Enter Sheriff and County Attorney. Mrs. Peters rises.*)

COUNTY ATTORNEY (*as one turning from serious thing to little pleasantries*). Well, ladies, have you decided whether she was going to quilt it or knot it?

MRS. PETERS. We think she was going to—knot it.

COUNTY ATTORNEY. Well, that's interesting, I'm sure. (*seeing the birdcage*) Has the bird flown?

MRS. HALE (*putting more quilt pieces over the box*). We think the—cat got it.

COUNTY ATTORNEY (*preoccupied*). Is there a cat? (*Mrs. Hale glances in a quick, covert way at Mrs. Peters.*)

MRS. PETERS. Well, not now. They're superstitious, you know. They leave.

COUNTY ATTORNEY (*to Sheriff Peters, continuing an interrupted conversation*). No sign at all of anyone having come from the outside. Their own rope. Now let's go up again and go over it piece by piece. (*They start upstairs.*) It would have to have been someone who knew just the—(*Mrs. Peters sits down. The two women sit there not looking at one another, but as if peering into something and at the same time holding back. When they talk now, it is in the manner of feeling their way over strange ground, as if afraid of what they are saying, but as if they cannot help saying it.*)

MRS. HALE. She liked the bird. She was going to bury it in that pretty box.

MRS. PETERS (*in a whisper*). When I was a girl—my kitten—there was a boy took a hatchet, and before my eyes—and before I could get there—(*covers her face an instant*) If they hadn't held me back, I would have—(*catches herself, looks upstairs, where steps are heard; falters weakly*)—hurt him.

MRS. HALE (*with a slow look around her*). I wonder how it would seem never to have had any children around. (*pause*) No, Wright wouldn't like the bird—a thing that sang. She used to sing. He killed that, too.

MRS. PETERS (*moving uneasily*). We don't know who killed the bird.

MRS. HALE. I knew John Wright.

MRS. PETERS. It was an awful thing was done in this house that night, Mrs. Hale. Killing a man while he slept, slipping a rope around his neck that choked the life out of him.

MRS. HALE. His neck. Choked the life out of him. (*Her hand goes out and rests on the birdcage.*)

MRS. PETERS (*with a rising voice*). We don't know who killed him. We don't know.

MRS. HALE (*her own feeling not interrupted*). If there'd been years and years of nothing, then a bird to sing to you, it would be awful—still, after the bird was still.

MRS. PETERS (*something within her speaking*). I know what stillness is. When we homesteaded in Dakota, and my first baby died—after he was two years old, and me with no other then—

MRS. HALE (*moving*). How soon do you suppose they'll be through, looking for evidence?

MRS. PETERS. I know what stillness is. (*pulling herself back*) The law has got to punish crime, Mrs. Hale.

MRS. HALE (*not as if answering that*). I wish you'd seen Minnie Foster when she wore a white dress with blue ribbons and stood up there in the choir and sang. (*a look around the room*) Oh, I wish I'd come over here once in a while! That was a crime! That was a crime! Who's going to punish that?

MRS. PETERS (*looking upstairs*). We mustn't—take on.

MRS. HALE. I might have known she needed help! I know how things can be—for women. I tell you, it's queer, Mrs. Peters. We live close together and we live far apart. We all go through the same things—it's all just a different kind of the same thing. (*brushes her eyes, noticing the bottle of fruit, reaches out for it*) If I was you, I wouldn't

17

tell her her fruit was gone. Tell her it ain't. Tell her it's all right. Take this in to prove it to her. She—she may never know whether it was broke or not.

MRS. PETERS (*takes the bottle; looks about for something to wrap it in; takes petticoat from the clothes brought from the other room, very nervously begins winding this around the bottle. In a false voice*). My, it's a good thing the men couldn't hear us. Wouldn't they just laugh! Getting all stirred up over a little thing like a—dead canary. As if that could have anything to do with—with—wouldn't they laugh! (*The men are heard coming downstairs.*)

MRS. HALE (*under her breath*). Maybe they would—maybe they wouldn't.

COUNTY ATTORNEY. No, Peters, it's all perfectly clear except a reason for doing it. But you know juries when it comes to women. If there was some definite thing—something to show—something to make a story about—a thing that would connect up with this strange way of doing it. (*The women's eyes meet for an instant. Enter Hale from outer door.*)

HALE. Well, I've got the team around. Pretty cold out there.

COUNTY ATTORNEY. I'm going to stay here awhile by myself. (*to the Sheriff*) You can send Frank out for me, can't you? I want to go over everything. I'm not satisfied that we can't do better.

SHERIFF. Do you want to see what Mrs. Peters is going to take in? (*The Lawyer goes to the table, picks up the apron, laughs.*)

COUNTY ATTORNEY. Oh I guess they're not very dangerous things the ladies have picked up. (*moves a few things about, disturbing the quilt pieces which cover the box. Steps back*) No, Mrs. Peters doesn't need supervising. For that matter, a sheriff's wife is married to the law. Ever think of it that way, Mrs. Peters?

MRS. PETERS. Not—just that way.

SHERIFF (*chuckling*). Married to the law. (*moves toward the other room*) I just want you to come in here a minute, George. We ought to take a look at these windows.

COUNTY ATTORNEY (*scoffingly*). Oh, windows!

SHERIFF. We'll be right out, Mr. Hale. (*Hale goes outside. The Sheriff follows the County Attorney into the other room. Then Mrs. Hale rises, hands tight together, looking intensely at Mrs. Peters, whose eyes take a slow turn, finally meeting Mrs. Hale's. A moment Mrs. Hale holds her, then her own eyes point the way to where the box is concealed. Suddenly Mrs. Peters throws back the quilt pieces and tries to put the box in the bag she is wearing. It is too big. She opens the box, starts to take the bird out, cannot touch it, goes to pieces, stands there helpless. Sound of a knob turning in the other room. Mrs. Hale snatches the box and puts it in the pocket of her big coat. Enter County Attorney and Sheriff.*)

COUNTY ATTORNEY (*facetiously*). Well, Henry, at least we found out that she was not going to quilt it. She was going to—what is it you call it, ladies!

MRS. HALE (*her hand against her pocket*). We call it—knot it, Mr. Henderson.

(CURTAIN)

PUBLISHED IN
1916

Reader's Response

1. Discuss your opinion about the ending of the play. Do the women do the right thing?

2. In your opinion, could the events in this play, which was written in 1916, happen in contemporary society? Explain your answer.

✑ CLOSE READING

1. Reread the following lines describing the opening scene. Underline the descriptive adjectives, nouns and verbs that create the mood of the setting. Then describe the setting (where and when the story takes place) and the mood (the atmosphere or ambiance) of the scene.

> Scene: *The kitchen in the now abandoned farmhouse of John Wright, a gloomy kitchen, and left without having been put in order—unwashed pans under the sink, a loaf of bread outside the breadbox, a dish towel on the table—other signs of uncompleted work. At the rear, the outer door opens, and the Sheriff comes in, followed by the county Attorney and Hale. The Sheriff and Hale are men in middle life, the county Attorney is a young man; all are much bundled up and go at once to the stove. They are followed by the two women— the Sheriff's Wife first; she is a slight wiry woman with a thin, nervous face. Mrs. Hale is larger and would ordinarily be called more comfortable looking, but she is disturbed now and looks fearfully about as she enters. The women have come in slowly and stand close together near the door.*

2. List the major characters and then state their relationship to each other.

Major Characters	Relationship to Each Other

3. What is significant about the rocking chair?

4. Why did Mr. Hale initially go to the farmhouse?

5. What do we understand to be Mr. Hale's opinions of John and Minnie Wright from his conversation with the Sheriff?

6. What was Mrs. Wright's emotional state when Mr. Hale found her? Cite Mrs. Wright's words and gestures to support your answer.

_____

_____

7. What had happened to Mr. Wright?

_____

_____

8. Describe Mrs. Hale's and Mrs. Peters' relationship with Mrs. Wright.

_____

9. How had Mrs. Wright's personality and lifestyle changed since her marriage?

_____

10. Describe the relationship between Mrs. Hale and Mrs. Peters. Does it change throughout the play?

_____

_____

✎ WRITER'S CRAFT

Symbolism

1. What is suggestive about Mrs. Wright's messy kitchen?

_____

2. John Wright does not want a telephone. What does this indicate about him?

_____

3. The quilt is a blanket made up of small pieces of cloth sewn together to create a design. Explain how the quilt symbolizes the discovery of the motive by Mrs. Hale and Mrs. Peters.

_____

4. Many symbols reveal the nature of the Wrights' personalities and marriage. Write an interpretation for each symbol below.

_____

a) Minnie Wright's girlhood clothes and her present clothes

_____

b) the dead canary

_____

c) the deterioration in the sewing of the quilt

_____

5. Find two other symbols used by the author to develop the other characters in the play.

   a) _____

   b) _____

6. What is the significance of the play's title?

   _____

Characterization: Male/Female Roles

1. What is the purpose of the male characters at the farmhouse?

   _____

2. Why are the female characters at the farmhouse?

   _____

3. Why does Mr. Hale say, "Well, women are used to worrying over trifles"?

   _____

   _____

4. Cite three examples from the play that reveal the men's attitude of superiority over the women.

   _____

   _____

   _____

5. How do the women react to the men's condescending attitude? Give examples of their verbal and physical responses.

   Verbal response: _____

   Physical response: _____

6. What is ironic about Mrs. Hale's statement ("We call it—knot it, Mr. Henderson.") at the end of the play?

   _____

   _____

Dramatic Structure

1. What information does the exposition of the play reveal?

   _____

2. What part of the play represents the rising action? What information does it provide?

   _____

3. What is the climax of the play?

_____

4. Give three examples of ways in which the characters' reactions or descriptions create moments of tension in the play.

_____

_____

_____

5. In your opinion, what is the theme of the play?

_____

VOCABULARY DEVELOPMENT
Standard versus Slang

Dialogue is crucial in a play. Characterization is depicted through dialogue. Read the following excerpts and answer the accompanying questions.

Excerpt 1

HALE. Why, I don't think she minded—one way or other. She didn't pay much attention. I said, "How do, Mrs. Wright, it's cold, ain't it?" And she said, "Is it?"—and went on kind of pleating at her apron. Well, I was surprised; she didn't ask me to come up to the stove, or to set down, but just sat there, not even looking at me, so I said, "I want to see John." And then she—laughed. I guess you would call it a laugh. I thought of Harry and the team outside, so I said a little sharp: "Can't I see John?" "No," she says, kind o' dull like. "Ain't he home?" says I. "Yes," says she, "he's home." "Then why can't I see him?" I asked her, out of patience. "'Cause he's dead," says she.

1. List the non-standard English words (slang) from the excerpt in the table below. Then write down the standard English words that correspond to them.

Non-Standard English Words	Standard English Words

Excerpt 2

COUNTY ATTORNEY. By the way, has anything been moved? Are things just as you left them yesterday?
[...]

Somebody should have been left here yesterday.

2. Is this excerpt an example of standard English or slang? Cite words from the excerpt to support your answer.

3. Write two dialogues on the subject of visiting friends. One dialogue should incorporate slang and the other should incorporate standard English. Your dialogues can be as funny or as serious as you wish.

Wrap-Up Activities

1. Write about John Wright's personality.

2. Write about the ironic symbolism of Mrs. Wright's name.

3. Discuss how the play exposes male-female stereotypes and gender biases.

4. Write about the effectiveness of the stage directions in maintaining tension throughout the play.

5. Explain how the character of Minnie Wright is developed throughout the play.

6. Discuss the significance of symbolism and imagery in the play.

7. Discuss the dramatic structure of the play.

8. Research further background about the author and write a literary essay using the biographical approach.

9. Examine the moral dilemma presented in the play and discuss its impact on the play's theme.

10. Discuss the pertinence of the play's title.

11. Read another play by Glaspell and compare it to *Trifles*.

12. Write a literary essay on *Trifles* from a feminist perspective.

My eLab ACTIVITY

My eLab

Read about the characteristics of radio plays and listen to excerpts of *Trifles* and from the most famous radio play of all, *War of the Worlds* by Orson Welles. Complete the proposed activities.

...munication skills are the

...to success both in your study

...and in your future career. Wri

...a clear, well-organized text

...will ensure that you communic

...your ideas accurately.

...Many students think that it is

...difficult to write well. However,

...this idea is not entirely true. Writing

...a skill, and like any skill, it improves

...with practice. Any student can become a

...better writer by paying careful

...attention to the writing process. Th

...book will provide you w...

...strategies for improving your writing and

...mastering the writing process

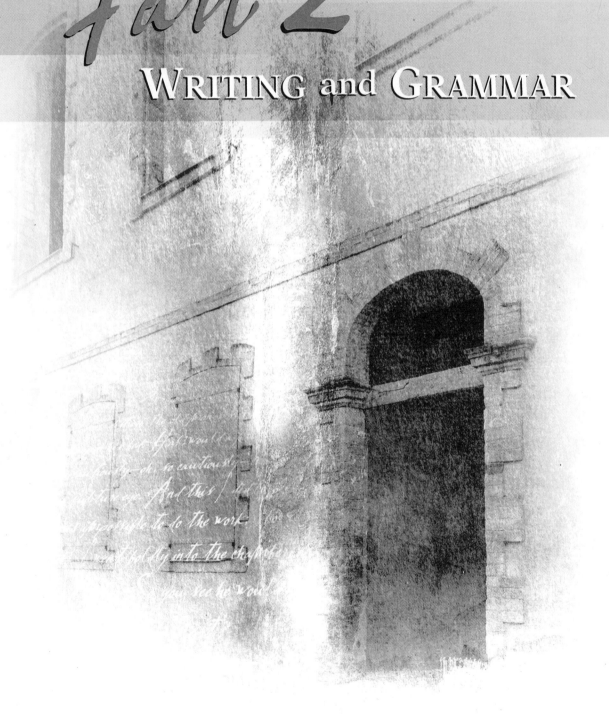

Part 2

WRITING and GRAMMAR

WRITING IN STYLE

Good communication skills are the key to success both in your studies and in your future career. Writing clear, well-organized texts will ensure that you communicate your ideas accurately.

Often students think that it is difficult to write well. However, this idea is not entirely true. Writing is a skill, and like any skill, it improves with practice. Any student can become a proficient writer by paying careful attention to the writing process. The following units will provide you with strategies for improving your writing skills and mastering the writing process.

Pre-Writing

You may feel panic when your instructor gives you a writing assignment. You may look at the blank sheet of paper and think that you have nothing to say. However, this sentiment is probably untrue. Following the next steps will help you discover ideas and topics to write about.

Pre-Writing

1. Identify your audience.
2. Determine your purpose.
3. Think of a topic.
4. Use pre-writing strategies:
   - Brainstorming
   - Free-writing

Identify Your Audience.

Ask yourself who is going to read your assignment. Is it your instructor, your classmates or someone else? Also ask yourself what information you should give your readers. Should your writing style be formal or informal? Take into consideration the tone and the level of language of your assignment.

For most writing assignments, your instructor will be your audience. Remember to include complete information in your assignment even if the audience for your work is your instructor. Never assume that there is no need to include complete information because your instructor knows more about the subject than you do. Most likely, your instructor is assessing your knowledge of the subject.

Determine Your Purpose.

What is your reason for writing the assignment? Your purpose may be to inform, to entertain, to argue, to analyze or to describe. Your writing may serve more than one purpose. For example, you may be writing both to inform and to entertain.

Think of a Topic.

Ask yourself what the subject of the assignment is. Your instructor may give you a topic, or you may have to develop your own. If the latter is the case, remember to choose a topic that you find interesting.

Use Pre-Writing Strategies.

After you have determined the audience, purpose and topic of your writing assignment, you may want to use pre-writing strategies in order to generate some ideas.

■ BRAINSTORMING

Make a list of ideas that come to mind regarding your topic. You do not have to write in complete sentences—you can just write down words or phrases.

Fairy tales

- written in the past
- live happily ever after
- witches
- prince and princess
- magic
- good versus evil
- talking animals
- sets of threes (three wishes, three brothers, three tasks)
- Charles Perrault
- Brothers Grimm

EXERCISE 1 Brainstorming

Brainstorm some ideas for the following topic.

Favourite books

_____

_____

_____

_____

■ FREE-WRITING

Write a paragraph without stopping and include all your ideas about the topic. Do not worry about spelling, punctuation or grammar. Even if you do not know what to write, keep on writing the ideas that come to mind.

Fairy tales

I liked to read them when I was young. So much fun. My favourite was *The Twelve Dancing Princesses*. The Brothers Grimm wrote fairy tales in Germany. They were also linguists. Common fairy tales all over the world. Sheherazad told wonderful fairy tales. The Arabian Nights. Tales with Ginns and magic lamps. What else? The princess was always pretty. The prince was always heroic. No such thing as a knight in shining armour. I don't know what else to say. Maybe this is enough. The princesses always married the handsome prince.

EXERCISE 2 Free-Writing

Free-write ideas for the following topic.

Books turned into films

_____

_____

_____

_____

Wrap-Up Activity

Use a pre-writing strategy to write about one of the following topics.

a) Children's books

b) A well-known author

c) My literature course

Writing Paragraphs

To write effectively, it is important to understand how to write in a well-organized fashion. In this unit, you will learn how to develop and organize your ideas. You will also write the first draft of your paragraph.

Paragraphs are units in which a writer expresses his or her ideas. They are groups of sentences containing a main idea. They have an inherent structure, containing a **topic sentence** and **supporting details**. A topic sentence introduces the subject of the paragraph and has a controlling idea. The supporting details consist of facts, examples or interpretations that give further information on the main idea or topic sentence. A paragraph sometimes ends with a concluding sentence.

Paragraph: group of sentences containing a main idea

Topic sentence: introduces the topic or main idea

- **Controlling idea:** expresses the writer's opinion, attitude or feeling

Support sentences: present facts, examples or interpretations that "prove" the main idea

Concluding sentence: reaffirms the main idea in a different way

---

There are certain points to remember when writing paragraphs.

- Paragraphs should only contain one main idea.
- All details should support the main idea of the paragraph.
- Paragraphs are generally six to ten sentences in length, containing 100 to 150 words in total. (This rule is, of course, very general and applies to descriptive and narrative forms rather than dialogue.) If a paragraph is too long, it is likely to contain more than one main idea and become unfocused. If a paragraph is too short, it is likely to contain insufficient information on the subject.
- Paragraphs should make an impact on the reader; they should be forceful and interesting.

Focus the Topic.

If your topic is too vague or general, it is necessary to focus it so that it can be supported in one paragraph. To focus your topic, use a pre-writing strategy, such as brainstorming or free-writing. After you have finished, underline some ideas that can be expanded into paragraphs.

EXAMPLE

Topic: Romance in literature

What shall I write about? Used to read romance literature as a teenager. The heroine was always beautiful. The man always rugged and handsome. <u>Very unrealistic</u>. Is there <u>any value to reading romance literature</u>? Great romances like *Anthony and Cleopatra*. Can't remember any others. The hero and heroine live happily ever after. <u>Similar to fairy tale endings</u>. The hero is always rich and intelligent. Never met anyone perfect like that. <u>Gives young girls the wrong ideas about life.</u> Shows that the man will always take care of the woman. Bad example for little girls.

EXERCISE **1** Focusing the Topic

The following topics are too broad for developing into one paragraph. Using a pre-writing strategy, such as free-writing or brainstorming, choose one topic and narrow it. Then underline some ideas that you might want to develop into paragraphs.

1. Horror literature

2. Literary heroes

3. Watching the film versus reading the book

_____

_____

_____

_____

_____

_____

_____

_____

_____

The Topic Sentence

The **topic sentence** introduces the topic of the paragraph and states the controlling idea. It is the focus of your paragraph or what your paragraph is about. It is followed by other sentences that provide supporting facts, examples or interpretations.

The **controlling idea** expresses the writer's opinion, attitude or feeling.

        topic   controlling idea

Edgar Allan Poe's story "The Tell-Tale Heart" <u>uses irony to release the dramatic tension</u>.

The **topic sentence** should
- be a complete sentence;

    Incorrect: The use of symbols in "The Mask of the Red Death" (lacks a controlling idea; has no verb)
    Correct: The use of symbols in "The Mask of the Red Death" adds depth to the story.

- not contain expressions such as "I believe," "I think" or "in my opinion ...";

> **Incorrect:** I think that Edgar Allan Poe's story "The Tell-Tale Heart" uses irony to release the dramatic tension. (redundant first-person announcement of opinion)
>
> **Correct:** Edgar Allan Poe's story "The Tell-Tale Heart" uses irony to release the dramatic tension.

- not be a factual statement;

> **Incorrect:** Edgar Allan Poe wrote short stories. (no opinion)
>
> **Correct:** Edgar Allan Poe wrote terrifying short stories.

- not be a direct question.

> **Incorrect:** Why do we study literature?
>
> **Correct:** We study literature to learn about the human condition.

Make an Impact on the Reader

Your topic sentence should be forceful and interesting. It should entice the reader into reading further.

EXERCISE 2 Analyzing Effective Topic Sentences

Choose the word from the list that best describes the problem associated with each topic sentence.

incomplete sentence	announcement of opinion	factual statement	question

1. I like to read horror stories.

Problem: _____

2. The Russian writer Leo Tolstoy wrote during the Golden Age of Russian literature.

Problem: _____

3. In the fairy tale "Little Red Riding Hood," why is the wolf a symbol of evil?

Problem: _____

4. Hemingway's writing style.

Problem: _____

5. John Steinbeck won the Nobel Prize for literature.

Problem: _____

6. My literature course.

Problem: _____

7. There are many children's books in English.

Problem: _____

8. In my opinion, romance literature has unrealistic endings.

   Problem: _____

9. What is great literature?

   Problem: _____

10. Robert Frost was a poet laureate of the United States.

   Problem: _____

EXERCISE 3 Identifying the Topic Sentence

Underline the topic sentence in the following paragraph. Remember that the topic sentence makes the main point of the paragraph. All other sentences must support the topic sentence.

> Many of us have read fairy tales in which a handsome prince rescues a beautiful girl from a wicked situation or person. Although on an initial reading most fairy-tale heroines seem weak and superficial, further analysis shows that they are very resilient. First, fairy-tale heroines such as Cinderella or Snow White work like slaves. Both heroines require much physical endurance to clean the house all day long. Cinderella has to look after the physical needs of her stepmother and stepsisters. Another example of a strong fairy-tale heroine is Rapunzel, who allows a prince to use her hair as a climbing rope. Furthermore, most fairy-tale princesses have to endure long years of emotional hardship. Snow White, Cinderella, Rapunzel and others suffer for many years at the hands of evil stepmothers and witches who want to isolate them or even kill them. These evil stepmothers prevent the young ladies from making friendships or finding romance. Yet the fairy-tale heroines never lose hope that someday they will be set free. Finally, fairy-tale heroines, far from being fragile, are tough and deserve the happy ending that they achieve at the end of the fairy tale.

EXERCISE 4 Writing Effective Topic Sentences

Choose two topics from Exercise 1 and write topic sentences for them.

1. **Topic:** _____

   Topic sentence: _____

   _____

2. **Topic:** _____

   Topic sentence: _____

   _____

Supporting Points

After you have written a topic sentence, you generate details for the paragraph. The details can be facts, examples or interpretations to support the topic sentence. There are two steps to follow when developing ideas for the body of your paragraph.

1. Developing supporting points

2. Choosing and organizing the best points

■ DEVELOPING YOUR SUPPORTING POINTS

You can make a list to generate supporting ideas.

EXAMPLE

Topic sentence: Symbols in the fairy tale "Little Red Riding Hood" offer important clues to the meaning of the story.

- Colour red denotes violence and blood.
- Original wolf was a werewolf.
- Young girl or female represents innocence.
- Male wolf symbolizes a predator, evil.
- Girl equals good while the wolf equals bad.
- Dark forest suggests isolation.
- Green of the forest suggests fertility.
- Hood symbolizes virginity.
- Grandmother is old.

■ ORGANIZING AND CHOOSING YOUR SUPPORTING POINTS

Organize your ideas logically. Ideas should flow sequentially. Choose the best ones and cross out any that do not belong. Then group any related ideas together and summarize the underlying connection.

EXAMPLE

First group of related points: Colours are symbolic.
- Colour red denotes violence and blood.
- Dark forest suggests isolation.
- Green of the forest suggests fertility.

Second group of related points: Female sex suggests weakness and innocence.
- Young girl or female represents innocence.
- Hood symbolizes virginity.

Third group of related points: Male sex suggests evil.
- Male wolf symbolizes a predator, evil.
- Girl equals good while the wolf equals bad.

Unrelated ideas
- ~~Original wolf was a werewolf.~~
- ~~Grandmother is old.~~

The Paragraph Plan

Once you have generated a list of supporting ideas, it is important to make a paragraph plan. In a paragraph plan, you write down the topic sentence and then the supporting ideas. Ask yourself if your supporting ideas are clear enough. To clarify your supporting ideas, add details (examples and interpretations) to your paragraph. Remember to organize your paragraph in a logical way.

EXAMPLE

Topic sentence: Symbols in the fairy tale "Little Red Riding Hood" offer important clues to the meaning of the story.

First supporting point: Colour is an important symbol found in the story.

Details:
- The colour red represents violence and blood.
- The girl's cape is red and she is a victim of violence.
- The forest is dark and isolated.
- Black symbolizes the sinister, scary and violent.

Second supporting point: The female sex suggests weakness.

Details:
- Little Red Riding Hood is a naive little girl.
- She wears a hood, which represents innocence and virginity.

Third supporting point: The male sex suggests the existence of evil.

Details:
- The wolf is male.
- He is the predator.
- He murders the grandmother and wants to eat Little Red Riding Hood.

EXERCISE 5 Deconstructing Your Paragraph

Read the following paragraph and make a paragraph plan. Write down the topic sentence followed by the supporting points and details. Follow the model above.

> The use of dialogue in drama is of primal importance. First, traditional drama contains very little narration or description, so communication between the characters is pivotal for moving the story forward. For example, dialogue illuminates the plot. Dialogue exposes the tensions and conflicts of characters such as in the famous balcony scene in Shakespeare's *Romeo and Juliet,* which introduces the audience to the next level of conflict. Next, dialogue is crucial to characterization. For instance, a character can be seen as educated, illiterate, selfish or selfless through dialogue. In Shakespeare's *King Richard III,* one of the most famous scenes clearly shows Richard's desperation when he cries, "A horse! A horse! My kingdom for a horse!" In fact, many of Shakespeare's characters are remembered through their conversations with others.

Topic sentence: _____

_____

First supporting point: _____

_____

Details: _____

_____

Second supporting point: _____

_____

Details: _____

_____

EXERCISE 6 Writing the Paragraph Plan

Choose one topic sentence from Exercise 4 or create your own topic sentence from a text with which you are already familiar. Write a paragraph plan. Make the topic sentence strong and interesting. The details should support the topic sentence.

Topic sentence: _____

_____

First supporting point: _____

_____

Details: _____

_____

Second supporting point: _____

_____

Details: _____

_____

Third supporting point: _____

_____

Details: _____

_____

Writing a Concluding Sentence

Not all paragraphs have a **concluding sentence**, but if you feel that your paragraph does not end smoothly, you should write one. There are several ways to write a concluding sentence.

- Reaffirm the topic sentence in another way.
- Make an unusual or interesting observation.

EXERCISE 7 Choosing the Best Concluding Sentence

In Exercise 5, the paragraph lacks a concluding sentence. Look at the following pair of sentences and choose the one that best concludes the paragraph.

a) Therefore, dialogue plays a crucial role in drama.

b) Therefore, the next time you see a play, listen carefully to the dialogue for insights into the plot and characterization.

EXERCISE 8 Writing a Concluding Sentence

Write a concluding sentence for the paragraph plan that you created in Exercise 6.

_____

_____

The First Draft

The next step after writing the paragraph plan is to write a first draft of the paragraph. In your first draft, you should include the topic sentence, the supporting points, the details and, if necessary, the concluding sentence.

Your first draft will probably contain some errors. You may not find the perfect word, or you may make some spelling mistakes. In fact, it is advisable to leave your first draft for a day or two after completing it. Then you can reread it with a new perspective in order to revise and edit it for errors.

As you read the example below, you will notice that it contains some errors. In the next unit, you will learn how to revise the paragraph and edit the errors.

EXAMPLE

Symbols in the fairy tale Little Red Riding Hood offer important clues to the meaning of the story. Colour is an important symbol in the story. The colour red is found throughout the narrative. Red is usualy a colour asociated with blood and violence. The little girls cape is red and she becomes a victim in the story. The female sex suggests weakness. Little Red Riding Hood is a naive young girl which becomes a victim of the wolf. She is clothed in a hood and cape, which suggests innocence. The male sex, by contrast, suggests violence and evil. The wolf is male. He is also the predator who murdered the grandmother and wants to kill the little girl.

EXERCISE 9 Writing the First Draft

Using your paragraph plan, write the first draft of your paragraph. Remember that at this stage, your paragraph does not have to be error-free. You will revise and edit it for errors in the next stage of your writing.

Wrap-Up Activity

Follow the guidelines below and write a paragraph.

- Choose a topic from a literary work that you have studied in class. Remember to use some pre-writing strategies to generate ideas.
- For writing ideas, refer to the suggested activities at the end of each short story chapter that you have studied.

Revising and Editing Paragraphs

Revising and editing are important steps in the writing process. You revise and edit a paragraph after you have written the first draft. When you revise, you improve the organization of the paragraph so that it is more logical and flows more smoothly. When you edit, you look for errors in grammar, spelling and punctuation.

Revising Paragraphs

■ REVISE FOR DEVELOPMENT.

A paragraph lacks adequate development if it does not contain enough details to support its topic sentence. When you write a paragraph, ensure that you include sufficient details to make your ideas clear to the reader. Add enough facts, examples and interpretations to your paragraph.

EXERCISE 1 Adding Adequate Details

The next paragraph is not well developed because it lacks adequate details.

> Fairy tales derive their meaning from specific universal characteristics. The setting of fairy tales is not specific, giving no date or place. The plot of most fairy tales has the same dramatic structure, and the protagonist must overcome great obstacles to achieve his or her goal. The main character is usually morally upstanding and is helped through magic or some other unnatural phenomena.

Add at least two details for each supporting idea to the paragraph plan of the preceding paragraph. Give specific examples of fairy tales or from fairy tales.

Paragraph Plan

Fairy tales derive their meaning from specific universal characteristics.

First supporting point: The setting of fairy tales is not specific, giving no date or place.
Details: _____

_____

Second supporting point: The plot of most fairy tales has the same dramatic structure, and the protagonist must overcome great obstacles to achieve his or her goal.

Details: _____

_____

Third supporting point: The main character is morally upstanding and is helped to achieve his or her goal through magic or some other unnatural phenomena.

Details: _____

_____

■ **REVISE FOR UNITY.**

To constitute a unified paragraph, all of the sentences should support the topic sentence. If some of the sentences do not support the topic sentence, the paragraph lacks unity.

EXERCISE 2 Revising for Unity

The next paragraph lacks unity. Underline the topic sentence and cross out any sentences that do not belong in the paragraph. (*Hint:* remember to ask yourself if each sentence supports the topic sentence.)

> Kate Chopin's story "Desirée's Baby" is a voice of social protest. First, Chopin criticizes male dominance prevalent in society during the nineteenth century. For example, Desirée gets her identity and status from her husband. Armand treats his wife as a slave, imposing his will over her. When Armand believes Desirée has shamed him, he forces his wife out of the house. Even today, many women are still struggling to achieve equality with men. Some women are hampered from promotions by the glass ceiling at work. Moreover, Chopin also judges the values of society of her time. Desirée, an orphan, does not know her racial roots. Thus, her social status is ambiguous, and she is considered lucky to be married to someone like Armand. Also, Armand puts much emphasis on class, wealth and racial purity and blames Desirée for their predicament. Thus, "Desirée's Baby" is a strong criticism of the social values of Chopin's world.

■ **REVISE FOR COHERENCE.**

To make your paragraph flow smoothly, ensure that you have organized your ideas logically. You can use transitional expressions to help guide the reader from one point to the next.

Common Transitional Expressions According to Usage	
Usage	**Expression**
Addition/ Sequence	above all, again, also, and, and then, besides, beyond that, equally important, finally, first, for one thing, further, furthermore, in addition, in fact, in the first place, last, moreover, next, second, still, then, to begin with, too
Cause/Effect/ Reason	accordingly, and so, as a result, because, consequently, for this purpose/reason, hence, otherwise, since, so, then, therefore, thus, to this end, with this object
Comparison	also, as well, both (neither), equally, in the same way, likewise, similarly, too
Concession of a point	certainly, granted that, of course, no doubt, to be sure
Conclusion/ Summary	all, all in all, altogether, as has been said, in brief, in any event, in conclusion, in other words, in short, in simpler terms, in summary, lastly, on the whole, that is, therefore, to summarize
Contrast	although, and yet, be that as it may, but, but at the same time, despite, even so, even though, for all that, however, in contrast, in one …/in other …, in spite of, instead, nevertheless, not only … but also, notwithstanding, on the contrary, on the other hand, regardless, still, though, yet, whereas

Common Transitional Expressions According to Usage (Continued)	
Usage	**Expression**
Example	as an illustration of, for example, for instance, in other words, in particular, in simpler terms, namely, one such, specifically, that is, to illustrate, yet another
Emphasis	above all, after all, especially, even, indeed, in fact, in particular, in all cases, indeed, it is true, most important, of course, surely, truly
Place/Position	above, adjacent to, below, beside, beyond, elsewhere, farther on, further, here, inside, near, nearby, next to, on the other/far side, opposite to, outside, there, to the north/east/south/west, to the right/left
Time	after a while, afterward, as long as, as soon as, at last, at length, at that time, at the moment, before, currently, during, earlier, eventually, firstly, formerly, gradually, immediately, in the meantime, in the past, lately, later, meanwhile, now, presently, shortly, simultaneously, since, so far, soon, subsequently, suddenly, then, thereafter, until, until now, when

EXERCISE 3 Adding Transitional Expressions

Exercise 1 contains no transitional expressions. To make the paragraph flow more smoothly, add appropriate transitions.

Fairy tales derive their meaning from specific universal characteristics. The setting

of fairy tales is not specific, giving no date or place. The plot of most fairy tales has

the same dramatic structure, and the protagonist must overcome great obstacles

to achieve his or her goal. The main character is usually morally upstanding and

is helped through magic or some other unnatural phenomena.

■ REVISE FOR STYLE.

Ensure that your paragraph contains exact language, sentence variety and parallel structure. For information on sentence variety, combining sentences and so on, see Unit 7.

Revising a Paragraph: An Example

The next paragraph is the first draft that was developed in the previous unit. Notice how the student has revised the paragraph for unity, support, coherence and style. The paragraph, however, still contains some grammar and punctuation errors. These errors will be edited in the next step of the writing process.

Symbols in the fairy tale Little Red Riding Hood offer important

                    First, colour

clues to the meaning of the story. ~~Colour~~ is an important symbol

      , and the

in the story. ~~The~~ colour red is found throughout. Red is usualy a

> Add transition

> Combine sentences

colour asociated with blood and violence. The little girls cape is

red and she becomes a victim in the story. ~~The~~ Next, the female sex suggests

| Add transition |

weakness. Little Red Riding Hood is a naive little girl which becomes

For example, she believe that the wolf is actually her grandmother.

a victim of the wolf. She is clothed in a hood and cape suggesting

| Add detail |

innocence. ~~The~~ Finally, the male sex, by contrast, suggests violence and evil. The

| Add transition |

wolf is male. He is also the predator who murdered the grandmother

and wants to kill the little girl. Clearly, the contrast between good and
evil adds to the meaning of the story.

| Add concluding sentence |

Editing for Errors: An Example

Editing is an important step in your writing. Before you write your final draft, proofread your work to ensure that it is free of grammar, spelling and punctuation errors. You can revise and edit at the same time. However, it might be easier to do these tasks in two different steps. Read the next edited paragraph.

Symbols in the fairy tale "Little Red Riding Hood" offer important

clues to the meaning of the story. First, colour is an important

| Add quotation marks |

symbol in the story, and the colour red is found throughout. Red is

is ~~usualy~~ usually a colour ~~asociated~~ associated with blood and violence. The little ~~girls~~ girl's

| Spelling |

cape is red, and she becomes a victim in the story. Next, the

female sex suggests weakness. Little Red Riding Hood is a naive

little girl ~~which~~ who becomes a victim of the wolf. For example, she

~~believe~~ believes that the wolf is actually her grandmother. She is clothed

| Relative pronoun/ subject/verb agreement |

in a hood and cape suggesting innocence. Finally, the male sex,

by contrast, suggests violence and evil. The wolf is a male. He is

also the predator who ~~murdered~~ murders the grandmother and wants to

| Tense shift |

kill the little girl. Clearly, the contrast between good and evil adds

to the meaning of the story.

EXERCISE 4 Writing Your Final Draft

Choose a paragraph that you have written and revise and edit it. Then write a final draft of the paragraph.

Literary Essays

One of the most common forms of writing you will produce as a student is the thesis/support essay. This is the type of essay that you will write for the purpose of literary analysis. Normally, a 500- to 800-word literary essay will have the following structure.

Literary Essay Structure

Introduction: presentation of the thesis statement in an interesting way

Body: three to five paragraphs, each built around one argument

Conclusion: reiteration of the thesis statement from a different angle, and the tying up of any loose ends

---

Longer papers deal with more complex analyses, but they are essentially structured in the same fashion. The body may consist of more than three or four arguments, or it may present the arguments in greater detail with more supporting facts.

Often, your instructor will provide you with a subject or topic that you must narrow into a thesis statement. A thesis statement is the main focus of the essay. The thesis statement contains the topic and the controlling idea. For example, "Hemingway's writing style" is a topic; "Hemingway's writing style differs radically from that of his predecessors," or "The writing styles of Ernest Hemingway and Ray Bradbury differ in many respects," are examples of **thesis statements**.

Next, you will be expected to think of at least three main arguments that can be used as support for your thesis. In subsequent steps, you will place your ideas in logical order and expand upon them, define your audience and choice of writing style, write an introduction and conclusion, and revise and edit your work.

The facts that you have learned about writing good paragraphs apply to the essay in an extended way. For example, the topic sentence of a paragraph has a role that is similar to that of the thesis statement of an essay. Both the body paragraphs and the thesis statement affirm the central idea of the piece and contain an opinion about the idea. In an essay, each of the supporting details, which you would otherwise find in the same paragraph, is expanded into an entire paragraph; likewise, the introduction and conclusion are longer in an essay.

Literary Analysis versus Book Report

Sometimes, when asked to produce a literary analysis, students write a descriptive text instead of an analytical one. A very effective way to understand the essence of a literary analysis is to contrast this type of essay with a book report.

EXERCISE 1 Understanding the Literary Essay

Read the two texts written by college students below. Then answer the questions that follow. (You do not have to read the literary works referred to in the essays in order to answer the questions.)

Essay 1

by Josée Bissonnette

1   *T**he Outsiders*** is a story written by S. E. Hinton. Recipient of the American Library Association Award, she wrote this first novel when she was sixteen. S. E. Hinton now lives with her husband and son in Tulsa, Oklahoma.

2   Ponyboy Michael Curtis is a fourteen-year-old orphan living with his two older brothers in a very tough neighbourhood. One day, his best friend, Johnny, kills Bob, a member of the rival gang. They must run away in order not to get caught. With Dallas Winston's help, they end up hiding in a Windrixville church. In that week, the friendship between Pony and Johnny becomes stronger. When Dally comes to take them home, the whole gang's life is about to be changed forever.

3   Ponyboy: He is the main character of the story. Ponyboy is a dreamer. He is often told that he doesn't use his head. He excels in school and is an excellent runner. He is the narrator of the story.

4   Johnny: Under his tough crust, Johnny is scared to death since he was beat up badly by the socs (rival gang). He killed Bob to save Ponyboy's life and to keep the Socs from beating him up again.

5   Dallas: Dally is the most bitter member of the gang. In New York, he blew off steam in gang fights. He became hard and cold at an early age. The only thing he loves is Johnny.

6   Soda: He is the middle child of the Curtis family. He is Ponyboy's favourite brother. Soda dropped out of school and works at a gas station.

7   Darry: He is Ponyboy's and Soda's older brother. Darry didn't go to college in order to give Ponyboy that chance. Instead he roofs houses to pay the bills.

8   Two-Bit: Two-Bit is the oldest member of the gang. He barely ever takes anything seriously, but he is a good friend to have.

9   Steve: Steve Randle knows everything there is to know about cars. He is Soda's best friend and they work at the same gas station.

10 The very first time I read *The Outsiders*, I decided that I had never read a better book. It is truly an extraordinary novel. It made me experience a thousand different emotions at the same time. The author allows the reader to see the world through the characters' eyes. *The Outsiders* is filled with thrilling action from beginning to end.

Essay 2

by Alexandra Deschamps-Sonsino

1   **B**ecause of the different times they lived in, Katherine Mansfield and John Steinbeck have very different ways of seeing the world that surrounds them. In "Miss Brill," Katherine Mansfield deals with issues such as aging and loneliness, whereas John Steinbeck captures feelings of sadness and weakness of the human soul in his short story, "The Chrysanthemums." By using imagery, characterization and various styles of narration, both authors portray two different women's reactions to their individual situation, and to allow their distressed emotions to come forward.

2   First, the use of imagery and the vocabulary in both stories help us to understand the world in which both Miss Brill and Elisa Allen live. In "Miss Brill," the use of imagery is particularly obvious. Miss Brill is an elderly woman who is in the habit of going to the Jardins Publiques every Sunday. The park is described as being "like white wine splashed over the Jardins Publiques," the air is "like a chill from a glass of iced water before you sip it," and the people around are as "still as statues." All of these comparisons and metaphors put the reader in an atmosphere of delighted perplexity, as the characters seem superficial, as if they are playing a part.

3   In "The Chrysanthemums," however, the Salinas Valley where Elisa Allen lives is described by the author with the use of very unusual vocabulary: "sharp and positive yellow leaves," "the air was cold and tender," "high grey-flannel fog," "like a lid on the mountains and made of the valley a closed pot." These descriptions, metaphors and oxymorons give the reader a very clear and yet childlike image of the setting in this story. As we have seen, in their own unique way, both authors communicate the setting of their story so as to give us a first impression of the events.

4   Furthermore, each character has her own personality, which is projected onto the reader in various ways. In "Miss Brill," there is no actual description of the main character, apart from the fact that she is wearing a stole made out of a fox fur and a red eiderdown. This description allows the reader to imagine Miss Brill's appearance based on the way she thinks and sees herself. In this story, the narrator has the ability to allow us to read Miss Brill's thoughts as she sits in the park.

5   On the other hand, John Steinbeck describes Elisa as being "thirty-five" with a "lean and strong" face and eyes that are "as clear as water" and a figure that looks "blocked and heavy." All of the adjectives used to describe her can also be used to describe a man, which indicates to the reader that she is a very powerful woman whose feminine side is hidden. However, these conclusions can only be made by the reader because there is no follow-up on Elisa's thoughts or feelings as there is for Miss Brill. Therefore, in both stories, the author does not reveal all so as not to make the characters seem too easy to figure out. The reader's imagination is required in order to make the characters seem more real.

6   Thirdly, in both stories, the characters' personality clashes with the way the world views them. In "Miss Brill," for example, the main character is convinced that the world she evolves in resembles a play in which everyone, including herself, plays an essential part. However, as soon as the young couple sits beside her, her view of the world is torn apart, and she is brought back to reality. She realizes that she is a lonely old woman with nothing to look forward to except a life of detail and habits. Just as she thought the others had come from "dark little rooms or even ... cupboards," she goes back to "her room like a cupboard." An object, which plays an important part in the story, is her stole, which is personified to express the repressed feelings that Miss Brill is experiencing. Although she took "it out of its box that afternoon, then out the moth powder ... and rubbed the life back into the dim little eyes," it is actually she who is getting ready to go out—old and rusty as she is.

7   In "The Chrysanthemums," Elisa's feelings about herself are present everywhere in the story as she is "over-eager, over-powerful" and her eyes are "hardened with resistance ... irritably." However, as soon as the man from the wagon talks to her about her flowers, her fear and anger disappear, which is expressed by the statement: "... the gloves were forgotten now." In other words, now she is naked in front of him, revealing her true nature and vulnerability.

8   In conclusion, various styles and methods of writing allow us to understand the characters that the authors put forward. Katherine Mansfield and John Steinbeck both allow us to dig deep inside their characters' emotions and—at the same time—into the human spirit in general.

1. Does Essay 1 have a thesis statement? If yes, what is it? (Write the first three words.)

_____

2. Does Essay 2 have a thesis statement? If yes, what is it? (Write the first three words.)

_____

3. a)   How are the characters presented in Essay 1?

_____

  b)   In a sentence or two, explain how this differs from the way they are presented in Essay 2.

_____

_____

4. a)   How does Essay 1 conclude?

_____

  b)   How is this type of conclusion different from the one in Essay 2?

_____

5. Which essay is a literary analysis?

_____

6. What are the major differences between the book report and the literary analysis presented here?

_____

7. Indicate whether each of the following statements is true or false.

  a)   A book report has a thesis statement.   **T   F**

  b)   A literary essay always contains short biographical information about the   **T   F**
       author.

  c)   A book report gives an in-depth analysis of a character.   **T   F**

  d)   A literary essay includes an analysis of elements and devices.   **T   F**

  e)   A book report contains the writer's personal opinion of the book.   **T   F**

  f)   A literary analysis contains the writer's personal opinion of the book.   **T   F**

  g)   A literary analysis gives a synopsis of the story.   **T   F**

  h)   A book report examines the theme of the story in detail.   **T   F**

Focus the Topic.

Sometimes, your instructor will provide you with a topic for writing an essay. Or you may be asked to find your own topic. When you are planning your essay, remember to keep in mind the audience and purpose. Although essays vary in length, instructors often ask students to write a five-paragraph essay. Therefore, it is necessary to take a broad or general subject and narrow it. Try some pre-writing strategies, such as brainstorming or free-writing, to focus your topic.

Remember that a five-paragraph essay (or longer) contains several points; therefore, your essay topic should be broad enough for developing at least three supporting points.

EXAMPLE

General topic: Romanticism

Focused topics:
- The Gothic novel
- The poetry of Keats
- The characteristics of Romanticism

EXERCISE 2 Focusing the Topic

The following topics are very general. Choose and narrow one of the topics so that it is appropriate for a short essay.

1. Children's literature

2. Suspense novels

3. Interesting literary characters

_____

_____

_____

The Thesis Statement

The thesis statement is the central focus of your essay. It tells the reader what you are going to analyze in your work. The thesis statement is similar to the topic sentence of the paragraph. However, since the essay is longer, the thesis statement is broader than the topic sentence.

The thesis statement states the topic and the controlling idea or the writer's point of view.

EXAMPLE

          topic   controlling idea

Emily Brontë's novel _Wuthering Heights_ is a scathing criticism of social hierarchy in nineteenth-century England.

The Thesis Statement

The thesis statement should
- be a complete sentence;
- express the writer's point of view;
- not contain the expressions "I think" or "I believe";
- be broad enough to be supported by at least three points;
- not be a question.

EXERCISE **3** Identifying Strong Thesis Statements

Read each of the following thesis statements and decide whether it is a strong (S) or a weak (W) thesis for a literary essay.

1. _____ I think that "Hills Like White Elephants" is a very good short story.

2. _____ Kate Chopin's use of irony in "The Story of an Hour" effectively heightens the dramatic structure of the story.

3. _____ Ray Bradbury's "The Veldt" contains descriptions of lions and vultures.

4. _____ Did you ever wonder why Edgar Allan Poe is considered to be the father of the modern short story?

5. _____ Louise Mallard, the protagonist of "The Story of an Hour," is a prototype of a feminist role model.

6. _____ The use of foreshadowing in "The Veldt" effectively creates a great deal of tension.

7. _____ Many twentieth-century writers experimented with style.

8. _____ "Hills Like White Elephants" was written by Ernest Hemingway in 1926.

EXERCISE **4** Writing Strong Thesis Statements

Choose one idea from the narrowed topics in Exercise 2 and write a thesis statement for it. Or choose a literary work that you have read, and after narrowing the topic, write a thesis statement for it.

Narrowed topic: _____

Thesis statement: _____

_____

The Supporting Points

To generate the supporting points for your essay, there are two steps that you should follow.

First, you should develop your points; second, you should organize them.

■ DEVELOPING YOUR POINTS

After you have narrowed your topic and written a thesis statement, you generate ideas to support the thesis statement. The supporting points can be examples and interpretations. Try using some pre-writing strategies to develop your supporting ideas.

EXAMPLE

Thesis statement: Emily Brontë's novel *Wuthering Heights* is a scathing criticism of social hierarchy in nineteenth-century England.

Supporting points:
- Heathcliff is described as a gypsy.
- Mr. Earnshaw went to Liverpool.
- Society treats him cruelly.
- The Lintons are of higher class than the Earnshaws.
- Catherine marries up.
- Catherine is better educated than Heathcliff.

- Ellen is of lower class than Catherine.
- Catherine thinks she is better than Heathcliff.
- The second generation reveals the same attitudes.
- Catherine's daughter laughs at illiterate Hareton.
- Heathcliff's son has a higher social status because of money.
- Heathcliff is dark-skinned and dark-haired.
- The Lintons look down on Heathcliff.

■ ORGANIZING YOUR POINTS

Organize your ideas by grouping related ideas together and summarize the underlying connection. Then cross out any ideas that do not belong.

EXAMPLE

First group of related points: Heathcliff inferior to Earnshaws
- Heathcliff is described as a gypsy.
- Heathcliff is dark-skinned and dark-haired.

Second group of related points: Lintons' attitude similar to society's opinions
- Society treats him cruelly.
- The Lintons are of higher class than the Earnshaws.
- The Lintons look down on Heathcliff.

Third group of related points: Catherine has superior attitude toward Heathcliff.
- Catherine thinks she is better than Heathcliff.
- Catherine is better educated than Heathcliff.
- Catherine marries up.

Fourth group of related points: Brontë's criticism of social prejudice
- The second generation reveals the same attitudes.
- Catherine's daughter laughs at illiterate Hareton.
- Heathcliff's son has a higher social status because of money.

Unrelated points
- ~~Mr. Earnshaw went to Liverpool.~~
- ~~Ellen is of lower class than Catherine.~~

EXERCISE 5 Developing Supporting Points

You wrote a thesis statement in the previous writing exercise. Now generate supporting ideas by using a pre-writing strategy. Remember to organize your ideas by grouping them together and crossing out any ideas that do not belong.

The Essay Plan

Writing an essay plan is a crucial step in the essay writing process. An essay plan will help you ensure that you have enough supporting details. An essay plan will also help you organize your ideas so that they are logical and flow smoothly. Look at the model of an essay plan in the following example. It contains the thesis statement and the supporting paragraphs. Each supporting paragraph consists of a topic sentence and details.

EXAMPLE

Thesis statement: Emily Brontë's novel *Wuthering Heights* is a scathing criticism of social hierarchy in nineteenth-century England.

First group of related ideas

Topic sentence: Right from the very beginning of the story, Heathcliff is characterized as inferior to the Earnshaws.
- Heathcliff is described as a gypsy.
- Heathcliff is dark-skinned and dark-haired.

Second group of related ideas

Topic sentence: Social attitudes toward class are seen through the Lintons' attitude toward Heathcliff.
- The Lintons are of higher class than the Earnshaws.
- The Lintons look down on Heathcliff.

Third group of related ideas

Topic sentence: Even members of the Earnshaw family hold a superior attitude toward Heathcliff.
- Catherine, her brother and Ellen think they are better than Heathcliff.
- Catherine is better educated than Heathcliff.
- Catherine marries up.

Fourth group of related ideas

Topic sentence: Brontë criticizes English society as being entrenched in its prejudices.
- The second generation reveals the same attitudes.
- Catherine's daughter laughs at illiterate Hareton.
- Heathcliff's son has a higher social status because of money.

EXERCISE 6 Making an Essay Plan

Read the next essay. Deconstruct the essay by making an essay plan in the space provided. Remember to write the thesis statement at the beginning of the plan. Each supporting idea should have a topic sentence and details.

A Comparison of "Finishing School" and "The Veldt"

by Alexandra Deschamps-Sonsino

1 For many writers, literature is the vehicle to expose social ills produced through human weaknesses. The fight between good and evil is one of the most fundamental themes in world literature. Evil has been portrayed through many faces. We search for evil most of our lives in order to know what we must fear. However, our understanding of evil changes as we grow. In the autobiographical short story "Finishing School" by Maya Angelou and "The Veldt" by Ray Bradbury, social evils are portrayed as a warning of what human beings must be aware of and overcome. If moral beings do not win this battle, there will be terrible consequences for mankind. "Finishing School" and "The Veldt" both expose the latent evil in society, which comes in the form of beliefs. Both stories provide strong commentaries on society, asserting that wrong beliefs such as racism and an over-reliance on technology, if not fought, will lead to human destruction.

2 First, in "Finishing School," Maya Angelou uses point of view in a very effective way to depict attitudes by having two different narrators tell a story about racism.

The author uses a first-person narrator to recount her autobiographical story about an encounter with social evil. Therefore, it is the author herself who is telling the tale, but at different ages of her life. The first-person narrator at the beginning of the story is the older Maya Angelou. She begins her story by writing, "Recently a white woman from Texas ... asked me about my hometown." The second narrator is the protagonist of the story, Margaret, who is the author at the age of ten. Her tale is recounted through a child-like naïveté which is seen, for example, when she describes her employer, Mrs. Cullinan, as "keeping herself embalmed" in order to explain Mrs. Cullinan's reasons for drinking alcohol. Both perspectives complement each other in that they show that racism affects anyone at any age.

3   The evil criticized in this story is racism directed toward blacks in the southern United States in the 1930s and 1940s. Margaret is black, and like other young black girls, she is being prepared to enter into society via a finishing school. This finishing school is the home of a white family, from whom Margaret learns to cook, clean and perform other duties that will be necessary for a girl of her status—in other words, a girl who will one day become a white family's maid. Mrs. Cullinan is Margaret's employer. She is white, middle-aged, fat and bourgeois. Mrs. Cullinan's opinion of Margaret and Miss Glory is not a good one. She has absolutely no respect for these women; she changes their names at will, calling Margaret "Mary." This name change horrifies the young Margaret, and in the manner of the older Margaret, she says:

4   It was dangerous practice to call a Negro anything that could be loosely construed as insulting because of the centuries of their having been called niggers, jigs, dingos, blackbirds, crows, boots and spooks.

5   The younger Margaret has similar thoughts: "Imagine letting some white woman rename you for her convenience." We can therefore see the cultural barriers that stand between Margaret and Mrs. Cullinan. Racism is such a strong force passed on from one generation to the next that in order to put it down and destroy the years of discrimination of blacks, Margaret symbolically breaks the family dishes, which were passed down from one generation to the next by Mrs. Cullinan's family. This action is a way of teaching people a lesson; to prove that if people's attitudes do not change, you have to make them change. There should be no acceptance of evil.

6   Moreover, in the short story "The Veldt," Bradbury presents a futuristic view of society in which technology originally meant for good uses can be turned into evil. Lydia and George Hadley live with their two children, Peter and Wendy, in a modern "Happy-life Home." This house is so technically advanced that it does everything for the family; the parents are no longer burdened with taking care of their children.

7   The central focus of the story is a nursery, which symbolizes the children's desires and dreams. The nursery, through its advanced technology, is also supposed to help the children work out their neuroses in a healthy way and enable them to overcome their psychological problems. The children come to depend on the nursery to meet their needs—a role that was formerly played by the parents. As Lydia states, "The house is wife and mother and now nursemaid." At the end of the story, when it is already too late, George and Lydia realize that their children, who symbolize innocence and unconditional love, have turned against them and used technology as a tool to commit acts of evil. The children eventually murder their parents.

8   In conclusion, we see that both authors warn humankind to be aware of the evils that exist in society.

Essay Plan

Thesis statement: _____

_____

Topic sentence 1: _____

_____

Details: _____

_____

Topic sentence 2: _____

_____

Details: _____

_____

Topic sentence 3: _____

_____

Details: _____

_____

Topic sentence 4: _____

_____

Details: _____

_____

Conclusion: _____

The Introduction

One of the most important parts of an essay is the introductory paragraph. A well-written introduction captures the reader's attention and makes him or her want to continue reading the essay. An introduction also sets the tone of the essay, allowing the reader to know immediately whether it is serious or humorous. The introduction further maintains the parameters for the essay by including the thesis statement. Generally, the thesis statement is the last sentence in the introductory paragraph. Although many professional writers do not follow this convention, it is advisable for you to place the thesis statement at the end of the introductory paragraph because you are beginning writers. Thus, it will be easier for you to find the thesis statement and ensure that it has been adequately supported.

Introductory Paragraph

Characteristics
- captures reader's attention
- sets the tone
- contains thesis statement often as last sentence

Possible beginnings
- a quotation
- a general background of the author or the work
- an interesting anecdote about the author or the work
- a definition that is central to the essay

No matter what type of introductory style you use, remember that since the thesis statement is the last sentence of the paragraph, it is important that the introduction and thesis statement be connected.

EXERCISE 7 Analyzing Introductory Styles

After reading the next paragraphs, determine which of the following introductory styles is used in each case.

quotation	general background	anecdote	definition

Prejudice is not just an ignorant attitude; it is an immoral intolerance. Prejudiced opinions have always existed throughout history. Even Emily Brontë's sweeping Romantic saga *Wuthering Heights* brings out the narrow-mindedness of society in her times. By showing the relationships and attitudes of her characters, Brontë brings to light the general opinions of society. Indeed, Emily Brontë's novel *Wuthering Heights* is a scathing criticism of social hierarchy in nineteenth-century England.

Introductory style: _____

Emily Brontë's passionate novel *Wuthering Heights* was surprisingly not well-received when it was first published in 1847. Critics thought it was too emotional. However, Charlotte Brontë, Emily Brontë's sister, wrote a foreword to the second edition, published in 1850. The novel has since become one of the most widely read novels of English literature. In fact, the novel's staying power rests in its universal themes, such as love, betrayal, vengeance and prejudice. Brontë's criticism of the small-mindedness of society in her times has been much analyzed. Indeed, Emily Brontë's novel *Wuthering Heights* is a scathing criticism of social hierarchy in nineteenth-century England.

Introductory style: _____

Emily Brontë was born in 1818 and lived a rather secluded life in Yorkshire. Her father was a parson, and Emily and her sisters and brother socialized mainly with each other. Given her lack of social contacts and the conventionality of her life, it is surprising that Brontë wrote such a passionate novel as *Wuthering Heights*. This novel reflects all the depths of human relationships. One of the most important aspects of human dynamics comes through in the attitudes of the characters. Indeed, Emily Brontë's novel *Wuthering Heights* is a scathing criticism of social hierarchy in nineteenth-century England.

Introductory style: _____

"My love for Heathcliff resembles the eternal rocks beneath: a source of little visible delight, but necessary," states Catherine, yet she marries Edgar Linton. She marries him because he is rich, and she will have social status. Status was very important to Victorian classes. Victorian society was rigid in its thinking and behaviour. Any actions out of the ordinary would lead to ostracism from society. Such social attitudes were confining even to those who led isolated lives such as Emily Brontë. Indeed, Emily Brontë's novel *Wuthering Heights* is a scathing criticism of social hierarchy in nineteenth-century England.

Introductory style: _____

The Conclusion

A conclusion brings your essay to a satisfactory close. In your concluding paragraph, always restate the thesis statement. Then summarize the essay's main points. There are a number of ways of ending your essay. As a final conclusion, you can end with a quotation, you can write about the broader implications deriving from the focus of your essay, or you can use an image or symbol that synthesizes your idea.

Concluding Paragraph

Characteristics
- brings the essay to a satisfactory close
- restates the thesis statement slightly differently
- summarizes the essay's main points

Possible endings
- a quotation
- writing about broader implications
- presenting an image or symbol that synthesizes your idea

EXERCISE 8 Writing a Conclusion

In the student essay "A comparison of 'Finishing School' and 'The Veldt,'" the thesis statement has been rephrased, but the essay has no concluding paragraph. Reread the essay and write an appropriate conclusion.

_____

_____

_____

_____

_____

The First Draft

You write the first draft after you have planned and organized your essay. At this stage of the writing process, your essay will contain mistakes. It is advisable to leave your work for a day or two and reread it later so that you can revise and edit it for errors.

EXERCISE 9 Writing the First Draft

Choose a topic from your literature course and follow the writing process. Focus your topic, write a thesis statement, generate supporting points and make an essay plan. Then write the first draft of your essay.

Revising and Editing Literary Essays

Revising and editing are critical steps in writing an essay before you hand in the final draft. When you revise your essay, you improve the organization of your ideas and ensure that you have adequately supported your thesis statement. When you edit, you verify that there are no errors in spelling, grammar or punctuation—the mechanics of written work.

Revising an Essay

There are four points to take into account when you revise your essay.

Revising an Essay

1. Development
- thesis statement is both broad enough and narrow enough for essay length
- body paragraphs have adequate examples

2. Unity
- topic sentences support thesis statement
- all sentences in a paragraph support topic sentence

3. Coherence
- links between thesis statement and topic sentences
- transitional words

4. Style
- varied sentence structure
- no clichés or slang
- exact language

■ REVISE FOR DEVELOPMENT.

When you revise for adequate development, consider the following:

- First, make sure that your thesis statement is broad enough to generate several supporting points, which will be developed into body paragraphs. If you do not have enough points to expand into body paragraphs, your essay will appear weak and unfocused. It may be necessary to revise your thesis statement to make certain that the scope of your focus is broad enough for the length of the essay you are required to write.

- Second, ensure that your body paragraphs provide adequate examples and interpretations to support their topic sentences. For further information on writing well-developed paragraphs, see Unit 3.

EXERCISE 1 Ensuring Adequate Supporting Details for Your Essays

Using a literary text with which you are familiar, write a thesis statement and develop at least three points that can be expanded into well-developed body paragraphs.

EXAMPLE

Thesis statement: Edgar Allan Poe uses important literary devices to create the tension in his short story "The Tell-Tale Heart."

Point 1: Poe relies heavily on imagery using sight and sound.

Point 2: Flashbacks and foreshadowing are important devices that create the tension.

Point 3: First-person narrator and subjective point of view also help to increase the tension.

Thesis statement: _____

Point 1: _____

Point 2: _____

Point 3: _____

EXERCISE 2 Analyzing a Well-Developed Body Paragraph

Read the next paragraph and underline the topic sentence. Then circle the specific examples that support the topic sentence.

According to Carl Jung, literature may be categorized into plot archetypes. For example, the love triangle plot is very common. In this type of plot, A loves B who loves C. In Shakespeare's play, *Othello*, there is a love conflict. Roderigo loves Desdemona who loves Othello. Another common plot archetype is the coming-of-age novel. In this type of plot, the main character goes through some personal or physical transformation to acquire maturity and understanding of the adult world. For instance, Holden Caulfield in *The Catcher in the Rye* by J. D. Salinger is a classic example of a transformation of a character. Furthermore, many literary works are built around the quest plot. In this type of plot, a hero overcomes many obstacles to achieve a goal or quest. At the end of his or her successful attempt, there is a large reward. The Harry Potter series is an example of the quest plot. In it, the hero, Harry, has to show his ingenuity and strength to reach his ultimate goal. Thus, literary achetypes are found throughout world literature.

EXERCISE 3 Writing a Well-Developed Body Paragraph

Using your answers from Exercise 1, choose one supporting point and write a well-developed body paragraph. Ensure that the paragraph contains adequate examples and interpretations.

▇ REVISE FOR UNITY.

When you check for unity in the body paragraphs of your essay, consider the following:

- Reread the topic sentences of every paragraph to verify that each one clearly supports the thesis statement.

- Reread each body paragraph to ensure that all sentences in the paragraph support the topic sentence. Two common problems may occur when you are writing the body of your essay: 1) there may be more than one idea in the body paragraphs; and 2) a body paragraph may be arbitrarily split and each smaller paragraph may lack a central focus. For more information on unity within body paragraphs, see Unit 3, "Revising and Editing Paragraphs."

EXERCISE 4 Verifying Unity in Body Paragraphs

The next thesis statements each have three points which may be developed into body paragraphs. Two points support the thesis statement, but one does not. Circle the letter of the point that offers no support.

1. **Thesis statement:** In the short story "The Lottery," the use of irony helps to create an element of surprise at the ending of the story.

   a) The title of the story distracts the reader.

   b) The author's reputation as an eccentric helps create a surprising ending.

   c) The setting of the story leads the readers astray.

3. **Thesis statement:** The Harry Potter series by J. K. Rowling is a modern version of traditional mythology.

   a) Traditional myths have main characters who are smart, intelligent and resourceful.

   b) In both traditional myths and Harry Potter books, the heroes have great powers.

   c) Myths do not have a known author, whereas modern mythological stories such as Harry Potter books have an identifiable author.

EXERCISE 5 Revising for Unity within Paragraphs

The following paragraph contains errors in unity. Revise the paragraph by crossing out any sentences that do not support the topic sentence.

In the novel *The Great Gatsby*, Fitzgerald shows the disintegration of values of American society. For example, Gatsby must resort to crime in order to achieve social status so that he can impress Daisy. Gatsby will do anything it takes to climb the social ladder. Next, Daisy uses her wealth and status to look down on people not considered "worthy" of her attention. Furthermore, both Daisy and Tom reject the family values of an earlier age when they embark on a series of affairs. The rules of society do not concern them because they think that they are above those rules. Today, there are many examples showing the disintegration of values. There is a rising divorce rate, and an overall perception that both governments and corporations want to acquire wealth at any cost. Thus, the acquisition of wealth at any price allows the characters of the novel to behave in a contemptuous manner toward others and to reject the rules of civilized society.

■ REVISE FOR COHERENCE.

Your essay will be much clearer to the reader if the ideas are organized and flow logically and smoothly. There are several ways to ensure that your essay is coherent.

1. Restate words, phrases or ideas from the thesis statement in the topic sentences of the body paragraphs.

    EXAMPLE

    Thesis statement: Emily Brontë's novel *Wuthering Heights* is a scathing <u>criticism of social hierarchy</u> in nineteenth-century England.

    Topic sentence 1: Right from the very beginning of the story, Heathcliff is characterized as <u>socially</u> <u>inferior</u> to the Earnshaws.

2. Connect an idea from the previous paragraph to the next paragraph.

    EXAMPLE

    Topic sentence 2: <u>Social attitudes</u> about class are seen through the Lintons' attitude toward Heathcliff.

    Topic sentence 3: Even members of the Earnshaw family hold a <u>superior attitude</u> toward Heathcliff.

3. Use transitional words or expressions to link ideas of body paragraphs. For a complete list of transitional words and expressions, see Unit 3.

    EXAMPLE

    Topic sentence 4: <u>Moreover</u>, Brontë criticizes English society as being entrenched in its prejudices.

EXERCISE 6 Revising for Coherence

Read the following outline for a literary essay. Underline all the words, phrases or ideas that link the thesis statement to the topic sentences and the topic sentences to each other.

Thesis statement: William Golding's novel *The Lord of the Flies* embraces Thomas Hobbes' philosophy, which asserts that man, in the state of nature, is nasty, brutish and short.

Topic sentence 1: First, the boys find themselves stranded on a deserted island, in a state of nature, with no rules or regulations.

Topic sentence 2: Next, despite their natural inclination to formulate rules, the boys soon embrace savagery and violence.

Topic sentence 3: Furthermore, savagery gains dominance over civilized behaviour as the boys' way of life.

Topic sentence 4: Finally, as the boys descend further and further into the state of nature, they lose the veneer of goodness and innocence that characterizes civilized society.

■ **REVISE FOR STYLE.**

Your essay will flow more smoothly if you vary your sentence structure and avoid clichés and slang expressions. It is important to use a dictionary and thesaurus so that you can check for the exact meaning of words and avoid repetitive words and phrases. For more information about exact language and sentence variety, refer to Units 6 and 7.

Editing for Errors

In the last stage of the writing process, you should reread your essay carefully and look for any errors in spelling, grammar and punctuation. For more information about spelling, grammar and punctuation, refer to Units 6 and 11.

Revising and Editing the Literary Essay: Example and Practice

Some students prefer to revise and edit their written work in two distinct steps. However, others find it easier to revise and edit simultaneously. Choose the method that is most convenient for you. Remember that the goal of this step is to produce a well-organized and error-free essay.

In the following paragraphs, all the key elements of an essay have been underlined. Notice how the text was revised and edited.

"Wuthering Heights," a Novel about Social Class

Prejudice is not just an ignorant attitude; it is an immoral intolerance.

> **Introduction by definition.**

Prejudiced opinions have always existed throughout history. Even Emily Brontë's sweeping Romantic saga *Wuthering Heights* brings out the narrow mindedness of society of her times. By showing the relationships and attitudes of her characters, Brontë brings to light the general opinions of society. **Indeed, Emily Brontë's novel** *Wuthering Heights* **is a scathing criticism of social hierarchy in nineteenth-century England.**

> **Thesis statement**

From
~~Right~~ from the ~~very~~ beginning of the story, Heathcliff is characterized as inferior to the Earnshaws. Indeed, he is described

> **Style: redundancy**
>
> **Topic sentence 1**

as a dark-haired and dark-skinned boy. Mr. Earnshaw ~~told~~ *tells* his

wife, "though it's as dark almost as if it came form the devil."(36)

| | Verb tense shift |
| Quotation |

By that comparison, Heathcliff is already something evil and

strange. He is not a good-standing member of society, but rather

something to look down on or be afraid of. Even Mrs. Earnshaw

categorizes him as a "gypsy brat" not fit to be a part of her family.

Futhermore, social
~~Social~~ attitudes toward class are seen through the Lintons'

attitude toward Heathcliff. The Lintons are conscious of their

| Add transition |
| Topic sentence 2 |

gentry status, their home, and their way of life. The Lintons belong

class, and Edgar
to the upper ~~class, Edgar~~ Linton is characterized as the perfect

| Run-on sentence |

gentleman in contrast to Heathcliff who does not know his parentage.

Linton does not treat Heathcliff as an equal. He does not offer him

friendship or respect. In fact, Edgar Linton observes that Heathcliff

is only a gentleman in "dress." ~~The Lintons are conscious of their~~

| Added example |
| Organization of ideas |

~~gentry status and their home and its furnishings and their way of~~

~~life reflect this attitude.~~

Even members of the Earnshaw family hold a superior attitude

toward Heathcliff. Catherine, her brother and Ellen Dean think

| Topic sentence 3 |

they are better than Heathcliff. Hindley terrorizes the young

, and he eventually
Heathcliff ~~. He~~ kicks him out of the house. Ellen, although a servant,

| Combining sentences |

knows that Heathcliff is not her master and treats him at times

disrespectfully. Catherine, although Heathcliff's best friend in

childhood, identifies herself in adulthood with the Lintons.

Indeed, Catherine seeks socially upward mobility when she rejects

Heathcliffe and marries Edgar Linton.

Moreover,

Brontë criticizes English society as being entrenched in its | **Transition**

prejudices. The second generation reveals some of the same attitudes | **Topic Sentence 4**

Catherine's

as the previous generation. ~~Catherines~~ daughter makes fun of an | **Spelling**

illiterate Hareton. She also looks down on Heathcliff, as well as on

weak

her own husband, the ~~week~~ Linton. Linton Heathcliff is sickly and | **Spelling**

unmanly, yet he has a higher social status because of his wealth.

Hareton Earnshaw is the second generation's Heathcliff, taunted

and neglected by all. When the young Catherine falls in love with

Hareton, the couple decide to leave Wuthering Heights and move

into the more elegant Thrushcross Grange, a symbol of class status.

Thus, Victorian society was a rigid system of class status, | **Restate thesis**

which Emily Brontë criticized in her novel *Wuthering Heights*.

was

Each of the characters ~~were~~ either trying to maintain their class | **Subject/verb agreement**

status or better it. Catherine and Heathcliff tried to climb the

social ladder while Edgar Linton tried to retain the status he was

born to. None of the characters were happy, and they all ultimately

led tragic lives.

EXERCISE 7 Revising and Editing a Literary Essay

The following student essay contains some errors. Read the essay and answer the questions that follow.

Othello and Lyra: Heroes for All Time

1   Throughout human history, the concept of heroism has permeated our collective culture. There are heroes found in religious texts, songs, plays, short stories and novels. There is also real-life heroes, whom we admire. Heroes have certain common caracteristics, such as courage, intelligence and determination.

Literature is full of heroes. William Shakespeare's character Othello in his play *Othello* and Phillip Pullman's Lyra Balcqua in *The Golden Compass* are protagonists of very different genres of literature. Yet they are both heroes.

2   First, Othello and Lyra are leaders. Othello is a general, he has led his soldiers into battle many times. He commands great respect from his soldiers. They show him loyalty and love by following him into dangerous situations. Othello is like many soldiers in present day. Soldiers are brave because they fight in very difficult situations. They risk their lives each time there is a war. As a young girl, Lyra led small battles against the other colleges. She takes charge of the children from the Oblation Board, leading them into the forest to be saved by the Gypsies. Thus, Lyra, like Othello, displays heroic qualities.

3   Othello and Lyra are courageous. Othello is introduced as a respected and successful military commander. At the begining of the novel, Lyra also proves that she knows no fear. She hides in a cupboard trying to find out information from Lord Asriel. She is brave when she gets kidnapped and when she talks to Iorek the Bear.

4   Moreover, neither Lyra nor Othello run away from detractors. Othello defends himself when accused of seducing Desdemona. He firmly speaks of his marriage to Desdemona with the Duke and Desdemona's father. Lyra, although at first scared of her father, Lord Asriel, speaks out against his research. And fights Mrs. Coulter, who kidnaps children for evil experiments.

5   Furthermore, Othello and Lyra show that they are determined characters. Othello is determined in his desire to achieve military victory against the Turks, he is also determined to marry Desdemona. Iago must manipulate events before Othello turns against Desdemona. Lyra also is a strong-willed girl. She is determined to go to the north and endure hardship to save her friend Roger. She overcomes many obstacles, such as the cold weather and enemies in the Oblation Board, to rescue her friend.

6   Although at first glance Othello and Lyra seem dissimilar, both have heroic qualities. They show leadership qualities, determination and are brave. Both have tragic lives, but their weaknesses are overshadowed by the heroic qualities they exhibit.

Questions

1. a) Which paragraph lacks unity? _____

    b) Cross out any sentences that do not support the topic sentence in that paragraph.

2. Which paragraph lacks adequate development? _____

3. a) Underline the thesis statement in the essay. Then underline all the words and phrases that link ideas from the topic sentences to the thesis statement.

    b) Which paragraph lacks a transitional expression? _____

    c) Add an appropriate transitional word or expression to the paragraph.

4. There are two run-on sentences in the essay (paragraphs 2 and 5). Correct the errors.

5. There are two subject/verb agreement errors in the essay (paragraphs 1 and 4). Correct the errors.

6. There are two spelling errors in the essay (paragraphs 1 and 3). Correct the errors.

7. There is one sentence fragment in the essay (paragraph 4). Correct the error.

8. There is one error in parallel construction in the essay (paragraph 6). Correct the error.

Essay-Writing Checklist

Before you hand in your essay, ask yourself the following questions. Make any changes necessary to ensure that you hand in a well-written essay.

USEFUL QUESTIONS FOR DEVELOPING AN ESSAY

- Did I write a thesis statement that introduces the topic and states the controlling idea?
- Did I support the thesis statement with facts, examples and interpretations?
- Did I write an essay plan to help organize the main and supporting ideas?
- Did I write a first draft?

USEFUL QUESTIONS FOR REVISING AND EDITING AN ESSAY

- Did I revise for unity?
- Did I revise for adequate development?
- Did I use transitional expressions to link ideas?
- Did I edit for errors in spelling, grammar and punctuation?

S E C T I O N 2

GRAMMAR

Spelling, grammar and punctuation are key elements in writing a text. If your writing contains errors in these elements, the meaning of your written text will be unclear. Therefore, it is important to become proficient at spelling, grammar and punctuation rules.

In the sections that follow, you will review the most important grammatical notions and rules you need to improve your written communication skills and writing style. In other words, mastery of this basic grammar will enable you to write with "style" and "substance." First you will look at words, which represent the basic elements of meaning, and then you will examine sentences and syntax—how words are put together—to express ideas.

Words

Connotations versus Denotations

Not only do words have meanings that you can look up in a dictionary, they also convey feelings and emotions in many instances. The **denotation** of a word refers to its dictionary meaning. The **connotation** refers to the emotional association that accompanies a word. The connotation can be positive, negative or neutral. For example, consider the words "house" and "home." Both can refer to the place where a person lives, but "home" conveys a feeling of warmth, security and intimacy that "house" does not convey. It is very important to pay close attention to word choice since synonyms can have the same denotation but very different connotations. When you are writing to convince the reader, you will want to use language that has the greatest impact.

EXERCISE 1 Determining Connotation

Look at the following groups of synonyms (or near synonyms) and indicate whether each word has a positive (+), negative (-) or neutral (0) connotation. If you are unsure, check your dictionary. (Choose the most common connotation.)

1. _____ thin   _____ lean   _____ wiry   _____ skinny   _____ cadaverous
   _____ slender

2. _____ guru   _____ pro   _____ expert   _____ teacher   _____ authority
   _____ mentor   _____ shark   _____ cunning person   _____ savant

3. _____ house   _____ home   _____ shack   _____ cottage   _____ dump
   _____ slum

4. _____ smart   _____ foxy   _____ brilliant _____ clever   _____ intelligent
   _____ gifted   _____ wily   _____ cagey   _____ shrewd   _____ learned

5. _____ great   _____ strong   _____ violent   _____ powerful   _____ warlike
   _____ influential

EXERCISE 2 Determining More Positive Emotional Impact

With a (+) sign, indicate which of the words in the following pairs has the most positive emotional impact.

1. _____ buyers   _____ clients   4. _____ educational   _____ pedantic

2. _____ famous   _____ notorious   5. _____ adds to   _____ complements

3. _____ genuine   _____ real   6. _____ yellow   _____ gold

Synonyms often have different **intensities**—that is, some are vivid and others are bland. Keeping this in mind when you write will help you choose more concrete and precise words.

EXERCISE 3 Determining Intensity

Underline the most intense word in each of the following pairs.

1.	cold	freezing	5.	hilarious	funny	9.	wet	soaking
2.	plummet	fall	6.	happy	ecstatic	10.	sadness	grief
3.	tired	exhausted	7.	terror	fear	11.	respect	veneration
4.	rage	anger	8.	interest	fascinate	12.	shout	say

Exact Language

■ AVOID WORDINESS AND REDUNDANCY

It is always better to be **brief** and **concise** than to be wordy and repetitious. For example, "beauty" is a better choice than "pulchritude" and "love" is a better choice than "amorousness" or "sentimental attachment."

Here are some common wordy expressions and the more concise substitutes you can use to replace them.

Substitutes for Wordy Expressions			
Wordy Expression	**Substitute**	**Wordy Expression**	**Substitute**
along the lines of	similar to	in spite of the fact	in spite of
at this/that point in time	then/at that time/presently	in the final analysis	finally/lastly
in close proximity	close or in proximity	in the habit of giving	gave
about the reason why	why	it did happen/it came about that	(nothing)
a difficult dilemma	a dilemma	it is in our own best interests	it is best
a true fact	a fact	it was because	because
concerning the matter of	about	owing/due to the fact that	because
exactly the same	the same	people from all walks of life	all, everyone
exceptions to the rule	exceptions	period of time	period
for the purpose of	for	regardless of the fact that	although
gave the appearance of being	looked like	spots and locations	places
great/few in number	great/few	still remain	remain
in order to	to	the point I am trying to make about why	I think (or nothing)

EXERCISE 4 Using Concise Language

Rewrite each of the following sentences to make it more concise. Change the word order if necessary.

> **EXAMPLE:** It is necessary that you study. ▸ You must study.

1. At this point in time it is in our own best interests to stop the investigation owing to the fact that no new evidence has been found.

   _____

   _____

2. Regardless of the fact that Samuel Clemens did not have a university education, it did happen that his work was highly regarded by people from all walks of life.

   _____

   _____

3. Due to the fact that Mark Twain was an entertaining storyteller, it came about that he was invited to many different spots and locations around the world, where he was in the habit of giving lectures along the lines of his short stories.

   _____

   _____

4. The point I am trying to make is about the reason why Mark Twain was appreciated worldwide; it was because of the universal appeal of his humour.

   _____

   _____

5. People from all walks of life talk about Mark Twain's vision of the truth.

   _____

   _____

■ AVOID VAGUE LANGUAGE

Always try to use **specific** words rather than vague ones.

Frequently Used Vague Words		
bad	man	school
big	nice	slowly
car	old	small
good	pretty	thing
happy	quickly	woman
house	sad	young
look	says	etc.

Use a thesaurus or a synonym dictionary to find substitutes for overly familiar and vague words. Choose the best synonym for the context.

EXERCISE 5 Using Specific Words

In the following sentences, replace the vague word (in italics) with a more concrete term from the box below.

EXAMPLE: I am *really hungry* _____*famished*_____.

perfect	stomped	dangerous	shone	repairs
well-written	rippled	examined	understanding	enraged
whispered	rotten	felt (2)		

1. She could do many *things* _____, such as install electrical wiring and fix a flat tire.

2. He is a *nice* _____ person. He always listens to people's problems.

3. The book was really *good* _____.

4. She *said* _____ to him in a very quiet voice that she had to leave early.

5. The child *was* _____ very frightened because he was in a *bad* _____ situation.

6. I *was* _____ so *mad* _____ that I *started walking* _____ angrily around the room.

7. The sun *was out* _____ and the breeze *was making little waves on* _____ the surface of the lake. It was a *nice* _____ day for a picnic.

8. The grapefruit tasted *bad* _____.

9. She *looked closely at* _____ the rip in his shirt.

Levels of Language

Good writers are aware of the appropriateness of the language they are using for their particular situation. For example, language choice differs between an academic report and a letter to a friend or between a business report and creative writing. It is important to understand what is being written and the intended audience. Language usage revolves around diction. Diction refers to the writer's choice of words. The following is a list of different types of usage.

Standard English
In general, it is the accepted form of usage in English-speaking countries. Although standard English can vary from one English-speaking country to another, it is the common language of the country and is used by government, business and educational institutions. It is considered to be formal language.

> **EXAMPLE:**   Hello. I am very pleased to meet you.

Colloquial English is common or informal language.

> **EXAMPLE:**   That boy keeps *hanging around* my girlfriend.

Slang is "very informal language that includes new and sometimes offensive words, and that is used especially only by people who belong to a particular group, such as young people or criminals" (*Longman Advanced American Dictionary* 1363).

> **EXAMPLE:**   My friends and I *pigged out* on pizza last night.

Dialect is a variety of English found in a particular geographic region and spoken by a particular group.

> **EXAMPLE:**   It was a good movie, *eh*? (This particular use of "eh" is found in the English-speaking parts of Canada.)

Jargon is language filled with unfamiliar terms, most of which are specific to a profession.

> **EXAMPLE:**   The *psychometric analysis* of the *standardized formative* test was used to determine the *intelligence quotient* of the *study group*.

Euphemism is vague or inoffensive language that is substituted for harsher words.

> **EXAMPLE:**   He works as a *sanitary engineer* (a garbage man in standard English).

In a literary essay or another type of academic essay, your audience is your instructor and the other students. In an academic setting, you should use a **formal style** and **standard language**.

Formal writing has:
- a serious tone;
- a conventional structure;
- correct grammar and spelling;
- a clear, precise word usage.

When writing for academic purposes, you should
- use standard English;
- find the precise meaning of words in a dictionary to make informed word choices;
- avoid informal language, such as "guys" or "gonna," slang, jargon, clichés, euphemisms and vague language. You should also avoid wordiness, triteness, redundancy, as well as any other sort of affected language.

Note: Check with your instructor concerning the use of the personal pronoun "I" and contractions. In some formal writing situations, it may be preferable not to use them.

Common Clichés to Avoid	
a drop in the bucket	crystal clear
as busy as a bee	easier said than done
as light as a feather	loss for words
as luck would have it	time and time again
axe to grind	top dog
better late than never	tried and true
between a rock and a hard place	under the weather
calm, cool and collected	work like a dog

EXERCISE 6 Avoiding Euphemisms and Clichés

Replace the euphemisms and clichés (in italics) with more direct language.

> **EXAMPLE:** After John's beloved wife *passed away* _____*died*_____, she was *laid to rest* _____*buried*_____ in Notre Dame Cemetery.

1. You must not drive *under the influence* _____. Otherwise, you will soon find yourself in the custody of a *law enforcement officer* _____.

2. Mark Twain *came into the world* _____ in Florida, Missouri, on November 30, 1835. It was *a red-letter day* _____ for readers.

3. *Last but not least* _____, do not forget to mention Twain's experience as a river pilot.

4. *Taking stock of* _____ the situation, the hero realized that *the writing was on the wall* _____.

5. The villain, however, *making the best of a bad situation* _____, decided to *turn over a new leaf* _____.

6. Several participants in the ride were *physically challenged* _____. Some suffered from *hearing impairment* _____ while others had *visual impairments* _____.

Common Slang to Avoid				
ace	downer	kick the bucket	pissed	weirdo
ain't	go bananas	kook	red-letter day	wimp
bad mouth	gonna	nerd	sweet tooth	wuss
chick	guts	outta here	top-notch	yucky
couch potato	guy	pass out	VIP	24/7
croak	jock	pig out	wanna	

EXERCISE 7 Using Standard Language to Replace Slang, Jargon and Wordiness

Replace the words in italics with a word (or words) conforming to standard English.

1. My neighbour is such a *jerk* _____. He blocks my driveway with his car.

2. The boy's *ego was battling with the forces of his id in order to gain hegemony of the conscious*
_____.

3. The new English teacher is *really* _____ *with it* _____.

4. That child *ain't never* _____ going to listen to his babysitter.

5. My mother *freaked out* _____ when I told her I *flunked*
_____ my math test.

EXERCISE 8 Transforming Informal Sentences into Formal Ones

Transform the following informal sentences into formal ones.

1. At the end of the story, there's this weird guy walkin' like he was pissed or something.
_____

2. I tell you that Hamlet ain't no wacko. He was just simulating and goofin' off so people'd leave him alone.
_____

3. Ray's a really top-notch writer. He's got a great writing style. Ray created some really neat VR (virtual reality) effects in "The Veldt."
_____

4. It was a real red-letter day for the hero of the story. The bad guy had croaked from a heart attack.
_____

5. The main character is this young kook who's gonna kill this old man because he has got the evil eye.

_____

6. In this story, a bunch of nerds face off with a bunch of jocks. It was pretty boring because all the characters were flat and you could guess what would happen.

_____

EXERCISE 9 Using Clear and Concise Language

The following paragraph contains vague words, clichés, slang and redundancies. Edit the paragraph to make it clear and concise by making any necessary changes.

Food has become the hot topic of conversation at this point in time. Everyone seems to be getting into the food and cooking frenzy for the purpose of cashing in on a hot market. And everyone seems to be as busy as a bee writing recipe books. I went to the bookstore the other day and I was at a loss for words. In the section on food, there were tons of books on how to cook food, how to eat food, how to loose weight, how to gain weight, how to get more nutritious food, how to avoid bad fat, how to get good fat, how to barbecue, how to be a vegetarian, etc. etc. etc. It seems that everybody has become an expert on food and nutrition, and everybody wants to tell everyone what he or she knows. In the final analysis, this is called imparting knowledge with a vengeance. I see this happening in many other areas as well, such as medicine and psychology. Everybody is an expert except me. In spite of the fact, I will remain in my den of ignorance, living, breathing, eating and enjoying my neuroses.

Spelling

■ SPELLING RULES

Different countries have different spelling systems. For example, the United States and Great Britain spell certain words differently. Canada also has its own spelling conventions. When you revise a text, pay close attention to spelling. Sometimes you will have a choice between certain alternatives, for example, writing *centre* or *center*. Be consistent. In other words, use the same spelling system throughout the text you are writing. Do not shift back and forth between two alternatives.

If you are using a word processor, you can do an automatic spell-check after choosing the appropriate dictionary (see American, Canadian and British spelling on page 328). If you are not using a computer, you should consult a dictionary when in doubt about the spelling of a word.

Here are a few basic rules that will help you spell certain words correctly.

---

1 Noun Plurals

Add -*s* to most nouns but -*es* to nouns ending in *s, sh, ch* and *x*.

> **EXAMPLES:**   car ▸ car**s**; church ▸ church**es**; box ▸ box**es**; bus ▸ bus**es**

In words that end in *f* or *fe*, change the *f* to *v* and add -*es.*

> **EXAMPLES:**   knife ▸ kni**ves**; half ▸ hal**ves**

> **EXCEPTIONS:** belief ▸ beliefs; roof ▸ roofs; chief ▸ chiefs
>
> man ▸ men; child ▸ children; woman ▸ women; mouse ▸ mice; goose ▸ geese; foot ▸ feet; tooth ▸ teeth
>
> Words borrowed from other languages: basis ▸ bases; hypothesis ▸ hypotheses; criterion ▸ criteria; medium ▸ media

2 Numbers

Spell out any numbers that begin a sentence.

> **EXAMPLE:**   **Five hundred** people attended the concert.

In essays that are **not** scientific, spell out numbers that do not consist of more than two words.

> **EXAMPLE:**   The narrator of the story had robbed **twenty-five** banks before being caught.

(You may wish to check with your instructor, as this rule does not always apply. In the field of psychology, for example, you may be expected to use numerals in your essays.)

Use numerals for time, dates, decimals, percentages, statistics, measurements and scores, exact amounts of money, addresses, chapters, pages, scenes and line numbers.

3 ie/ei

Use *i* before *e* except after *c*, or when the combination of *e* and *i* is sounded like *ay* as in "neighbour" and "weigh."

> **EXAMPLE:**   bel**ie**ve; repr**ie**ve; rec**ei**ve; sl**ei**gh

> **EXCEPTIONS:** science, species, height, either, neither, leisure, foreign, seize

4 Double Consonants

When adding a suffix, double the final consonant of one-syllable words when a vowel precedes the consonant.

> **EXAMPLES:**   sit ▸ si**tt**ing; hot ▸ ho**tt**est

In words with more than one syllable, double the final consonant when the last syllable of the word is stressed.

> **EXAMPLE:**   confer ▸ confe**rr**ed, but develop ▸ developed (the final syllable in *confer* is stressed, but the final syllable in *develop* is not.)

5 y/i

When adding the suffix **-s** or **-ed** to a word ending in a consonant followed by **y**, change **y** to **i** and add the suffix.

> **EXAMPLE:**   carr**y** ▸ carr**ies** or carr**ied** (note that "carrying" is used because in this word you hear two separate **i** sounds when it is pronounced).

When a vowel is followed by **y**, do not change **y**.

> **EXAMPLE:**   pla**y** ▸ plays, played, playing

> **EXCEPTIONS:** pay ▸ paid; say ▸ said; lay ▸ laid

6 Prefixes and Suffixes

When adding a prefix or a suffix to a word, various rules apply and you may need to check a dictionary. Often, however, the original base word does not change.

> **EXAMPLES:**   mature ▸ premature; awful ▸ awfully; change ▸ changeable

Commonly Misspelled English Words					
acceptable	definitely	exaggerate	humorous	personnel	separate
a lot	desperate	fiery	independent	precede	vacuum
argument	disappointed	foreign	its/it's	publicly	Wednesday
believe	dilemma	gauge	leisure	receive	weird
business	discipline	government	millennium	relevant	with
calendar	embarrass	guarantee	minuscule	rhyme	writer
changeable	environment	harass	misspell	rhythm	writing
conscious	especially	height	pastime	schedule	written

EXERCISE 10 Choosing Correct Spelling

Underline the correct spelling of each word pair below.

> **EXAMPLE:** percieve/<u>perceive</u>

1. thief/theif

2. iresponsible/irresponsible

3. foxes/foxs

4. beautifull/beautiful

5. soceity/society

6. berrys/berries

7. echos/echoes

8. knifes/knives

9. writing/writting

10. envyable/enviable

11. cleaner/cleanner

12. argument/arguement

13. enterring/entering

14. truly/truely

15. unnatural/unatural

16. wierd/weird

EXERCISE 11 Correcting Misspelled Words

Underline all the incorrectly spelled words and write the correct spelling above them.
(*Hint:* there are three mistakes in each sentence.)

1. The girl shreiked with delight at her loss in wieght of eigth pounds.

2. I refered her to the manager because she sayd that she had already paied her bill.

3. When he played football, he usualy carryed the ball a full twenty-five yards before the guards stoped him.

4. Those hypotheses do not meat whit the women's approuval.

5. Last nite, we purchased two boxs of special envelopes and sent the films to be developed.

EXERCISE 12 Correcting Misspelled Words in a Paragraph

Edit the following paragraph for spelling. There are ten errors.

The short story "The Lottery" creates alot of controversy. Many poeple are shocked by the ending because they expect that a lottery would bring happyness to the winner. The story seems inocent at the begining. It is a sunny summer day, and all the villagers are gathering in the town square at ten in the morning. First the childrens come and than the adults. They collect stones in piles, but the audience does not realy think about this action. The narrator tells the reader that everything will be over before noon dinner and describes the setting and the pro-cedure very objectivly. There is no emotion as the story unfolds, and the reader wonders what is going to happen. Suddenly the end arrives, and it is truely terrible. The reader just cannot believe it.

■ AMERICAN, CANADIAN AND BRITISH ENGLISH

Canadian English is influenced by both the language of our British forebears and that of our American neighbours. Sometimes Canadians prefer a particular spelling, but other times they can choose between two spellings. Usually a Canadian dictionary, such as the *The Canadian Oxford Dictionary*, will list the preferred spelling first. Vocabulary sometimes varies too; British and American meanings can be very different. As mentioned above, when you have a choice, it is important to be consistent throughout your essay.

6

Spelling Variants		
British	**American**	**Canadian**
-yse, -ise: analyse, criticise	**-yze, -ize**: analyze, criticize	*-yze, -ize* preferred
-our: colour, flavour	**-or**: color, flavor	*-our* preferred
-re: centre, theatre	**-er**: center, theater	*-re* preferred
-xion: reflexion, connexion	**-ction**: reflection, connection	*-ction*
-ight: night, light	**-ite**: nite, lite (slang)	*-ight*
cheque	check	cheque
defence	defense	defence
dialogue	dialog or dialogue	dialogue
programme	program	program preferred
judgement	judgment	judgment preferred
double consonant s, l: travelled, marvellous, focussed	**single consonant s, l:** traveled, marvelous, focused	double consonant *s, l*

In Canada, words doubling as nouns and verbs end in *-ice* or *-ise*: *-ice* is used in the noun form and *-ise* in the verb form.

EXAMPLES: Every day, I practise piano.
I enjoy doing piano practice.

Vocabulary Variants		
British	**American**	**Canadian**
lorry	truck	truck
lift	elevator	elevator
boot	trunk	trunk
flat	apartment	apartment (flat: rented, self-contained part of a house)
tap	faucet	tap
holiday	vacation	holiday, vacation
biscuits	cookies	biscuits, cookies
veranda	porch	veranda, porch
purse	pocketbook	purse
pictures	movies	movies
telly	TV	TV
pudding	dessert	dessert

Original Canadian Words

Some typical Canadian words that do not come from either British or American English include muskeg, tuque, bush pilot, electoral riding, chinook, the Prairies, portage.

Canadian Pronunciation

Differing from British pronunciation, Canadian pronunciation is similar to the pronunciation existing in certain regions of the United States. However, Canadians always give themselves away when they pronounce words ending in **-out**, such as "out" and "about," when they pronounce the last letter of the alphabet as "zed" (instead of the American "zee") and when they finish a sentence with "eh?"

EXERCISE 13 Knowing Canadian Spelling

Using the previous list, write the words in the blanks using the appropriate spelling. Write "either" if both forms are appropriate for Canadian spelling.

	EXAMPLE:	*Canadian*	*British*	*American*
		program	*programme*	*program*

	Canadian	*British*	*American*
1.	_____	_____	color
2.	_____	travelled	_____
3.	_____	judgement	_____
4.	_____	defense	_____
5.	honour	_____	_____
6.	_____	centre	_____
7.	_____	cheque	_____
8.	_____	_____	dialog
9.	reflection	_____	_____
10.	neighbour	_____	_____

EXERCISE 14 Knowing Canadian Vocabulary

The following sentences include either British or American vocabulary. Edit them using Canadian vocabulary.

   EXAMPLE: Last night, we went to the ~~pictures.~~ *movies*

1. The boot of our car is full of old books.

2. The faucet in our bathroom is leaking.

3. Every Thursday night, I watch reality shows on the telly.

4. Yesterday, a thief stole my sister's pocketbook.

5. Would you like some pudding? It's chocolate cake.

Homonyms and Other Commonly Confused Words

Words in English that are pronounced the same but have different spellings and meanings are called **homonyms**, for example, "I" and "eye."

Common Homonyms		
ad/add	fir/fur	red/read
allowed/aloud	for/four/fore	sail/sale
altar/alter	gorilla/guerrilla	sea/see
ascent/assent	hi/high	sight/site/cite
ate/eight	hole/whole	some/sum
bear/bare	its/it's	stationary/stationery
be/bee	lead/led	steal/steel
brake/break	lie/lye	sundae/Sunday
by/bye/buy	meet/meat	their/they're/there
cell/sell	miner/minor	to/two/too
cent/scent/sent	new/knew	waste/waist
cereal/serial	no/know	war/wore
complement/compliment	one/won	weather/whether
council/counsel	or/oar/ore	weak/week
die/dye	peace/piece	which/witch
disc/disk	principal/principle	who's/whose
eye/I	rain/rein/reign	wood/would
fare/fair	read/reed	your/you're

Many other English words often cause confusion because they have similar spellings but different pronunciations and meanings, for example "though," "tough" and "thought."

EXERCISE 15 Writing Sentences Using Homonyms and Other Commonly Confused Words

Find the definition for the following words in your dictionary. Then write a sentence using each word correctly.

1. ascent _____

   assent _____

2. accept _____

   except _____

3. affect _____

effect _____

4. altar _____

alter _____

5. cite _____

site _____

sight _____

6. complement _____

compliment _____

7. council _____

counsel _____

8. principle _____

principal _____

9. weather _____

whether _____

10. who's _____

whose _____

EXERCISE 16 Correcting Errors in Spelling and Meaning

The following paragraph contains fifteen errors. Underline each mistake and write the correct word above it.

My to friends and I decided to go on vacation. We where really excited about the

idea, but could not agree upon the destination. Should it be near the see, or should

it be in the mountains? We just could not decide. After much debate, we finally

agreed too go to the beach on a tropical island. I thought that this idea was great,

and my friends taught so to. With hour suitcases packed, we met one our before

our plane was to take off. Their was so much noise and excitement at the airport.

It was filled with tourists going on wonderful vacations, and leaving there worries

behind them. At the check-in counter, we all put our bags on the belt and their was

some confusion about who's bag was whose. We finally arrived at hour destination

and now we our looking forward to a hole weak of rest and relaxation.

Sentences

Not only is it important for people to write correct sentences when they communicate with others in a formal capacity—for example, in academic or workplace situations—it is also necessary to write with a certain amount of style. To do this, you must be able to recognize the difference between what constitutes a correct sentence and what is unacceptable. You must also work at developing style in your writing through the use of various types of sentences.

Read the explanations that follow and complete the accompanying exercises.

Useful Definitions: Sentences, Clauses and Phrases

- A **sentence** is a group of words containing a subject and a verb and expressing a complete thought.

    EXAMPLE: My brother lives in New Zealand.

Sentences can be affirmative, negative, interrogative, imperative or exclamatory.

    EXAMPLE: The sky is blue. (affirmative)
    The sky is not blue. (negative)
    Is the sky blue? (interrogative)
    Look! The sky is so blue! (exclamatory)
    Look at the sky. (imperative)
    Come. Go. Look. Stop. (imperative) *

* These verbs in the imperative are actually sentences because the subject "you" is understood.

- A **clause** is also a group of words containing a subject and a verb. However, a clause does not necessarily express a complete thought. Clauses are **independent** (or **principal** or **main**) if they express a complete thought.

    EXAMPLE: My brother lives in New Zealand.

Clauses are **dependent** (or **subordinate**) if they do not express a complete thought.

    EXAMPLE: which I am wearing

Note: An independent clause is the same as a simple sentence. However, sentences often contain more than one clause. Sentences are classified according to how many independent and dependent clauses they contain (see "Types of Sentences," p. 335).

- A **phrase** is a group of words that fit together. Phrases often begin with a preposition or a participle.

    EXAMPLE: to the store
    sitting next to me

Phrases never have a fully conjugated verb.

EXERCISE 1 Identifying Different Types of Clauses

Underline the independent clauses and place parentheses around the dependent clauses in the following sentences.

EXAMPLE: <u>The man</u> (who is driving the red car) <u>is my father.</u>

1. Jennifer especially likes books that are written by feminist authors.

2. My friend Raffi, whom you met at the party last week, is leaving for Paris tomorrow.

3. In the early morning light, Hoa noticed flocks of parrots in the trees.

4. Nancy decided that she would apply for the job.

5. Do you know who is going to see the film with us?

6. Students who are in their final year are looking forward to graduating.

7. Do you think that the professor will ask us difficult questions?

8. We know that Avi has worked very hard.

9. The car, which is red, belongs to my brother.

10. Dorothy, who is standing by the door, is my teacher.

EXERCISE 2 Distinguishing between Clauses and Phrases

Write *P* or *C* in the blanks to indicate if the expressions are phrases or clauses.

1. _____ in the morning

2. _____ because he wanted to go

3. _____ if you were a millionaire

4. _____ to the library

5. _____ until yesterday

6. _____ a healthy appetite

7. _____ usually each week

8. _____ although she tried

9. _____ when it happened

10. _____ a great book

Types of Sentences

There are four types of sentences: simple, compound, complex and compound-complex.

---

1 A **simple** sentence contains one independent clause, with one principal verb and one complete idea.

>   **EXAMPLE:** The children play in the park every day.

2 A **compound** sentence contains two independent clauses joined together by a coordinate conjunction (*and, or, nor, but, yet, so*), with two principal verbs and two complete ideas.

>   **EXAMPLE:** The children play in the park every day, and their parents go to work.

3 A **complex** sentence contains one independent clause, with one principal verb and one complete idea, and at least one dependent clause, with one or more subordinate verbs and one or more incomplete ideas.

>   **EXAMPLE:** The children, who are eight years old, play in the park every day.
>   Independent clause: The children play in the park every day.
>   Dependent clause: who are eight years old

4 A **compound-complex** sentence contains at least two independent clauses, with two or more principal verbs and two or more complete ideas, and at least one dependent clause, with one or more subordinate verbs and one or more incomplete ideas.

>   **EXAMPLE:** The children, who are eight years old, play in the park every day, but their parents go to work.

EXERCISE 3 Identifying Different Types of Sentences

Indicate whether the following sentences are simple (*S*), compound (*C*), complex (*CX*) or compound-complex (*CC*).

1. _____ Paula, who has been studying medicine for the last three years, will work at the new clinic next summer.

2. _____ Next week, Ron is leaving for Haiti.

3. _____ Jonathan promised me that he would look after the hotel reservations, and I told him that I would buy the plane tickets.

4. _____ Do you know which novel we will read for the course?

5. _____ Marilyn wants to get married next year, but Fred would like to finish his studies first.

6. _____ Where are you going?

7. _____ Jung Chang wrote a memoir about China that has become very popular, and she has recently written a biography about Mao.

8. _____ Mandarin is the official language in China, but there are also many other Chinese languages.

9. _____ My friends recently travelled to Peru, where they saw many interesting places.

10. _____ Do you like to read?

Sentence Variety

If you write a paragraph or essay using sentences that are the same length and structure, it can become monotonous to read. A passage written using only simple sentences is not smooth. To improve your writing style, try using a variety of sentence types. Sentence variety means that your sentences use different lengths and stylistic patterns. You can vary your sentence style and structure in the following way.

■ COMBINING SENTENCES USING COORDINATION AND SUBORDINATION

When you write a text, you are arranging a series of ideas that you wish to communicate to a reader. Some, but not all, of these ideas are of equal importance, so that is where the concepts of coordination and subordination come into play.

Coordination and Subordination

- **Coordination** joins two or more ideas of equal importance.
  Look at the following pair of simple sentences.

     My sister lives in Toronto. My brother lives in New York.

  Both of the sentences contain ideas of equal importance. An appropriate coordinating conjunction (and, or, nor, but or so) can be used to transform these simple sentences into a compound sentence. A comma is placed before the conjunction when it joins two complete sentences.

     EXAMPLE: My sister lives in Toronto, and my brother lives in New York.

- **Subordination** shows a relation between ideas having different degrees of importance—that is, a less important idea is subordinated to a more important one.
  Read the following pair of simple sentences.

     My sister lives in Toronto. She found a job there.

  Since these sentences are too short, we will transform them into a complex sentence. First we must choose the idea that we want to emphasize. For the purpose of this example, we will say it is the fact that my sister lives in Toronto. This idea will be expressed in the independent (or main) clause.

  Now we must find a way to express the less important idea in a dependent or subordinate clause linked to the main clause by choosing an appropriate subordinating conjunction (see the table on page 337 for an extensive list). For the purpose of this example, we will use the subordinate clause to show the reason that my sister lives in Toronto.

     EXAMPLE: My sister lives in Toronto because she found a job there.

When using subordinate clauses, you must be careful that the underlying logic of your complex sentence is correct. Compare the following sentences.

John ate a pizza because he was hungry.
John was hungry because he ate a pizza.

You can see that the second sentence contains faulty logic.

Relative pronouns (who, whose, whom, which, that) are also used to show subordination.

> **EXAMPLE:** The book that I just read was fantastic.
> Independent/main clause: The book was fantastic.
> Dependent/subordinate clause: that I just read
> Relative pronoun: that

Subordinating Conjunctions		
Usage	**Subordinating Conjunctions**	**Example**
To indicate time (When?)	after, before, when, since, as soon as, while	As soon as he arrives, we'll leave.
To indicate place (Where?)	where, wherever	I don't know where he went.
To show cause/purpose (Why?)	because, so, in order that, as, since	He doesn't eat dessert because he doesn't want to gain weight.
To show manner (How?)	as, as if, as though	Tom walks as though he's wounded.
To show contrast/ comparison/concession	although, even if, though, than, whereas	Although she was late, she won the prize. Sarah ran faster than I thought.
To indicate condition	if, whether, unless	She doesn't know whether he will come.

EXERCISE 4 Building Better Sentences

Transform two simple sentences into an appropriate simple, compound or complex sentence. Sometimes you will use a coordinating or subordinating conjunction or a relative pronoun, but you may also simply delete and rearrange words.

> **EXAMPLE:** The girls rapidly ate the pizza. They were famished.
>
> → The girls rapidly ate the pizza because they were famished.
> or
> The famished girls rapidly ate the pizza.

1. Edgar Allan Poe was primarily an author of horror stories. There are also some poems that he wrote that are excellent, such as "The Raven."

_____

_____

_____

2. Edgar Allan Poe was very interested in the mysterious and strange aspects of life. He died in a mysterious and strange way.

_____

_____

_____

3. *The Silence of the Lambs* is one of the most macabre movies ever produced. I wonder if Poe would have written a story like it if he were alive today.

_____

_____

_____

4. Brently Mallard arrived home in a hurry. It was as if he were dying to see his wife.

_____

_____

_____

5. In "The Masque of the Red Death," Prince Prospero allows only his friends into his castle. This cruel prince abandons most of his subjects to an atrocious death outside the castle walls.

_____

_____

_____

EXERCISE 5 Combining Sentences

Combine the important elements of several short sentences into a longer, better one. Use subordinating conjunctions, delete repetitions, rearrange elements and shorten phrasing to accomplish the task.

> **EXAMPLE:** Menka plays the piano. She is my piano teacher, and she is excellent. Menka has been playing piano for many years. She is an outstanding performer. She practises a lot.
>
> ➤ Menka, my excellent piano teacher, has been playing piano for many years, and she is also an outstanding performer because she practises a lot.

1. Edgar Allan Poe wrote many short stories. They always deal with death and the supernatural. His short stories are very scary. This is the reason that I really appreciate them. Poe is one of my favourite authors.

_____

_____

_____

2. Kate Chopin lived at a time when women were oppressed. She didn't let this stop her. She expressed her views on sexuality and marriage anyway. She also demanded freedom for women.

_____

_____

_____

3. "The Tell-Tale Heart" is a short story written by Edgar Allan Poe. It is narrated by a man who seems crazy and obsessed. This man decides to kill an old man. It's because of the old man's blue eyes. The narrator thinks that the old man's eyes have the supernatural power of the evil eye.

_____

_____

_____

4. I read another short story by Kate Chopin. In this story, the main character is a woman who is oppressed by her marriage. This makes her seem only half alive. Then she discovers passionate love outside of marriage.

_____

_____

_____

5. Literature can be read on many different levels. Some people read it only for the storyline. Others appreciate the fictional world the author has created. Others like to delve deeper. They are searching for universal truths and meanings.

_____

_____

_____

■ OTHER TRANSITIONAL WORDS AND EXPRESSIONS

Transitional expressions make connections between sentences and paragraphs and serve to make your writing flow smoothly. In the following example, transitional expressions appear in italics.

> *Indeed*, Edgar Allan Poe was truly a master of suspense, as "The Tell-Tale Heart" most certainly shows. *In this story*, Poe exploits several literary devices to their fullest in order to create tension. *For instance*, he uses imagery appealing to both hearing and sight to great effect. The striking contrasts between dark and light, *such as* the black of night in contrast to the lantern's beam of light and the reassuring light of day, create tension in the reader's mind. *Similarly*, the interplay of quiet and loud sounds, *for example*, the sound of a mouse crossing the floor in contrast to the old man's shriek or the louder and louder beating of the dead heart, makes the suspense rise to an almost intolerable level. *In brief*, these few examples serve to confirm Poe's reputation as a master of suspense.

You can find a chart of common transitional expressions on pages 293-294 in Unit 3. You will notice that the coordinating and subordinating conjunctions are also included in this chart.

EXERCISE 6 Identifying Transitional Expressions

The following paragraph contains eleven transitional expressions. Identify them and indicate their usage. For example, "furthermore" is a transitional expression that indicates sequence. Be sure to check the chart on pages 293-294.

> Philosophy cafés are one of the most popular trends in France at the moment. Indeed, they are sprouting up all over Paris. Every Sunday, many people converge on one of several cafés to discuss burning philosophical questions. In one very popular café in particular, a different philosopher leads the discussion each week. However, in other cafés, the same philosopher leads the weekly discussions and proposes the following week's topic. Then people have a week to read up and gather their thoughts on it. Moreover, in all cases, would-be participants have to arrive early to get a seat inside the café. Not only are the budding philosophers excited and enthusiastic about their weekly philosophy sessions, but café owners are also delighted because of the crowds thronging to their establishments. In fact, they are amazed that philosophy has become such a popular Sunday pastime.

Transitional Expression	Usage	Transitional Expression	Usage
_____	_____	_____	_____
_____	_____	_____	_____
_____	_____	_____	_____
_____	_____	_____	_____
_____	_____	_____	_____
_____	_____		

■ VARYING THE BEGINNING OF SENTENCES

When you write, try varying the beginning of your sentences.

- Use an adverb.

    EXAMPLE: <u>Usually</u>, I drive my car to work.
    <u>Surprisingly</u>, the reporter didn't ask any personal questions.

- Use a prepositional phrase.

    EXAMPLE: <u>In the afternoon</u>, Ms. Strongly went to get the information that she had requested.
    <u>On weekends</u>, we often go hiking.

- Use an appositive.

    EXAMPLE: My professor, <u>Yasmine Kahn</u>, is an expert criminologist.
    Blue Rodeo, <u>a band from Toronto</u>, is going to play here next week.

- Use a present or past participle.

    EXAMPLE: <u>Writing</u> a novel critical of the Russian government, Alexander Solzhenitsyn knew he was engaging in dangerous activities.
    <u>Deported</u> by the Russian government, Alexander Solzhenitsyn eventually immigrated to the United States.

EXERCISE 7 Writing Varied Sentences

1. Write a sentence that begins with an adverb.

    _____

2. Write a sentence that begins with a prepositional phrase.

    _____

3. Write a sentence that begins with an appositive.

    _____

4. Write a sentence that begins with a present or past participle.

    _____

▨ USING INTERESTING SENTENCES

Try the following strategies to add variety to your written work.

- Add a quotation.

    EXAMPLE: Mark Twain declared, "I have never let my schooling interfere with my education."

- Ask and answer a question.

    EXAMPLE: Is Margaret Atwood really a great writer?

- Use an exclamation.

    EXAMPLE: Canadian writers always write about dreary winters!

EXERCISE 8 Editing for Sentence Variety

The next paragraph lacks sentence variety. Using the methods that you have learned, edit the paragraph and write at least five varied sentences.

> George Orwell is a very famous writer. His real name is Arthur Eric Blair. He was born in 1903. He was born in India. He died in 1950. He did his studies in Britain. He worked for the Indian Imperial Police in Burma. He developed a hatred for imperialism. He returned to Britain. He started his writing career. He wrote many novels. He also wrote many essays. His works were usually based on his personal experience. His best-known works are *Animal Farm* and *Nineteen Eighty-Four*. Both of these books are allegories. Orwell was critical of Stalin. He was also critical of totalitarianism. Orwell died in London of tuberculosis.

Capitalization and Punctuation

Capitalization

Here are the basic rules for capitalization in English.

You should capitalize:

- the first word in a sentence or direct quotation, the pronoun *I* and proper nouns;

    EXAMPLES: the Milky Way
    What is the most interesting book that you have ever read?

- certain abbreviations, especially titles;

    EXAMPLES: Dr. Guy Charpentier, Line Pritchard, M.D., Sen. Pat McDonald, 2:00 A.M., 2005 AD

- full and official names of organizations, official titles of people, titles of documents and literary works;

    EXAMPLE: Prime Minister Jean Chrétien

- the words in the title of a book, film, song or work of art, excluding articles and conjunctions;

    EXAMPLE: *One Flew Over the Cuckoo's Nest*

- the names of days, months and holidays;

    EXAMPLES: Monday, March, Easter

- the names of historical periods, events and eras;

    EXAMPLES: the Renaissance, the Second World War

- the name of religions, nationalities, tribes, races, and specific languages;

    EXAMPLES: Hinduism, Spanish, Mohawk, a Chinese restaurant
    God, Buddha, Islam, Jesus and His apostles

- the points of the compass when they refer to a specific region;

    EXAMPLE: in the West

- the names of specific school courses;

    EXAMPLE: English 101

- personifications.

    EXAMPLE: "O Love, thou wond'rous thing!"

EXERCISE 1 Practising Capitalization

Place capital letters where they are needed in the following sentences.

1. former british prime minister margaret thatcher and former american president ronald reagan seemed to agree on many important points.

2. he and i will arrive on easter monday, which occurs on april 26th this year.

3. did you enjoy reading *the catcher in the rye* by j. d. salinger, or did you prefer *for whom the bell tolls* by ernest hemingway?

4. during the renaissance, many christians believed that god ruled his people with an iron hand.

5. i'm going to the west next summer to work at chateau lake louise near banff, alberta.

6. my parents have a cottage near lake simcoe.

7. next saturday, i'm going to eat at a japanese restaurant.

8. my history class went to the museum of natural history last february.

Punctuation

The role of punctuation is to make writing easier to understand. Most punctuation marks serve to separate or end the writer's thoughts.

Punctuation for Separating Parts of Sentences

Comma (,)

Use a comma

- to separate items in a series;
- after introductory words, phrases or clauses;
- to set off words of direct address or mild interjections;
- to mark off additional but not essential information (for example, non-restrictive clauses, phrases and appositives) about a noun preceding the information;
- before and after parenthetical expressions (words that interrupt the thought);
- before a conjunction (*and, or, nor, but, so*) linking two independent clauses in a compound sentence;
- between the day (or month) and the year and after the year if the sentence continues;
- before examples introduced by *such as* or *especially*;
- after identification of the speaker and before a direct quotation.

    EXAMPLE: I bought a carton of milk, a dozen eggs and a box of cookies at the store.
    My brother's son was born April 1, 1975.

Semi-colon (;)

Use a semi-colon

- to connect two closely related thoughts (independent clauses) in place of a conjunction;

- before conjunctive adverbs, such as *nevertheless*, *however*, *consequently*, etc., in two closely related clauses;
- between items in a series that already contain commas.

> **EXAMPLE:** There was a tornado warning yesterday; the meteorologists had seen funnel clouds forming.
> There was a tornado warning yesterday; however, many people were unaware of anything unusual.

Colon (:)
Use a colon

- to introduce a list (but not after a verb or preposition);
- to introduce a formal quotation that is a complete sentence or a block quotation;
- after the salutation in a business letter;
- when using numerals to express time;
- in plays and biblical and volume references;
- to introduce a subtitle.

> **EXAMPLE:** Please bring the following for the camping trip: a tent, a sleeping bag, warm clothes and food for three days.

Parentheses ( )
Use parentheses around supplementary explanations and comments.

> **EXAMPLE:** The supermodel (a snob) looked down her nose at the reporter during the interview.

Dash (—)
Use a dash

- before and/or after parenthetical elements;
- before a sudden break in thought;
- to show unfinished or interrupted dialogue;
- after a statement in order to explain or expand upon it.

> **EXAMPLE:** Shakespeare's Hamlet—the protagonist—is a very indecisive character.

Brackets ([ ])
Use brackets to insert an editorial comment in quoted material.

> **EXAMPLE:** The film critic gave the movie a great review, saying that "it [*Casablanca*] was the best romance" he had seen in a long time.

Terminal Punctuation

Period (.)
Use a period

- at the end of statements or commands;
- after certain abbreviations (but not metric symbols);
- to indicate decimals.

Question Mark (?)
Use a question mark after a direct question or to indicate uncertainty.

Exclamation Point (!)
Use an exclamation point after a sentence that expresses strong emotion or surprise.

Other Punctuation

Hyphen (-)
Use a hyphen

- in certain compound words (nouns, adjectives, phrases);
- to divide a word at the end of a line if you are not using a word processor;
- in compound numbers (twenty-one to ninety-nine);
- with fractions used as adjectives.

> **EXAMPLE:** The recipe said to include a half-kilo of sugar.

Apostrophe (')
Use an apostrophe

- to show possession (mainly for animate objects);
  - before the *s* for singular and plural nouns not ending in *-s*;
  - after the *s* for plural nouns ending in *-s*;
- in contractions.

> **EXAMPLES:** My younger brother's girlfriend is an opera singer.
> My two older brothers' girlfriends are scientists. I'm a poor singer.

Quotation Marks (" ")
Use quotation marks

- around direct speech and dialogue;
- around short quotations in essays, term papers and so on;
- around titles of short stories, poems, songs, essays, book chapters and magazine and newspaper articles;
- to emphasize specific words in a text;
- around definitions.

> **EXAMPLE:** The prisoner said, "I'm not guilty of murdering that man."

Ellipsis Marks (...)
Use ellipsis marks

- to indicate that something has been omitted in quoted material;
- to show a pause in dialogue.

> **EXAMPLE:** Juliet declares her love for Romeo on the balcony when she says, "O Romeo, O Romeo ..."

Italics (*italics*)
Although italics are not a form of punctuation, they are important in texts written on a computer.

Use italics

- for titles of books, magazines, newspapers, plays and movies;
- for names of ships or planes;
- to emphasize specific words in a text;
- for foreign words or phrases in a text.

> **EXAMPLE:** My favourite play is *Man and Superman* by George Bernard Shaw.

Note: If you are writing a text by hand, use underlining instead of italics.

EXERCISE 2 Punctuating Correctly

Add the necessary punctuation to the following sentences.

1. The flight attendant said Dont let the dog out of its cage until an American inspector has approved it

2. Twenty five students were present for the final review nevertheless six of them failed the exam

3. Do you know if each of us must bring the following two pens three pencils a notebook and a dictionary

4. The singular is child the plural is children

5. Wow You won the Chalmers Award I exclaimed Which prize did he win

6. She got up at 6 am and started studying the period from the second century BC to the end of first century AD

7. Mark Twain he was an amazing speaker gave lectures as far away as Australia

8. The boys jackets were left on the chairs and someone stole six of them

9. Who was it who said One small step for mankind when he first stepped onto the moons surface

10. One of my favourite short stories is The Lottery by Shirley Jackson and one of my favourite books is The Bone People by Keri Hulme

EXERCISE 3 Practising Capitalization and Punctuation

Correct eleven capitalization errors and add four commas in the following text.

Mark twain born samuel l. clemens led a very interesting and varied life. His father, judge John Clemens, brought the family up first in a town called florida. In 1839, the family moved to Hannibal near the Mississippi river, a river that was to influence Clemens enormously. Young Samuel left school at the age of twelve after learning the printers trade and, a few years later, worked at setting the type at a newspaper called the hannibal journal. It was here that samuel first began to write. At the age of eighteen, he left for New york, Philadelphia, Cincinnati and New orleans, where he completed his studies to become a river pilot. The american civil war broke out and clemens served for two weeks as a confederate soldier before leaving for California where he tried his hand at gold mining. Soon he started to write articles and stories which he now signed Mark Twain, a term used by river pilots to indicate the depth of the water. He became increasingly popular, but he still did not make much money. Since he was an excellent storyteller a friend suggested that he become a lecturer. His lectures were so popular that he ended up touring the world.

Common Sentence Errors

There are two types of common sentence errors: sentence fragments and run-ons. When you edit your work, make sure that your writing is free of these errors.

Sentence Fragments

A **sentence fragment** is part of a sentence (an incomplete idea) that has been given the punctuation marks of a full sentence. A sentence fragment lacks either a main subject or a main verb. Most sentence fragments are unacceptable in formal writing—unless they are part of dialogue.

> **EXAMPLE:** Because the car was fast. (This is an incomplete idea because we do not know the result of the car's being fast.)

> **COMPLETE SENTENCE:** The teenager wanted to buy the car because it was fast.

■ TYPES OF FRAGMENTS

There are four types of sentence fragments. Read the following explanations to be able to identify and correct fragments.

Phrase Fragments

Phrase fragments are groups of words that are missing a subject or a verb. The fragments are italicized.

No verb: . *The hot weather.* It is causing a drought.
No subject: *In the afternoon.* The children were having fun.

To correct phrase fragments, add the missing subject or verb by joining the fragment to the sentence.

One sentence: The hot weather. ~~It~~ is causing a drought.

                            it was hot.
Two sentences: In the afternoon. The children were having fun.

Fragments with *-ing* and *to*

Some fragments begin with a present participle (*-ing*) or an infinite (*to*). Usually these types of fragments appear beside a sentence that contains the subject. Avoid making this error.

-ing fragment: Planning my weekend. I realized that I wanted to do nothing at all.
to fragment: Carlyle Jones wanted to go home. To see his mother.

To correct an *-ing* or *to* fragment, join the fragment to the sentence to form one sentence, or add words to form two separate sentences.

One sentence: Planning my weekend. I realized that I wanted to do nothing at all.

                            to
Carlyle Jones wanted to go home. ~~To~~ see his mother.

was easy.

Two sentences: Planning my weekend. I realized that I wanted to do nothing at all.

He wanted to

Carlyle Jones wanted to go home. ~~To~~ see his mother.

Explanatory Fragments

Some fragments are **explanatory fragments.** Such fragments usually provide an explanation but lack a subject or verb. These types of fragments begin with one of the following words.

as well as	except	for instance
like	such as	also
for example	including	particularly

EXAMPLES: There are many different prizes for literature. Such as the Man Booker Prize.

My sister is a writer. Also a poet.

To correct explanatory fragments, join the fragment to the sentence to form one sentence, or add words to form two separate sentences.

, such

One sentence: There are many different prizes for literature. ~~Such~~ as the Man Booker Prize.

She is also

Two sentences: My sister is a writer. ~~Also~~ a poet.

Dependent-Clause Fragments

Dependent-clause fragments begin with subordinating conjunctions or relative pronouns. The following table contains some of the most common words that begin dependent clauses.

Common Subordinating Conjunctions				Relative Pronouns
after	before	though	whenever	that
although	even though	unless	where	which
as	if	until	whereas	who(m)
because	since	what	whether	whose

EXAMPLES: Unless you finish reading *War and Peace.* You won't be able to pass the exam.

My favourite author is Leo Tolstoy. Whose books are very long but very interesting.

To correct dependent-clause fragments, join the fragment to a complete sentence, or add the necessary words to make it a complete idea. You can also delete the subordinating conjunction.

, you

One sentence: Unless you finish reading *War and Peace.* ~~You~~ won't be able to pass the exam.

His

Two sentences: My favourite author is Leo Tolstoy. ~~Whose~~ books are very long but very interesting.

EXERCISE **1** Correcting Sentence Fragments

Correct the fragments in the following sentences.

1. People eating popcorn and slurping soft drinks. These sounds disturb my enjoyment of a movie.

2. I love to dive into the refreshing water of a lake after a strenuous bike ride. Whenever it is possible.

3. To be able to earn my living as an artist. It is a major goal in my life.

4. Some driving laws are too permissive. For example, letting people talk on cellphones and drive at the same time.

5. Since her parents died within months of each other. I've noticed that Joan hasn't been the same.

6. Jack did well on all his final exams. Except for mathematics which he has decided to take over during the summer term.

7. Global warming. This is the greatest problem facing our planet.

Run-Ons and Comma Splices

A **run-on sentence** consists of two or more independent clauses that have no punctuation or too many conjunctions between them.

> **EXAMPLES:** The students studied for the exam they all passed.

A **comma splice** results from punctuating two independent clauses incorrectly with a comma.

> **EXAMPLE:**   The students studied for the exam, they all passed.

There are four ways to correct run-on sentences and comma splices.

---

1 Write separate sentences using proper terminal punctuation.

> **EXAMPLE:** The students studied for the exam. They all passed.

2 Place a semicolon between two independent clauses that are closely connected.

> **EXAMPLE:** The students studied for the exam; they all passed.

3 Place a comma and a coordinating conjunction between the two independent clauses.

> **EXAMPLE:** The students studied for the exam, and they all passed.

4 Make one of the independent clauses dependent.

> **EXAMPLE:** The students who studied for the exam all passed.

EXERCISE 2 Correcting Sentence Errors

Read the following sentences. In the blanks provided, write *C* for sentences that are correct, *R* for run-on sentences or comma splices and *F* for sentence fragments. Then correct the errors. Note that you may have to change the word order or add some words.

1. _____ Because I had to give my little brother a present.

2. _____ He travelled to Tibet, he had to get special permission from the government for his trip.

3. _____ My mother comes from Brazil, people speak Portuguese there.

4. _____ Can you tell me what homework we have to do for tonight I didn't hear what the teacher said.

5. _____ That she was praised for her singing abilities, which made her very happy.

6. _____ Can you tell me where Lacombe Street is?

7. _____ Who knows?

8. _____ The restaurant which is Italian and very expensive and the service is slow and inefficient.

9. _____ Eat!

10. _____ I am very happy that we could get the hockey tickets for tonight.

EXERCISE 3 Correcting Common Sentence Errors

Make any necessary corrections in the following paragraph in order to form complete sentences. There are six sentence errors.

Scary and spooky novels have become very popular in the current culture, we know this by the sales and box office successes of books and films whose subject matter is meant to scare the reader or audience. Why are we so fascinated by scary subject matter? It is meant to create terror in our minds. Horror and terror raise our emotions to an adrenaline-driven peak. Which we find both stimulating and necessary. We like the rush that we get after a good scare it is for this reason that we try extreme sports or choose risky professions; we like the effects of the adrenaline on our bodies.

I avoid watching anything that scares me. Or causes me highly anxious moments. Scary movies keep me awake all night I am very tired the next day. I guess I will never dive with sharks, go piranha watching in the Amazon or climb Mount Everest. I just like to sit in my armchair. And read about such adventures.

Parallel Construction

Parallel construction refers to the use of similar grammatical structures. For example, words, clauses and phrases should be in similar grammatical form in sentences.

EXAMPLES: **Incorrect:** The speaker at the protest rally *waved* his arms, *jumped* up and down and *was screaming* in a very loud voice to the audience.
Correct: The speaker at the protest rally *waved* his arms, *jumped* up and down and *screamed* in a very loud voice to the audience.

Incorrect: *Actors*, *singers* and *people who write* attended the ceremonies.
Correct: *Actors*, *singers* and *writers* attended the ceremonies.

Incorrect: You will find the library *on* the second floor, *in* the new section and *go near* the elevators.
Correct: You will find the library *on* the second floor, *in* the new section, *near* the elevators.

EXERCISE 4 Correcting Errors in Parallel Construction

Correct the parallel construction errors in the following sentences.

1. Successful students organize their time, study on a regular basis and are trying to get enough sleep.

   _____

2. The astronauts were conducting experiments, such as studying the effects of gravity, looked at behavioural problems and examining the effects of isolation on people.

   _____

3. The people in the next apartment sang loudly, melodically and with enthusiasm.

   _____

4. The science professor gave a lecture about volcanoes, earthquakes and waves that are huge.

   _____

5. We ran across the street, through the neighbourhood and went up the hill for the marathon.

   _____

Redundancy and Restatements

Restating words and ideas in a paragraph can have either a **positive** or **negative** effect.

Sometimes stating the same word or idea can be redundant and ineffective. In English, subjects and objects in clauses are not repeated.

> **EXAMPLES:** **Incorrect:** The *girl, she* is my sister. In this example, the subject *girl* is repeated with the pronoun *she*.
> **Correct:** The *girl* is my sister.
>
> **Incorrect:** The *basketball player, whom* the movie was based on *him,* played in the NBA.
> **Correct:** The *basketball player, whom* the movie was based on, played in the NBA.

However, sometimes repetition can be used very effectively. Restating key ideas in a paragraph can emphasize their importance. The following paragraph contains repetitions of key words.

> **EXAMPLE:** The interest generated by the events in **World War II** has been re-exploited recently by the film industry. This **war** was one of the bloodiest, long-term **battles** of the twentieth century. During the **war**, many people died as a result of political decisions. Such events make great subjects for films, such as **futility of war**, the exploitation of the weak by the strong and the heroic efforts of many individuals. Films such as *Saving Private Ryan, Schindler's List* and *The Thin Red Line* are the results of recent experimentation on an old subject, **World War II**.

EXERCISE 5 Correcting Errors in Parallel Construction and Redundancy

The following paragraph contains five errors in parallel construction. It also has two examples of redundancy. Find and correct the errors.

> **EXAMPLE:** The cat, she was beautiful.

The cat was lying in the sun, and she licked her paws. She was a white Persian cat with silky soft fur that shone. Her eyes, they were blue-grey. Those eyes were shiny, bright and with sparkles. She watched the birds in the trees, quietly, pensively and with attention. The birds suddenly became nervous. They were becoming aware of the cat, which they had been watching it. The cat slowly twitched her tail, watched the birds and was purring quietly. The birds shrieked and flew away. The cat continued washing herself.

Agreement

■ SUBJECTS AND VERBS

A verb must agree with its subject. A singular subject must take a singular verb and a plural subject must take a plural verb.

- Nouns and verbs must agree in person and number.
- Third-person singular verbs end in -s or -es.

  EXAMPLE: The boy *sings* in the choir.

  Verbs with plural subjects do not take -s or -es.

  EXAMPLE: The boys *sing* in the choir.

- Compound subjects joined by *and* take plural verbs.

  EXAMPLE: My sister and her friend *play* on a soccer team.

  However, if compound subjects joined with *and* express a singular idea, they take a singular verb.

  EXAMPLE: The stormy sea's rocking and moving *becomes* calm in fair weather.

- Verbs that follow compound subjects joined with *or, nor, either ... or, neither ... nor* and *not ... but* agree with the closest subject.

  EXAMPLE: Neither your sister nor your *brothers play* hockey.
  Neither your brothers nor *your sister plays* hockey.

- Collective nouns, such as *family, jury, salt, furniture*, etc., take singular verbs.

  EXAMPLE: The furniture in the house *is* new.

- Nouns that are singular in meaning but plural in form, such as *politics, economics, news*, etc., take a singular verb.

  EXAMPLE: Mathematics *is* a difficult subject.

- Words that refer to portions, such as *some, most, all, part, half* and fractions, take either a singular or plural verb, depending on the noun that follows *of the*.

  EXAMPLE: All of the *banana is* ripe.
  All of the *bananas are* ripe.

- Indefinite determiners, such as *each* and *every*, and indefinite pronouns, such as *anybody, somebody, anyone, everyone* and *something*, take singular verbs.

  EXAMPLES: Each student *speaks* for five minutes.
  Everyone *is* responsible for bringing his or her own lunch.
  Somebody always *forgets* the key to the office.

- In sentences containing *there* or *here*, the verb *to be* is either singular or plural depending on the noun that follows.

  EXAMPLE: There *is* one *window* in the kitchen.
  There *are* two *windows* in the kitchen.

- A title takes a singular verb.

  EXAMPLE: *The Boats of the Oceans* is a funny book.

EXERCISE 6 Making Subjects and Verbs Agree

Underline the correct answer.

1. There (was/were) lots of people at the concert last night.

2. Each of the girls (want/wants) to buy a new car.

3. News of the accident (was/were) broadcast on all the channels.

4. My homework (is/are) not complicated this week.

5. Mathematics (is/are) my favourite subject.

6. Everyone (is/are) happy that he got the job.

7. There (is/are) cake, pie or ice cream for dessert.

8. Neither of them (want/wants) to go to court next week.

9. One of the girls (has/have) forgotten her books.

10. Both the cat and the dog (has/have) a special place to sleep.

11. If you want to make lunch, there (is/are) cheese, ham, lettuce, mustard and fresh bread.

12. (Has/Have) anybody remembered to bring the tablecloth?

13. The information (was/were) very important.

14. (Don't/Doesn't) either of you remember his phone number?

15. Some members of the baseball team (was/were) late for the game.

16. Five dollars (is/are) the price of the ticket to the game.

17. On the desk (is/are) several letters from his friend.

18. Either he or his brother (is/are) going on the trip.

19. A period of two weeks (is/are) not long enough to appreciate that country.

20. The winner of the race, surrounded by his fans, (was/were) trying to catch the attention of his family.

■ PRONOUN/ANTECEDENT AND PRONOUN/CASE AGREEMENT

Pronouns are words that replace nouns. Used to avoid repetition, they should agree in person, number and gender with their antecedents.

> **EXAMPLES:** At that college, *every student* must bring *his or her* own portable computer to class.
> At that college, *students* must bring *their* own portable computers to class.

> **INCORRECT USAGE:** Every student must bring their own computer to class.

Pronouns must also agree in case. This refers to whether they function as a subject or an object.

> **EXAMPLES:** Both you and I are excited about the award.
> He gave a copy of the document to both you and me.

The next chart illustrates the pronoun cases.

			Possessive	
Pronouns				
	Subjective	Objective	Possessive Adjective	Possessive Pronoun
Singular 1st person 2nd person 3rd person	I you he, she, it, who, whoever	me you him, her, it whom, whomever	my your his, her, its, whose	mine yours his, hers
Plural 1st person 2nd person 3rd person	we you they	us you them	our your their	ours yours theirs

Problems with Pronoun Case

Making Comparisons with *Than* or *As*
Use the correct pronoun case in comparisons with *than* or *as*. If you use the incorrect pronoun case, the meaning of the sentence will change. Compare the next sentences.

> I like to read more than (he/him).

If you use the subjective case (he), your sentence will have a different meaning than if you use the objective case (him).

> I like to read more than he (likes to read).
> I like to read more than (I like) him.

Using Prepositional Phrases
When you use a prepositional phrase, the words after the preposition are always the objects of the preposition. Thus, you must always use the object pronoun after the preposition.

> **EXAMPLES:** *Between you and me*, the students don't participate enough.
> They gave their congratulations *to us*.

Selecting *His* or *Her*

In English, the possessive pronoun reflects the gender of the possessor, not that of the object. If something belongs to a female, use *her*. If something belongs to a male, use *his*. If something belongs to an inanimate object, use *its*.

EXAMPLES: My <u>mother</u> took *her* umbrella.
My <u>brother</u> told *his* friend about the film.
The <u>book</u> had *its* cover ripped.

EXERCISE 7 Making Pronouns Agree with Antecedents or Case

Underline the correct pronouns in the sentences below.

1. Between you and (me/I), I think that he should have won the debate.

2. (We/Us) people are all in agreement on that question.

3. John and (he/him) make a great duo.

4. I think it was (she/her) who started the fight.

5. May Jim and (I/me) borrow your typewriter tomorrow?

6. Janice is a better writer than (she/her).

7. I do not approve of (you/your) smoking in my office.

8. Did you ask (he/him) to arrive early?

9. To (we/us) teachers, the remarks were very interesting.

10. It must have been (they/them) who sent the message.

EXERCISE 8 Correcting Errors in Agreement

Correct the errors in the following sentences by crossing out the incorrect words and writing the correct replacements above them.

1. My father and mother is going to the cinema tonight.

2. The boy told her own mother that he preferred the pies she baked for him to the ones at school.

3. Each one will bring their own car.

4. Economics are my favourite course, but I like math, too.

5. Neither Jane nor Jim want to go to Janet's party.

6. All of the pie were eaten.

7. Many people prefer rock to classical music, but him and me prefer classical.

8. None of them is going to the dance.

9. Five dollars are a lot of money for a nine-year-old.

10. Anyone would know their own address.

11. The news were terrible last night.

12. My dad loves his new car, but he lent her to my mum for the evening.

13. She likes sugar in his tea but not in his coffee.

14. One of them are lying.

15. Jack and John are lending his books to Rhonda before the exam.

16. Everyone are shouting at me that mathematics are the best subject.

17. Have anyone called?

18. Neither Bob nor John have completed the homework.

19. Do he has any information about the trip to Europe next summer?

20. The women was walking while her children was running.

Shifts in Verbs and Pronouns

■ SHIFT IN VERBS

If there is an incorrect shift in tense or voice, the point of view of the text becomes illogical. There are two main problem areas related to shifting points of view.

---

1 Past to Present Tense

Avoid shifting verb tenses. It is important to be consistent in your use of verb tenses. A shift from present to past or from past to present should be avoided unless there is a concrete change in time in the narrative or setting of the text.

> EXAMPLE: **Incorrect:** The night **was** dark and there **is** no moon.
> **Correct:** The night **was** dark and there **was** no moon.

2 Passive to Active Voice

Avoid shifting from passive voice to active voice and vice versa.

> EXAMPLE: **Incorrect:** The tree was cut by the man and he chopped the branches.
> **Correct:** The man cut the tree and chopped the branches.

EXERCISE 9 Correcting Shifts in Verb Tenses

Correct the errors in the sentences below.

1. Yesterday I went to an outdoor concert, but it was so crowded that it is impossible to hear the music.

    _____

2. The car was being inspected by the mechanic because somebody bought it.

    _____

3. I swim every morning. However, this morning, I don't go swimming because I don't feel well.

    _____

4. Do you see the accident when you went to work yesterday?

    _____

5. The boy was sick and is being examined by the nurse on duty.

    _____

■ SHIFT IN PRONOUNS

Avoid shifting pronouns in a text. Keep a consistent point of view. In other words, do not start with *we* and shift to *you* or *he* or *she*.

> **EXAMPLE:** **Incorrect:** When we reread this story, you realize that the author included a lot of foreshadowing.
> **Correct:** When we reread this story, we realize that the author included a lot of foreshadowing.

EXERCISE 10 Correcting Errors in Pronoun Shift

Correct the errors in pronoun shift in each of the following sentences.

> **EXAMPLE:** We enjoy going to see films, but ~~they~~ we are always disappointed that the film isn't more like the book.

1. Our literature professor, Dr. Komo, gave us a surprise test. We were all worried because you knew that the test was going to be difficult.

2. I was taking a course in Mandarin this semester. I was really surprised when I realized you could actually have a five-minute discussion on the weather.

3. My classmate, Ahmed, did a presentation on Hemingway's style. He was very nervous during the presentation, but then, they always get very nervous.

4. You will be a successful student if one has good study habits.

5. My favourite book is *The Lord of the Rings* by Tolkien. Since it is so long, I read it last summer, but it still took you two weeks to complete.

EXERCISE 11 Correcting Errors in Verb and Pronoun Shift

Underline and correct nine verb tense and pronoun shift errors.

The girl is about ten years old. Presently, she and her family lived in a cottage by the lake. Everyday, the girl and her younger brother, a boy of eight or so, were taught to fish by the neighbour, a teenager who was around sixteen years old. The teenager's name was Lisa, and she seemed to be a proficient fisher.

Yesterday morning, the three fishers go out on the lake in a canoe. They knew that one had to wear warm clothes because it was windy.

It was very early in the morning. They stopped the canoe in the middle of the lake, and you could see the fish circling beneath them in the clear water. They fished all morning, but you didn't think they were successful because I didn't see any fish when they came back to shore.

Misplaced and Dangling Modifiers

Many people do not use modifiers correctly when they write an essay. **Modifiers** are words, phrases or clauses that describe or qualify another element in the sentence, usually a noun, pronoun or verb. They are often left dangling (with no word to modify), or they are placed too far from the word they describe.

EXAMPLES: The young boy lives next door. (The adjective *young* modifies *boy* and *next* modifies *door*.)

The girl wearing the red dress is my niece. (The phrase *wearing the red dress* modifies *girl* and *red* modifies *dress*.)

She is walking quickly. (The adverb *quickly* modifies *walking*.)

Here are two important rules to follow.

Rule 1: A modifier must always have a word to modify.

Incorrect: Coming home late, dinner had already been served in the dining room.

Coming home late does not modify any other word or phrase in the sentence. It is not *dinner* that has come home late but some unmentioned person. *Coming home late* has been left dangling with nothing to modify and is therefore termed a "dangling modifier."

Correct: Coming home late, he rushed into the dining room, where dinner had already been served.

Coming home late modifies *he* not *dinner*. It is he who has come home late.

Rule 2: It is usually very important to place a modifier close to the word or phrase it is modifying.

In the next example, we want to indicate that a plane that is on fire is landing on a tarmac.

Incorrect: The plane finally landed on the tarmac, now on fire.

This sentence is incorrect because *now on fire* modifies *tarmac* in this particular context. However, we want to say that the plane is on fire.

Correct: The plane, now on fire, finally landed on the tarmac.
Now on fire, the plane finally landed on the tarmac.

EXERCISE 12 Identifying Dangling and Misused Modifiers

Choose the correct sentence in the pairs below.

1. _____ **a)** Inventing a new recipe, the meal was truly a success.
b) Inventing a new recipe, the chef made a truly successful meal.

2. _____ **a)** After failing to brake, John had an accident when his car swerved out of control.
b) After failing to brake, the car swerved out of control and John had an accident.

3. _____ **a)** Stuck outside the house, the phone rang and rang, and I couldn't answer it.
b) Stuck outside the house, I couldn't answer the phone, and it rang and rang.

4. _____ **a)** Mother talked on the phone while I was trying to study in a very animated way.
b) Mother talked on the phone in a very animated way while I was trying to study.

5. _____ **a)** When writing the book, Frederick often used to gaze at the river for inspiration.
b) When writing the book, the river provided plenty of inspiration for Frederick.

EXERCISE 13 Correcting Dangling and Misplaced Modifiers

Rewrite the following sentences in order to correct the dangling and misplaced modifiers.

> **EXAMPLE:** Going to the gym regularly, John's weight loss was substantial. (incorrect)
> Going to the gym regularly, John lost a substantial amount of weight. (correct)

1. After turning out the lights, the book was very hard to read in the dark room.

   _____

2. When on a hunting safari, the animals were lucky to escape Hemingway's sharp aim.

   _____

3. An "honest charlatan," I have a very high opinion of the Amazing Randi.

   _____

4. When finished college, my parents took me out for a huge celebration.

   _____

5. Screaming loudly and recognizing their own screams, the hungry lions savagely ate the children's parents.

   _____

6. The ending of "The Veldt" was shocking for me, while reading this popular short story by Ray Bradbury.

   _____

7. Thinking himself very different from the others, the banquet seemed very boring to Tom.

   _____

Quoting, Paraphrasing, Summarizing

Paragraphs and essays require supporting details. These details can take the form of statistics, data, examples or information from secondary sources. Whichever type of detail you use, it is important to know how to use ideas and information from secondary sources correctly. Incorrect usage of secondary source material can result in plagiarism.

Quoting

■ DIRECT QUOTATIONS

Direct quotations are repetitions of the exact words a person has said or written. They must be placed in double quotation marks.

Punctuation in Direct Quotations

- Put a colon after a complete sentence that introduces a direct quotation.

    EXAMPLE: Hemingway often uses action to emphasize a character trait: "The girl looked at the bead curtain, put her hand out and took hold of two of the strings of beads."

- Put a comma after a phrase that introduces a quotation. The first word of the quotation must be capitalized.

    EXAMPLE: In Hemingway's story, the girl said, "They look like white elephants."

- End the quotation with a comma if your sentence continues after the end of the quotation.

    EXAMPLE: The female character in Hemingway's short story remarked, "Everything tastes of licorice," revealing her attitude toward life.

- Punctuate an incomplete quotation within a sentence only with quotation marks.

    EXAMPLE: Hemingway's short story takes place at a railway station "between two lines of rails in the sun."

- Use ellipsis marks to indicate that words are missing from a quotation.

    EXAMPLE: The story starts in the following way: "The hills across the valley of the Ebro ..."

- Use single quotation marks to indicate a quotation within a quotation.

    EXAMPLE: Hemingway wrote, "'They're lovely hills,' she said. 'They don't really look like white elephants.'"

Note: The use of quotations differs in English-speaking countries. The style used above is American and this style also predominates in Canada. However, the British tend to use single quotation marks around direct speech and double quotation marks for quotations within

---

quotations (in other words, the opposite of American usage). American usage requires most end punctuation (commas, periods, etc.) to be placed inside the final quotation marks whereas British usage requires it to be placed outside the final quotation marks.

> **EXAMPLE:** Here is how the above example would appear according to British usage.
> Hemingway wrote, '"They're lovely hills," she said. "They don't really look like white elephants"'.

■ INDIRECT QUOTATIONS

When the exact words of a speaker or writer are not used, quotation marks are unnecessary. Compare the first two examples (a statement and indirect speech) with the third (direct speech).

> **EXAMPLES:** In Hemingway's work, the girl compares the hills to white elephants.
> I told him that in Hemingway's work, the girl compares the hills to white elephants.
> John answered, "In Hemingway's work, the girl compares the hills to white elephants."

■ LONGER QUOTATIONS

For quotations exceeding three lines of a text, do not use quotation marks. Indent the entire length of the quotation and use double-space.

> **EXAMPLE:** The Hills across the valley of the Ebro were long and white. On this side there was no shade and no trees and the station was between two lines of rails in the sun. Close against the side of the station there was the warm shadow of the building and a curtain, made of strings of bamboo beads, hung across the open door into the bar, to keep out flies ...

EXERCISE 1 Punctuating Direct Quotations

Add the correct punctuation to the following quotations (the quotations are in italics).

> **EXAMPLE:** Playwright George Bernard Shaw said, *"Youth is wasted upon the young."*

1. The German writer, Goethe, declared *when ideas fail, words come in handy*.

2. *A poem begins in delight and ends in wisdom* stated Robert Frost.

3. Oscar Wilde proclaimed *I can resist anything but temptation*.

4. The French author Voltaire made a profound observation *those who make you believe absurdities can make you commit atrocities*.

5. *Action is eloquence* said William Shakespeare.

EXERCISE 2 Transforming Direct Quotations into Indirect Quotations

Change each of the direct quotations below into indirect quotations.

> **EXAMPLE:** Alfred, Lord Tennyson wrote, "Knowledge comes, but wisdom lingers."
> Alfred, Lord Tennyson wrote that knowledge comes, but wisdom lingers.

1. Kingsley Amis once wrote, "If you can't annoy somebody, there's little point in writing."

   _____

2. G. K. Chesterton stated, "A good novel tells us the truth about its hero; but a bad novel tells us the truth about its author."

   _____

3. Winston Churchill declared, "History will be kind to me because I intend to write it."

   _____

4. The poet e.e. cummings observed, "The most wasted of all days is one without laughter."

   _____

5. W. Somerset Maugham said, "The ability to quote is a serviceable substitute for wit."

   _____

Paraphrasing

Paraphrasing refers to restating, in your own words, someone else's ideas or texts. A paraphrase must indicate the source of the original idea in order to avoid plagiarism. Each sentence of the original text must be paraphrased in the order in which it originally appears. Here is an example using an excerpt from "The Lady or the Tiger" by Frank Stockton.

> In the very olden time there lived a semi-barbaric king, whose ideas, though somewhat polished and sharpened by the progressiveness of distant Latin neighbours, were still large and unrestricted, as became the half of him, which was barbaric. He was a man of exuberant fancy, and of such irresistible authority that, at his will, he turned his varied fancies into facts.

Paraphrase:

In "The Lady or the Tiger," Frank Stockton tells the story of a half-barbaric, half-civilized king who lived a very long time ago. Like himself, his unrestrained ideas were only half-civilized from contact with his more progressive but far-off "Latin neighbours." He had a wild imagination but such strong authority that he was able to turn his wildest ideas into reality whenever he wanted.

How to Paraphrase

- Underline the key ideas in the original text. Read the text carefully because you will have to restate the main ideas and supporting details when you paraphrase them.
- Restate the ideas of the original text in your own words. Use a dictionary and thesaurus to find exact meanings and synonyms of words in the original text.
- Mention the source of the original text, stating its author and title.

- Put quotation marks around any words or phrases used from the original source (sometimes, it is impossible to substitute your own words for the phrases mentioned in the original text).
- Use approximately the same number of words as the original passage (a paraphrase is a restated version of the original text).
- Do not change the meaning of the original text (do not include your own opinions).
- Verify that your paraphrase expresses the ideas and intent of the author of the original text.

10

EXERCISE 3 Paraphrasing Sentences

Rewrite the following sentences in your own words. Remember to keep the original meaning of the ideas. (You do not need to mention the source in this exercise.)

1. The animal kingdom reflects a predominance of male-female work stereotypes.

   _____

   _____

2. Male birds have very colourful plumage, while female birds look more ordinary; the reason for this difference is that females need to be camouflaged in order to hatch their eggs while being safe from predators.

   _____

   _____

3. Male birds also use their colourful appearance as a way to court females.

   _____

   _____

4. The Black Widow spider is aptly named because the females eat the males after mating.

   _____

   _____

5. Male lions are truly the lords of their domain; the females hunt for food, look after the young and allow the males to eat first.

   _____

   _____

EXERCISE 4 Paraphrasing Direct Quotations

Paraphrase the following direct quotations from the stories you have read.

1. "When she abandoned herself, a little whispered word escaped her slightly parted lips. She said it over and over under her breath: 'Free, free, free!'" ("The Story of an Hour")

   _____

   _____

10

2. "I've never seen one," the man drank his beer.

"No, you wouldn't have."

"I might have," the man said. "Just because you say I wouldn't have doesn't prove anything." ("Hills Like White Elephants")

_____

_____

3. "When I had waited a long time, very patiently, without hearing him lie down, I resolved to open a little—a very, very little crevice in the lantern. So I opened it—you cannot imagine how stealthily, stealthily—until, at length a simple dim ray, like the thread of the spider, shot from out the crevice and fell full upon the vulture eye." ("The Tell-Tale Heart")

_____

_____

EXERCISE 5 Paraphrasing a Paragraph

The next excerpt is taken from George Orwell's "Shooting an Elephant." Read the paragraph carefully and paraphrase it in your own words.

> One day something happened which in a roundabout way was enlightening. It was a tiny incident in itself, but it gave me a better glimpse than I had had before of the real nature of imperialism—the real motives for which despotic governments act. Early one morning the sub-inspector at a police station at the other end of the town rang me up on the phone and said that an elephant was ravaging the bazaar. Would I please come and do something about it? I did not know what I could do, but I wanted to see what was happening, and I got on to a pony and started out. I took my rifle, an old .44 Winchester and much too small to kill an elephant, but I thought the noise might be useful _in terrorem_. Various Burmans stopped me on the way and told me about the elephant's doings. It was not, of course, a wild elephant, but a tame one which had gone "must." It had been chained up, as tame elephants always are when their attack of "must" is due, but on the previous night it had broken its chain and escaped. Its mahout, the only person who could manage it when it was in that state, had set out in pursuit, but had taken the wrong direction and was now twelve hours' journey away, and in the morning the elephant had suddenly reappeared in the town. The Burmese population had no weapons and were quite helpless against it. It had already destroyed somebody's bamboo hut, killed a cow and raided some fruit-stalls and devoured the stock; also it had met the municipal rubbish van and, when the driver jumped out and took to his heels, had turned the van over and inflicted violences upon it.

_____

_____

_____

_____

_____

_____

_____

10

Summarizing

Summarizing refers to restating in your own words the general ideas of a text. A summary focuses on the main ideas and does not contain specific details or examples. It should be brief and concise. Here is an example.

> Joining us most nights were the Nahumovskys. They attended the same English classes and travelled with my parents on the same bus. Rita Nahumovsky was a beautician, her face spackled with makeup, and Misha Nahumovsky was a tool-and-die maker. They came from Minsk and didn't know a soul in Canada. With abounding enthusiasm, they incorporated themselves into our family. My parents were glad to have them. Our life was tough, we had it hard—but the Nahumovskys had it harder. They were alone, they were older, they were stupefied by the demands of language. Being essentially helpless themselves, my parents found it gratifying to help the more helpless Nahumovskys. (From "Tapka" by David Bezmozgis.)

> **Summary**
> In his short story "Tapka," David Bezmozgis wrote that every night the narrator's parents helped an older Russian couple learn English.

How to Summarize

- Read the original text carefully.
- Underline the key ideas in the text. In a summary, you will only present the main ideas of a text.
- Ask yourself who, what, when, where, why and how questions about the text. These questions will help you to synthesize the ideas of the original text.
- Restate the essential ideas in your own words. Refer to a dictionary and thesaurus for meanings of words and useful synonyms.
- Maintain the original meaning of the text. Do not include your own opinions. Mention the source (author and title) of the original text.
- Reread your summary and verify that you have explained the critical message of the text.

EXERCISE 6 Summarizing a Short Text

Write a summary of the following excerpt from "The Yellow Sweater" by Hugh Garner.

> He stepped on the gas when he reached the edge of town. The big car took hold of the pavement and began to eat up the miles on the straight, almost level, highway. With his elbow stuck through the open window he stared ahead at the shimmering greyness of the road. He felt heavy and pleasantly satiated after his good small town breakfast, and he shifted his bulk in the seat, at the same time brushing some cigar ash from the front of his salient vest. In another four hours he would be home—a day ahead of himself this trip, but with plenty to show the office for last week's work. He unconsciously patted the wallet resting in the inside pocket of his jacket as he thought of the orders he had taken.

> Four thousand units to Slanders ... his second best line too ... four thousand at twelve percent ... four hundred and eighty dollars! He rolled the sum over in his mind as if tasting it, enjoying its tartness like a kid with a gumdrop.

10

_____

_____

_____

_____

_____

_____

EXERCISE 7 Summarizing Dialogue

Write a summary of the following excerpt from "The Veldt" by Ray Bradbury.

"George, I wish you'd look at the nursery."

"What's wrong with it?"

"I don't know."

"Well, then."

"I just want you to look at it, is all, or call a psychologist in to look at it."

"What would a psychologist want with a nursery?"

"You know very well what he'd want." His wife paused in the middle of the kitchen and watched the stove busy humming to itself, making supper for four.

"It's just that the nursery is different now than it was."

"All right, let's have a look."

They walked down the hall of their soundproofed, Happylife Home, which had cost them thirty thousand dollars installed, this house which clothed and fed and rocked them to sleep and played and sang and was good to them. Their approach sensitized a switch somewhere and the nursery light flicked on when they came within ten feet of it. Similarly, behind them, in the halls, lights went on and off as they left them behind, with a soft automaticity.

"Well," said George Hadley.

They stood on the thatched floor of the nursery. It was forty feet across by forty feet long and thirty feet high—it had cost half again as much as the rest of the house. "But nothing's too good for our children," George had said.

_____

_____

_____

_____

_____

_____

_____

Citing Sources (MLA)

Many colleges require students to follow the rules and conventions established by the Modern Language Association (MLA) for acknowledging sources of quoted, paraphrased and summarized texts. Citations in the text of an academic essay or research paper are usually followed by parentheses containing the name of the author and a page number.

EXAMPLE: Some critics have said that Ray Bradbury "fears and distrusts science" (Knight 4).

At the end of the essay or paper, the cited sources are listed in a **Works Cited** list. The entries are listed in alphabetical order according to the author's last name (or the title of the work if no author is named). Quotation marks are placed around the titles of articles, and the name of the publisher is mentioned as briefly as possible.

Generic MLA Format for Common Sources

Book: Author. Title of book. City of publication: publisher, year.

Anthology: Author of story. "Title of story." Title of book. Name of editor. Edition (if given). City of publication: publisher, year. Page numbers.

Encyclopedia: Author of article (if given). "Article title." Title of book. City of publication: publisher, year.

Article: Author. "Title of article." Title of magazine Date: page(s).

Website: Title of the website. Editor. Date and/or version number. Name of sponsoring institution. Date of access <URL>.

EXAMPLE: Works Cited

"Analysis of Ray Bradbury's Writing." Planet Papers. 5 November 2006 <http://www.planetpapers.com/Assets/5150.php>

Bradbury, Ray. Classic Stories 1. New York: Bantam, 1990.

Knight, Damon. "When I Was in Kneepants: Ray Bradbury." Modern Critical Views: Ray Bradbury. Ed. Harold Bloom. Philadelphia: Chelsea House Publishers, 2001. 3–8.

Web English Teacher. Carla Beard. 15 April 2006. Web English Teacher. 5 November 2006 http://www.webenglishteacher.com/bradbury.html

---

For more detailed information on using MLA style for citing sources, consult the following websites.

http://kclibrary.nhmccd.edu/mlastyle.htm

http://owl.english.purdue.edu/owl/resource/557/01/

http://www.library.cornell.edu/newhelp/res_strategy/citing/mla.html

EXERCISE 8 Using MLA Format for Citing Works

Rewrite the following sources in MLA format for a Works Cited list.

Book: *Science Fiction: History, Science, Vision* by Robert Scholes and Eric Rabkin, published by Oxford University Press in 1977 in New York.

Article: "C. S. Lewis' Trilogy: A Cosmic Romance" by Kathryn Hume. It appeared in *Modern Fiction Studies 20* on pages 505–517 in 1974–75.

Website: By Don Bassie in <u>Made in Canada</u>, 2000. "Timothy Findlay." Accessed July 17th, 2006. <http://www.geocities.com/canadian_sf/pages/authors/findlay.htm>

Encyclopedia: 1993, Tenth edition of the <u>World Book Encyclopedia:</u> "Columbus Discovers America."

Anthology: "There Will Come Soft Rains" by Ray Bradbury in the anthology titled <u>21 Great Stories</u>. Edited by Abraham H. Lass and Norma L. Tasman in New York and published by the New American Library in 1969. Pages 75–88.

EXERCISE 9 Creating an MLA Works Cited List

Number the following fifteen works in the proper order to create an MLA Works Cited list.

———— Hillegas, Mark R. "Science Fiction and the Idea of Progress." *Extrapolation* 1 May 1960: 25–28.

———— Heinlein, Robert A. "Science Fiction: Its Nature, Faults, and Virtues." In Damon Knight, ed., *Turning Points: Essays on the Art of Science Fiction*. New York: Harper and Row, 1977. 3-28.

———— "The Fireman." *Galaxy Science Fiction* 1.5 (1951). Reprinted in *Science Fiction Origins*, ed. William F. Nolan & Martin H. Greenburg. New York: Popular Library, 1980.

———— Pohl, Frederik. "Tunnel Under the World." *Galaxy Science Fiction* January 1955.

———— Landsburg, Steven E. "Who Shall Inherit the Earth?" *Slate.* 1 May 1997. 2 May 1997 <http://www.slate.com>.

———— Markoff, John. "The Voice on the Phone Is Not Human, but It's Helpful." *New York Times on the Web* 21 June 2002.

———— Fishkin, Fred. "Privacy and the Net." *Boot Camp*. CBS Radio. WCBS, New York. 21 Aug. 2002. 15 Feb. 2004 <http://newsradio88.com/bootcamp-cgi/story=123445>

———— Smith, John. "Fresco." *Encyclopedia Britannica Online*. 2003. Encyclopedia Britannica. 29 Mar. 2005 <http://www.eb.com/>

———— Bois, Danuta. *The Glory of the Temple and the State: Henry Purcell 1659–1695*. 1997. The British Library. 7 May 2004 <http://www.bl.uk.whatson/ purcell.html>

———— Tully, Shawn. "The Universal Teenager." *Fortune* 4 Apr. 1994: 14–16.

_____ Smith, John and Susan Lane. *Space Travel in the 21st Century*. New York: Random House, 2002. 103.

_____ "Columbus Discovers America." *World Book Encyclopedia*. 10th ed. 1993.

_____ Green, Bob. "Heartland AIDS Ride." *Chicago Tribune* 27 July 2002, Final Edition, Tempo section: 14–15.

_____ O'Connor, Flannery. "Everything That Rises Must Converge." *Mirrors: An Introduction to Literature*. Ed. John R. Knott, Jr. and Christopher R. Reaske. 2nd ed. San Francisco: Canfield, 1975. 58–67.

_____ Cohen, Henning and Tristram Peter Coffin. *America Celebrates*. Detroit: Visible Ink Press, 1991.

10

Practice and Review

Short Exercises

EXERCISE 1 Correcting Mixed Errors

The following items were taken from introductory compositions by college students. Each phrase or sentence contains at least one error. Correct the error(s).

1. Their's ways to see life that is generally not told are talk about.

   _____

2. but i know i can make a real good presentation ...

   _____

3. In my last course, I learn all king of way to rite texts.

   _____

4. Sometime little detail is importants like ...

   _____

5. In 1998, I've taken a course in ... but I've never read or wrote much in english.

   _____

6. Reading worths a lot ... but I did'nt read a lot in english before this course.

   _____

7. I'm reading a lot of English books at home, but I'm not writing that much.

   _____

8. Each person as is own personnality.

   _____

9. I live in St. Jerome since august, before, I lived in St. Agathe where is my family.

   _____

10. Does it worth it to do a job wich you do'nt appreciate?

   _____

EXERCISE 2 Correcting Mixed Errors

The following items were taken from paragraphs by students. Each phrase or sentence contains at least one error. Correct the error(s).

1. The General Zaroff is a man who have hunted every kind of animals. He dosen't have ...

   _____

2. Equality shouln't be base on materials possesions.

   _____

3. That what show us that the sentense is ...

   _____

4. Everybody have a life and do what he want's with it.

   _____

5. ... its true than in those days society there is not much place for the weaks ...

   _____

6. Divorce wasn't existing at that time.

   _____

7. These storys often have ironics element witch brings an unexpected ending to them.

   _____

8. She feel free because she have learn ...

   _____

9. Its about human beings taking advantage of it's supremacy toward's other animal's.

   _____

10. The author present the same theme in this storie.

    _____

Editing Two Student Essays with Grammar Errors

EXERCISE 3 Editing an Essay with Annotated Errors

A student wrote the following essay. Edit the essay for grammatical errors. The type of error is indicated in parentheses after the underlined error.

Key

cap ➡ capitalization	sp ➡ spelling	agr ➡ agreement
poss ➡ possessive form	verb ➡ verb form	ww ➡ wrong word
frag ➡ sentence fragment	run-on ➡ run-on sentence	// ➡ parallel structure

Romantic Elements in Edgar Allan Poe's "The Tell-Tale Heart"
by Virginie Lachapelle

Edgar Allan Poe (1809–1849) is known as the father of the modern short story in <u>english</u> (*cap*). Furthermore, he is also known as a great <u>writter</u> (*sp*) of horror stories. One of his most popular short stories is "The Tell-Tale Heart." This short story captures the imagination of the reader because it is filled with horror elements influenced by Romanticism.

First, "The Tell-Tale Heart" <u>reflect</u> (*agr*) a popular theme present in literary works during Romanticism: the theme of madness. The protagonist of the story is mad. His madness is <u>show</u> (*verb*) by the evolution in the <u>killers</u> (*poss*) mind. For example, the murderer pretends to be very calm while he begins to tell his story: "How, then, am I mad? Hearken and observe how healthily—how calmly I can tell you the whole story." (lines 4-5) At the <u>begining</u> (*sp*) of the story, the killer really seems to be calm, but as the story progresses, the tension increases in the killer's mind: "I have told you that I am nervous: so I am." (line 72) The killer is losing both patience and control. By the end of the story, when the tension is at <u>it's</u> (*ww*) peak, the murderer truly <u>looses</u> (*ww*) control. He confesses that he is the killer: "I admit the deed!" (line 124) The protagonist behaves insanely as the story unfolds.

Moreover, the protagonist of "The Tell-Tale Heart" is a marginalized member of society. First, he has an obsession with the old man's eye. This obsession causes him to <u>comitt</u> (*sp*) actions that are condemned by society. The obsession also leads to his eventual insanity. Second, he is ostracized from society. <u>Because he is a murderer</u> (*frag*). So the protagonist fulfills the criterion of Romantic literature. Such characters did not belong to mainstream society, but rather to the peripheral world. <u>Like Poe's protagonist in "The Tell-Tale Heart."</u> (*frag*)

Finally, Poe's "Tell-Tale Heart" also reflects other characteristics of Romanticism, such as hallucinations and the dream world. Indeed, the protagonist lives in a world full of hallucinations. As he realizes the severity of his action, <u>he is starting</u> (*//*) to hear sounds that eventually drive him insane. As his sense of guilt increases, his hallucinations also <u>increases</u> (*agr*). He hears the beating of his victim's heart. He shrieks, "It is the beating of his hideous heart!" (line 124) <u>Poe increases the tension in this story by describing the sounds the protagonist hears as his guilt grows for murdering the old man, it is this sense of guilt that finally betrays him.</u> (*run-on*)

Thus, Poe's "The Tell-Tale Heart" contains many elements found in Romantic literature. Poe's ability to create tension by using an insane protagonist and <u>describes</u> (*//*) his hallucinations makes it a chilling story to read.

EXERCISE 4 Editing an Essay

A student wrote the following essay. Find and correct the errors in spelling, grammar and punctuation. There are fifteen errors in total.

The Role of Women in Kate Chopin's "The Story of an Hour"
by Nancy Jalbert

On november 19, 1863, Abraham Lincoln delivered the Gettysburg Address in which he proclaimed "that all men are created equal." Indeed, the american Civil War was fought for this principle. Lincoln and his supporters taught that slavery was inhuman and therefore had to be abolished. As noble as this cause was, the president and others forgot that one half of the population did not enjoy the equal rights of the other half. Women did not have the same freedom as men. Kate Chopin, a largely forgotten writter until recently, was ahead of her time, her literary works discussed the conditions of women. In "The Story of an Hour," she reflects the reality that women of the nineteenth century were forced to endure.

Chopin argues that marriage may not be the ultimate goal for women. Life for Mrs. Mallard is not as happy as she would have wanted. She feels her life is so painful. That she almost prefers to die young. Before she hears the news of her husband's death, she want to die. "It was only yesterday she had thought with a shudder that life might be long." (line 60) However, since she hears about the death of her husband, she feels a sense of relief. She is convinced that she will have a better life. Mrs. Mallard's ironic reaction to the news of his husband's death shows that she considers him an obstacle to her joy of living. His absence changes her perceptions about life.

Furthermore, Chopin also reveal how many women deceive themselves about the state of there happiness. Mrs. Mallard has been married for many years. Up until now, she has been discontent. Chopin describes Mrs. Mallard's face, "whose lines bespoke a certain repression." (line 24) Her husband is not an abusive or violent man. She has no specific reason for her discontentment. In fact, she loves her husband and feels a certain amount of grief: "...she saw the kind, tender hands folded in death; the face that had never looked save with love upon her ..." (line 41). Yet she feels this overwhelming sense of freedom at the prospect of living as a single woman. Women of Chopins time never expressed such a sentiment, and men probably never imagined that a woman could be discontent with a loving husband.

Moreover, Chopin exposes all the social hypocrisy regarding a wife's role. In a marriage. In the immediate moments after the news, Mrs. Mallard visualizes the coming years with joy. She thinks, "she would live for herself." (line 44) The reader can only assume that in her marriage, Mrs. Mallard is oppressed. Yet social norms of the time presumed that all women wanted to get married and that all marriages were happy. Mrs. Mallard's secret thoughts-she probably cannot expose them in public-we're different from social expectations.

Thus, in the short story "The Story of an Hour," Chopin depicts societys expectation of the role women had to play. Mrs. Mallard's reactions to her husband's death are totally unexpected. Yet she is probably not alone in feeling such sentiments. Chopin just gives a stage. To the repressed voices of women.

Glossary of Literary Terms

Abstractness — not perceptible by the senses, only by the mind

Action — events taking place in a dramatic or narrative work

Acts — divisions of a play that usually last from thirty minutes to an hour

Alliteration — repetition of initial consonant sounds in a series of words

Anapestic foot — a poetic foot of three syllables, the first two unaccented and the last accented

Antagonist — a person or a power who opposes the main character in a narrative or dramatic work

Antithesis — a linking of two contrasting ideas to make each more vivid

Apostrophe — an address to a dead person as if he or she were still alive

Assonance — repetition of the same vowel sounds in a series of words

Atmosphere — the emotional mood of the text

Blank verse — a poem that has meter but no rhymes

Cacaphony — unpleasant or harsh combination of sounds

Character — all persons found in a dramatic or narrative work

Climax — the point of maximum conflict of the story

Comparison — looking at similarities and differences between two or more items

Complication — a conflict in the story

Concreteness — perceptible by the senses

Conflict — the opposition of different characters or forces in a dramatic or narrative work; may be either internal or external

Connotation — the secondary level of a meaning or association of a word

Consonance — repetition of consonants in a series of words

Couplet — two lines that rhyme

Dactylic foot — poetic foot of three syllables, the first stressed, the last two unstressed

Denotation — the literal or dictionary meaning of a word

Denouement — the conclusion of the story

Dialogue — conversation in written form

Dimeter — poetic meter of two feet

Dramatic structure — the beginning, middle, and end of a dramatic or narrative work

Drama — a theatrical production intended for actors to present a story through dialogue and action

Epic — a long poem about someone heroic or noble who performs an important action

Essay — short written composition on a single subject

Euphony — pleasant or harmonious sounds

Exposition — what happens at the beginning of the story

External action — action that happens outside the body of the characters in a narrative or dramatic work

Fiction — all written works that are created in the imagination of the author

First-person narrator — the story told through the perspective of an "I"

First-person plural narrators — the story told through different narrators such as the "I" or "We"

Flashback — an event that happened before the story takes place

Flat character — a minor character, who does not develop throughout the narrative or dramatic work

Foreshadowing — clue as to what will happen in the future of the narrative or dramatic work

Formal speech — language of the educated, standard usage of language; more regimented rules of language

Formal structure — how a poem is divided into stanzas and how the stanzas fit together

Free verse — poetry that does not conform to any fixed pattern (in either rhyme or rhythm)

Hyperbole — an exaggerated statement

Heptameter — verse or line consisting of seven poetic feet

Hexameter — verse or line consisting of six poetic feet, also called Alexandrian

Iamb — a poetic foot containing an unstressed followed by a stressed syllable

Iambic pentameter — five poetic feet consisting of alternating unstressed and stressed syllables

Imagery — descriptions that create mental images aroused by the perception of the five senses: sight, sound, smell, taste and touch.

Informal speech — language that is more familiar; less regimented

Internal action — action that happens in the minds of the characters in a narrative or dramatic work

Intrusive point of view — third-person point of view, which is judgmental

Irony — the opposite of that which is expected

Limited point of view — third-person point of view, which narrates from the perspective of one character

Lyric — a short poem that expresses an idea or emotion

Metaphor — describing one thing as if it were something else

Meter — rhythmical pattern of words in poetry based on the accented and non-accented syllables of speech

Mood — the emotions of a narrative or dramatic work

Narration — recounting of a sequence of events in a narrative work

Narrative structure — the way in which the sequence of events in a narrative or dramatic work is ordered; they can be chronological or non-chronological

Nonfiction — prose works that are not fiction, such as biographies, historical accounts, scientific articles, etc.

Novel — a long work of prose fiction with many characters and plots

Objective point of view — third-person narration from a non-judgmental point of view

Octave — stanza of eight lines

Ode — a serious, longer poem on a noble subject

Omniscient point of view — third-person narration that is all-seeing and all-knowing; describes thoughts and actions of all characters

Onomatopoeia — words that resemble sounds associated with the object

Oxymoron — a figure of speech containing a contradiction

Paradox — a statement that seems to be contradictory and true at the same time

Pentameter — a verse or line consisting of five poetic feet

Poetic feet — a measure of rhythm that groups stressed and unstressed syllables in English poetry

Persona — the voice or character of the speaker (or narrator) in a literary piece

Personification — giving human qualities to non-human objects or animals

Play — a story written for the stage; story performed on stage

Plot — what happens in the story

Poem — a poetic work

Poetry — literary genre that is qualified by its rhythmic language

Point of view — the eye and voice through which a story is told

Prose — ordinary speech or writing with no specific rhythmical pattern (meter), as opposed to poetry

Protagonist — the main character in a narrative work; does not have to have heroic qualities

Pun — a play on words creating two different meanings simultaneously

Quatrain — a stanza containing four lines of sometimes alternating rhymes

Refrain — a line, phrase, or stanza that is repeated at intervals throughout a poem or song

Repetition — a sound, word, sentence or image seen over and over throughout a narrative or dramatic work

Rhyme — repetition of stressed sounds in words usually found at the end of lines of verse

Round character — usually a major character who changes or develops in a narrative or dramatic work — one of the major genres of fiction; one play

Scenes — sections of acts in a play that can last from a few seconds to thirty minutes

Sestet — last six lines of a sonnet

Setting — where and when the story takes place

Short story — compact narrative fiction; usually can be read in one sitting

Simile — a comparison using "like" or "as"

Slang — language that is not accepted as the standard

Sonnet — a poem of Italian origin with 14 lines in iambic pentameter

Spondaic foot — a poetic foot of two syllables, both stressed

Stage directions — instructions to the actors and the director about the set, gestures, entrances and exits, etc.

Stanza — a group of lines (usually four) which rhyme

Static character — a secondary character; one who does not change

Stereotype — a character who is ordinary; a type or mould

Stream of consciousness — an unbroken flow of thought in a character's mind

Style — the way in which a narrative or dramatic or poetic work is written; the types of words, sentences, and structures used for an overall effect

Suggestion — the choice of evocative words to enhance meaning

Symbol — a particular word or idea that may represent something else

Syntax — the grammatical order in which sentences are organized

Tercet — three lines that rhyme; a triplet

Tetrameter — a verse or line consisting of four poetic feet

Theme — the central concept in a narrative, dramatic, or poetic work

Third-person narrator — story told through a "he" or "she" who is not part of the events

Tone — the technique in which an attitude is revealed

Tragedy — a drama in which the outcome is serious

Trimeter — a verse or line consisting of three poetic feet

Trochaic foot — a poetic foot of two syllables, the first stressed, the second unstressed

Verse — a line of poetry that usually has a particular rhythm

Voice (or persona or narrator) — the person who speaks in a literary piece

Glossary of Grammatical Terms

Clause — a group of words containing a subject and a verb, but not necessarily expressing a complete idea

Comma splice — incorrect use of comma between two independent clauses when a period should be used

Complex sentence — one independent clause and at least one dependent clause

Compound sentence — two independent clauses joined by a coordinate such as "and," "but," "or," etc.

Compound-complex sentence — two independent clauses and at least one dependent clause

Co-ordinating conjunction — words such as "and," "but," "or," "so," "yet"

Co-ordination — joining of two or more ideas of equal importance

Dependent or subordinate clause — a clause which expresses an incomplete idea

Direct quotation — repetition of the exact words a person has said or written using double quotation marks

Homographs — words that are spelled the same but pronounced differently and with different meaning

Homonyms — words that are pronounced the same but have diferent spellings and meanings

Independent clause or principal clause — clause that expresses a complete idea

Indirect quotation — stating indirectly what someone else has said

Modifiers — words, phrases, or clauses that describe or qualify another element in the sentence

Paragraph — a group of sentences with one main idea

Parallel construction — refers to the use of similar grammatical structures

Paraphrasing — restating in your own words someone else's ideas or texts

Phrase — a group of words often beginning with a preposition or participle

Prefix — additions affixed "in front of" root words in order to give them particular meanings

Principal or main clause — *see* independent clause

Pronouns — words that replace nouns

Quotations — repetitions of exact words a person has said or written

Relative pronouns — words such as "who," "whom," "whose," "which" and "that," which are used to show subordination

Run-on sentence — at two or more independent clauses with no punctuation or too many conjunctions (and, but, or) between them

Sentence — a group of words containing a subject and a verb and expressing a complete idea

Sentence fragment — an incomplete idea; part of a sentence punctuated as if it were a complete sentence

Simple sentence — one independent clause

Subordination — shows a relation between ideas having different degrees of importance; less important idea is subordinated to a more important idea

Subordinating conjunction — words such as "because," "when," "since," etc. (See Unit 2 for a complete list.)

Suffix — additions affixed "after" root words in order to give them particular meanings

Summarizing — restating in your own words briefly and concisely the general ideas of a text

Supporting details — facts, examples or statistics that give further information on the central idea of a paragraph

Thesis statement — a sentence that contains the essential opinion or gist about the subject of an essay

Topic sentence — introduces the subject of the paragraph and contains an opinion about the subject

Transitional expressions — words and expressions that make connections between sentences and paragraphs in order to link ideas such as "furthermore," "moreover," etc. (See Unit 3 for a complete list.)

Index

The index lists key words and expressions pertaining to literary terms or periods, authors featured in the book or in My eLab, and writing or grammatical terms. Page numbers printed in bold indicate that this is where the word or expression has either been defined or has received its fullest treatment.

euphemism 322
euphony **183**
Euripides 250
exact language **319**
exposition 4
external action 5
external (setting) 6

F

feminist (approach) **220**
feminist (elements) 244
Ferlinghetti, Lawrence 191
Fielding, Henry 252
figurative devices/language **12**, 49
figures of speech **12**, **13**, **181**
first draft 291
first-person narration/narrator 10, **11**, 24
first-person plural narrator 10,
flashback 5, 71-72
flat character 7, **8**, 62
foreign words **84**
foreshadowing 5, **60**, 71, 107, 136, 262
form 185, 230, 233, 239, 241, 244
formal speech **12**, **13**
formal (elements) **222**, 224
formalist (approach) **219**
fragment 347-349
free verse **184**
free-writing 283
Freud, Sigmund 139
Frost, Robert 191, **223**

G

Garner, Hugh **127**
Gay, John 252
general setting 6
gerunds 165
Ginsberg, Allen 191
Glaspell, Susan **264**
Golden Age 189
Goldsmith, Oliver 252
Gothic (elements, novel) 17, 19, 63
Graves, Robert 191
Gray, Thomas 190

H

hard consonant sounds **183**
Hemingway, Ernest **76**, 192
heptameter **184**
hexameter **184**
historical **219**, **222**, 231, 241
histories 255
history—English language **27**
history—poetry **188**
history—short story **16**
homonyms 331
Housman, A. E. 179
Hughes, Ted 191
Humanism 251
hyperbole **181**

I

iamb, iambic foot **184**
iambic pentameter **184**
imagery **12**, **13**, 25, **50**, 60, 70, 71, 95, 146,
    163, 173, **181**, 224, 233, 236, 244, 246

informal speech **12**, **13**
interior monologue 148
internal action 5
internal (setting) 6
introduction 305
introductory paragraph/style 305, 306
intrusive 10, **11**
irony **12**, **13**, 26, **27**, 51, 58, 59-60, 107, 146,
    147, **181**, 262

J

Jackson, Shirley 99
jargon 322
Johnson, Ben 255
Judeo-Christian 138

K

Keats, John 190
King Solomon 192
Kyd, Thomas 251

L

Larkin, Philip 191
Layton, Irving 191, **232**
Leacock, Stephen **167**
level of abstractness **13**
level of concreteness **13**
level(s) of language **12**, **13**, **181**, 321
Lewis, C. D. 191
limited point of view 10, **11**
lines **178**, **254**
literary analysis 296
literary essay(s) **296**
lyric **183**

M

main clause 333
Mallarmé, Stéphane 191
Marlowe, Christopher 251, 255
Marxist (approach) **220**
Maugham, Somerset 252
McCrae, Dr. John **220**
metaphor(s) **12**, **13**, **26**, 173, **181**, 262
metaphysical 189
meter **183**, **184**
Milton, John 189
minority (approach) **220**
misplaced modifier 359
Molière 252
monometer **184**
mood 6, **12**, 26, **178**
morality play 250
multiple narrators **11**

N

narration **10**, **11**, 37, 59, 108, 161, 238
narrative structure **4**, 5, 71, 97, 122
narrator **11**

O

objective **11**
objectively 10
octaves **182**
ode **183**
omniscient point of view 10, **11**
one-act play 264
O'Neill, Eugene 264